Modeling in Transport Phenomena

A Conceptual Approach

Second Edition

Modeling in Transport Phenomena

A Conceptual Approach

Second Edition

İsmail Tosun

Department of Chemical Engineering
Middle East Technical University
Ankara, Turkey

ELSEVIER

Amsterdam • Boston • Heidelberg • London • New York • Oxford • Paris
San Diego • San Francisco • Singapore • Sydney • Tokyo

Elsevier
Radarweg 29, PO Box 211, 1000 AE Amsterdam, The Netherlands
The Boulevard, Langford Lane, Kidlington, Oxford OX5 1GB, UK

First edition 2002
Second edition 2007

Library of Congress Cataloging-in-Publication Data
A catalog record for this book is available from the Library of Congress

British Library Cataloguing in Publication Data
A catalogue record for this book is available from the British Library

ISBN-13: 978-0-444-53021-9

For information on all Elsevier publications
visit our website at books.elsevier.com

Printed and bound by CPI Group (UK) Ltd, Croydon, CR0 4YY

Transferred to Digital Print 2012

Working together to grow
libraries in developing countries

www.elsevier.com | www.bookaid.org | www.sabre.org

ELSEVIER   BOOK AID
             International   Sabre Foundation

To Ayşe

To Ayşe

CONTENTS

PREFACE TO THE SECOND EDITION

While the main skeleton of the first edition is preserved, Chapters 10 and 11 have been rewritten and expanded in this new edition. The number of example problems in Chapters 8–11 has been increased to help students to get a better grasp of the basic concepts. Many new problems have been added, showing step-by-step solution procedures. The concept of time scales and their role in attributing a physical significance to dimensionless numbers are introduced in Chapter 3.

Several of my colleagues and students helped me in the preparation of this new edition. I thank particularly Dr. Ufuk Bakır, Dr. Ahmet N. Eraslan, Dr. Yusuf Uludağ, and Meriç Dalgıç for their valuable comments and suggestions. I extend my thanks to Russell Fraser for reading the whole manuscript and improving its English.

İSMAİL TOSUN
(itosun@metu.edu.tr)

Ankara, Turkey
October 2006

The Solutions Manual is available for instructors who have adopted this book for their course. Please contact the author to receive a copy, or visit http://textbooks.elsevier.com/9780444530219

PREFACE TO THE FIRST EDITION

During their undergraduate education, students take various courses on fluid flow, heat transfer, mass transfer, chemical reaction engineering, and thermodynamics. Most of them, however, are unable to understand the links between the concepts covered in these courses and have difficulty in formulating equations, even of the simplest nature. This is a typical example of not seeing the forest for the trees.

The pathway from the real problem to the mathematical problem has two stages: perception and formulation. The difficulties encountered at both of these stages can be easily resolved if students recognize the forest first. Examination of the trees one by one comes at a later stage.

In science and engineering, the forest is represented by the **basic concepts**, i.e., conservation of chemical species, conservation of mass, conservation of momentum, and conservation of energy. For each one of these conserved quantities, the following inventory rate equation can be written to describe the transformation of the particular conserved quantity φ:

$$\begin{pmatrix} \text{Rate of} \\ \varphi \text{ in} \end{pmatrix} - \begin{pmatrix} \text{Rate of} \\ \varphi \text{ out} \end{pmatrix} + \begin{pmatrix} \text{Rate of } \varphi \\ \text{generation} \end{pmatrix} = \begin{pmatrix} \text{Rate of } \varphi \\ \text{accumulation} \end{pmatrix}$$

in which the term φ may stand for chemical species, mass, momentum, or energy.

My main purpose in writing this textbook is to show students how to translate the inventory rate equation into mathematical terms at both the macroscopic and microscopic levels. It is not my intention to exploit various numerical techniques to solve the governing equations in momentum, energy, and mass transport. The emphasis is on obtaining the equation representing a physical phenomenon and its interpretation.

I have been using the draft chapters of this text in my third year *Mathematical Modelling in Chemical Engineering* course for the last two years. It is intended as an undergraduate textbook to be used in an (Introduction to) Transport Phenomena course in the junior year. This book can also be used in unit operations courses in conjunction with standard textbooks. Although it is written for students majoring in chemical engineering, it can also be used as a reference or supplementary text in environmental, mechanical, petroleum, and civil engineering courses.

An overview of the manuscript is shown schematically in the figure below.

Chapter 1 covers the basic concepts and their characteristics. The terms appearing in the inventory rate equation are discussed qualitatively. Mathematical formulations of the "rate of input" and "rate of output" terms are explained in Chapters 2, 3, and 4. Chapter 2 indicates that the total flux of any quantity is the sum of its molecular and convective fluxes. Chapter 3 deals with the formulation of the inlet and outlet terms when the transfer of matter takes place through the boundaries of the system by making use of the transfer coefficients, i.e., friction factor, heat transfer coefficient, and mass transfer coefficient. The correlations available in the literature to evaluate these transfer coefficients are given in Chapter 4. Chapter 5 briefly talks about the rate of generation in transport of mass, momentum, and energy.

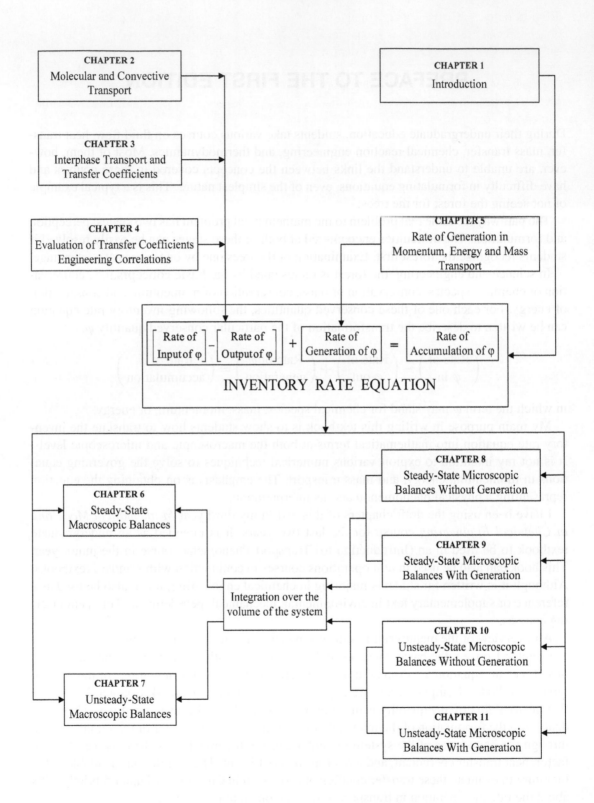

Traditionally, the development of the microscopic balances precedes that of the macroscopic balances. However, it is my experience that students grasp the ideas better if the reverse pattern is followed. Chapters 6 and 7 deal with the application of the inventory rate equations at the macroscopic level.

The last four chapters cover the inventory rate equations at the microscopic level. Once the velocity, temperature, or concentration distributions are determined, the resulting equations are integrated over the volume of the system to obtain the macroscopic equations covered in Chapters 6 and 7.

I had the privilege of having Professor Max S. Willis of the University of Akron as my PhD supervisor, who introduced me to the real nature of transport phenomena. All that I profess to know about transport phenomena is based on the discussions with him as a student, a colleague, a friend, and a mentor. His influence is clear throughout this book. Two of my colleagues, Güniz Gürüz and Zeynep Hiçşaşmaz Katnaş, kindly read the entire manuscript and made many helpful suggestions. My thanks are also extended to the members of the Chemical Engineering Department for their many discussions with me and especially to Timur Doğu, Türker Gürkan, Gürkan Karakaş, Önder Özbelge, Canan Özgen, Deniz Üner, Levent Yılmaz, and Hayrettin Yücel. I appreciate the help provided by my students, Gülden Camçı, Yeşim Güçbilmez, and Özge Oğuzer, for proofreading and checking the numerical calculations.

Finally, without the continuous understanding, encouragement and tolerance shown by my wife Ayşe and our children Çiğdem and Burcu, this book could not have been completed and I am particularly grateful to them.

Suggestions and criticisms from instructors and students using this book will be appreciated.

İSMAİL TOSUN
(itosun@metu.edu.tr)

Ankara, Turkey
March 2002

1

INTRODUCTION

1.1 BASIC CONCEPTS

A concept is a unit of thought. Any part of experience that we can organize into an idea is a concept. For example, man's concept of cancer is changing all the time as new medical information is gained as a result of experiments.

Concepts or ideas that are the basis of science and engineering are *chemical species*, *mass*, *momentum*, and *energy*. These are all conserved quantities. A *conserved* quantity is one that can be transformed. However, transformation does not alter the total amount of the quantity. For example, money can be transferred from a checking account to a savings account but the transfer does not affect the total assets.

For any quantity that is conserved, an *inventory rate equation* can be written to describe the transformation of the conserved quantity. Inventory of the conserved quantity is based on a specified unit of time, which is reflected in the term *rate*. In words, this rate equation for any conserved quantity φ takes the form

$$\begin{pmatrix} \text{Rate of} \\ \text{input of } \varphi \end{pmatrix} - \begin{pmatrix} \text{Rate of} \\ \text{output of } \varphi \end{pmatrix} + \begin{pmatrix} \text{Rate of} \\ \text{generation of } \varphi \end{pmatrix} = \begin{pmatrix} \text{Rate of} \\ \text{accumulation of } \varphi \end{pmatrix} \quad (1.1\text{-}1)$$

Basic concepts upon which the technique for solving engineering problems is based are the rate equations for the

- Conservation of chemical species,
- Conservation of mass,
- Conservation of momentum,
- Conservation of energy.

The entropy inequality is also a basic concept but it only indicates the feasibility of a process and, as such, is not expressed as an inventory rate equation.

A rate equation based on the conservation of the value of money can also be considered as a basic concept, i.e., economics. Economics, however, is outside the scope of this text.

1.1.1 Characteristics of the Basic Concepts

The basic concepts have certain characteristics that are always taken for granted but seldom stated explicitly. The basic concepts are

- Independent of the level of application,
- Independent of the coordinate system to which they are applied,
- Independent of the substance to which they are applied.

1

Table 1.1. Levels of application of the basic concepts

Level	Theory	Experiment
Microscopic	Equations of Change	Constitutive Equations
Macroscopic	Design Equations	Process Correlations

The basic concepts are applied at both the microscopic and the macroscopic levels as shown in Table 1.1.

At the microscopic level, the basic concepts appear as partial differential equations in three independent space variables and time. Basic concepts at the microscopic level are called the *equations of change*, i.e., conservation of chemical species, mass, momentum, and energy.

Any mathematical description of the response of a material to spatial gradients is called a *constitutive equation*. Just as the reaction of different people to the same joke may vary, the response of materials to the variable condition in a process differs. Constitutive equations are postulated and cannot be derived from the fundamental principles[1]. The coefficients appearing in the constitutive equations are obtained from experiments.

Integration of the equations of change over an arbitrary engineering volume exchanging mass and energy with the surroundings gives the basic concepts at the macroscopic level. The resulting equations appear as ordinary differential equations, with time as the only independent variable. The basic concepts at this level are called the *design equations* or *macroscopic balances*. For example, when the microscopic level mechanical energy balance is integrated over an arbitrary engineering volume, the result is the macroscopic level engineering Bernoulli equation.

Constitutive equations, when combined with the equations of change, may or may not comprise a determinate mathematical system. For a determinate mathematical system, i.e., the number of unknowns is equal to the number of independent equations, the solutions of the equations of change together with the constitutive equations result in the velocity, temperature, pressure, and concentration profiles within the system of interest. These profiles are called *theoretical* (or *analytical*) *solutions*. A theoretical solution enables one to design and operate a process without resorting to experiments or scale-up. Unfortunately, the number of such theoretical solutions is small relative to the number of engineering problems that must be solved.

If the required number of constitutive equations is not available, i.e., the number of unknowns is greater than the number of independent equations, then the mathematical description at the microscopic level is indeterminate. In this case, the design procedure appeals to an experimental information called *process correlation* to replace the theoretical solution. All process correlations are limited to a specific geometry, equipment configuration, boundary conditions, and substance.

1.2 DEFINITIONS

The functional notation

$$\varphi = \varphi(t, x, y, z) \tag{1.2-1}$$

---

[1]The mathematical form of a constitutive equation is constrained by the *second law of thermodynamics* so as to yield a positive entropy generation.

indicates that there are three *independent space variables*, x, y, z, and one *independent time variable*, t. The φ on the right side of Eq. (1.2-1) represents the functional form, and the φ on the left side represents the value of the dependent variable, φ.

1.2.1 Steady-State

The term steady-state means that at a particular location in space the dependent variable does not change as a function of time. If the dependent variable is φ, then

$$\left(\frac{\partial \varphi}{\partial t}\right)_{x,y,z} = 0 \qquad (1.2\text{-}2)$$

The partial derivative notation indicates that the dependent variable is a function of more than one independent variable. In this particular case, the independent variables are (x, y, z) and t. The specified location in space is indicated by the subscripts (x, y, z), and Eq. (1.2-2) implies that φ is not a function of time, t. When an ordinary derivative is used, i.e., $d\varphi/dt = 0$, then this implies that φ is a constant. It is important to distinguish between partial and ordinary derivatives because the conclusions are very different.

Example 1.1 A Newtonian fluid with constant viscosity μ and density ρ is initially at rest in a very long horizontal pipe of length L and radius R. At $t = 0$, a pressure gradient, $|\Delta P|/L$, is imposed on the system and the volumetric flow rate, $\mathcal{Q}$, is expressed as

$$\mathcal{Q} = \frac{\pi R^4 |\Delta P|}{8\mu L} \left[1 - 32 \sum_{n=1}^{\infty} \frac{\exp(-\lambda_n^2 \tau)}{\lambda_n^4} \right]$$

where τ is the dimensionless time defined by

$$\tau = \frac{\mu t}{\rho R^2}$$

and $\lambda_1 = 2.405$, $\lambda_2 = 5.520$, $\lambda_3 = 8.654$, etc. Determine the volumetric flow rate under steady conditions.

Solution

Steady-state solutions are independent of time. To eliminate time from the unsteady-state solution, we have to let $t \rightarrow \infty$. In that case, the exponential term approaches zero and the resulting steady-state solution is given by

$$\mathcal{Q} = \frac{\pi R^4 |\Delta P|}{8\mu L}$$

which is known as the Hagen-Poiseuille law.

Comment: If time appears in the exponential term, then the term must have a negative sign to ensure that the solution does not blow as $t \rightarrow \infty$.

Example 1.2 A cylindrical tank is initially half full with water. The water is fed into the tank from the top and it leaves the tank from the bottom. The inlet and outlet volumetric flow rates are different from each other. The differential equation describing the time rate of change of water height is given by

$$\frac{dh}{dt} = 6 - 8\sqrt{h}$$

where h is the height of water in meters. Calculate the height of water in the tank under steady conditions.

Solution

Under steady conditions dh/dt must be zero. Then

$$0 = 6 - 8\sqrt{h}$$

or,

$$h = 0.56 \text{ m}$$

1.2.2 Uniform

The term *uniform* means that at a particular instant in time, the dependent variable is not a function of position. This requires that all three of the partial derivatives with respect to position be zero, i.e.,

$$\left(\frac{\partial \varphi}{\partial x}\right)_{y,z,t} = \left(\frac{\partial \varphi}{\partial y}\right)_{x,z,t} = \left(\frac{\partial \varphi}{\partial z}\right)_{x,y,t} = 0 \tag{1.2-3}$$

The variation of a physical quantity with respect to position is called *gradient*. Therefore, the gradient of a quantity must be zero for a uniform condition to exist with respect to that quantity.

1.2.3 Equilibrium

A system is in *equilibrium* if both steady-state and uniform conditions are met simultaneously. An equilibrium system does not exhibit any variation with respect to position or time. The state of an equilibrium system is specified completely by the non-Euclidean coordinates[2] (P, V, T). The response of a material under equilibrium conditions is called *property correlation*. The ideal gas law is an example of a thermodynamic property correlation that is called an *equation of state*.

1.2.4 Flux

The flux of a certain quantity is defined by

$$\text{Flux} = \frac{\text{Flow of a quantity/Time}}{\text{Area}} = \frac{\text{Flow rate}}{\text{Area}} \tag{1.2-4}$$

where area is normal to the direction of flow. The units of momentum, energy, mass, and molar fluxes are Pa (N/m^2, or kg/m·s^2), W/m^2 (J/m^2·s), kg/m^2·s, and kmol/m^2·s, respectively.

---

[2]A *Euclidean* coordinate system is one in which length can be defined. The coordinate system (P, V, T) is *non-Euclidean*.

1.3 MATHEMATICAL FORMULATION OF THE BASIC CONCEPTS

In order to obtain the mathematical description of a process, the general inventory rate equation given by Eq. (1.1-1) should be translated into mathematical terms.

1.3.1 Inlet and Outlet Terms

A quantity may enter or leave the system by two means: (*i*) by inlet and/or outlet streams, (*ii*) by exchange of a particular quantity between the system and its surroundings through the boundaries of the system. In either case, the rate of input and/or output of a quantity is expressed by using the flux of that particular quantity. The flux of a quantity may be constant or dependent on position. Thus, the rate of a quantity can be determined as

$$
\text{Inlet/Outlet rate} =
\begin{cases}
(\text{Flux})(\text{Area}) & \text{if flux is constant} \\
\iint\limits_A \text{Flux} \, dA & \text{if flux is position dependent}
\end{cases}
\tag{1.3-1}
$$

where A is the area perpendicular to the direction of the flux. The differential areas in cylindrical and spherical coordinate systems are given in Section A.1 in Appendix A.

Example 1.3 Velocity can be interpreted as the volumetric flux ($m^3/m^2 \cdot s$). Therefore, volumetric flow rate can be calculated by the integration of velocity distribution over the cross-sectional area that is perpendicular to the flow direction. Consider the flow of a very viscous fluid in the space between two concentric spheres as shown in Figure 1.1. The velocity distribution is given by Bird *et al.* (2002) as

$$
v_\theta = \frac{R \, |\Delta P|}{2\mu E(\varepsilon) \sin\theta} \left[\left(1 - \frac{r}{R} \right) + \kappa \left(1 - \frac{R}{r} \right) \right]
$$

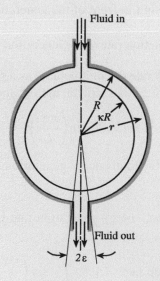

Figure 1.1. Flow between concentric spheres.

where

$$E(\varepsilon) = \ln\left(\frac{1 + \cos\varepsilon}{1 - \cos\varepsilon}\right)$$

Calculate the volumetric flow rate, $\mathcal{Q}$.

Solution

Since the velocity is in the θ-direction, the differential area that is perpendicular to the flow direction is given by Eq. (A.1-9) in Appendix A as

$$dA = r\sin\theta\, dr\, d\phi \tag{1}$$

Therefore, the volumetric flow rate is

$$\mathcal{Q} = \int_0^{2\pi}\int_{\kappa R}^R v_\theta\, r\sin\theta\, dr\, d\phi \tag{2}$$

Substitution of the velocity distribution into Eq. (2) and integration give

$$\mathcal{Q} = \frac{\pi R^3(1-\kappa)^3}{6\mu E(\varepsilon)}\,|\Delta P| \tag{3}$$

1.3.2 Rate of Generation Term

The generation rate per unit volume is denoted by $\mathfrak{R}$ and it may be constant or dependent on position. Thus, the generation rate is expressed as

$$\text{Generation rate} = \begin{cases} (\mathfrak{R})(\text{Volume}) & \text{if } \mathfrak{R} \text{ is constant} \\[2ex] \iiint\limits_V \mathfrak{R}\, dV & \text{if } \mathfrak{R} \text{ is position dependent} \end{cases} \tag{1.3-2}$$

where V is the volume of the system in question. It is also possible to have the depletion of a quantity. In that case, the plus sign in front of the generation term must be replaced by the minus sign, i.e.,

$$\text{Depletion rate} = -\text{ Generation rate} \tag{1.3-3}$$

Example 1.4 Energy generation rate per unit volume as a result of an electric current passing through a rectangular plate of cross-sectional area A and thickness L is given by

$$\mathfrak{R} = \mathfrak{R}_o \sin\left(\frac{\pi x}{L}\right)$$

where $\mathfrak{R}$ is in W/m^3. Calculate the total energy generation rate within the plate.

Solution

Since $\mathfrak{R}$ is dependent on position, energy generation rate is calculated by integration of $\mathfrak{R}$ over the volume of the plate, i.e.,

$$\text{Energy generation rate} = A\,\mathfrak{R}_o \int_0^L \sin\left(\frac{\pi x}{L}\right) dx = \frac{2AL\,\mathfrak{R}_o}{\pi}$$

1.3.3   Rate of Accumulation Term

The rate of accumulation of any quantity φ is the time rate of change of that particular quantity within the volume of the system. Let ρ be the mass density and $\widehat{\varphi}$ be the quantity per unit mass. Thus,

$$\text{Total quantity of } \varphi = \iiint\limits_{V} \rho\widehat{\varphi}\, dV \tag{1.3-4}$$

and the rate of accumulation is given by

$$\text{Accumulation rate} = \frac{d}{dt}\left(\iiint\limits_{V} \rho\widehat{\varphi}\, dV\right) \tag{1.3-5}$$

If $\widehat{\varphi}$ is independent of position, then Eq. (1.3-5) simplifies to

$$\text{Accumulation rate} = \frac{d}{dt}(m\,\widehat{\varphi}) \tag{1.3-6}$$

where m is the total mass within the system.

The accumulation rate may be positive or negative depending on whether the quantity is increasing or decreasing with time within the volume of the system.

1.4   SIMPLIFICATION OF THE RATE EQUATION

In this section, the general rate equation given by Eq. (1.1-1) will be simplified for two special cases: (*i*) steady-state transport without generation, (*ii*) steady-state transport with generation.

1.4.1   Steady-State Transport Without Generation

For this case Eq. (1.1-1) reduces to

$$\text{Rate of input of } \varphi = \text{Rate of output of } \varphi \tag{1.4-1}$$

Equation (1.4-1) can also be expressed in terms of flux as

$$\iint\limits_{A_{in}} (\text{Inlet flux of } \varphi)\, dA = \iint\limits_{A_{out}} (\text{Outlet flux of } \varphi)\, dA \tag{1.4-2}$$

For constant inlet and outlet fluxes Eq. (1.4-2) reduces to

$$\begin{pmatrix} \text{Inlet flux} \\ \text{of } \varphi \end{pmatrix} \begin{pmatrix} \text{Inlet} \\ \text{area} \end{pmatrix} = \begin{pmatrix} \text{Outlet flux} \\ \text{of } \varphi \end{pmatrix} \begin{pmatrix} \text{Outlet} \\ \text{area} \end{pmatrix} \tag{1.4-3}$$

If the inlet and outlet areas are equal, then Eq. (1.4-3) becomes

$$\text{Inlet flux of } \varphi = \text{Outlet flux of } \varphi \tag{1.4-4}$$

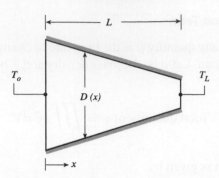

Figure 1.2. Heat transfer through a solid circular cone.

It is important to note that Eq. (1.4-4) is valid as long as the areas perpendicular to the direction of flow at the inlet and outlet of the system are equal to each other. The variation of the area in between does not affect this conclusion. Equation (1.4-4) obviously is not valid for the transfer processes taking place in the radial direction in cylindrical and spherical coordinate systems. In this case either Eq. (1.4-2) or Eq. (1.4-3) should be used.

Example 1.5 Consider a solid cone of circular cross-section whose lateral surface is well insulated as shown in Figure 1.2. The diameters at $x = 0$ and $x = L$ are 25 cm and 5 cm, respectively. If the heat flux at $x = 0$ is 45 W/m^2 under steady conditions, determine the heat transfer rate and the value of the heat flux at $x = L$.

Solution

For steady-state conditions without generation, the heat transfer rate is constant and can be determined from Eq. (1.3-1) as

$$\text{Heat transfer rate} = (\text{Heat flux})_{x=0}(\text{Area})_{x=0}$$

Since the cross-sectional area of the cone is $\pi D^2/4$, then

$$\text{Heat transfer rate} = (45)\left[\frac{\pi (0.25)^2}{4}\right] = 2.21 \text{ W}$$

The value of the heat transfer rate is also 2.21 W at $x = L$. However, the heat flux does depend on position and its value at $x = L$ is

$$(\text{Heat flux})_{x=L} = \frac{2.21}{[\pi(0.05)^2/4]} = 1126 \text{ W/m}^2$$

Comment: Heat flux values are different from each other even though the heat flow rate is constant. Therefore, it is important to specify the area upon which a given heat flux is based when the area changes as a function of position.

1.4.2 Steady-State Transport with Generation

For this case Eq. (1.1-1) reduces to

$$\left(\begin{array}{c} \text{Rate of} \\ \text{input of } \varphi \end{array}\right) + \left(\begin{array}{c} \text{Rate of} \\ \text{generation of } \varphi \end{array}\right) = \left(\begin{array}{c} \text{Rate of} \\ \text{output of } \varphi \end{array}\right) \qquad (1.4\text{-}5)$$

Equation (1.4-5) can also be written in the form

$$\iint_{A_{in}} (\text{Inlet flux of } \varphi)\, dA + \iiint_{V_{sys}} \Re\, dV = \iint_{A_{out}} (\text{Outlet flux of } \varphi)\, dA \qquad (1.4\text{-}6)$$

where $\Re$ is the generation rate per unit volume. If the inlet and outlet fluxes together with the generation rate are constant, then Eq. (1.4-6) reduces to

$$\left(\begin{array}{c} \text{Inlet flux} \\ \text{of } \varphi \end{array}\right)\left(\begin{array}{c} \text{Inlet} \\ \text{area} \end{array}\right) + \Re \left(\begin{array}{c} \text{System} \\ \text{volume} \end{array}\right) = \left(\begin{array}{c} \text{Outlet flux} \\ \text{of } \varphi \end{array}\right)\left(\begin{array}{c} \text{Outlet} \\ \text{area} \end{array}\right) \qquad (1.4\text{-}7)$$

Example 1.6 An exothermic chemical reaction takes place in a 20 cm thick slab and the energy generation rate per unit volume is 1×10^6 W/m^3. The steady-state heat transfer rate into the slab at the left-hand side, i.e., at $x = 0$, is 280 W. Calculate the heat transfer rate to the surroundings from the right-hand side of the slab, i.e., at $x = L$. The surface area of each face is 40 cm^2.

Solution

At steady-state, there is no accumulation of energy and the use of Eq. (1.4-5) gives

$$(\text{Heat transfer rate})_{x=L} = (\text{Heat transfer rate})_{x=0} + \Re\,(\text{Volume})$$

$$= 280 + (1 \times 10^6)(40 \times 10^{-4})(20 \times 10^{-2}) = 1080 \text{ W}$$

The values of the heat fluxes at $x = 0$ and $x = L$ are

$$(\text{Heat flux})_{x=0} = \frac{280}{40 \times 10^{-4}} = 70 \times 10^3 \text{ W/m}^2$$

$$(\text{Heat flux})_{x=L} = \frac{1080}{40 \times 10^{-4}} = 270 \times 10^3 \text{ W/m}^2$$

Comment: Even though the steady-state conditions prevail, neither the heat transfer rate nor the heat flux are constant. This is due to the generation of energy within the slab.

REFERENCE

Bird, R.B., W.E. Stewart and E.N. Lightfoot, 2002, Transport Phenomena, 2nd Ed., Wiley, New York.

SUGGESTED REFERENCES FOR FURTHER STUDY

Brodkey, R.S. and H.C. Hershey, 1988, Transport Phenomena: A Unified Approach, McGraw-Hill, New York.
Fahien, R.W., 1983, Fundamentals of Transport Phenomena, McGraw-Hill, New York.
Felder, R.M. and R.W. Rousseau, 2000, Elementary Principles of Chemical Processes, 3rd Ed., Wiley, New York.
Incropera, F.P. and D.P. DeWitt, 2002, Fundamentals of Heat and Mass Transfer, 5th Ed., Wiley, New York.

PROBLEMS

1.1 One of your friends writes down the inventory rate equation for money as

$$\begin{pmatrix} \text{Change in amount} \\ \text{of dollars} \end{pmatrix} = (\text{Interest}) - \begin{pmatrix} \text{Service} \\ \text{charge} \end{pmatrix} + \begin{pmatrix} \text{Dollars} \\ \text{deposited} \end{pmatrix} - \begin{pmatrix} \text{Checks} \\ \text{written} \end{pmatrix}$$

Identify the terms in the above equation.

1.2 Determine whether steady- or unsteady-state conditions prevail for the following cases:

a) The height of water in a dam during heavy rain,
b) The weight of an athlete during a marathon,
c) The temperature of an ice cube as it melts.

1.3 What is the form of the function $\varphi(x, y)$ if $\partial^2\varphi/\partial x \partial y = 0$?

(**Answer:** $\varphi(x, y) = f(x) + h(y) + C$, where C is a constant)

1.4 Steam at a temperature of 200 °C flows through a pipe of 5 cm inside diameter and 6 cm outside diameter. The length of the pipe is 30 m. If the steady rate of heat loss per unit length of the pipe is 2 W/m, calculate the heat fluxes at the inner and outer surfaces of the pipe.

(**Answer:** 12.7 W/m^2 and 10.6 W/m^2)

1.5 Dust evolves at a rate of 0.3 kg/h in a foundry of dimensions 20 m × 8 m × 4 m. According to ILO (International Labor Organization) standards, the dust concentration should not exceed 20 mg/m^3 to protect workers' health. Determine the volumetric flow rate of ventilating air to meet the standards of ILO.

(**Answer:** 15, 000 m^3/h)

1.6 An incompressible Newtonian fluid flows in the z-direction in space between two parallel plates that are separated by a distance $2B$ as shown in Figure 1.3(a). The length and the width of each plate are L and W, respectively. The velocity distribution under steady conditions is given by

$$v_z = \frac{|\Delta P|B^2}{2\mu L}\left[1 - \left(\frac{x}{B}\right)^2\right]$$

a) For the coordinate system shown in Figure 1.3(b), show that the velocity distribution takes the form

$$v_z = \frac{|\Delta P|B^2}{2\mu L}\left[2\left(\frac{x}{B}\right) - \left(\frac{x}{B}\right)^2\right]$$

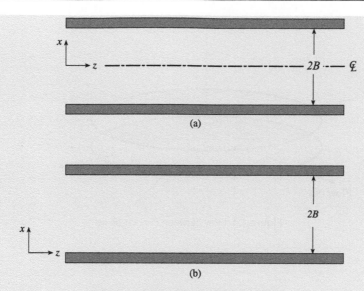

Figure 1.3. Flow between parallel plates.

b) Calculate the volumetric flow rate by using the velocity distributions given above. What is your conclusion?

$$\left(\textbf{Answer:} \text{ b) For both cases } Q = \frac{2\,|\Delta P|\,B^3 W}{3\mu L}\right)$$

1.7 An incompressible Newtonian fluid flows in the z-direction through a straight duct of triangular cross-sectional area, bounded by the plane surfaces $y = H$, $y = \sqrt{3}\,x$ and $y = -\sqrt{3}\,x$. The velocity distribution under steady conditions is given by

$$v_z = \frac{|\Delta P|}{4\mu L H}(y - H)\left(3x^2 - y^2\right)$$

Calculate the volumetric flow rate.

$$\left(\textbf{Answer:} \ Q = \frac{\sqrt{3}\,H^4\,|\Delta P|}{180\mu L}\right)$$

1.8 For radial flow of an incompressible Newtonian fluid between two parallel circular disks of radius R_2 as shown in Figure 1.4, the steady-state velocity distribution is (Bird *et al.*, 2002)

$$v_r = \frac{b^2|\Delta P|}{2\mu r \ln(R_2/R_1)}\left[1 - \left(\frac{z}{b}\right)^2\right]$$

where R_1 is the radius of the entrance hole. Determine the volumetric flow rate.

$$\left(\textbf{Answer:} \ Q = \frac{4}{3}\frac{\pi b^3\,|\Delta P|}{\ln(R_2/R_1)}\right)$$

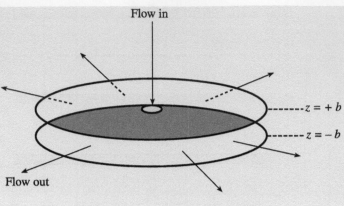

Figure 1.4. Flow between circular disks.

2

MOLECULAR AND CONVECTIVE TRANSPORT

The total flux of any quantity is the sum of the molecular and convective fluxes. The fluxes arising from potential gradients or driving forces are called *molecular fluxes*. Molecular fluxes are expressed in the form of *constitutive* (or *phenomenological*) *equations* for momentum, energy, and mass transport. Momentum, energy, and mass can also be transported by bulk fluid motion or bulk flow, and the resulting flux is called *convective flux*. This chapter deals with the formulation of molecular and convective fluxes in momentum, energy, and mass transport.

2.1 MOLECULAR TRANSPORT

Substances may behave differently when subjected to the same gradients. *Constitutive equations* identify the characteristics of a particular substance. For example, if the gradient is momentum, then the viscosity is defined by the constitutive equation called *Newton's law of viscosity*. If the gradient is energy, then the thermal conductivity is defined by *Fourier's law of heat conduction*. If the gradient is concentration, then the diffusion coefficient is defined by *Fick's first law of diffusion*. Viscosity, thermal conductivity, and diffusion coefficient are called *transport properties*.

2.1.1 Newton's Law of Viscosity

Consider a fluid contained between two large parallel plates of area A, separated by a very small distance Y. The system is initially at rest but at time $t = 0$ the lower plate is set in motion in the x-direction at a constant velocity V by applying a force F in the x-direction while the upper plate is kept stationary. The resulting velocity profiles are shown in Figure 2.1 for various times. At $t = 0$, the velocity is zero everywhere except at the lower plate, which has a velocity V. Then the velocity distribution starts to develop as a function of time. Finally, at steady-state, a linear velocity distribution is obtained.

Experimental results show that the force required to maintain the motion of the lower plate per unit area (or momentum flux) is proportional to the velocity gradient, i.e.,

$$\underbrace{\frac{F}{A}}_{\substack{\text{Momentum} \\ \text{flux}}} = \underbrace{\mu}_{\substack{\text{Transport} \\ \text{property}}} \underbrace{\frac{V}{Y}}_{\substack{\text{Velocity} \\ \text{gradient}}} \qquad (2.1\text{-}1)$$

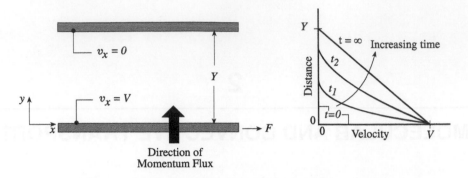

Figure 2.1. Velocity profile development in flow between parallel plates.

and the proportionality constant, μ, is the *viscosity*. Equation (2.1-1) is a macroscopic equation. The microscopic form of this equation is given by

$$\tau_{yx} = -\mu \frac{dv_x}{dy} = -\mu \dot{\gamma}_{yx} \tag{2.1-2}$$

which is known as *Newton's law of viscosity* and any fluid obeying Eq. (2.1-2) is called a *Newtonian fluid*. The term $\dot{\gamma}_{yx}$ is called *rate of strain*[1] or *rate of deformation* or *shear rate*. The term τ_{yx} is called *shear stress*. It contains two subscripts: x represents the direction of force, i.e., F_x, and y represents the direction of the normal to the surface, i.e., A_y, on which the force is acting. Therefore, τ_{yx} is simply the force per unit area, i.e., F_x/A_y. It is also possible to interpret τ_{yx} as the flux of x-momentum in the y-direction.

Since the velocity gradient is negative, i.e., v_x decreases with increasing y, a negative sign is introduced on the right-hand side of Eq. (2.1-2) so that the stress in tension is positive.

In SI units, shear stress is expressed in N/m^2 (Pa) and velocity gradient in (m/s)/m. Thus, the examination of Eq. (2.1-1) indicates that the units of viscosity in SI units are

$$\mu = \frac{N/m^2}{(m/s)/m} = Pa \cdot s = \frac{N \cdot s}{m^2} = \frac{(kg \cdot m/s^2) \cdot s}{m^2} = \frac{kg}{m \cdot s}$$

Most viscosity data in the cgs system are usually reported in g/(cm·s), known as a poise (P), or in centipoise (1 cP = 0.01 P), where

$$1 \, Pa \cdot s = 10 \, P = 10^3 \, cP$$

Viscosity varies with temperature. While liquid viscosity decreases with increasing temperature, gas viscosity increases with increasing temperature. Concentration also affects viscosity for solutions or suspensions. Viscosity values of various substances are given in Table D.1 in Appendix D.

Example 2.1 A Newtonian fluid with a viscosity of 10 cP is placed between two large parallel plates. The distance between the plates is 4 mm. The lower plate is pulled in the positive x-direction with a force of 0.5 N, while the upper plate is pulled in the negative

---

[1]Strain is defined as deformation per unit length. For example, if a spring of original length L_o is stretched to a length L, then the strain is $(L - L_o)/L_o$.

x-direction with a force of 2 N. Each plate has an area of 2.5 m^2. If the velocity of the lower plate is 0.1 m/s, calculate:

a) The steady-state momentum flux,
b) The velocity of the upper plate.

Solution

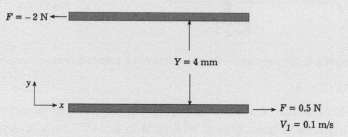

$$F = -2\,\text{N} \longleftarrow$$

$$Y = 4\ \text{mm}$$

$$F = 0.5\ \text{N}$$
$$V_1 = 0.1\ \text{m/s}$$

a) The momentum flux (or force per unit area) is

$$\tau_{yx} = \frac{F}{A} = \frac{0.5 + 2}{2.5} = 1\ \text{Pa}$$

b) Let V_2 be the velocity of the upper plate. From Eq. (2.1-2)

$$\tau_{yx} \int_0^Y dy = -\mu \int_{V_1}^{V_2} dv_x \quad \Rightarrow \quad V_2 = V_1 - \frac{\tau_{yx} Y}{\mu} \tag{1}$$

Substitution of the values into Eq. (1) gives

$$V_2 = 0.1 - \frac{(1)(4 \times 10^{-3})}{10 \times 10^{-3}} = -0.3\ \text{m/s} \tag{2}$$

The minus sign indicates that the upper plate moves in the negative x-direction. Note that the velocity gradient is $dv_x/dy = -100\ \text{s}^{-1}$.

2.1.2   Fourier's Law of Heat Conduction

Consider a slab of solid material of area A between two large parallel plates of a distance Y apart. Initially the solid material is at temperature T_o throughout. Then the lower plate is suddenly brought to a slightly higher temperature, T_1, and maintained at that temperature. The second law of thermodynamics states that heat flows spontaneously from the higher temperature T_1 to the lower temperature T_o. As time proceeds, the temperature profile in the slab changes, and ultimately a linear steady-state temperature is attained as shown in Figure 2.3.

Experimental measurements made at steady-state indicate that the rate of heat flow per unit area is proportional to the temperature gradient, i.e.,

$$\underbrace{\frac{\dot{Q}}{A}}_{\substack{\text{Energy} \\ \text{flux}}} = \underbrace{k}_{\substack{\text{Transport} \\ \text{property}}} \underbrace{\frac{T_1 - T_o}{Y}}_{\substack{\text{Temperature} \\ \text{gradient}}} \tag{2.1-3}$$

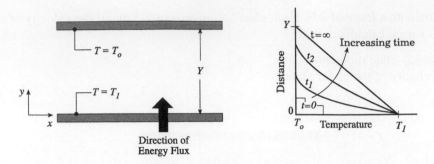

Figure 2.3. Temperature profile development in a solid slab between two plates.

The proportionality constant, k, between the energy flux and the temperature gradient is called *thermal conductivity*. In SI units, $\dot{Q}$ is in W(J/s), A in m^2, dT/dx in K/m, and k in W/m·K. The thermal conductivity of a material is, in general, a function of temperature. However, in many engineering applications the variation is sufficiently small to be neglected. Thermal conductivity values for various substances are given in Table D.2 in Appendix D.

The microscopic form of Eq. (2.1-3) is known as *Fourier's law of heat conduction* and is given by

$$q_y = -k \frac{dT}{dy} \tag{2.1-4}$$

in which the subscript y indicates the direction of the energy flux. The negative sign in Eq. (2.1-4) indicates that heat flows in the direction of decreasing temperature.

Example 2.2 One side of a copper slab receives a net heat input at a rate of 5000 W due to radiation. The other face is held at a temperature of 35 °C. If steady-state conditions prevail, calculate the surface temperature of the side receiving radiant energy. The surface area of each face is 0.05 m^2, and the slab thickness is 4 cm.

Solution

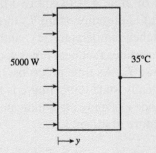

Physical Properties

For copper: $k = 398$ W/m·K

Analysis

System: Copper slab

Under steady conditions with no internal generation, the conservation statement for energy reduces to

$$\text{Rate of energy in} = \text{Rate of energy out} = 5000 \text{ W}$$

Since the slab area across which heat transfer takes place is constant, the heat flux through the slab is also constant, and is given by

$$q_y = \frac{5000}{0.05} = 100{,}000 \text{ W/m}^2$$

Therefore, the use of Fourier's law of heat conduction, Eq. (2.1-4), gives

$$100{,}000 \int_0^{0.04} dy = -398 \int_{T_o}^{35} dT \quad \Rightarrow \quad T_o = 45.1\,^\circ\text{C}$$

2.1.3 Fick's First Law of Diffusion

Consider two large parallel plates of area A. The lower one is coated with a material, A, which has a very low solubility in the stagnant fluid B filling the space between the plates. Suppose that the saturation concentration of A is ρ_{A_o} and A undergoes a rapid chemical reaction at the surface of the upper plate and its concentration is zero at that surface. At $t = 0$ the lower plate is exposed to B and, as time proceeds, the concentration profile develops as shown in Figure 2.4. Since the solubility of A is low, an almost linear distribution is reached under steady conditions.

Experimental measurements indicate that the mass flux of A is proportional to the concentration gradient, i.e.,

$$\underbrace{\frac{\dot{m}_A}{A}}_{\substack{\text{Mass} \\ \text{flux of } A}} = \underbrace{\mathcal{D}_{AB}}_{\substack{\text{Transport} \\ \text{property}}} \underbrace{\frac{\rho_{A_o}}{Y}}_{\substack{\text{Concentration} \\ \text{gradient}}} \tag{2.1-5}$$

where the proportionality constant, $\mathcal{D}_{AB}$, is called the *binary molecular mass diffusivity* (or *diffusion coefficient*) of species A through B. The microscopic form of Eq. (2.1-5) is known

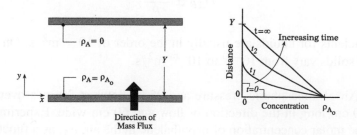

Figure 2.4. Concentration profile development between parallel plates.

as *Fick's first law of diffusion* and is given by

$$
j_{A_y} = -\mathcal{D}_{AB}\rho\,\frac{d\omega_A}{dy} \tag{2.1-6}
$$

where j_{A_y} and ω_A represent the molecular mass flux of species A in the y-direction and mass fraction of species A, respectively. If the total density, ρ, is constant, then the term $\rho(d\omega_A/dy)$ can be replaced by $d\rho_A/dy$ and Eq. (2.1-6) becomes

$$
j_{A_y} = -\mathcal{D}_{AB}\,\frac{d\rho_A}{dy} \qquad \rho = \text{constant} \tag{2.1-7}
$$

To measure $\mathcal{D}_{AB}$ experimentally, it is necessary to design an experiment (like the one given above) in which the convective mass flux is almost zero.

In mass transfer calculations, it is sometimes more convenient to express concentrations in molar units rather than in mass units. In terms of molar concentration, Fick's first law of diffusion is written as

$$
J_{A_y}^* = -\mathcal{D}_{AB}\,c\,\frac{dx_A}{dy} \tag{2.1-8}
$$

where $J_{A_y}^*$ and x_A represent the molecular molar flux of species A in the y-direction and the mole fraction of species A, respectively. If the total molar concentration, c, is constant, then the term $c(dx_A/dy)$ can be replaced by dc_A/dy, and Eq. (2.1-8) becomes

$$
J_{A_y}^* = -\mathcal{D}_{AB}\,\frac{dc_A}{dy} \qquad c = \text{constant} \tag{2.1-9}
$$

The diffusion coefficient has the dimensions of m^2/s in SI units. Typical values of $\mathcal{D}_{AB}$ are given in Appendix D. Examination of these values indicates that the diffusion coefficient of gases has an order of magnitude of 10^{-5} m^2/s under atmospheric conditions. Assuming ideal gas behavior, the pressure and temperature dependence of the diffusion coefficient of gases may be estimated from the relation

$$
\mathcal{D}_{AB} \propto \frac{T^{3/2}}{P} \tag{2.1-10}
$$

Diffusion coefficients for liquids are usually in the order of 10^{-9} m^2/s. On the other hand, $\mathcal{D}_{AB}$ values for solids vary from 10^{-10} to 10^{-14} m^2/s.

Example 2.3 Air at atmospheric pressure and $95\,°C$ flows at 20 m/s over a flat plate of naphthalene 80 cm long in the direction of flow and 60 cm wide. Experimental measurements report the molar concentration of naphthalene in the air, c_A, as a function of distance x from the plate as follows:

x (cm)	c_A (mol/m^3)
0	0.117
10	0.093
20	0.076
30	0.063
40	0.051
50	0.043

Determine the molar flux of naphthalene from the plate surface under steady conditions.

Solution

Physical properties

Diffusion coefficient of naphthalene ($\mathcal{A}$) in air ($\mathcal{B}$) at 95 °C (368 K) is

$$(\mathcal{D}_{AB})_{368} = (\mathcal{D}_{AB})_{300} \left(\frac{368}{300}\right)^{3/2} = (0.62 \times 10^{-5}) \left(\frac{368}{300}\right)^{3/2} = 0.84 \times 10^{-5} \text{ m}^2/\text{s}$$

Assumptions

1. The total molar concentration, c, is constant.
2. Naphthalene plate is also at a temperature of 95 °C.

Analysis

The molar flux of naphthalene transferred from the plate surface to the flowing stream is determined from

$$J_{A_x}^*\big|_{x=0} = -\mathcal{D}_{AB} \left(\frac{dc_A}{dx}\right)_{x=0} \tag{1}$$

It is possible to calculate the concentration gradient on the surface of the plate by using one of the several methods explained in Section A.5 in Appendix A.

Graphical method

The plot of c_A versus x is given in Figure 2.5. The slope of the tangent to the curve at $x = 0$ is -0.0023 (mol/m^3)/cm.

Curve fitting method

From semi-log plot of c_A versus x, shown in Figure 2.6, it appears that a straight line represents the data fairly well. The equation of this line can be determined by the method of least squares in the form

$$y = mx + b \tag{2}$$

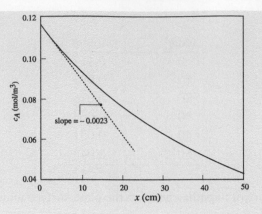

Figure 2.5. Concentration of species $\mathcal{A}$ as a function of position.

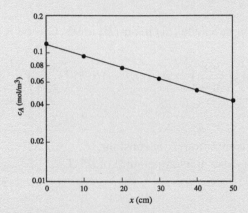

Figure 2.6. Concentration of species $\mathcal{A}$ as a function of position.

where

$$y = \log c_A \qquad (3)$$

To determine the values of m and b from Eqs. (A.6-10) and (A.6-11) in Appendix A, the required values are calculated as follows:

y_i	x_i	$x_i y_i$	x_i^2
-0.932	0	0	0
-1.032	10	-10.32	100
-1.119	20	-22.38	400
-1.201	30	-36.03	900
-1.292	40	-51.68	1600
-1.367	50	-68.35	2500
$\sum y_i = -6.943$	$\sum x_i = 150$	$\sum x_i y_i = -188.76$	$\sum x_i^2 = 5500$

The values of m and b are

$$m = \frac{(6)(-188.76) - (150)(-6.943)}{(6)(5500) - (150)^2} = -0.0087$$

$$b = \frac{(-6.943)(5500) - (150)(-188.76)}{(6)(5500) - (150)^2} = -0.94$$

Therefore, Eq. (2) takes the form

$$\log c_A = -0.087x - 0.94 \quad \Rightarrow \quad c_A = 0.115 e^{-0.02x} \tag{4}$$

Differentiation of Eq. (4) gives the concentration gradient on the surface of the plate as

$$\left(\frac{dc_A}{dx}\right)_{x=0} = -(0.115)(0.02) = -0.0023 \,(\text{mol/m}^3)/\text{cm} = -0.23 \,\text{mol/m}^4$$

Substitution of the numerical values into Eq. (1) gives the molar flux of naphthalene from the surface as

$$J_{A_x}^*\big|_{x=0} = (0.84 \times 10^{-5})(0.23) = 19.32 \times 10^{-7} \,\text{mol/m}^2{\cdot}\text{s}$$

2.2 DIMENSIONLESS NUMBERS

Newton's "law" of viscosity, Fourier's "law" of heat conduction, and Fick's first "law" of diffusion, in reality, **are not laws** but defining equations for viscosity, μ, thermal conductivity, k, and diffusion coefficient, $\mathcal{D}_{AB}$. The fluxes (τ_{yx}, q_y, j_{A_y}) and the gradients (dv_x/dy, dT/dy, $d\rho_A/dy$) must be known or measurable for the experimental determination of μ, k, and $\mathcal{D}_{AB}$.

Newton's law of viscosity, Eq. (2.1-2), Fourier's law of heat conduction, Eq. (2.1-4), and Fick's first law of diffusion, Eqs. (2.1-7) and (2.1-9), can be generalized as

$$\boxed{\begin{pmatrix} \text{Molecular} \\ \text{flux} \end{pmatrix} = \begin{pmatrix} \text{Transport} \\ \text{property} \end{pmatrix} \begin{pmatrix} \text{Gradient of} \\ \text{driving force} \end{pmatrix}} \tag{2.2-1}$$

Although the constitutive equations are similar, they are not completely analogous because the transport properties (μ, k, $\mathcal{D}_{AB}$) have different units. These equations can also be expressed in the following forms:

$$\tau_{yx} = -\frac{\mu}{\rho}\frac{d}{dy}(\rho v_x) \qquad \rho = \text{constant} \qquad \rho v_x = \text{momentum/volume} \tag{2.2-2}$$

$$q_y = -\frac{k}{\rho \widehat{C}_P}\frac{d}{dy}(\rho \widehat{C}_P T) \qquad \rho \widehat{C}_P = \text{constant} \qquad \rho \widehat{C}_P T = \text{energy/volume} \tag{2.2-3}$$

$$j_{A_y} = -\mathcal{D}_{AB}\frac{d\rho_A}{dy} \qquad \rho = \text{constant} \qquad \rho_A = \text{mass of } \mathcal{A}/\text{volume} \tag{2.2-4}$$

The term μ/ρ in Eq. (2.2-2) is called *momentum diffusivity* or *kinematic viscosity*, and the term $k/\rho\widehat{C}_P$ in Eq. (2.2-3) is called *thermal diffusivity*. Momentum and thermal diffusivities

Table 2.1. Analogous terms in constitutive equations for momentum, energy, and mass (or mole) transfer in one-dimension

	Momentum	Energy	Mass	Mole
Molecular flux	τ_{yx}	q_y	j_{A_y}	$J_{A_y}^*$
Transport property	μ	k	$\mathcal{D}_{AB}$	$\mathcal{D}_{AB}$
Gradient of driving force	$\dfrac{dv_x}{dy}$	$\dfrac{dT}{dy}$	$\dfrac{d\rho_A}{dy}$	$\dfrac{dc_A}{dy}$
Diffusivity	ν	α	$\mathcal{D}_{AB}$	$\mathcal{D}_{AB}$
Quantity/Volume	ρv_x	$\rho \widehat{C}_P T$	ρ_A	c_A
Gradient of Quantity/Volume	$\dfrac{d(\rho v_x)}{dy}$	$\dfrac{d(\rho \widehat{C}_P T)}{dy}$	$\dfrac{d\rho_A}{dy}$	$\dfrac{dc_A}{dy}$

are designated by ν and α, respectively. Note that the terms ν, α, and $\mathcal{D}_{AB}$ all have the same units, m^2/s, and Eqs. (2.2-2)–(2.2-4) can be expressed in the general form as

$$\boxed{\left(\begin{array}{c}\text{Molecular}\\ \text{flux}\end{array}\right) = (\text{Diffusivity})\left(\begin{array}{c}\text{Gradient of}\\ \text{Quantity/Volume}\end{array}\right)} \qquad (2.2\text{-}5)$$

The quantities that appear in Eqs. (2.2-1) and (2.2-5) are summarized in Table 2.1.

Since the terms ν, α, and $\mathcal{D}_{AB}$ all have the same units, the ratio of any two of these diffusivities results in a dimensionless number. For example, the ratio of momentum diffusivity to thermal diffusivity gives the *Prandtl number*, Pr:

$$\text{Prandtl number} = \text{Pr} = \frac{\nu}{\alpha} = \frac{\widehat{C}_P \mu}{k} \qquad (2.2\text{-}6)$$

The Prandtl number is a function of temperature and pressure. However, its dependence on temperature, at least for liquids, is much stronger. The order of magnitude of the Prandtl number for gases and liquids can be estimated as

$$\text{Pr} = \frac{(10^3)(10^{-5})}{10^{-2}} = 1 \qquad \text{for gases}$$

$$\text{Pr} = \frac{(10^3)(10^{-3})}{10^{-1}} = 10 \qquad \text{for liquids}$$

The *Schmidt number* is defined as the ratio of the momentum to mass diffusivities:

$$\text{Schmidt number} = \text{Sc} = \frac{\nu}{\mathcal{D}_{AB}} = \frac{\mu}{\rho \mathcal{D}_{AB}} \qquad (2.2\text{-}7)$$

The order of magnitude of the Schmidt number for gases and liquids can be estimated as

$$\text{Sc} = \frac{10^{-5}}{(1)(10^{-5})} = 1 \qquad \text{for gases}$$

$$\text{Sc} = \frac{10^{-3}}{(10^3)(10^{-9})} = 10^3 \qquad \text{for liquids}$$

Finally, the ratio of α to $\mathcal{D}_{AB}$ gives the *Lewis number*, Le:

$$\text{Lewis number} = \text{Le} = \frac{\alpha}{\mathcal{D}_{AB}} = \frac{k}{\rho \, \widehat{C}_P \mathcal{D}_{AB}} = \frac{\text{Sc}}{\text{Pr}} \qquad (2.2\text{-}8)$$

2.3 CONVECTIVE TRANSPORT

Convective flux or bulk flux of a quantity is expressed as

$$\boxed{\begin{pmatrix} \text{Convective} \\ \text{flux} \end{pmatrix} = (\text{Quantity/Volume}) \begin{pmatrix} \text{Characteristic} \\ \text{velocity} \end{pmatrix}} \qquad (2.3\text{-}1)$$

When air is pumped through a pipe, it is considered a single phase and a single component system. In this case, there is no ambiguity in defining the characteristic velocity. However, if the oxygen in the air were reacting, then the fact that air is composed predominantly of two species, O_2 and N_2, would have to be taken into account. Hence, air should be considered a single phase, binary component system. For a single phase system composed of n components, the general definition of a characteristic velocity is given by

$$v_{ch} = \sum_i^n \beta_i \, v_i \qquad (2.3\text{-}2)$$

where β_i is the weighting factor and v_i is the velocity of a constituent. The three most common characteristic velocities are listed in Table 2.2. The term $\overline{V}_i$ in the definition of the volume average velocity represents the partial molar volume of a constituent. The molar average velocity is equal to the volume average velocity when the total molar concentration, c, is constant. On the other hand, the mass average velocity is equal to the volume average velocity when the total mass density, ρ, is constant.

The choice of a characteristic velocity is arbitrary. For a given problem, it is more convenient to select a characteristic velocity that will make the convective flux zero and thus yield a simpler problem. In the literature, it is common practice to use the molar average velocity for dilute gases, i.e., $c = \text{constant}$, and the mass average velocity for liquids, i.e., $\rho = \text{constant}$.

It should be noted that the molecular mass flux expression given by Eq. (2.1-6) represents the molecular mass flux with respect to the mass average velocity. Therefore, in the equation representing the total mass flux, the characteristic velocity in the convective mass flux term is taken as the mass average velocity. On the other hand, Eq. (2.1-8) is the molecular molar flux with respect to the molar average velocity. Therefore, the molar average velocity is considered the characteristic velocity in the convective molar flux term.

Table 2.2. Common characteristic velocities

Characteristic Velocity	Weighting Factor	Formulation
Mass average	Mass fraction (ω_i)	$v = \sum_i \omega_i v_i$
Molar average	Mole fraction (x_i)	$v^* = \sum_i x_i v_i$
Volume average	Volume fraction ($c_i \overline{V}_i$)	$v^\blacksquare = \sum_i c_i \overline{V}_i v_i$

2.4 TOTAL FLUX

Since the total flux of any quantity is the sum of its molecular and convective fluxes, then

$$
\begin{pmatrix} \text{Total} \\ \text{flux} \end{pmatrix} = \underbrace{\begin{pmatrix} \text{Transport} \\ \text{property} \end{pmatrix} \begin{pmatrix} \text{Gradient of} \\ \text{driving force} \end{pmatrix}}_{\text{Molecular flux}} + \underbrace{\begin{pmatrix} \text{Quantity} \\ \overline{\text{Volume}} \end{pmatrix} \begin{pmatrix} \text{Characteristic} \\ \text{velocity} \end{pmatrix}}_{\text{Convective flux}} \quad (2.4\text{-}1)
$$

or,

$$
\begin{pmatrix} \text{Total} \\ \text{flux} \end{pmatrix} = \underbrace{(\text{Diffusivity}) \begin{pmatrix} \text{Gradient of} \\ \text{Quantity/Volume} \end{pmatrix}}_{\text{Molecular flux}} + \underbrace{\begin{pmatrix} \text{Quantity} \\ \overline{\text{Volume}} \end{pmatrix} \begin{pmatrix} \text{Characteristic} \\ \text{velocity} \end{pmatrix}}_{\text{Convective flux}} \quad (2.4\text{-}2)
$$

The quantities that appear in Eqs. (2.4-1) and (2.4-2) are given in Table 2.3.

The general flux expressions for momentum, energy, and mass transport in different coordinate systems are given in Appendix C.

From Eq. (2.4-2), the ratio of the convective flux to the molecular flux is given by

$$
\frac{\text{Convective flux}}{\text{Molecular flux}} = \frac{(\text{Quantity/Volume})(\text{Characteristic velocity})}{(\text{Diffusivity})(\text{Gradient of Quantity/Volume})} \quad (2.4\text{-}3)
$$

Table 2.3. Analogous terms in flux expressions for various types of transport in one-dimension

Type of Transport	Total Flux	Molecular Flux	Convective Flux	Constraint
Momentum	π_{yx}	$-\mu \dfrac{dv_x}{dy}$	$(\rho v_x) v_y$	None
		$-\nu \dfrac{d(\rho v_x)}{dy}$		$\rho = \text{const.}$
Energy	e_y	$-k \dfrac{dT}{dy}$	$(\rho \widehat{C}_P T) v_y$	None
		$-\alpha \dfrac{d(\rho \widehat{C}_P T)}{dy}$		$\rho \widehat{C}_P = \text{const.}$
Mass	$\mathcal{W}_{A_y}$	$-\rho \mathcal{D}_{AB} \dfrac{d\omega_A}{dy}$	$\rho_A v_y$	None
		$-\mathcal{D}_{AB} \dfrac{d\rho_A}{dy}$		$\rho = \text{const.}$
Mole	N_{A_y}	$-c \mathcal{D}_{AB} \dfrac{dx_A}{dy}$	$c_A v_y^*$	None
		$-\mathcal{D}_{AB} \dfrac{dc_A}{dy}$		$c = \text{const.}$

Since the gradient of a quantity represents the variation of that particular quantity over a characteristic length, the "Gradient of Quantity/Volume" can be expressed as

$$\text{Gradient of Quantity/Volume} = \frac{\text{Difference in Quantity/Volume}}{\text{Characteristic length}} \qquad (2.4\text{-}4)$$

The use of Eq. (2.4-4) in Eq. (2.4-3) gives

$$\frac{\text{Convective flux}}{\text{Molecular flux}} = \frac{(\text{Characteristic velocity})(\text{Characteristic length})}{\text{Diffusivity}} \qquad (2.4\text{-}5)$$

The ratio of the convective flux to the molecular flux is known as the *Peclet number*, Pe. Therefore, Peclet numbers for heat and mass transfers are

$$\text{Pe}_\text{H} = \frac{v_{ch} L_{ch}}{\alpha} \qquad (2.4\text{-}6)$$

$$\text{Pe}_\text{M} = \frac{v_{ch} L_{ch}}{\mathcal{D}_{AB}} \qquad (2.4\text{-}7)$$

Hence, the total flux of any quantity is given by

$$\text{Total flux} = \begin{cases} \text{Molecular flux} & \text{Pe} \ll 1 \\ \text{Molecular flux} + \text{Convective flux} & \text{Pe} \simeq 1 \\ \text{Convective flux} & \text{Pe} \gg 1 \end{cases} \qquad (2.4\text{-}8)$$

2.4.1   Rate of Mass Entering and/or Leaving the System

The mass flow rate of species i entering and/or leaving the system, $\dot{m}_i$, is expressed as

$$\dot{m}_i = \left[\underbrace{\left(\begin{array}{c}\text{Mass} \\ \text{Diffusivity}\end{array}\right)\left(\begin{array}{c}\text{Gradient of} \\ \text{Mass of } i/\text{Volume}\end{array}\right)}_{\text{Molecular mass flux of species } i} + \underbrace{\left(\begin{array}{c}\text{Mass of } i \\ \text{Volume}\end{array}\right)\left(\begin{array}{c}\text{Characteristic} \\ \text{velocity}\end{array}\right)}_{\text{Convective mass flux of species } i} \right] \left(\begin{array}{c}\text{Flow} \\ \text{area}\end{array}\right)$$

$$(2.4\text{-}9)$$

In general, the mass of species i may enter and/or leave the system by two means:

- Entering and/or leaving conduits,
- Exchange of mass between the system and its surroundings through the boundaries of the system, i.e., interphase transport.

When a mass of species i enters and/or leaves the system by a conduit(s), the characteristic velocity is taken as the average velocity of the flowing stream and it is usually large enough to neglect the molecular flux compared to the convective flux, i.e., $\text{Pe}_\text{M} \gg 1$. Therefore, Eq. (2.4-9) simplifies to

$$\dot{m}_i = \left(\frac{\text{Mass of } i}{\text{Volume}}\right)\left(\begin{array}{c}\text{Average} \\ \text{velocity}\end{array}\right)\left(\begin{array}{c}\text{Flow} \\ \text{area}\end{array}\right) \qquad (2.4\text{-}10)$$

or,

$$\boxed{\dot{m}_i = \rho_i \langle v \rangle A = \rho_i \mathcal{Q}} \qquad (2.4\text{-}11)$$

Summation of Eq. (2.4-11) over all species leads to the total mass flow rate, $\dot{m}$, entering and/or leaving the system by a conduit in the form

$$\dot{m} = \rho \langle v \rangle A = \rho \mathcal{Q}$$ (2.4-12)

On a molar basis, Eqs. (2.4-11) and (2.4-12) take the form

$$\dot{n}_i = c_i \langle v \rangle A = c_i \mathcal{Q}$$ (2.4-13)

$$\dot{n} = c \langle v \rangle A = c \mathcal{Q}$$ (2.4-14)

On the other hand, when a mass of species i enters and/or leaves the system as a result of interphase transport, the flux expression to be used is dictated by the value of the Peclet number as shown in Eq. (2.4-8).

Example 2.4 Liquid $\mathcal{B}$ is flowing over a vertical plate as shown in Figure 2.7. The surface of the plate is coated with a material, $\mathcal{A}$, which has a very low solubility in liquid $\mathcal{B}$. The concentration distribution of species $\mathcal{A}$ in the liquid is given by Bird *et al.* (2002) as

$$\frac{c_A}{c_{A_o}} = \frac{1}{\Gamma(4/3)} \int_\eta^\infty e^{-u^3} \, du$$

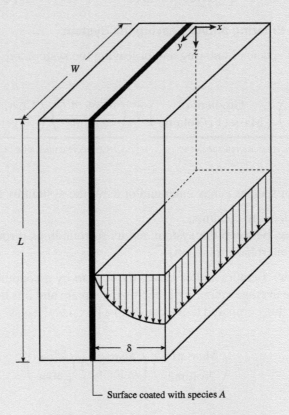

Figure 2.7. Solid dissolution into a falling film.

where c_{A_o} is the solubility of $\mathcal{A}$ in $\mathcal{B}$, η is the dimensionless parameter defined by

$$\eta = x \left(\frac{\rho g \delta}{9 \mu \mathcal{D}_{AB} z} \right)^{1/3}$$

and $\Gamma(4/3)$ is the gamma function defined by

$$\Gamma(n) = \int_0^\infty \beta^{n-1} e^{-\beta} \, d\beta \quad n > 0$$

Calculate the rate of transfer of species $\mathcal{A}$ into the flowing liquid.

Solution

Assumptions

1. The total molar concentration in the liquid phase is constant.
2. In the x-direction, the convective flux is small compared to the molecular flux.

Analysis

The molar rate of transfer of species $\mathcal{A}$ can be calculated from the expression

$$\dot{n}_A = \int_0^W \int_0^L N_{A_x} \big|_{x=0} \, dz \, dy \tag{1}$$

where the total molar flux of species $\mathcal{A}$ at the interface, $N_{A_x}|_{x=0}$, is given by

$$N_{A_x} \big|_{x=0} = J_{A_x}^* \big|_{x=0} = -\mathcal{D}_{AB} \left(\frac{\partial c_A}{\partial x} \right)_{x=0} \tag{2}$$

By the application of the chain rule, Eq. (2) takes the form

$$N_{A_x} \big|_{x=0} = -\mathcal{D}_{AB} \frac{\partial \eta}{\partial x} \left(\frac{d c_A}{d \eta} \right)_{\eta=0} \tag{3}$$

The term $\partial \eta / \partial x$ is

$$\frac{\partial \eta}{\partial x} = \left(\frac{\rho g \delta}{9 \mu \mathcal{D}_{AB} z} \right)^{1/3} \tag{4}$$

On the other hand, the term $d c_A / d \eta$ can be calculated by the application of the Leibnitz formula, i.e., Eq. (A.4-3) in Appendix A, as

$$\frac{d c_A}{d \eta} = -\frac{c_{A_o}}{\Gamma(4/3)} e^{-\eta^3} \tag{5}$$

Substitution of Eqs. (4) and (5) into Eq. (3) yields

$$N_{A_x} \big|_{x=0} = \frac{\mathcal{D}_{AB} \, c_{A_o}}{\Gamma(4/3)} \left(\frac{\rho g \delta}{9 \mu \mathcal{D}_{AB} z} \right)^{1/3} \tag{6}$$

Finally, the use of Eq. (6) in Eq. (1) gives the molar rate of transfer of species $\mathcal{A}$ as

$$\dot{n}_A = \frac{1}{2} \frac{W c_{A_o}}{\Gamma(4/3)} \left(\frac{3 \rho g \delta}{\mu} \right)^{1/3} (\mathcal{D}_{AB} L)^{2/3} \tag{7}$$

2.4.2 Rate of Energy Entering and/or Leaving the System

The rate of energy entering and/or leaving the system, $\dot{E}$, is expressed as

$$\dot{E} = \left[\underbrace{\left(\begin{array}{c} \text{Thermal} \\ \text{diffusivity} \end{array} \right) \left(\begin{array}{c} \text{Gradient of} \\ \text{Energy/Volume} \end{array} \right)}_{\text{Molecular energy flux}} + \underbrace{\left(\begin{array}{c} \text{Energy} \\ \text{Volume} \end{array} \right) \left(\begin{array}{c} \text{Characteristic} \\ \text{velocity} \end{array} \right)}_{\text{Convective energy flux}} \right] \left(\begin{array}{c} \text{Flow} \\ \text{area} \end{array} \right)$$

(2.4-15)

As in the case of mass, energy may enter or leave the system by two means:

- By inlet and/or outlet streams,
- By exchange of energy between the system and its surroundings through the boundaries of the system in the form of heat and work.

When energy enters and/or leaves the system by a conduit(s), the characteristic velocity is taken as the average velocity of the flowing stream and it is usually large enough to neglect the molecular flux compared to the convective flux, i.e., $Pe_H \gg 1$. Therefore, Eq. (2.4-15) simplifies to

$$\dot{E} = \left(\frac{\text{Energy}}{\text{Volume}} \right) \left(\begin{array}{c} \text{Average} \\ \text{velocity} \end{array} \right) \left(\begin{array}{c} \text{Flow} \\ \text{area} \end{array} \right)$$

(2.4-16)

Energy per unit volume, on the other hand, is expressed as the product of energy per unit mass, $\widehat{E}$, and mass per unit volume, i.e., density, such that Eq. (2.4-16) becomes

$$\dot{E} = \left(\frac{\text{Energy}}{\text{Mass}} \right) \underbrace{\left(\frac{\text{Mass}}{\text{Volume}} \right) \left(\begin{array}{c} \text{Average} \\ \text{velocity} \end{array} \right) \left(\begin{array}{c} \text{Flow} \\ \text{area} \end{array} \right)}_{\text{Mass flow rate}} = \widehat{E} \dot{m}$$

(2.4-17)

NOTATION

A	area, m^2
$\widehat{C}_P$	heat capacity at constant pressure, kJ/kg·K
c	total concentration, kmol/m^3
c_i	concentration of species i, kmol/m^3
$\mathcal{D}_{AB}$	diffusion coefficient for system $\mathcal{A}$-$\mathcal{B}$, m^2/s
$\dot{E}$	rate of energy, W
e	total energy flux, W/m^2
F	force, N
J^*	molecular molar flux, kmol/m^2·s
j	molecular mass flux, kg/m^2·s
k	thermal conductivity, W/m·K
$\dot{m}$	total mass flow rate, kg/s
$\dot{m}_i$	mass flow rate of species i, kg/s
N	total molar flux, kmol/m^2·s

$\dot{n}$	total molar flow rate, kmol/s
$\dot{n}_i$	molar flow rate of species i, kmol/s
P	pressure, Pa
$\dot{Q}$	heat transfer rate, W
Q	volumetric flow rate, m^3/s
q	heat flux, W/m^2
T	temperature, °C or K
t	time, s
V	volume, m^3
$\overline{V}_i$	partial molar volume of species i, m^3/kmol
v	velocity, m/s
v^*	molar average velocity, m/s
$v^{\blacksquare}$	volume average velocity, m/s
$\mathcal{W}$	total mass flux, kg/m$^2\cdot$s
x	rectangular coordinate, m
x_i	mole fraction of species i
y	rectangular coordinate, m
α	thermal diffusivity, m^2/s
$\dot{\gamma}$	rate of strain, 1/s
μ	viscosity, kg/m·s
ν	kinematic viscosity (or momentum diffusivity), m^2/s
π	total momentum flux, N/m^2
ρ	total density, kg/m^3
ρ_i	density of species i, kg/m^3
τ_{yx}	flux of x-momentum in the y-direction, N/m^2
ω_i	mass fraction of species i

Overlines

$\frown$	per unit mass
–	partial molar

Bracket

$\langle a \rangle$	average value of a

Superscript

sat	saturation

Subscripts

A, B	species in binary systems
ch	characteristic
i	species in multicomponent systems

Dimensionless Numbers

Le	Lewis number
Pe_H	Peclet number for heat transfer
Pe_M	Peclet number for mass transfer
Pr	Prandtl number
Sc	Schmidt number

REFERENCES

Bird, R.B., W.E. Stewart and E.N. Lightfoot, 2002, Transport Phenomena, 2nd Ed., Wiley, New York.
Kelvin, W.T., 1864, The secular cooling of the earth, Trans. Roy. Soc. Edin. 23, 157.

SUGGESTED REFERENCES FOR FURTHER STUDY

Brodkey, R.S. and H.C. Hershey, 1988, Transport Phenomena – A Unified Approach, McGraw-Hill, New York.
Cussler, E.L., 1997, Diffusion – Mass Transfer in Fluid Systems, 2nd Ed., Cambridge University Press, Cambridge.
Fahien, R.W., 1983, Fundamentals of Transport Phenomena, McGraw-Hill, New York.

PROBLEMS

2.1 Show that the force per unit area can be interpreted as the momentum flux.

2.2 A Newtonian fluid with a viscosity of 50 cP is placed between two large parallel plates separated by a distance of 8 mm. Each plate has an area of 2 m^2. The upper plate moves in the positive x-direction with a velocity of 0.4 m/s while the lower plate is kept stationary.

a) Calculate the steady force applied to the upper plate.
b) The fluid in part (a) is replaced with another Newtonian fluid of viscosity 5 cP. If the steady force applied to the upper plate is the same as that of part (a), calculate the velocity of the upper plate.

(**Answer:** a) 5 N   b) 4 m/s)

2.3 Three parallel flat plates are separated by two fluids as shown in the figure below. What should be the value of Y_2 so as to keep the plate in the middle stationary?

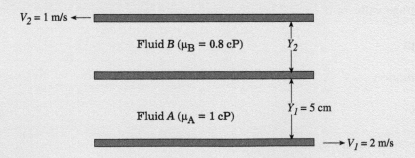

(**Answer:** 2 cm)

2.4   The steady rate of heat loss through a plane slab, which has a surface area of 3 m^2 and is 7 cm thick, is 72 W. Determine the thermal conductivity of the slab if the temperature distribution in the slab is given as

$$T = 5x + 10$$

where T is temperature in °C and x is the distance measured from one side of the slab in cm.

(**Answer:** 0.048 W/m·K)

2.5   The inner and outer surface temperatures of a 20 cm thick brick wall are 30 °C and -5 °C, respectively. The surface area of the wall is 25 m^2. Determine the steady rate of heat loss through the wall if the thermal conductivity is 0.72 W/m·K.

(**Answer:** 3150 W)

2.6   Energy is generated uniformly in a 6 cm thick wall. The steady-state temperature distribution is

$$T = 145 + 3000z - 1500z^2$$

where T is temperature in °C and z is the distance measured from one side of the wall in meters. Determine the rate of heat generation per unit volume if the thermal conductivity of the wall is 15 W/m·K.

(**Answer:** 45 kW/m^3)

2.7   The temperature distribution in a one-dimensional wall of thermal conductivity 20 W/m·K and thickness 60 cm is

$$T = 80 + 10e^{-0.09t} \sin(\pi \xi)$$

where T is temperature in °C, t is time in hours, $\xi = z/L$ is the dimensionless distance with z being a coordinate measured from one side of the wall, and L is the wall thickness in meters. Calculate the total amount of heat transferred in half an hour if the surface area of the wall is 15 m^2.

(**Answer:** 15,360 J)

2.8   The steady-state temperature distribution within a plane wall 1 m thick with a thermal conductivity of 8 W/m·K is measured as a function of position as follows:

z (m)	0	0.1	0.2	0.3	0.4	0.5	0.6	0.7	0.8	0.9	1.0
T (°C)	30	46	59	70	79	85	89	90	89	86	80

where z is the distance measured from one side of the wall. Determine the uniform rate of energy generation per unit volume within the wall.

(**Answer:** 1920 W/m^3)

2.9   The *geothermal gradient* is the rate of increase of temperature with depth in the earth's crust.

a) If the average geothermal gradient of the earth is about 25 °C/km, estimate the steady rate of heat loss from the surface of the earth.

b) One of your friends claims that the amount of heat escaping from 1 m^2 in 4 days is enough to heat a cup of coffee. Do you agree? Justify your answer.

Take the diameter and the thermal conductivity of the earth as 1.27×10^4 km and 3 W/m·K, respectively.

(**Answer:** a) 38×10^9 kW)

2.10   Estimate the earth's age by making use of the following assumptions:

(*i*) Neglecting the curvature, the earth may be assumed to be a semi-infinite plane that began to cool from an initial molten state of $T_o = 1200$ °C. Taking the interface temperature at $z = 0$ to be equal to zero, the corresponding temperature distribution takes the form

$$T = T_o \operatorname{erf}\left(\frac{z}{2\sqrt{\alpha t}}\right) \tag{1}$$

where $\operatorname{erf}(x)$ is the *error function,* defined by

$$\operatorname{erf}(x) = \frac{2}{\sqrt{\pi}} \int_0^x e^{-u^2}\, du \tag{2}$$

(*ii*) The temperature gradient at $z = 0$ is equal to the geothermal gradient of the earth, i.e., 25 °C/km.

(*iii*) The thermal conductivity, the density and the heat capacity of the earth are 3 W/m·K, 5500 kg/m^3 and 2000 J/kg·K, respectively.

Estimation of the age of the earth, based on the above model, was first used by Lord Kelvin (1864). However, he knew nothing about radioactivity or heating of the earth's crust by radioactive decay at that time. As a result, his estimates, ranging from 20 to 200 million years, were completely wrong. Today, geologists generally accept the age of the earth as 4.55 billion years.

(**Answer:** 85.3×10^6 year)

2.11   A slab is initially at a uniform temperature T_o and occupies the space from $z = 0$ to $z = \infty$. At time $t = 0$, the temperature of the surface at $z = 0$ is suddenly changed to T_1 ($T_1 > T_o$) and maintained at that temperature for $t > 0$. Under these conditions the temperature distribution is given by

$$\frac{T_1 - T}{T_1 - T_o} = \operatorname{erf}\left(\frac{z}{2\sqrt{\alpha t}}\right) \tag{1}$$

If the surface area of the slab is A, determine the amount of heat transferred into the slab as a function of time.

$$\left(\textbf{Answer: } Q = \frac{2kA(T_1 - T_o)}{\sqrt{\pi\alpha}}\sqrt{t}\right)$$

2.12 Air at $20\,°C$ and 1 atm pressure flows over a porous plate that is soaked in ethanol. The molar concentration of ethanol in the air, c_A, is given by

$$c_A = 4e^{-1.5z}$$

where c_A is in $kmol/m^3$ and z is the distance measured from the surface of the plate in meters. Calculate the molar flux of ethanol from the plate.

(**Answer:** 0.283 $kmol/m^2{\cdot}h$)

2.13 The formal definition of the partial molar volume is given by

$$\overline{V}_i = \left(\frac{\partial V}{\partial n_i}\right)_{T,P,n_{j\neq i}} \tag{1}$$

Substitute

$$V = \frac{n}{c} \tag{2}$$

into Eq. (1) and show that the volume fraction is equal to the mole fraction for constant total molar concentration, c, i.e.,

$$c_i \overline{V}_i = x_i \tag{3}$$

This further implies that the molar average velocity is equal to the volume average velocity when the total molar concentration is constant.

2.14 For a gas at constant pressure, why does the Schmidt number usually remain fairly constant over a large temperature range, while the diffusion coefficient changes markedly?

2.15 Gas $\mathcal{A}$ dissolves in liquid $\mathcal{B}$ and diffuses into the liquid phase. As it diffuses, species $\mathcal{A}$ undergoes an irreversible chemical reaction as shown in the figure below. Under steady conditions, the resulting concentration distribution in the liquid phase is given by

$$\frac{c_A}{c_{A_o}} = \frac{\cosh\left\{\Lambda\left[1 - \left(\frac{z}{L}\right)\right]\right\}}{\cosh\Lambda}$$

in which

$$\Lambda = \sqrt{\frac{kL^2}{\mathcal{D}_{AB}}}$$

where c_{A_o} is the surface concentration, k is the reaction rate constant and $\mathcal{D}_{AB}$ is the diffusion coefficient.

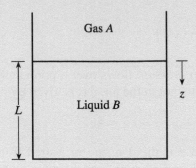

a) Determine the rate of moles of $\mathcal{A}$ entering the liquid phase if the cross-sectional area of the tank is A.

b) Determine the molar flux at $z = L$. What is the physical significance of this result?

$$\left(\textbf{Answer: a) } \dot{n}_A = \frac{A\mathcal{D}_{AB}c_{A_o}\Lambda \tanh \Lambda}{L} \quad \text{b) } 0 \right)$$

3

INTERPHASE TRANSPORT AND TRANSFER COEFFICIENTS

In engineering calculations, we are interested in the determination of the rate of momentum, heat, and mass transfer from one phase to another across the phase interface. This can be achieved by integrating the flux expression over the interfacial area. Equation (2.4-2) gives the value of the flux at the interface as

$$
\begin{pmatrix} \text{Interphase} \\ \text{flux} \end{pmatrix} = \left[(\text{Diffusivity}) \begin{pmatrix} \text{Gradient of} \\ \text{Quantity/Volume} \end{pmatrix} \right.
$$
$$
\left. + \begin{pmatrix} \text{Quantity} \\ \text{Volume} \end{pmatrix} \begin{pmatrix} \text{Characteristic} \\ \text{velocity} \end{pmatrix} \right]_{\text{interface}}
$$

Note that the determination of the interphase flux requires the values of the *quantity/volume* and its gradient to be known at the interface. Therefore, equations of change must be solved to obtain the distribution of *quantity/volume* as a function of position. These analytical solutions, however, are not possible most of the time. In that case we resort to experimental data and correlate the results by the transfer coefficients, namely, the friction factor, the heat transfer coefficient, and the mass transfer coefficient. The resulting correlations are then used in designing equipment.

This chapter deals with the physical significance of these three transfer coefficients. In addition, the relationships between these transfer coefficients will be explained by using dimensionless numbers and analogies.

3.1 FRICTION FACTOR

Let us consider a flat plate of length L and width W suspended in a uniform stream having an approach velocity v_∞ as shown in Figure 3.1.

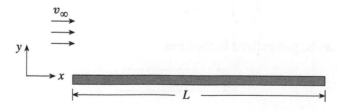

Figure 3.1. Flow on a flat plate.

As engineers, we are interested in the determination of the total drag force, i.e., the component of the force in the direction of flow, exerted by the flowing stream on the plate. This force can be calculated by integrating the total momentum flux at the wall over the surface area. The total momentum flux at the wall, $\pi_{yx}|_{y=0}$, is

$$\pi_{yx}\big|_{y=0} = \tau_{yx}\big|_{y=0} + (\rho v_x v_y)\big|_{y=0} \tag{3.1-1}$$

where $\tau_{yx}|_{y=0}$ is the value of the shear stress at the wall. Since the plate is stationary, the fluid in contact with the plate is also stagnant[1] and both v_x and v_y are zero at $y = 0$. Therefore, Eq. (3.1-1) reduces to

$$\pi_{yx}\big|_{y=0} = \tau_{yx}\big|_{y=0} = \tau_w = \mu \frac{\partial v_x}{\partial y}\bigg|_{y=0} \tag{3.1-2}$$

Note that the minus sign is omitted in Eq. (3.1-2) since the value of v_x increases as the distance y increases. The drag force, F_D, on one side of the plate is calculated from

$$F_D = \int_0^W \int_0^L \tau_w \, dx \, dz \tag{3.1-3}$$

Evaluation of the integral in Eq. (3.1-3) requires the value of the velocity gradient at the wall to be known as a function of position. Obtaining analytical expressions for the velocity distribution from the solution of the equations of change, however, is almost impossible in most cases. Thus, it is customary in engineering practice to replace τ_w with a dimensionless term called the *friction factor*, f, such that

$$\boxed{\tau_w = \frac{1}{2}\rho v_\infty^2 f} \tag{3.1-4}$$

Substitution of Eq. (3.1-4) into Eq. (3.1-3) gives

$$F_D = \frac{1}{2}\rho v_\infty^2 \int_0^W \int_0^L f \, dx \, dz = (WL)\left(\frac{1}{2}\rho v_\infty^2\right)\langle f \rangle \tag{3.1-5}$$

where $\langle f \rangle$ is the friction factor averaged over the area of the plate[2], i.e.,

$$\langle f \rangle = \frac{\displaystyle\int_0^W \int_0^L f \, dx \, dz}{\displaystyle\int_0^W \int_0^L dx \, dz} = \frac{1}{WL}\int_0^W \int_0^L f \, dx \, dz \tag{3.1-6}$$

Equation (3.1-5) can be generalized in the form

$$\boxed{F_D = A_{ch} K_{ch} \langle f \rangle} \tag{3.1-7}$$

---

[1]This is known as the *no-slip boundary condition*.

[2]See Section A.2 in Appendix A.

in which the terms A_{ch}, characteristic area, and K_{ch}, characteristic kinetic energy, are defined by

$$A_{ch} = \begin{cases} \text{Wetted surface area} & \text{for flow in conduits} \\ \text{Projected area} & \text{for flow around submerged objects} \end{cases} \qquad (3.1\text{-}8)$$

$$K_{ch} = \frac{1}{2}\rho v_{ch}^2 \qquad (3.1\text{-}9)$$

where v_{ch} is the characteristic velocity.

Power, $\dot{W}$, is defined as the rate at which work is done. Therefore,

$$\text{Power} = \frac{\text{Work}}{\text{Time}} = \frac{(\text{Force})(\text{Distance})}{\text{Time}} = (\text{Force})(\text{Velocity}) \qquad (3.1\text{-}10)$$

or,

$$\boxed{\dot{W} = F_D v_{ch}} \qquad (3.1\text{-}11)$$

Example 3.1 Advertisements for cars in magazines give a complete list of their features, one of which is the friction factor (or drag coefficient), based on the frontal area. Sports cars, such as the Toyota Celica, usually have a friction factor of around 0.24. If the car has a width of 2 m and a height of 1.5 m,

a) Determine the power consumed by the car when it is going at 100 km/h.
b) Repeat part **(a)** if the wind blows at a velocity of 30 km/h opposite to the direction of the car.
c) Repeat part **(a)** if the wind blows at a velocity of 30 km/h in the direction of the car.

Solution

Physical properties

For air at 20 °C (293 K): $\rho = 1.2 \text{ kg/m}^3$

Assumption

1. Air is at 20 °C.

Analysis

a) The characteristic velocity is

$$v_{ch} = (100)\left(\frac{1000}{3600}\right) = 27.78 \text{ m/s}$$

The drag force can be calculated from Eq. (3.1-7) as

$$F_D = A_{ch}\left(\frac{1}{2}\rho v_{ch}^2\right)\langle f\rangle = (2 \times 1.5)\left[\frac{1}{2}(1.2)(27.78)^2\right](0.24) = 333.4 \text{ N}$$

The use of Eq. (3.1-11) gives the power consumed as

$$\dot{W} = F_D v_{ch} = (333.4)(27.78) = 9262 \text{ W}$$

b) In this case the characteristic velocity is

$$v_{ch} = (100 + 30)\left(\frac{1000}{3600}\right) = 36.11 \text{ m/s}$$

Therefore, the drag force and the power consumed are

$$F_D = (2 \times 1.5)\left[\frac{1}{2}(1.2)(36.11)^2\right](0.24) = 563.3 \text{ N}$$

$$\dot{W} = (563.3)(36.11) = 20{,}341 \text{ W}$$

c) In this case the characteristic velocity is

$$v_{ch} = (100 - 30)\left(\frac{1000}{3600}\right) = 19.44 \text{ m/s}$$

Therefore, the drag force and the power consumed are

$$F_D = (2 \times 1.5)\left[\frac{1}{2}(1.2)(19.44)^2\right](0.24) = 163.3 \text{ N}$$

$$\dot{W} = (163.3)(19.44) = 3175 \text{ W}$$

3.1.1 Physical Interpretation of Friction Factor

Combination of Eqs. (3.1-2) and (3.1-4) leads to

$$\frac{1}{2} f = \frac{\mu}{\rho v_\infty^2} \left.\frac{\partial v_x}{\partial y}\right|_{y=0} \tag{3.1-12}$$

The friction factor can be determined from Eq. (3.1-12) if the physical properties of the fluid (viscosity and density), the approach velocity of the fluid, and the velocity gradient at the wall are known. Since the calculation of the velocity gradient requires the velocity distribution in the fluid phase to be known, the actual case is idealized as shown in Figure 3.2.

The entire resistance to momentum transport is assumed to be due to a laminar film of thickness δ next to the wall. The velocity gradient in the film is constant and is equal to

$$\left.\frac{\partial v_x}{\partial y}\right|_{y=0} = \frac{v_\infty}{\delta} \tag{3.1-13}$$

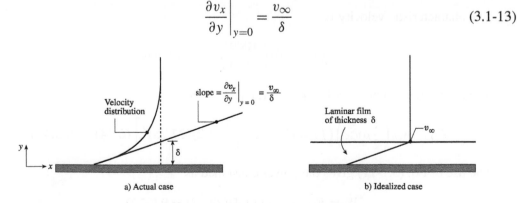

a) Actual case   b) Idealized case

Figure 3.2. The film model for momentum transfer.

Substitution of Eq. (3.1-13) into Eq. (3.1-12) and multiplication of the resulting equation by the characteristic length, L_{ch}, yield

$$\boxed{\frac{1}{2} f \, Re = \frac{L_{ch}}{\delta}} \tag{3.1-14}$$

where the dimensionless term Re is the *Reynolds number*, defined by

$$Re = \frac{L_{ch} v_{\infty} \rho}{\mu} \tag{3.1-15}$$

Equation (3.1-14) indicates that the product of the friction factor with the Reynolds number is directly proportional to the characteristic length and inversely proportional to the thickness of the momentum boundary layer.

3.2   HEAT TRANSFER COEFFICIENT

3.2.1   Convection Heat Transfer Coefficient

Let us consider a flat plate suspended in a uniform stream of velocity v_{∞} and temperature T_{∞} as shown in Figure 3.3. The temperature at the surface of the plate is kept constant at T_w.

As engineers, we are interested in the total rate of heat transfer from the plate to the flowing stream. This can be calculated by integrating the total energy flux at the wall over the surface area. The total energy flux at the wall, $e_y|_{y=0}$, is

$$e_y\big|_{y=0} = q_y\big|_{y=0} + \left(\rho \widehat{C}_P T v_y\right)\big|_{y=0} \tag{3.2-1}$$

where $q_y|_{y=0}$ is the molecular (or conductive) energy flux at the wall. As a result of the no-slip boundary condition at the wall, the fluid in contact with the plate is stagnant and heat is transferred by pure conduction through the fluid layer immediately adjacent to the plate. Therefore, Eq. (3.2-1) reduces to

$$e_y\big|_{y=0} = q_y\big|_{y=0} = q_w = -k \left.\frac{\partial T}{\partial y}\right|_{y=0} \tag{3.2-2}$$

The rate of heat transfer, $\dot{Q}$, from one side of the plate to the flowing stream is calculated from

$$\dot{Q} = \int_0^W \int_0^L q_w \, dx \, dz \tag{3.2-3}$$

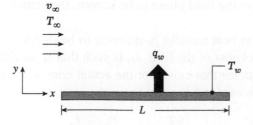

Figure 3.3. Flow over a flat plate.

Evaluation of the integral in Eq. (3.2-3) requires the temperature gradient at the wall to be known as a function of position. However, the fluid motion makes the analytical solution of the temperature distribution impossible to obtain in most cases. Hence, we usually resort to experimentally determined values of the energy flux at a solid-fluid boundary in terms of the *convection heat transfer coefficient*, *h*, as

$$q_w = h\,(T_w - T_\infty) \qquad (3.2\text{-}4)$$

which is known as *Newton's law of cooling*. The convection heat transfer coefficient, *h*, has the units of $W/m^2 \cdot K$. It depends on the fluid flow mechanism, fluid properties (density, viscosity, thermal conductivity, heat capacity) and flow geometry.

Substitution of Eq. (3.2-4) into Eq. (3.2-3) gives the rate of heat transfer as

$$\dot{Q} = (T_w - T_\infty) \int_0^W \int_0^L h\,dx\,dz = (WL)\langle h \rangle (T_w - T_\infty) \qquad (3.2\text{-}5)$$

where $\langle h \rangle$ is the heat transfer coefficient averaged over the area of the plate and is defined by

$$\langle h \rangle = \frac{\displaystyle \int_0^W \int_0^L h\,dx\,dz}{\displaystyle \int_0^W \int_0^L dx\,dz} = \frac{1}{WL} \int_0^W \int_0^L h\,dx\,dz \qquad (3.2\text{-}6)$$

Equation (3.2-5) can be generalized in the form

$$\dot{Q} = A_H \langle h \rangle (\Delta T)_{ch} \qquad (3.2\text{-}7)$$

where A_H is the heat transfer area and $(\Delta T)_{ch}$ is the characteristic temperature difference.

3.2.1.1 *Physical interpretation of heat transfer coefficient*   Combination of Eqs. (3.2-2) and (3.2-4) leads to

$$h = -\frac{k}{T_w - T_\infty} \left. \frac{\partial T}{\partial y} \right|_{y=0} \qquad (3.2\text{-}8)$$

The convection heat transfer coefficient can be determined from Eq. (3.2-8) if the thermal conductivity of the fluid, the overall temperature difference, and the temperature gradient at the wall are known. Since the calculation of the temperature gradient at the wall requires the temperature distribution in the fluid phase to be known, the actual case is idealized as shown in Figure 3.4.

The entire resistance to heat transfer is assumed to be due to a stagnant film in the fluid next to the wall. The thickness of the film, δ_t, is such that it provides the same resistance to heat transfer as the resistance that exists for the actual convection process. The temperature gradient in the film is constant and is equal to

$$\left. \frac{\partial T}{\partial y} \right|_{y=0} = \frac{T_\infty - T_w}{\delta_t} \qquad (3.2\text{-}9)$$

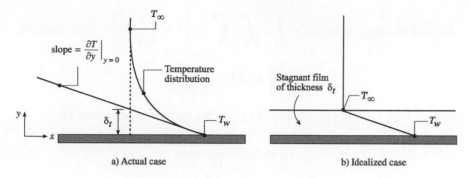

Figure 3.4. The film model for energy transfer.

Substitution of Eq. (3.2-9) into Eq. (3.2-8) gives

$$h = \frac{k}{\delta_t}$$

(3.2-10)

Equation (3.2-10) indicates that the thickness of the film, δ_t, determines the value of h. For this reason the term h is frequently referred to as the *film heat transfer coefficient*.

Example 3.2 Energy generation rate per unit volume as a result of fission within a spherical reactor of radius R is given as a function of position as

$$\Re = \Re_o \left[1 - \left(\frac{r}{R} \right)^2 \right]$$

where r is the radial distance measured from the center of the sphere. Cooling fluid at a temperature of T_∞ flows over the reactor. If the average heat transfer coefficient $\langle h \rangle$ at the surface of the reactor is known, determine the surface temperature of the reactor at steady-state.

Solution

System: Reactor

Analysis

The inventory rate equation for energy becomes

$$\text{Rate of energy out} = \text{Rate of energy generation}$$

(1)

The rate at which energy leaves the sphere by convection is given by Newton's law of cooling as

$$\text{Rate of energy out} = (4\pi R^2)\langle h \rangle (T_w - T_\infty)$$

(2)

where T_w is the surface temperature of the sphere. The rate of energy generation can be determined by integrating $\Re$ over the volume of the sphere. The result is

$$\text{Rate of energy generation} = \int_0^{2\pi} \int_0^{\pi} \int_0^R \Re_o \left[1 - \left(\frac{r}{R} \right)^2 \right] r^2 \sin\theta \, dr d\theta d\phi$$

$$= \frac{8\pi}{15} \Re_o R^3 \tag{3}$$

Substitution of Eqs. (2) and (3) into Eq. (1) gives the surface temperature as

$$T_w = T_\infty + \frac{2}{15} \frac{\Re_o R}{\langle h \rangle} \tag{4}$$

3.2.2 Radiation Heat Transfer Coefficient

The heat flux due to radiation, q^R, from a small object to the surroundings wall is given as

$$q^R = \varepsilon \sigma \left(T_1^4 - T_2^4 \right) \tag{3.2-11}$$

where ε is the emissivity of the small object, σ is the *Stefan-Boltzmann constant* (5.67×10^{-8} W/m^2·K^4), and T_1 and T_2 are the temperatures of the small object and the wall in degrees Kelvin, respectively.

In engineering practice, Eq. (3.2-11) is written in a fashion analogous to Eq. (3.2-4) as

$$q^R = h^R (T_1 - T_2) \tag{3.2-12}$$

where h^R is the *radiation heat transfer coefficient*. Comparison of Eqs. (3.2-11) and (3.2-12) gives

$$h^R = \frac{\varepsilon \sigma \left(T_1^4 - T_2^4 \right)}{T_1 - T_2} \simeq 4\varepsilon \sigma \langle T \rangle^3 \tag{3.2-13}$$

provided that $\langle T \rangle \gg (T_1 - T_2)/2$, where $\langle T \rangle = (T_1 + T_2)/2$.

3.3 MASS TRANSFER COEFFICIENT

Let us consider a flat plate suspended in a uniform stream of fluid (species $\mathcal{B}$) having a velocity v_∞ and species $\mathcal{A}$ concentration c_{A_∞} as shown in Figure 3.5. The surface of the plate is also coated with species $\mathcal{A}$ with concentration c_{A_w}.

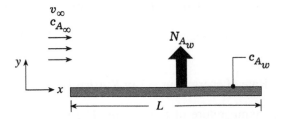

Figure 3.5. Flow over a flat plate.

As engineers, we are interested in the total number of moles of species A transferred from the plate to the flowing stream. This can be calculated by integrating the total molar flux at the wall over the surface area. The total molar flux at the wall, $N_{A_y}|_{y=0}$, is

$$N_{A_y}\big|_{y=0} = J_{A_y}^*\big|_{y=0} + \left(c_A v_y^*\right)\big|_{y=0} \tag{3.3-1}$$

where $J_{A_y}^*|_{y=0}$ is the molecular (or diffusive) molar flux at the wall. For low mass transfer rates Eq. (3.3-1) can be simplified to[3]

$$N_{A_y}\big|_{y=0} = N_{A_w} \simeq J_{A_y}^*\big|_{y=0} = -\mathcal{D}_{AB} \frac{\partial c_A}{\partial y}\bigg|_{y=0} \tag{3.3-2}$$

and the rate of moles of species A transferred, $\dot{n}_A$, from one side of the plate to the flowing stream is

$$\dot{n}_A = \int_0^W \int_0^L N_{A_w} \, dx \, dz \tag{3.3-3}$$

Evaluation of the integral in Eq. (3.3-3) requires the value of the concentration gradient at the wall to be known as a function of position. Since this is almost impossible to obtain in most cases, in a manner analogous to the definition of the heat transfer coefficient, the *convection mass transfer coefficient*, k_c, is defined by the following expression

$$\boxed{N_{A_w} = k_c(c_{A_w} - c_{A_\infty})} \tag{3.3-4}$$

which may be called *Newton's law of mass transfer* as suggested by Slattery (1999). The mass transfer coefficient has the units of m/s. It depends on the fluid flow mechanism, fluid properties (density, viscosity, diffusion coefficient) and flow geometry.

Substitution of Eq. (3.3-4) into Eq. (3.3-3) gives the rate of moles of species A transferred as

$$\dot{n}_A = (c_{A_w} - c_{A_\infty}) \int_0^W \int_0^L k_c \, dx \, dz = (WL)\langle k_c \rangle (c_{A_w} - c_{A_\infty}) \tag{3.3-5}$$

where $\langle k_c \rangle$ is the mass transfer coefficient averaged over the area of the plate and is defined by

$$\langle k_c \rangle = \frac{\displaystyle\int_0^W \int_0^L k_c \, dx \, dz}{\displaystyle\int_0^W \int_0^L dx \, dz} = \frac{1}{WL} \int_0^W \int_0^L k_c \, dx \, dz \tag{3.3-6}$$

---

[3]Note that v_y^* is the molar average velocity defined by

$$v_y^* = \frac{c_A v_{A_y} + c_B v_{B_y}}{c}$$

At the wall, i.e., $y = 0$, $v_{B_y} = 0$ due to the no-slip boundary condition. However, $v_{A_y} \neq 0$ as a result of the transfer of species A from the surface to the flowing stream. Therefore, $v_y^*|_{y=0} \neq 0$.

Equation (3.3-5) can be generalized in the form

$$\dot{n}_A = A_M \langle k_c \rangle (\Delta c_A)_{ch} \tag{3.3-7}$$

where A_M is the mass transfer area and $(\Delta c_A)_{ch}$ is the characteristic concentration difference.

3.3.1 Physical Interpretation of Mass Transfer Coefficient

Combination of Eqs. (3.3-2) and (3.3-4) leads to

$$k_c = -\frac{\mathcal{D}_{AB}}{c_{A_w} - c_{A_\infty}} \left. \frac{\partial c_A}{\partial y} \right|_{y=0} \tag{3.3-8}$$

The convection mass transfer coefficient can be determined from Eq. (3.3-8) if the diffusion coefficient, the overall concentration difference, and the concentration gradient at the wall are known. Since the calculation of the concentration gradient at the wall requires the concentration distribution to be known, the actual case is idealized as shown in Figure 3.6.

The entire resistance to mass transfer is due to a stagnant film in the fluid next to the wall. The thickness of the film, δ_c, is such that it provides the same resistance to mass transfer by molecular diffusion as the resistance that exists for the actual convection process. The concentration gradient in the film is constant and equal to

$$\left. \frac{\partial c_A}{\partial y} \right|_{y=0} = \frac{c_{A_\infty} - c_{A_w}}{\delta_c} \tag{3.3-9}$$

Substitution of Eq. (3.3-9) into Eq. (3.3-8) gives

$$k_c = \frac{\mathcal{D}_{AB}}{\delta_c} \tag{3.3-10}$$

Equation (3.3-10) indicates that the mass transfer coefficient is directly proportional to the diffusion coefficient and inversely proportional to the thickness of the concentration boundary layer.

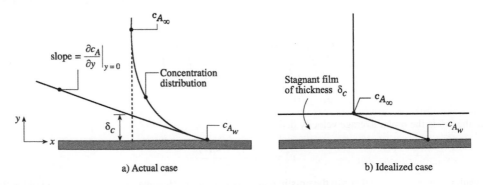

a) Actual case   b) Idealized case

Figure 3.6. The film model for mass transfer.

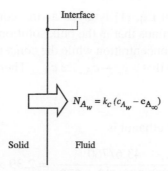

Figure 3.7. Transfer of species $\mathcal{A}$ from the solid to the fluid phase.

3.3.2   Concentration at the Phase Interface

Consider the transfer of species $\mathcal{A}$ from the solid phase to the fluid phase through a flat interface as shown in Figure 3.7. The molar flux of species $\mathcal{A}$ is expressed by Eq. (3.3-4). In the application of this equation to practical problems of interest, there is no difficulty in defining the concentration in the bulk fluid phase, c_{A_∞}, since this can be measured experimentally. However, to estimate the value of c_{A_w}, one has to make an assumption about the conditions at the interface. It is generally assumed that the two phases are in equilibrium with each other at the solid-fluid interface. If T_w represents the interface temperature, the value of c_{A_w} is given by

$$c_{A_w} = \begin{cases} P_A^{sat}/\mathcal{R}T & \text{(Assuming ideal gas behavior)} & \text{fluid} = \text{gas} \\ \text{Solubility of solid in liquid at } T_w & & \text{fluid} = \text{liquid} \end{cases} \qquad (3.3\text{-}11)$$

The *Antoine equation* is widely used to estimate vapor pressures and it is given in Appendix D.

Example 3.3 0.5 L of ethanol is poured into a cylindrical tank of 2 L capacity and the top is quickly sealed. The total height of the cylinder is 1 m. Calculate the mass transfer coefficient if the ethanol concentration in the air reaches 2% of its saturation value in 5 minutes. The cylinder temperature is kept constant at 20 °C.

Solution

Physical properties

For ethanol ($\mathcal{A}$) at 20 °C (293 K): $\begin{cases} \rho = 789 \text{ kg/m}^3 \\ \mathcal{M} = 46 \\ P_A^{sat} = 43.6 \text{ mmHg} \end{cases}$

Assumption

1. Ideal gas behavior.

Analysis

The mass transfer coefficient can be calculated from Eq. (3.3-4), i.e.,

$$N_{A_w} = k_c(c_{A_w} - c_{A_\infty}) \qquad (1)$$

The concentration difference in Eq. (1) is given as the concentration of ethanol vapor at the surface of the liquid, c_{A_w}, minus that in the bulk solution, c_{A_∞}. The concentration at the liquid surface is the saturation concentration while the concentration in the bulk is essentially zero at relatively short times so that $c_{A_w} - c_{A_\infty} \simeq c_{A_w}$. Therefore Eq. (1) simplifies to

$$N_{A_w} = k_c c_{A_w} \tag{2}$$

The saturation concentration of ethanol is

$$c_{A_w} = \frac{P_A^{sat}}{\mathcal{R}T} = \frac{43.6/760}{(0.08205)(20+273)} = 2.39 \times 10^{-3} \text{ kmol/m}^3 \tag{3}$$

Since the ethanol concentration within the cylinder reaches 2% of its saturation value in 5 minutes, the moles of ethanol evaporated during this period are

$$n_A = (0.02)(2.39 \times 10^{-3})(1.5 \times 10^{-3}) = 7.17 \times 10^{-8} \text{ kmol} \tag{4}$$

where 1.5×10^{-3} m^3 is the volume of the air space in the tank. Therefore, the molar flux at 5 minutes can be calculated as

$$N_{A_w} = \frac{\text{Number of moles of species } \mathcal{A}}{(\text{Area})(\text{Time})} \tag{5}$$

$$= \frac{7.17 \times 10^{-8}}{(2 \times 10^{-3}/1)(5 \times 60)} = 1.2 \times 10^{-7} \text{ kmol/m}^2 \cdot \text{s} \tag{3.1}$$

Substitution of Eqs. (3) and (5) into Eq. (2) gives the mass transfer coefficient as

$$k_c = \frac{1.2 \times 10^{-7}}{2.39 \times 10^{-3}} = 5 \times 10^{-5} \text{ m/s} \tag{6}$$

3.4 DIMENSIONLESS NUMBERS

Rearrangement of Eqs. (3.1-4), (3.2-4) and (3.3-4) gives

$$\tau_w = \frac{1}{2} f v_{ch} \Delta(\rho v_{ch}) \qquad \Delta(\rho v_{ch}) = \rho v_\infty - 0 \tag{3.4-1}$$

$$q_w = \frac{h}{\rho \widehat{C}_P} \Delta(\rho \widehat{C}_P T) \qquad \Delta(\rho \widehat{C}_P T) = \rho \widehat{C}_P T_w - \rho \widehat{C}_P T_\infty \tag{3.4-2}$$

$$N_{A_w} = k_c \Delta c_A \qquad \Delta c_A = c_{A_w} - c_{A_\infty} \tag{3.4-3}$$

Note that Eqs. (3.4-1)–(3.4-3) have the general form

$$\begin{pmatrix} \text{Interphase} \\ \text{flux} \end{pmatrix} = \begin{pmatrix} \text{Transfer} \\ \text{coefficient} \end{pmatrix} \begin{pmatrix} \text{Difference in} \\ \text{Quantity/Volume} \end{pmatrix} \tag{3.4-4}$$

and the terms $f v_{ch}/2$, $h/\rho \widehat{C}_P$, and k_c all have the same units, m/s. Thus, the ratio of these quantities must yield dimensionless numbers:

$$\text{Heat transfer Stanton number} = \text{St}_\text{H} = \frac{h}{\rho \widehat{C}_P v_{ch}} \qquad (3.4\text{-}5)$$

$$\text{Mass transfer Stanton number} = \text{St}_\text{M} = \frac{k_c}{v_{ch}} \qquad (3.4\text{-}6)$$

Since the term $f/2$ is dimensionless itself, it is omitted in Eqs. (3.4-5) and (3.4-6).

Dimensionless numbers can also be obtained by taking the ratio of the fluxes. For example, when the concentration gradient is expressed in the form

$$\text{Gradient of Quantity/Volume} = \frac{\text{Difference in Quantity/Volume}}{\text{Characteristic length}} \qquad (3.4\text{-}7)$$

the expression for the molecular flux, Eq. (2.2-5), becomes

$$\text{Molecular flux} = \frac{(\text{Diffusivity})\,(\text{Difference in Quantity/Volume})}{\text{Characteristic length}} \qquad (3.4\text{-}8)$$

Therefore, the ratio of the total interphase flux, Eq. (3.4-4), to the molecular flux, Eq. (3.4-8), is

$$\frac{\text{Interphase flux}}{\text{Molecular flux}} = \frac{(\text{Transfer coefficient})\,(\text{Characteristic length})}{\text{Diffusivity}} \qquad (3.4\text{-}9)$$

The quantities in Eq. (3.4-9) for various transport processes are given in Table 3.1.

The dimensionless terms representing the ratio of the interphase flux to the molecular flux in Table 3.1 are defined in terms of the dimensionless numbers as

$$\frac{1}{2} f \frac{\rho \, v_{ch} L_{ch}}{\mu} = \frac{1}{2} f \, \text{Re} \qquad (3.4\text{-}10)$$

$$\frac{h \, L_{ch}}{k} = \text{Nu} \qquad (3.4\text{-}11)$$

$$\frac{k_c L_{ch}}{\mathcal{D}_{AB}} = \text{Nu}_\text{M} = \text{Sh} \qquad (3.4\text{-}12)$$

Table 3.1. Transfer coefficient, diffusivity and flux ratio for the transport of momentum, energy and mass

Process	Transfer Coefficient	Diffusivity	Interphase Flux / Molecular Flux
Momentum	$\frac{1}{2} f v_{ch}$	$\frac{\mu}{\rho}$	$\frac{1}{2} f \frac{\rho \, v_{ch} L_{ch}}{\mu}$
Energy	$\frac{h}{\rho \widehat{C}_P}$	$\frac{k}{\rho \widehat{C}_P}$	$\frac{h L_{ch}}{k}$
Mass	k_c	$\mathcal{D}_{AB}$	$\frac{k_c L_{ch}}{\mathcal{D}_{AB}}$

Table 3.2. Analogous dimensionless numbers in energy and mass transfer

Energy	Mass
$\mathrm{Pr} = \dfrac{\nu}{\alpha} = \dfrac{\mu \widehat{C}_P}{k}$	$\mathrm{Sc} = \dfrac{\nu}{\mathcal{D}_{AB}} = \dfrac{\mu}{\rho \mathcal{D}_{AB}}$
$\mathrm{Nu} = \dfrac{h L_{ch}}{k}$	$\mathrm{Nu_M} = \mathrm{Sh} = \dfrac{k_c L_{ch}}{\mathcal{D}_{AB}}$
$\mathrm{St_H} = \dfrac{\mathrm{Nu}}{\mathrm{Re}\,\mathrm{Pr}} = \dfrac{h}{\rho \widehat{C}_P v_{ch}}$	$\mathrm{St_M} = \dfrac{\mathrm{Sh}}{\mathrm{Re}\,\mathrm{Sc}} = \dfrac{k_c}{v_{ch}}$

where Nu is the *heat transfer Nusselt number* and $\mathrm{Nu_M}$ is the *mass transfer Nusselt number*. The mass transfer Nusselt number is generally called the *Sherwood number*, Sh. Equations (3.4-10)–(3.4-12) indicate that the product $(f\,\mathrm{Re}/2)$ is more closely analogous to the Nusselt and Sherwood numbers than f is itself. A summary of the analogous dimensionless numbers for energy and mass transfer covered so far is given in Table 3.2. The Stanton numbers for heat and mass transfer are designated by $\mathrm{St_H}$ and $\mathrm{St_M}$, respectively.

3.4.1 Dimensionless Numbers and Time Scales

A characteristic time is the time over which a given process takes place. Consider, for example, the free fall of a stone of mass 0.5 kg from the top of a skyscraper. If the height, L, of the building is 250 m, how long does it take for the stone to reach the ground? Since the acceleration of gravity, i.e., $g = 9.8$ m/s^2, is responsible for the falling process, then the characteristic time representing the free fall of a stone is given by

$$t_{ch} = \sqrt{\frac{L\,(\mathrm{m})}{g\,(\mathrm{m/s^2})}} \tag{3.4-13}$$

$$= \sqrt{\frac{250}{9.8}} = 5.1 \text{ s}$$

From physics, the actual time of fall can be calculated from the formula

$$L = \frac{1}{2} g t^2 \tag{3.4-14}$$

or,

$$t = \sqrt{\frac{(2)(250)}{9.8}} = 7.1 \text{ s}$$

which is different from 5.1 s. It should be kept in mind that the time scale gives a rough estimate, or order-of-magnitude, of the characteristic time of a given process. As far as the order-of-magnitude is concerned, the values 5.1 s and 7.1 s are almost equivalent.

Diffusivities (ν, α, $\mathcal{D}_{AB}$) all have the same units, m^2/s. Therefore, the characteristic time (or time scale) for molecular transport is given by

$$(t_{ch})_{mol} = \frac{(\text{Characteristic Length})^2}{\text{Diffusivity}} \tag{3.4-15}$$

Table 3.3. Time scales for different transport mechanisms

Type of Transport	Molecular Time Scale	Convective Time Scale
Momentum	$\dfrac{L_{ch}^2}{\nu}$	$\dfrac{L_{ch}}{v_{ch}}$
Heat	$\dfrac{L_{ch}^2}{\alpha}$	$\dfrac{L_{ch}}{h/\rho \widehat{C}_P}$
Mass	$\dfrac{L_{ch}^2}{\mathcal{D}_{AB}}$	$\dfrac{L_{ch}}{k_{ch}}$

Note that each process experiences an unsteady-state period before reaching steady-state conditions. Thus, Eq. (3.4-15) gives an idea of the time it takes for a given process to reach steady-state.

Transfer coefficients ($f v_{ch}/2$, $h/\rho \widehat{C}_P$, and k_c) all have the same units, m/s. Therefore, the characteristic time (or time scale) for convective transport is given by

$$(t_{ch})_{conv} = \frac{\text{Characteristic Length}}{\text{Transfer Coefficient}} \qquad (3.4\text{-}16)$$

Table 3.3 summarizes the molecular and convective time scales for the transport of momentum, heat, and mass. The tricky issue in the estimation of order of magnitude is how to identify the characteristic length. In general, the characteristic length used in the molecular time scale may be different from that used in the convective time scale.

Since the $f/2$ term is dimensionless itself, it is omitted from the convective time scale for momentum. Note that the convective time scale for momentum transport, L_{ch}/v_{ch}, is the time it takes for the fluid to move through the system, also known as the *residence time*.

It is possible to redefine the dimensionless numbers in terms of the time scales as follows:

$$\text{Pr} = \frac{\text{Conductive time scale}}{\text{Viscous time scale}} = \frac{\nu}{\alpha} \qquad (3.4\text{-}17)$$

$$\text{Sc} = \frac{\text{Diffusive time scale}}{\text{Viscous time scale}} = \frac{\nu}{\mathcal{D}_{AB}} \qquad (3.4\text{-}18)$$

$$\text{Le} = \frac{\text{Diffusive time scale}}{\text{Conductive time scale}} = \frac{\alpha}{\mathcal{D}_{AB}} \qquad (3.4\text{-}19)$$

$$\text{Pe}_H = \frac{\text{Conductive time scale}}{\text{Convective time scale for momentum transport}} = \frac{v_{ch} L_{ch}}{\alpha} \qquad (3.4\text{-}20)$$

$$\text{Pe}_M = \frac{\text{Diffusive time scale}}{\text{Convective time scale for momentum transport}} = \frac{v_{ch} L_{ch}}{\mathcal{D}_{AB}} \qquad (3.4\text{-}21)$$

3.5 TRANSPORT ANALOGIES

Existing analogies in various transport processes depend on the relationship between the dimensionless numbers defined by Eqs. (3.4-10)–(3.4-12). In Section 3.1.1 we showed that

$$\frac{1}{2} f \, \text{Re} = \frac{L_{ch}}{\delta} \qquad (3.5\text{-}1)$$

On the other hand, substitution of Eqs. (3.2-10) and (3.3-10) into Eqs. (3.4-11) and (3.4-12), respectively, gives

$$\text{Nu} = \frac{L_{ch}}{\delta_t} \qquad (3.5\text{-}2)$$

and

$$\text{Sh} = \frac{L_{ch}}{\delta_c} \qquad (3.5\text{-}3)$$

Examination of Eqs. (3.5-1)–(3.5-3) indicates that

$$\frac{\text{Interphase flux}}{\text{Molecular flux}} = \frac{\text{Characteristic length}}{\text{Effective film thickness}} \qquad (3.5\text{-}4)$$

Comparison of Eqs. (3.4-9) and (3.5-4) implies that

$$\text{Effective film thickness} = \frac{\text{Diffusivity}}{\text{Transfer coefficient}} \qquad (3.5\text{-}5)$$

Note that the *effective film thickness* is the thickness of a fictitious film that would be required to account for the entire resistance if only molecular transport were involved.

Using Eqs. (3.5-1)–(3.5-3), it is possible to express the characteristic length as

$$L_{ch} = \frac{1}{2} f \, \text{Re} \, \delta = \text{Nu} \, \delta_t = \text{Sh} \, \delta_c \qquad (3.5\text{-}6)$$

Substitution of $\text{Nu} = \text{St}_H \, \text{Re} \, \text{Pr}$ and $\text{Sh} = \text{St}_M \, \text{Re} \, \text{Sc}$ into Eq. (3.5-6) gives

$$\frac{1}{2} f \delta = \text{St}_H \, \text{Pr} \, \delta_t = \text{St}_M \, \text{Sc} \, \delta_c \qquad (3.5\text{-}7)$$

3.5.1 The Reynolds Analogy

Similarities between the transport of momentum, energy, and mass were first noted by Reynolds in 1874. He proposed that the effective film thicknesses for the transfer of momentum, energy, and mass are equal, i.e.,

$$\delta = \delta_t = \delta_c \qquad (3.5\text{-}8)$$

Therefore, Eq. (3.5-7) becomes

$$\frac{f}{2} = \text{St}_H \, \text{Pr} = \text{St}_M \, \text{Sc} \qquad (3.5\text{-}9)$$

Reynolds further assumed that $\text{Pr} = \text{Sc} = 1$. Under these circumstances Eq. (3.5-9) reduces to

$$\boxed{\frac{f}{2} = \text{St}_H = \text{St}_M} \qquad (3.5\text{-}10)$$

which is known as the *Reynolds analogy*. Physical properties in Eq. (3.5-10) must be evaluated at $T = (T_w + T_\infty)/2$.

The Reynolds analogy is reasonably valid for gas systems but should not be considered for liquid systems.

3.5.2 The Chilton-Colburn Analogy

In the Chilton-Colburn analogy the relationships between the effective film thicknesses are expressed as

$$\frac{\delta}{\delta_t} = \text{Pr}^{1/3} \qquad \frac{\delta}{\delta_c} = \text{Sc}^{1/3} \tag{3.5-11}$$

Substitution of Eq. (3.5-11) into Eq. (3.5-7) yields

$$\frac{f}{2} = \text{St}_\text{H}\, \text{Pr}^{2/3} \equiv j_H \tag{3.5-12}$$

and

$$\frac{f}{2} = \text{St}_\text{M}\, \text{Sc}^{2/3} \equiv j_M \tag{3.5-13}$$

where j_H and j_M are the *Colburn j-factors* for heat and mass transfer, respectively. Physical properties in Eqs. (3.5-12) and (3.5-13) must be evaluated at $T = (T_w + T_\infty)/2$. Note that Eqs. (3.5-12) and (3.5-13) reduce to the Reynolds analogy, Eq. (3.5-10), for fluids with $\text{Pr} = 1$ and $\text{Sc} = 1$.

The Chilton-Colburn analogy is valid when $0.6 \leqslant \text{Pr} \leqslant 60$ and $0.6 \leqslant \text{Sc} \leqslant 3000$. However, even if these criteria are satisfied, the use of the Chilton-Colburn analogy is restricted by the flow geometry. The validity of the Chilton-Colburn analogy for flow in different geometries is given in Table 3.4.

Examination of Table 3.4 indicates that the term $f/2$ is not equal to the Colburn j-factors in the case of flow around cylinders and spheres. The drag force is the component of the force in the direction of mean flow and both viscous and pressure forces contribute to this force[4]. For flow over a flat plate, the pressure always acts normal to the surface of the plate and the component of this force in the direction of mean flow is zero. Thus, only viscous force contributes to the drag force. In the case of curved surfaces, however, the component of normal force to the surface in the direction of mean flow is not necessarily zero as shown

Table 3.4. Validity of the Chilton-Colburn analogy for various geometries

Flow Geometry	Chilton-Colburn Analogy
Flow over a flat plate	$\dfrac{f}{2} = j_H = j_M$
Flow over a cylinder	$j_H = j_M$
Flow over a sphere	$j_H = j_M$ if $\begin{cases} \text{Nu} \gg 2 \\ \text{Sh} \gg 2 \end{cases}$
Flow in a pipe	$\dfrac{f}{2} = j_H = j_M$ if $\text{Re} > 10{,}000$ (Smooth pipe)

---

[4]The drag force arising from viscous and pressure forces is called *friction* (or *skin*) *drag* and *form drag*, respectively.

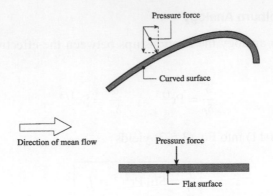

Figure 3.8. Pressure force acting on curved and flat surfaces.

in Figure 3.8. Therefore, the friction factor for flow over flat plates and for flow inside circular ducts includes only friction drag, whereas the friction factor for flow around cylinders, spheres, and other bluff objects includes both friction and form drags. As a result, the $f/2$ term for flow around cylinders and spheres is greater than the j-factors.

Example 3.4 Water evaporates from a wetted surface of rectangular shape when air at 1 atm and 35 °C is blown over the surface at a velocity of 15 m/s. Heat transfer measurements indicate that for air at 1 atm and 35 °C the average heat transfer coefficient is given by the following empirical relation

$$\langle h \rangle = 21 v_\infty^{0.6}$$

where $\langle h \rangle$ is in W/m²·K and v_∞, air velocity, is in m/s. Estimate the mass transfer coefficient and the rate of evaporation of water from the surface if the area is 1.5 m².

Solution

Physical properties

For water at 35 °C (308 K): $P^{sat} = 0.0562$ bar

For air at 35 °C (308 K): $\begin{cases} \rho = 1.1460 \text{ kg/m}^3 \\ v = 16.47 \times 10^{-6} \text{ m}^2/\text{s} \\ \widehat{C}_P = 1.005 \text{ kJ/kg·K} \\ \text{Pr} = 0.711 \end{cases}$

Diffusion coefficient of water ($\mathcal{A}$) in air ($\mathcal{B}$) at 35 °C (308 K) is

$$(\mathcal{D}_{AB})_{308} = (\mathcal{D}_{AB})_{313}\left(\frac{308}{313}\right)^{3/2} = (2.88 \times 10^{-5})\left(\frac{308}{313}\right)^{3/2} = 2.81 \times 10^{-5} \text{ m}^2/\text{s}$$

The Schmidt number is

$$\text{Sc} = \frac{v}{\mathcal{D}_{AB}} = \frac{16.47 \times 10^{-6}}{2.81 \times 10^{-5}} = 0.586$$

Assumption

1. Ideal gas behavior.

Analysis

The use of the Chilton-Colburn analogy, $j_H = j_M$, gives

$$\langle k_c \rangle = \frac{\langle h \rangle}{\rho \widehat{C}_P} \left(\frac{\mathrm{Pr}}{\mathrm{Sc}} \right)^{2/3} = \frac{21 v_\infty^{0.6}}{\rho \widehat{C}_P} \left(\frac{\mathrm{Pr}}{\mathrm{Sc}} \right)^{2/3} \tag{1}$$

Substitution of the values into Eq. (1) gives the average mass transfer coefficient as

$$\langle k_c \rangle = \frac{(21)(15)^{0.6}}{(1.1460)(1005)} \left(\frac{0.711}{0.586} \right)^{2/3} = 0.105 \ \mathrm{m/s}$$

Saturation concentration of water is

$$c_{A_w} = \frac{P_A^{sat}}{\mathcal{R} T} = \frac{0.0562}{(8.314 \times 10^{-2})(35 + 273)} = 2.19 \times 10^{-3} \ \mathrm{kmol/m^3}$$

Therefore, the evaporation rate of water from the surface is

$$\dot{n}_A = A \langle k_c \rangle (c_{A_w} - c_{A_\infty}) = (1.5)(0.105)(2.19 \times 10^{-3} - 0) = 3.45 \times 10^{-4} \ \mathrm{kmol/s}$$

NOTATION

A	area, m^2
A_H	heat transfer area, m^2
A_M	mass transfer area, m^2
$\widehat{C}_P$	heat capacity at constant pressure, kJ/kg·K
c_i	concentration of species i, kmol/m^3
$\mathcal{D}_{AB}$	diffusion coefficient for system $\mathcal{A}$-$\mathcal{B}$, m^2/s
F_D	drag force, N
f	friction factor
h	heat transfer coefficient, W/m^2·K
j_H	Chilton-Colburn j-factor for heat transfer
j_M	Chilton-Colburn j-factor for mass transfer
K	kinetic energy per unit volume, J/m^3
k	thermal conductivity, W/m·K
k_c	mass transfer coefficient, m/s
L	length, m
$\mathcal{M}$	molecular weight, kg/kmol
N	total molar flux, kmol/m^2·s
$\dot{n}_i$	molar flow rate of species i, kmol/s
P	pressure, Pa
$\dot{Q}$	heat transfer rate, W

q heat flux, W/m^2

q^R heat flux due to radiation, W/m^2

$\mathcal{R}$ gas constant, J/mol·K

$\mathfrak{R}$ energy generation rate per unit volume, W/m^3

T temperature, °C or K

t time, s

v velocity, m/s

$\dot{W}$ rate of work, W

x rectangular coordinate, m

y rectangular coordinate, m

z rectangular coordinate, m

α thermal diffusivity, m^2/s

Δ difference

δ fictitious film thickness for momentum transfer, m

δ_c fictitious film thickness for mass transfer, m

δ_t fictitious film thickness for heat transfer, m

ε emissivity

μ viscosity, kg/m·s

ν kinematic viscosity (or momentum diffusivity), m^2/s

π total momentum flux, N/m^2

ρ density, kg/m^3

σ Stefan-Boltzmann constant, W/m^2·K^4

τ_{yx} flux of x-momentum in the y-direction, N/m^2

Bracket

$\langle a \rangle$ average value of a

Superscript

sat   saturation

Subscripts

A, B species in binary systems

ch characteristic

i species in multicomponent systems

w surface or wall

∞ free-stream

Dimensionless Numbers

Nu$_H$ Nusselt number for heat transfer

Nu$_M$ Nusselt number for mass transfer

Pr Prandtl number
Re Reynolds number
Sc Schmidt number
Sh Sherwood number
St_H Stanton number for heat transfer
St_M Stanton number for mass transfer

REFERENCE

Slattery, J.C., 1999, Advanced Transport Phenomena, Cambridge University Press, Cambridge.

SUGGESTED REFERENCES FOR FURTHER STUDY

Bird, R.B., W.E. Stewart and E.N. Lightfoot, 2002, Transport Phenomena, 2nd Ed., Wiley, New York.
Cussler, E.L., 1984, How we make mass transfer seem difficult, Chem. Eng. Ed. 18 (3), 124.
Fahien, R.W., 1983, Fundamentals of Transport Phenomena, McGraw-Hill, New York.

PROBLEMS

3.1 Your friend claims that humid air causes an increase in the gas consumption of cars. Do you agree?

3.2 Air at 20 °C flows over a flat plate of dimensions 50 cm × 25 cm. If the average heat transfer coefficient is 250 W/m^2·K, determine the steady rate of heat transfer from one side of the plate to air when the plate is maintained at 40 °C.

(**Answer:** 625 W)

3.3 Air at 15 °C flows over a spherical LPG tank of radius 4 m. The outside surface temperature of the tank is 4 °C. If the steady rate of heat transfer from the air to the storage tank is 62,000 W, determine the average heat transfer coefficient.

(**Answer:** 28 W/m^2·K)

3.4 The volumetric heat generation in a hollow aluminum sphere of inner and outer radii of 20 cm and 50 cm, respectively, is given by

$$\Re = 4.5 \times 10^4 (1 + 0.6r^2)$$

in which $\Re$ is in W/m^3 and r is the radial coordinate measured in meters. The inner surface of the sphere is subjected to a uniform heat flux of 15,000 W/m^2, while heat is dissipated by convection to an ambient air at 25 °C through the outer surface with an average heat transfer coefficient of 150 W/m^2·K. Determine the temperature of the outer surface under steady conditions.

(**Answer:** 92.3 °C)

3.5   In the system shown below, the rate of heat generation is 800 W/m^3 in Region A, which is perfectly insulated on the left-hand side. Given the conditions indicated in the figure, calculate the heat flux and temperature at the right-hand side, i.e., at $x = 100$ cm, under steady-state conditions.

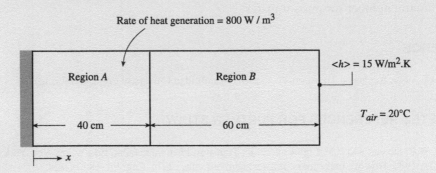

(**Answer:** 320 W/m^2, 41.3 °C)

3.6   Uniform energy generation rate per unit volume at $\Re = 2.4 \times 10^6$ W/m^3 is occurring within a spherical nuclear fuel element of 20 cm diameter. Under steady conditions the temperature distribution is given by

$$T = 900 - 10,000r^2$$

where T is in degrees Celsius and r is in meters.

a) Determine the thermal conductivity of the nuclear fuel element.
b) What is the average heat transfer coefficient at the surface of the sphere if the ambient temperature is 35 °C?

(**Answer:** a) 40 W/m·K   b) 104.6 W/m^2·K)

3.7   A plane wall, with a surface area of 30 m^2 and a thickness of 20 cm, separates a hot fluid at a temperature of 170 °C from a cold fluid at 15 °C. Under steady-state conditions, the temperature distribution across the wall is given by

$$T = 150 - 600x - 50x^2$$

where x is the distance measured from the hot wall in meters and T is the temperature in degrees Celsius. If the thermal conductivity of the wall is 10 W/m·K:

a) Calculate the average heat transfer coefficients at the hot and cold surfaces.
b) Determine the rate of energy generation within the wall.

(**Answer:** a) $\langle h \rangle_{hot} = 300$ W/m^2·K, $\langle h \rangle_{cold} = 477$ W/m^2·K   b) 6000 W)

3.8   Derive Eq. (3.2-13).

(**Hint:** Express T_1 and T_2 in terms of $\langle T \rangle$.)

3.9 It is also possible to interpret the Nusselt and Sherwood numbers as dimensionless temperature and concentration gradients, respectively. Show that the Nusselt and Sherwood numbers can be expressed as

$$Nu = \frac{-(\partial T/\partial y)_{y=0}}{(T_w - T_\infty)/L_{ch}}$$

and

$$Sh = \frac{-(\partial c_A/\partial y)_{y=0}}{(c_{A_w} - c_{A_\infty})/L_{ch}}$$

4

EVALUATION OF TRANSFER COEFFICIENTS: ENGINEERING CORRELATIONS

Since most engineering problems do not have theoretical solutions, a large portion of engineering analysis is concerned with experimental information, which is usually expressed in terms of engineering correlations. These correlations, however, are limited to a specific geometry, equipment configuration, boundary conditions, and substance. As a result, the values obtained from correlations are not exact and it is possible to obtain two different answers from two different correlations for the same problem. Therefore, one should keep in mind that the use of a correlation introduces an error in the order of $\pm 25\%$.

Engineering correlations are given in terms of dimensionless numbers. For example, the correlations used to determine the friction factor, heat transfer coefficient, and mass transfer coefficient are generally expressed in the form

$$f = f(\text{Re})$$

$$\text{Nu} = \text{Nu}(\text{Re}, \text{Pr})$$

$$\text{Sh} = \text{Sh}(\text{Re}, \text{Sc})$$

In this chapter, some of the available correlations for momentum, energy, and mass transport in different geometries will be presented. Emphasis will be placed on the calculations of force (or rate of work), heat transfer rate, and mass transfer rate under steady conditions.

4.1 REFERENCE TEMPERATURE AND CONCENTRATION

The evaluation of the dimensionless numbers that appear in the correlation requires the physical properties of the fluid to be known or estimated. These properties, such as density and viscosity, depend on temperature and/or concentration. Temperature and concentration, on the other hand, vary as a function of position. Two commonly used reference temperatures and concentrations are the *bulk temperature* or *concentration* and the *film temperature* or *concentration*.

4.1.1 Bulk Temperature and Concentration

For flow inside pipes, the *bulk temperature* or *concentration* at a particular location in the pipe is the average temperature or concentration if the fluid were thoroughly mixed, sometimes called the *mixing-cup temperature* or *concentration*. The bulk temperature and the bulk concentration are denoted by T_b and c_b, respectively, and are defined by

$$T_b = \frac{\iint_A v_n T \, dA}{\iint_A v_n \, dA} \qquad \text{and} \qquad c_b = \frac{\iint_A v_n c \, dA}{\iint_A v_n \, dA} \tag{4.1-1}$$

where v_n is the component of velocity in the direction of mean flow.

For the case of flow past bodies immersed in an infinite fluid, the bulk temperature and bulk concentration become the free stream temperature and free stream concentration, respectively, i.e.,

$$\left. \begin{array}{l} T_b = T_\infty \\ c_b = c_\infty \end{array} \right\} \text{ For flow over submerged objects} \tag{4.1-2}$$

4.1.2 Film Temperature and Concentration

The *film temperature*, T_f, and the *film concentration*, c_f, are defined as the arithmetic average of the bulk and surface values, i.e.,

$$T_f = \frac{T_b + T_w}{2} \qquad \text{and} \qquad c_f = \frac{c_b + c_w}{2} \tag{4.1-3}$$

where subscript w represents the conditions at the surface or the wall.

4.2 FLOW PAST A FLAT PLATE

Let us consider a flat plate suspended in a uniform stream of velocity v_∞ and temperature T_∞ as shown in Figure 3.1. The length of the plate in the direction of flow is L and its width is W. The local values of the friction factor, the Nusselt number, and the Sherwood number are given in Table 4.1 for both laminar and turbulent flow conditions. The term Re_x is the Reynolds number based on the distance x, and defined by

$$\mathrm{Re}_x = \frac{x v_\infty \rho}{\mu} = \frac{x v_\infty}{\nu} \tag{4.2-1}$$

The expression for the friction factor under laminar flow conditions, Eq. (A) in Table 4.1, can be obtained analytically from the solution of the equations of change. Blausius (1908) was

Table 4.1. The local values of the friction factor, the Nusselt number, and the Sherwood number for flow over a flat plate

	Laminar		Turbulent	
f_x	$0.664\,\mathrm{Re}_x^{-1/2}$	(A)	$0.0592\,\mathrm{Re}_x^{-1/5}$	(D)
Nu_x	$0.332\,\mathrm{Re}_x^{1/2}\,\mathrm{Pr}^{1/3}$	(B)	$0.0296\,\mathrm{Re}_x^{4/5}\,\mathrm{Pr}^{1/3}$	(E)
Sh_x	$0.332\,\mathrm{Re}_x^{1/2}\,\mathrm{Sc}^{1/3}$	(C)	$0.0296\,\mathrm{Re}_x^{4/5}\,\mathrm{Sc}^{1/3}$	(F)
	$\mathrm{Re}_x \leqslant 500{,}000$		$5 \times 10^5 < \mathrm{Re}_x < 10^7$	
	$0.6 \leqslant \mathrm{Pr} \leqslant 60$		$0.6 \leqslant \mathrm{Sc} \leqslant 3000$	

the first to obtain this solution using a mathematical technique called the *similarity solution* or the *method of combination of variables*. Note that Eqs. (B) and (C) in Table 4.1 can be obtained from Eq. (A) by using the Chilton-Colburn analogy. Since analytical solutions are impossible for turbulent flow, Eq. (D) in Table 4.1 is obtained experimentally. The use of this equation in the Chilton-Colburn analogy yields Eqs. (E) and (F).

The average values of the friction factor, the Nusselt number, and the Sherwood number can be obtained from the local values by the application of the mean value theorem. In many cases, however, the transition from laminar to turbulent flow will occur on the plate. In this case, both the laminar and turbulent flow regions must be taken into account in calculating the average values. For example, if the transition takes place at x_c, where $0 < x_c < L$, then the average friction factor is given by

$$\langle f \rangle = \frac{1}{L} \left[\int_0^{x_c} (f_x)_{lam}\, dx + \int_{x_c}^{L} (f_x)_{turb}\, dx \right] \tag{4.2-2}$$

Change of variable from x to Re_x reduces Eq. (4.2-2) to

$$\langle f \rangle = \frac{1}{\mathrm{Re}_L} \left[\int_0^{\mathrm{Re}_c} (f_x)_{lam}\, d\,\mathrm{Re}_x + \int_{\mathrm{Re}_c}^{\mathrm{Re}_L} (f_x)_{turb}\, d\,\mathrm{Re}_x \right] \tag{4.2-3}$$

where Re_c, the Reynolds number at the point of transition, and Re_L, the Reynolds number based on the length of the plate, are defined by

$$\mathrm{Re}_c = \frac{x_c v_\infty}{\nu} \tag{4.2-4}$$

$$\mathrm{Re}_L = \frac{L v_\infty}{\nu} \tag{4.2-5}$$

Substitution of Eqs. (A) and (D) in Table 4.1 into Eq. (4.2-3) gives

$$\langle f \rangle = \frac{0.074}{\mathrm{Re}_L^{1/5}} + \frac{1.328\,\mathrm{Re}_c^{1/2} - 0.074\,\mathrm{Re}_c^{4/5}}{\mathrm{Re}_L} \tag{4.2-6}$$

Taking $\mathrm{Re}_c = 500{,}000$ results in

$$\langle f \rangle = \frac{0.074}{\mathrm{Re}_L^{1/5}} - \frac{1743}{\mathrm{Re}_L} \tag{4.2-7}$$

The average values of the friction factor, the Nusselt number, and the Sherwood number can be calculated in a similar way for a variety of flow conditions. The results are given in Table 4.2. In these correlations all physical properties must be evaluated at the film temperature.

Once the average values of the Nusselt and Sherwood numbers are determined, the average values of the heat and mass transfer coefficients are calculated from

$$\langle h \rangle = \frac{\langle \mathrm{Nu} \rangle k}{L} \tag{4.2-8}$$

$$\langle k_c \rangle = \frac{\langle \mathrm{Sh} \rangle \mathcal{D}_{AB}}{L} \tag{4.2-9}$$

Table 4.2. Correlations for flow past a flat plate

		Laminar		Laminar and Turbulent		Turbulent
$\langle f \rangle$	(A)	$1.328\,\mathrm{Re}_L^{-1/2}$	(D)	$0.074\,\mathrm{Re}_L^{-1/5} - 1743\,\mathrm{Re}_L^{-1}$	(G)	$0.074\,\mathrm{Re}_L^{-1/5}$
$\langle \mathrm{Nu} \rangle$	(B)	$0.664\,\mathrm{Re}_L^{1/2}\,\mathrm{Pr}^{1/3}$	(E)	$(0.037\,\mathrm{Re}_L^{4/5} - 871)\,\mathrm{Pr}^{1/3}$	(H)	$0.037\,\mathrm{Re}_L^{4/5}\,\mathrm{Pr}^{1/3}$
$\langle \mathrm{Sh} \rangle$	(C)	$0.664\,\mathrm{Re}_L^{1/2}\,\mathrm{Sc}^{1/3}$	(F)	$(0.037\,\mathrm{Re}_L^{4/5} - 871)\,\mathrm{Sc}^{1/3}$	(I)	$0.037\,\mathrm{Re}_L^{4/5}\,\mathrm{Sc}^{1/3}$
		$\mathrm{Re}_L \leq 500{,}000$		$5 \times 10^5 < \mathrm{Re}_L < 10^8$		$\mathrm{Re}_L > 10^8$
		$0.6 \leqslant \mathrm{Pr} \leqslant 60$		$0.6 \leqslant \mathrm{Sc} \leqslant 3000$		

On the other hand, the rate of momentum transfer, i.e., the drag force, the rate of heat transfer, and the rate of mass transfer of species $\mathcal{A}$ from one side of the plate are calculated as

$$F_D = (WL)\left(\frac{1}{2}\rho v_\infty^2\right)\langle f \rangle \tag{4.2-10}$$

$$\dot{Q} = (WL)\langle h \rangle |T_w - T_\infty| \tag{4.2-11}$$

$$\dot{n}_A = (WL)\langle k_c \rangle |c_{A_w} - c_{A_\infty}| \tag{4.2-12}$$

Engineering problems associated with the flow of a fluid over a flat plate are classified as follows:

- Calculate the transfer rate; given the physical properties, the velocity of the fluid, and the dimensions of the plate.
- Calculate the length of the plate in the direction of flow; given the physical properties, the velocity of the fluid, and the transfer rate.
- Calculate the fluid velocity; given the dimensions of the plate, the transfer rate, and the physical properties of the fluid.

Example 4.1 Water at 20 °C flows over a 2 m long flat plate with a velocity of 3 m/s. The width of the plate is 1 m. Calculate the drag force on one side of the plate.

Solution

Physical properties

For water at 20 °C (293 K): $\begin{cases} \rho = 999 \text{ kg/m}^3 \\ \mu = 1001 \times 10^{-6} \text{ kg/m·s} \end{cases}$

Assumption

1. Steady-state conditions prevail.

Analysis

To determine which correlation to use for calculating the average friction factor $\langle f \rangle$, we must first determine the Reynolds number:

$$\text{Re}_L = \frac{L v_\infty \rho}{\mu} = \frac{(2)(3)(999)}{1001 \times 10^{-6}} = 6 \times 10^6$$

Therefore, both laminar and turbulent flow regions exist on the plate. The use of Eq. (D) in Table 4.2 gives the friction factor as

$$\langle f \rangle = \frac{0.074}{\text{Re}_L^{1/5}} - \frac{1743}{\text{Re}_L} = \frac{0.074}{(6 \times 10^6)^{1/5}} - \frac{1743}{6 \times 10^6} = 3 \times 10^{-3}$$

The drag force can then be calculated from Eq. (4.2-10) as

$$F_D = (WL)\left(\frac{1}{2}\rho v_\infty^2\right)\langle f \rangle = (1 \times 2)\left[\frac{1}{2}(999)(3)^2\right](3 \times 10^{-3}) = 27 \text{ N}$$

Example 4.2 Air at a temperature of 25 °C flows over a 30 cm wide electric resistance flat plate heater with a velocity of 13 m/s. The heater dissipates energy into the air at a constant rate of 2730 W/m^2. How long must the heater be in the direction of flow for the surface temperature not to exceed 155 °C?

Solution

Physical properties

The film temperature is $(25 + 155)/2 = 90$ °C.

For air at 90 °C (363 K) and 1 atm: $\begin{cases} \nu = 21.95 \times 10^{-6} \text{ m}^2/\text{s} \\ k = 30.58 \times 10^{-3} \text{ W/m·K} \\ \text{Pr} = 0.704 \end{cases}$

Assumptions

1. Steady-state conditions prevail.
2. Both laminar and turbulent flow regions exist over the plate.

Analysis

The average convection heat transfer coefficient can be calculated from Newton's law of cooling as

$$\langle h \rangle = \frac{q_w}{T_w - T_\infty} = \frac{2730}{155 - 25} = 21 \text{ W/m}^2\text{·K} \tag{1}$$

To determine which correlation to use, it is necessary to calculate the Reynolds number. However, the Reynolds number cannot be determined a priori since the length of the heater is unknown. Therefore, a trial-and-error procedure must be used. Since we assumed that both laminar and turbulent flow regions exist over the heater, the use of Eq. (E) in Table 4.2 gives

$$\langle \text{Nu} \rangle = \frac{\langle h \rangle L}{k} = \left(0.037 \, \text{Re}_L^{4/5} - 871 \right) \text{Pr}^{1/3}$$

$$\frac{(21)L}{30.58 \times 10^{-3}} = \left\{ 0.037 \left[\frac{(13)L}{21.95 \times 10^{-6}} \right]^{4/5} - 871 \right\} (0.704)^{1/3} \tag{2}$$

Simplification of Eq. (2) yields

$$F(L) = L - 1.99 \, L^{4/5} + 1.13 = 0 \tag{3}$$

The length of the heater can be determined from Eq. (3) by using one of the numerical methods for root finding given in Section A.7.2 in Appendix A. The iteration scheme given by Eq. (A.7-25) is expressed as

$$L_k = L_{k-1} - \frac{0.02 L_{k-1} F(L_{k-1})}{F(1.01 L_{k-1}) - F(0.99 L_{k-1})} \tag{4}$$

Assuming $L^{4/5} \simeq L$, a starting value can be estimated as $L_o = 1.141$. The iterations are given in the table below:

k	L_k
0	1.141
1	1.249
2	1.252
3	1.252

Thus, the length of the plate is approximately 1.25 m. Now it is necessary to check the validity of the second assumption:

$$\mathrm{Re}_L = \frac{(1.25)(13)}{21.95 \times 10^{-6}} = 7.4 \times 10^5 \quad \Rightarrow \quad \text{Checks!}$$

Example 4.3 A water storage tank open to the atmosphere is 12 m in length and 6 m in width. The water and the surrounding air are at a temperature of 25 °C, and the relative humidity of the air is 60%. If the wind blows at a velocity of 2 m/s along the long side of the tank, what is the steady rate of water loss due to evaporation from the surface?

Solution

Physical properties

For air at 25 °C (298 K): $\nu = 15.54 \times 10^{-6}$ m^2/s

Diffusion coefficient of water ($\mathcal{A}$) in air ($\mathcal{B}$) at 25 °C (298 K):

$$(\mathcal{D}_{AB})_{298} = (\mathcal{D}_{AB})_{313}\left(\frac{298}{313}\right)^{3/2} = (2.88 \times 10^{-5})\left(\frac{298}{313}\right)^{3/2} = 2.79 \times 10^{-5} \text{ m}^2/\text{s}$$

The Schmidt number is

$$\mathrm{Sc} = \frac{\nu}{\mathcal{D}_{AB}} = \frac{15.54 \times 10^{-6}}{2.79 \times 10^{-5}} = 0.56$$

For water at 25 °C (298 K): $P^{sat} = 0.03165$ bar

Assumptions

1. Steady-state conditions prevail.
2. Ideal gas behavior.

Analysis

To determine which correlation to use, we must first calculate the Reynolds number:

$$\mathrm{Re}_L = \frac{L v_\infty}{\nu} = \frac{(12)(2)}{15.54 \times 10^{-6}} = 1.54 \times 10^6$$

Since both laminar and turbulent conditions exist, the use of Eq. (F) in Table 4.2 gives

$$\langle \mathrm{Sh} \rangle = \left(0.037\,\mathrm{Re}_L^{4/5} - 871\right)\mathrm{Sc}^{1/3} = \left[0.037(1.54 \times 10^6)^{4/5} - 871\right](0.56)^{1/3} = 2000$$

Therefore, the average mass transfer coefficient is

$$\langle k_c \rangle = \frac{\langle \text{Sh} \rangle \mathcal{D}_{AB}}{L} = \frac{(2000)(2.79 \times 10^{-5})}{12} = 4.65 \times 10^{-3} \text{ m/s}$$

The number of moles of H_2O ($\mathcal{A}$) evaporated in unit time is

$$\dot{n}_A = A \langle k_c \rangle \left[c_A^{sat} - c_A(air) \right] = A \langle k_c \rangle \left(c_A^{sat} - 0.6 c_A^{sat} \right) = 0.4 A \langle k_c \rangle c_A^{sat}$$

The saturation concentration of water, c_A^{sat}, is

$$c_A^{sat} = \frac{P_A^{sat}}{\mathcal{R}T} = \frac{0.03165}{(8.314 \times 10^{-2})(25 + 273)} = 1.28 \times 10^{-3} \text{ kmol/m}^3$$

Hence, the rate of water loss is

$$\dot{m}_A = \dot{n}_A \mathcal{M}_A = 0.4 A \langle k_c \rangle c_A^{sat} \mathcal{M}_A$$

$$= (0.4)(12 \times 6)\left(4.65 \times 10^{-3}\right)\left(1.28 \times 10^{-3}\right)(18)(3600) = 11.1 \text{ kg/h}$$

4.3   FLOW PAST A SINGLE SPHERE

Consider a single sphere immersed in an infinite fluid. We may consider two exactly equivalent cases: (*i*) the sphere is stagnant, the fluid flows over the sphere, (*ii*) the fluid is stagnant, the sphere moves through the fluid.

According to Newton's second law of motion, the balance of forces acting on a single spherical particle of diameter D_P, falling in a stagnant fluid with a constant terminal velocity v_t, is expressed in the form

$$\text{Gravitational force} = \text{Buoyancy} + \text{Drag force} \tag{4.3-1}$$

or,

$$\left(\frac{\pi D_P^3}{6} \right) \rho_P g = \left(\frac{\pi D_P^3}{6} \right) \rho g + \left(\frac{\pi D_P^2}{4} \right) \left(\frac{1}{2} \rho v_t^2 \right) f \tag{4.3-2}$$

where ρ_P and ρ represent the densities of the particle and fluid, respectively. In the literature, the friction factor f is also called the *drag coefficient* and is denoted by C_D. Simplification of Eq. (4.3-2) gives

$$f v_t^2 = \frac{4}{3} \frac{g D_P (\rho_P - \rho)}{\rho} \tag{4.3-3}$$

Equation (4.3-3) can be rearranged in dimensionless form as

$$\boxed{f \, \text{Re}_P^2 = \frac{4}{3} \text{Ar}} \tag{4.3-4}$$

where the Reynolds number, Re_P, and the Archimedes number, Ar, are defined by

$$\text{Re}_P = \frac{D_P v_t \rho}{\mu} \tag{4.3-5}$$

$$\text{Ar} = \frac{D_P^3 g \rho (\rho_P - \rho)}{\mu^2} \tag{4.3-6}$$

Engineering problems associated with the motion of spherical particles in fluids are classified as follows:

- Calculate the terminal velocity, v_t; given the viscosity of fluid, μ, and the particle diameter, D_P.
- Calculate the particle diameter, D_P; given the viscosity of the fluid, μ, and the terminal velocity, v_t.
- Calculate the fluid viscosity, μ; given the particle diameter, D_P, and the terminal velocity, v_t.

The difficulty in these problems arises from the fact that the friction factor f in Eq. (4.3-4) is a complex function of the Reynolds number and the Reynolds number cannot be determined *a priori*.

4.3.1   Friction Factor Correlations

For flow of a sphere through a stagnant fluid, Lapple and Shepherd (1940) presented their experimental data in the form of f versus Re_P. Their data can be approximated as

$$f = \frac{24}{\text{Re}_P} \qquad \text{Re}_P < 2 \tag{4.3-7}$$

$$f = \frac{18.5}{\text{Re}_P^{0.6}} \qquad 2 \leqslant \text{Re}_P < 500 \tag{4.3-8}$$

$$f = 0.44 \qquad 500 \leqslant \text{Re}_P < 2 \times 10^5 \tag{4.3-9}$$

Equations (4.3-7) and (4.3-9) are generally referred to as Stokes' law and Newton's law, respectively.

In recent years, efforts have been directed to obtain a single comprehensive equation for the friction factor that covers the entire range of Re_P. Turton and Levenspiel (1986) proposed the following five-constant equation, which correlates the experimental data for $\text{Re}_P \leqslant 2 \times 10^5$:

$$\boxed{f = \frac{24}{\text{Re}_P}\left(1 + 0.173\,\text{Re}_P^{0.657}\right) + \frac{0.413}{1 + 16{,}300\,\text{Re}_P^{-1.09}}} \tag{4.3-10}$$

4.3.1.1 *Solutions to the engineering problems* Solutions to the engineering problems described above can now be summarized as follows:

■ **Calculate v_t; given μ and D_P**

Substitution of Eq. (4.3-10) into Eq. (4.3-4) gives

$$\mathrm{Ar} = 18\big(\mathrm{Re}_P + 0.173\,\mathrm{Re}_P^{1.657}\big) + \frac{0.31\,\mathrm{Re}_P^2}{1 + 16{,}300\,\mathrm{Re}_P^{-1.09}} \qquad (4.3\text{-}11)$$

Since Eq. (4.3-11) expresses the Archimedes number as a function of the Reynolds number, calculation of the terminal velocity for a given particle diameter and fluid viscosity requires an iterative solution. To circumvent this problem, it is necessary to express the Reynolds number as a function of the Archimedes number. The following explicit expression relating the Archimedes number to the Reynolds number is proposed by Turton and Clark (1987):

$$\boxed{\mathrm{Re}_P = \frac{\mathrm{Ar}}{18}\big(1 + 0.0579\,\mathrm{Ar}^{0.412}\big)^{-1.214}} \qquad (4.3\text{-}12)$$

The procedure to calculate the terminal velocity is as follows:

a) Calculate the Archimedes number from Eq. (4.3-6),
b) Substitute the Archimedes number into Eq. (4.3-12) and determine the Reynolds number,
c) Once the Reynolds number is determined, the terminal velocity can be calculated from the equation

$$v_t = \frac{\mu\,\mathrm{Re}_P}{\rho D_P} \qquad (4.3\text{-}13)$$

Example 4.4 Calculate the velocities at which a drop of water, 5 mm in diameter, would fall in air at 20 °C and the same size air bubble would rise through water at 20 °C.

Solution

Physical properties

For water at 20 °C (293 K): $\begin{cases} \rho = 999 \ \mathrm{kg/m^3} \\ \mu = 1001 \times 10^{-6} \ \mathrm{kg/m \cdot s} \end{cases}$

For air at 20 °C (293 K): $\begin{cases} \rho = 1.2047 \ \mathrm{kg/m^3} \\ \mu = 18.17 \times 10^{-6} \ \mathrm{kg/m \cdot s} \end{cases}$

Analysis

Water droplet falling in air

To determine the terminal velocity of water, it is necessary to calculate the Archimedes number using Eq. (4.3-6):

$$\mathrm{Ar} = \frac{D_P^3 g \rho (\rho_P - \rho)}{\mu^2} = \frac{(5 \times 10^{-3})^3 (9.8)(1.2047)(999 - 1.2047)}{(18.17 \times 10^{-6})^2} = 4.46 \times 10^6$$

The Reynolds number is calculated from Eq. (4.3-12):

$$\mathrm{Re}_P = \frac{\mathrm{Ar}}{18}(1 + 0.0579\,\mathrm{Ar}^{0.412})^{-1.214}$$

$$= \frac{4.46 \times 10^6}{18}\left[1 + 0.0579(4.46 \times 10^6)^{0.412}\right]^{-1.214} = 3581$$

Hence, the terminal velocity is

$$v_t = \frac{\mu\,\mathrm{Re}_P}{\rho D_P} = \frac{(18.17 \times 10^{-6})(3581)}{(1.2047)(5 \times 10^{-3})} = 10.8 \text{ m/s}$$

Air bubble rising in water

In this case, the Archimedes number is

$$Ar = \frac{D_P^3 g \rho(\rho_P - \rho)}{\mu^2} = \frac{(5 \times 10^{-3})^3(9.8)(999)(1.2047 - 999)}{(1001 \times 10^{-6})^2} = -1.219 \times 10^6$$

The minus sign indicates that the motion of the bubble is in the direction opposite to gravity, i.e., it is rising. The Reynolds number and the terminal velocity are

$$\mathrm{Re}_P = \frac{\mathrm{Ar}}{18}(1 + 0.0579\,\mathrm{Ar}^{0.412})^{-1.214}$$

$$= \frac{1.219 \times 10^6}{18}\left[1 + 0.0579(1.219 \times 10^6)^{0.412}\right]^{-1.214} = 1825$$

$$v_t = \frac{\mu\,\mathrm{Re}_P}{\rho D_P} = \frac{(1001 \times 10^{-6})(1825)}{(999)(5 \times 10^{-3})} = 0.37 \text{ m/s}$$

■ **Calculate D_P; given μ and v_t**

In this case, Eq. (4.3-4) must be rearranged such that the particle diameter is eliminated. If both sides of Eq. (4.3-4) are divided by Re_P^3, the result is

$$\frac{f}{\mathrm{Re}_P} = Y \tag{4.3-14}$$

where Y, which is independent of D_P, is a dimensionless number defined by

$$Y = \frac{4}{3}\frac{g(\rho_P - \rho)\mu}{\rho^2 v_t^3} \tag{4.3-15}$$

Substitution of Eq. (4.3-10) into Eq. (4.3-14) yields

$$Y = \frac{24}{\mathrm{Re}_P^2}\left(1 + 0.173\,\mathrm{Re}_P^{0.657}\right) + \frac{0.413}{\mathrm{Re}_P + 16{,}300\,\mathrm{Re}_P^{-0.09}} \tag{4.3-16}$$

Since Eq. (4.3-16) expresses Y as a function of the Reynolds number, calculation of the particle diameter for a given terminal velocity and fluid viscosity requires an iterative solution. To circumvent this problem, the following explicit expression relating Y to the Reynolds number is proposed by Tosun and Akşahin (1992) as

$$\boxed{\mathrm{Re}_P = \frac{\Psi(Y)}{(6\,Y^{13/20} - Y^{6/11})^{17/20}}} \tag{4.3-17}$$

where $\Psi(Y)$ is given by

$$\Psi(Y) = \exp\left(3.15 + \frac{0.052}{Y^{1/4}} + \frac{0.007}{Y^{1/2}} - \frac{0.00019}{Y^{3/4}}\right) \tag{4.3-18}$$

The procedure to calculate the particle diameter is as follows:

a) Calculate Y from Eq. (4.3-15),
b) Substitute Y into Eqs. (4.3-17) and (4.3-18) and determine Re_P,
c) Once the Reynolds number is determined, the particle diameter can be calculated from the equation

$$D_P = \frac{\mu\,\mathrm{Re}_P}{\rho v_t} \tag{4.3-19}$$

Example 4.5 A gravity settling chamber is one of the diverse range of equipment used to remove particulate solids from gas streams. In a settling chamber, the entering gas stream encounters a large and abrupt increase in cross-sectional area as shown in the figure below. As a result of the sharp decrease in the gas velocity, the solid particles settle down with gravity. In practice, the gas velocity through the chamber should be kept below 3 m/s to prevent the re-entrainment of the settled particles.

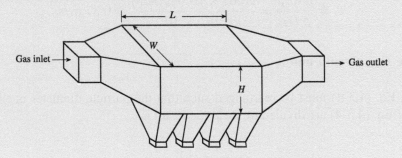

Spherical dust particles having a density of 2200 kg/m³ are to be separated from an air stream at a temperature of 25 °C. Determine the diameter of the smallest particle that can be removed in a settling chamber 7 m long, 2 m wide, and 1 m high.

Solution

Physical properties

For air at 25 °C (298 K): $\begin{cases} \rho = 1.1845\ \text{kg/m}^3 \\ \mu = 18.41 \times 10^{-6}\ \text{kg/m·s} \end{cases}$

Analysis

For the minimum particle size that can be removed with 100% efficiency, the time required for this particle to fall a distance H must be equal to the time required to move this particle horizontally a distance L, i.e.,

$$t = \frac{H}{v_t} = \frac{L}{\langle v \rangle} \quad \Rightarrow \quad v_t = \langle v \rangle \left(\frac{H}{L} \right)$$

where $\langle v \rangle$ represents the average gas velocity in the settling chamber. Taking $\langle v \rangle = 3$ m/s, the settling velocity of the particles can be calculated as

$$v_t = (3)\left(\frac{1}{7} \right) = 0.43 \text{ m/s}$$

The value of Y is calculated from Eq. (4.3-15) as

$$Y = \frac{4}{3} \frac{g(\rho_P - \rho)\mu}{\rho^2 v_t^3} = \frac{4}{3} \frac{(9.8)(2200 - 1.1845)(18.41 \times 10^{-6})}{(1.1845)^2(0.43)^3} = 4.74$$

Substitution of the value of Y into Eq. (4.3-18) gives

$$\Psi(Y) = \exp\left(3.15 + \frac{0.052}{Y^{1/4}} + \frac{0.007}{Y^{1/2}} - \frac{0.00019}{Y^{3/4}} \right)$$

$$= \exp\left[3.15 + \frac{0.052}{(4.74)^{1/4}} + \frac{0.007}{(4.74)^{1/2}} - \frac{0.00019}{(4.74)^{3/4}} \right] = 24.3$$

Therefore, the Reynolds number and the particle diameter are

$$\text{Re}_P = \frac{\Psi(Y)}{(6Y^{13/20} - Y^{6/11})^{17/20}} = \frac{24.3}{[6(4.74)^{13/20} - (4.74)^{6/11}]^{17/20}} = 2.55$$

$$D_P = \frac{\mu \, \text{Re}_P}{\rho \, v_t} = \frac{(18.41 \times 10^{-6})(2.55)}{(1.1845)(0.43)} = 92 \times 10^{-6} \text{ m}$$

■ **Calculate μ; given D_P and v_t**

In this case, Eq. (4.3-4) must be rearranged so that the fluid viscosity can be eliminated. If both sides of Eq. (4.3-4) are divided by Re_P^2, the result is

$$f = X \tag{4.3-20}$$

where X, which is independent of μ, is a dimensionless number defined by

$$X = \frac{4}{3} \frac{g D_P (\rho_P - \rho)}{\rho v_t^2} \tag{4.3-21}$$

Substitution of Eq. (4.3-10) into Eq. (4.3-20) gives

$$X = \frac{24}{\text{Re}_P}\left(1 + 0.173\,\text{Re}_P^{0.657}\right) + \frac{0.413}{1 + 16,300\,\text{Re}_P^{-1.09}} \tag{4.3-22}$$

Since Eq. (4.3-22) expresses X as a function of the Reynolds number, calculation of the fluid viscosity for a given terminal velocity and particle diameter requires an iterative solution. To circumvent this problem, the following explicit expression relating X to the Reynolds number is proposed by Tosun and Akşahin (1992):

$$\boxed{\text{Re}_P = \frac{24}{X}(1 + 120X^{-20/11})^{4/11}} \qquad X \geqslant 0.5 \qquad (4.3\text{-}23)$$

The procedure to calculate the fluid viscosity is as follows:

a) Calculate X from Eq. (4.3-21),
b) Substitute X into Eq. (4.3-23) and determine the Reynolds number,
c) Once the Reynolds number is determined, the fluid viscosity can be calculated from the equation

$$\mu = \frac{D_P v_t \rho}{\text{Re}_P} \qquad (4.3\text{-}24)$$

Example 4.6 One way of measuring fluid viscosity is to use a falling ball viscometer in which a spherical ball of known density is dropped into a fluid-filled graduated cylinder and the time of fall for the ball for a specified distance is recorded.

A spherical ball, 5 mm in diameter, has a density of 1000 kg/m^3. It falls through a liquid of density 910 kg/m^3 at 25 °C and travels a distance of 10 cm in 1.8 min. Determine the viscosity of the liquid.

Solution

The terminal velocity of the sphere is

$$v_t = \frac{\text{Distance}}{\text{Time}} = \frac{10 \times 10^{-2}}{(1.8)(60)} = 9.26 \times 10^{-4} \text{ m/s}$$

The value of X is calculated from Eq. (4.3-21) as

$$X = \frac{4}{3} \frac{g D_P (\rho_P - \rho)}{\rho v_t^2} = \frac{4}{3} \frac{(9.8)(5 \times 10^{-3})(1000 - 910)}{(910)(9.26 \times 10^{-4})^2} = 7536$$

Substitution of the value of X into Eq. (4.3-23) gives the Reynolds number as

$$\text{Re}_P = \frac{24}{X}(1 + 120X^{-20/11})^{4/11} = \frac{24}{7536}\left[1 + 120(7536)^{-20/11}\right]^{4/11} = 3.2 \times 10^{-3}$$

Hence, the viscosity of the fluid is

$$\mu = \frac{D_P v_t \rho}{\text{Re}_P} = \frac{(5 \times 10^{-3})(9.26 \times 10^{-4})(910)}{3.2 \times 10^{-3}} = 1.32 \text{ kg/m·s}$$

4.3.1.2   *Deviations from ideal behavior*   It should be noted that Eqs. (4.3-4) and (4.3-10) are only valid for a single spherical particle falling in an unbounded fluid. The presence of container walls and other particles as well as any deviations from spherical shape affect the terminal velocity of particles. For example, as a result of the upflow of displaced fluid in a suspension of uniform particles, the settling velocity of particles in suspension is slower than the terminal velocity of a single particle of the same size. The most general empirical equation relating the settling velocity to the volume fraction of particles, ω, is given by

$$\frac{v_t(\text{suspension})}{v_t(\text{single sphere})} = (1 - \omega)^n \qquad (4.3\text{-}25)$$

where the exponent n depends on the Reynolds number based on the terminal velocity of a particle in an unbounded fluid. In the literature, values of n are reported as

$$n = \begin{cases} 4.65 - 5.00 & \text{Re}_P < 2 \\ 2.30 - 2.65 & 500 \leqslant \text{Re}_P \leqslant 2 \times 10^5 \end{cases} \qquad (4.3\text{-}26)$$

Particle shape is another factor affecting terminal velocity. The terminal velocity of a non-spherical particle is less than that of a spherical one by a factor of *sphericity*, ϕ, i.e.,

$$\frac{v_t(\text{non-spherical})}{v_t(\text{spherical})} = \phi < 1 \qquad (4.3\text{-}27)$$

Sphericity is defined as the ratio of the surface area of a sphere having the same volume as the non-spherical particle to the actual surface area of the particle.

4.3.2   Heat Transfer Correlations

When a sphere is immersed in an infinite stagnant fluid, the analytical solution for steady-state conduction is possible[1] and the result is expressed in the form

$$\text{Nu} = 2 \qquad (4.3\text{-}28)$$

In the case of fluid motion, the contribution of the convective mechanism must be included in Eq. (4.3-28). Correlations for including convective heat transfer are as follows:

Ranz-Marshall correlation

Ranz and Marshall (1952) proposed the following correlation for constant surface temperature:

$$\boxed{\text{Nu} = 2 + 0.6\,\text{Re}_P^{1/2}\,\text{Pr}^{1/3}} \qquad (4.3\text{-}29)$$

All properties in Eq. (4.3-29) must be evaluated at the film temperature.

---

[1] See Example 8.12 in Chapter 8.

Whitaker correlation

Whitaker (1972) considered heat transfer from the sphere to be a result of two parallel processes occurring simultaneously. He assumed that the laminar and turbulent contributions are additive and proposed the following equation:

$$\mathrm{Nu} = 2 + \left(0.4\,\mathrm{Re}_P^{1/2} + 0.06\,\mathrm{Re}_P^{2/3}\right)\mathrm{Pr}^{0.4}(\mu_\infty/\mu_w)^{1/4} \tag{4.3-30}$$

All properties except μ_w should be evaluated at T_∞. Equation (4.3-30) is valid for

$$3.5 \leqslant \mathrm{Re}_P \leqslant 7.6 \times 10^4 \qquad 0.71 \leqslant \mathrm{Pr} \leqslant 380 \qquad 1.0 \leqslant \mu_\infty/\mu_w \leqslant 3.2$$

4.3.2.1   *Calculation of the heat transfer rate*   Once the average heat transfer coefficient is estimated by using correlations, the rate of heat transferred is calculated as

$$\dot{Q} = \left(\pi D_P^2\right)\langle h\rangle |T_w - T_\infty| \tag{4.3-31}$$

Example 4.7 An instrument is enclosed in a protective spherical shell, 5 cm in diameter, and submerged in a river to measure the concentrations of pollutants. The temperature and the velocity of the river are 10 °C and 1.2 m/s, respectively. To prevent any damage to the instrument as a result of the low river temperature, the surface temperature is kept constant at 32 °C by installing electrical heaters in the protective shell. Calculate the electrical power dissipated under steady conditions.

Solution

Physical properties

For water at 10 °C (283 K): $\begin{cases} \rho = 1000 \text{ kg/m}^3 \\ \mu = 1304 \times 10^{-6} \text{ kg/m·s} \\ k = 587 \times 10^{-3} \text{ W/m·K} \\ \mathrm{Pr} = 9.32 \end{cases}$

For water at 32 °C (305 K): $\mu = 769 \times 10^{-6}$ kg/m·s

Analysis

System: Protective shell

Under steady conditions, the electrical power dissipated is equal to the rate of heat loss from the shell surface to the river. The rate of heat loss is given by

$$\dot{Q} = \left(\pi D_P^2\right)\langle h\rangle (T_w - T_\infty) \tag{1}$$

To determine $\langle h\rangle$, it is necessary to calculate the Reynolds number

$$\mathrm{Re}_P = \frac{D_P v_\infty \rho}{\mu} = \frac{(5 \times 10^{-2})(1.2)(1000)}{1304 \times 10^{-6}} = 4.6 \times 10^4 \tag{2}$$

The Whitaker correlation, Eq. (4.3-30), gives

$$Nu = 2 + \left(0.4\,Re_P^{1/2} + 0.06\,Re_P^{2/3}\right) Pr^{0.4}(\mu_\infty/\mu_w)^{1/4}$$

or,

$$Nu = 2 + \left[0.4(4.6 \times 10^4)^{1/2} + 0.06(4.6 \times 10^4)^{2/3}\right](9.32)^{0.4}$$

$$\times \left(\frac{1304 \times 10^{-6}}{769 \times 10^{-6}}\right)^{1/4} = 456 \tag{3}$$

The average heat transfer coefficient is

$$\langle h \rangle = Nu\left(\frac{k}{D_P}\right) = (456)\left(\frac{587 \times 10^{-3}}{5 \times 10^{-2}}\right) = 5353 \ W/m^2{\cdot}K \tag{4}$$

Therefore, the rate of heat loss is calculated from Eq. (1) as

$$\dot{Q} = \left[\pi(5 \times 10^{-2})^2\right](5353)(32 - 10) = 925 \ W \tag{5}$$

4.3.3 Mass Transfer Correlations

When a sphere is immersed in an infinite stagnant fluid, the analytical solution for steady-state diffusion is possible[2] and the result is expressed in the form

$$Sh = 2 \tag{4.3-32}$$

In the case of fluid motion, the contribution of convection must be taken into consideration. Correlations for convective mass transfer are as follows:

Ranz-Marshall correlation

For constant surface composition and low mass transfer rates, Eq. (4.3-29) may be applied to mass transfer problems simply by replacing Nu and Pr with Sh and Sc, respectively, i.e.,

$$\boxed{Sh = 2 + 0.6\,Re_P^{1/2}\,Sc^{1/3}} \tag{4.3-33}$$

Equation (4.3-33) is valid for

$$2 \leqslant Re_P \leqslant 200 \qquad 0.6 \leqslant Sc \leqslant 2.7$$

Frossling correlation

Frossling (1938) proposed the following correlation:

$$\boxed{Sh = 2 + 0.552\,Re_P^{1/2}\,Sc^{1/3}} \tag{4.3-34}$$

Equation (4.3-34) is valid for

$$2 \leqslant Re_P \leqslant 800 \qquad 0.6 \leqslant Sc \leqslant 2.7$$

---

[2]See Example 8.19 in Chapter 8.

Steinberger and Treybal (1960) modified the Frossling correlation as

$$\text{Sh} = 2 + 0.552\,\text{Re}_P^{0.53}\,\text{Sc}^{1/3} \qquad (4.3\text{-}35)$$

which is valid for

$$1500 \leqslant \text{Re}_P \leqslant 12,000 \qquad 0.6 \leqslant \text{Sc} \leqslant 1.85$$

Steinberger-Treybal correlation

The correlation originally proposed by Steinberger and Treybal (1960) includes a correction term for natural convection. The lack of experimental data, however, makes this term very difficult to calculate in most cases. The effect of natural convection becomes negligible when the Reynolds number is high, and the Steinberger-Treybal correlation reduces to

$$\text{Sh} = 0.347\,\text{Re}_P^{0.62}\,\text{Sc}^{1/3} \qquad (4.3\text{-}36)$$

Equation (4.3-36) is recommended for liquids when

$$2000 \leqslant \text{Re}_P \leqslant 16,900$$

4.3.3.1 *Calculation of the mass transfer rate*   Once the average mass transfer coefficient is estimated by using correlations, the rate of mass of species $\mathcal{A}$ transferred is calculated as

$$\dot{m}_A = \left(\pi D_P^2\right)\langle k_c \rangle |c_{A_w} - c_{A_\infty}|\mathcal{M}_A \qquad (4.3\text{-}37)$$

Example 4.8 A solid sphere of benzoic acid ($\rho = 1267$ kg/m^3) with a diameter of 12 mm is dropped into a long cylindrical tank filled with pure water at 25 °C. If the height of the tank is 3 m, determine the amount of benzoic acid dissolved from the sphere when it reaches the bottom of the tank. The saturation solubility of benzoic acid in water is 3.412 kg/m^3.

Solution

Physical properties

For water ($\mathcal{B}$) at 25 °C (298 K): $\begin{cases} \rho = 1000 \text{ kg/m}^3 \\ \mu = 892 \times 10^{-6} \text{ kg/m·s} \\ \mathcal{D}_{AB} = 1.21 \times 10^{-9} \text{ m}^2/\text{s} \end{cases}$

The Schmidt number is

$$\text{Sc} = \frac{\mu}{\rho\mathcal{D}_{AB}} = \frac{892 \times 10^{-6}}{(1000)(1.21 \times 10^{-9})} = 737$$

Assumptions

1. Initial acceleration period is negligible and the sphere reaches its terminal velocity instantaneously.
2. Diameter of the sphere does not change appreciably. Thus, the Reynolds number and the terminal velocity remain constant.

3. Steady-state conditions prevail.
4. Physical properties of water do not change as a result of mass transfer.

Analysis

To determine the terminal velocity of the benzoic acid sphere, it is necessary to calculate the Archimedes number using Eq. (4.3-6):

$$Ar = \frac{D_P^3 g \rho (\rho_P - \rho)}{\mu^2} = \frac{(12 \times 10^{-3})^3 (9.8)(1000)(1267 - 1000)}{(892 \times 10^{-6})^2} = 5.68 \times 10^6$$

The Reynolds number is calculated from Eq. (4.3-12):

$$Re_P = \frac{Ar}{18}(1 + 0.0579\, Ar^{0.412})^{-1.214}$$

$$= \frac{5.68 \times 10^6}{18}\left[1 + 0.0579(5.68 \times 10^6)^{0.412}\right]^{-1.214} = 4056$$

Hence, the terminal velocity is

$$v_t = \frac{\mu\, Re_P}{\rho\, D_P} = \frac{(892 \times 10^{-6})(4056)}{(1000)(12 \times 10^{-3})} = 0.3 \text{ m/s}$$

Since the benzoic acid sphere falls the distance of 3 m with a velocity of 0.3 m/s, the falling time is

$$t = \frac{\text{Distance}}{\text{Time}} = \frac{3}{0.3} = 10 \text{ s}$$

The Sherwood number is calculated from the Steinberger-Treybal correlation, Eq. (4.3-36), as

$$Sh = 0.347\, Re_P^{0.62}\, Sc^{1/3} = 0.347(4056)^{0.62}(737)^{1/3} = 541$$

The average mass transfer coefficient is

$$\langle k_c \rangle = Sh\left(\frac{\mathcal{D}_{AB}}{D_P}\right) = (541)\left(\frac{1.21 \times 10^{-9}}{12 \times 10^{-3}}\right) = 5.46 \times 10^{-5} \text{ m/s}$$

The rate of transfer of benzoic acid (species $\mathcal{A}$) to water is calculated by using Eq. (4.3-37):

$$\dot{m}_A = \left(\pi D_P^2\right)\langle k_c \rangle (c_{A_w} - c_{A_\infty})\mathcal{M}_A = \left(\pi D_P^2\right)\langle k_c \rangle (\rho_{A_w} - \rho_{A_\infty})$$

$$= \left[\pi(12 \times 10^{-3})^2\right](5.46 \times 10^{-5})(3.412 - 0) = 8.43 \times 10^{-8} \text{ kg/s}$$

The amount of benzoic acid dissolved in 10 s is

$$M_A = \dot{m}_A\, t = (8.43 \times 10^{-8})(10) = 8.43 \times 10^{-7} \text{ kg}$$

Verification of assumption # 2

The initial mass of the benzoic acid sphere, M_o, is

$$M_o = \left[\frac{\pi (12 \times 10^{-3})^3}{6} \right] (1267) = 1.146 \times 10^{-3} \, \text{kg}$$

The percent decrease in the mass of the sphere is given by

$$\left(\frac{8.43 \times 10^{-7}}{1.146 \times 10^{-3}} \right) \times 100 = 0.074\%$$

Therefore, the assumed constancy of D_P and v_t is justified.

4.4　FLOW NORMAL TO A SINGLE CYLINDER

4.4.1　Friction Factor Correlations

For cross flow over an infinitely long circular cylinder, Lapple and Shepherd (1940) presented their experimental data in the form of f versus Re_D, the Reynolds number based on the diameter of the cylinder. Their data can be approximated as

$$f = \frac{6.18}{\text{Re}_D^{8/9}} \qquad \text{Re}_D < 2 \tag{4.4-1}$$

$$f = 1.2 \qquad 10^4 \leqslant \text{Re}_D \leqslant 1.5 \times 10^5 \tag{4.4-2}$$

The friction factor f in Eqs. (4.4-1) and (4.4-2) is based on the projected area of a cylinder, i.e., diameter times length, and Re_D is defined by

$$\text{Re}_D = \frac{D v_\infty \rho}{\mu} \tag{4.4-3}$$

Tosun and Akşahin (1992) proposed the following single equation for the friction factor that covers the entire range of the Reynolds number in the form

$$\boxed{f = \frac{6.18}{\text{Re}_D^{8/9}} \left(1 + 0.36 \, \text{Re}_D^{5/9} \right)^{8/5}} \qquad \text{Re}_D \leqslant 1.5 \times 10^5 \tag{4.4-4}$$

Once the friction factor is determined, the drag force is calculated from

$$\boxed{F_D = (DL) \left(\frac{1}{2} \rho v_\infty^2 \right) f} \tag{4.4-5}$$

Example 4.9 A distillation column has an outside diameter of 80 cm and a height of 10 m. Calculate the drag force exerted by air on the column if the wind speed is 2.5 m/s.

Solution

Physical properties

For air at 25 °C (298 K): $\begin{cases} \rho = 1.1845 \text{ kg/m}^3 \\ \mu = 18.41 \times 10^{-6} \text{ kg/m·s} \end{cases}$

Assumption

1. Air temperature is 25 °C.

Analysis

From Eq. (4.4-3) the Reynolds number is

$$\text{Re}_D = \frac{D v_\infty \rho}{\mu} = \frac{(0.8)(2.5)(1.1845)}{18.41 \times 10^{-6}} = 1.29 \times 10^5$$

The use of Eq. (4.4-4) gives the friction factor as

$$f = \frac{6.18}{\text{Re}_D^{8/9}} \left(1 + 0.36 \text{Re}_D^{5/9}\right)^{8/5}$$

$$= \frac{6.18}{(1.29 \times 10^5)^{8/9}} \left[1 + 0.36(1.29 \times 10^5)^{5/9}\right]^{8/5} = 1.2$$

Therefore, the drag force is calculated from Eq. (4.4-5) as

$$F_D = (DL)\left(\frac{1}{2}\rho v_\infty^2\right) f = (0.8 \times 10)\left[\frac{1}{2}(1.1845)(2.5)^2\right](1.2) = 35.5 \text{ N}$$

4.4.2 Heat Transfer Correlations

As stated in Section 4.3.2, the analytical solution for steady-state conduction from a sphere to a stagnant medium gives $\text{Nu} = 2$. Therefore, the correlations for heat transfer in spherical geometry require that $\text{Nu} \to 2$ as $\text{Re} \to 0$. In the case of a single cylinder, however, no solution for the case of steady-state conduction exists. Hence, it is required that $\text{Nu} \to 0$ as $\text{Re} \to 0$. The following heat transfer correlations are available in this case:

Whitaker correlation

Whitaker (1972) proposed a correlation in the form

$$\boxed{\text{Nu} = \left(0.4 \text{Re}_D^{1/2} + 0.06 \text{Re}_D^{2/3}\right) \text{Pr}^{0.4} (\mu_\infty/\mu_w)^{1/4}} \tag{4.4-6}$$

in which all properties except μ_w are evaluated at T_∞. Equation (4.4-6) is valid for

$$1.0 \leqslant \text{Re}_D \leqslant 1.0 \times 10^5 \qquad 0.67 \leqslant \text{Pr} \leqslant 300 \qquad 0.25 \leqslant \mu_\infty/\mu_w \leqslant 5.2$$

Table 4.3. Constants of Eq. (4.4-7) for the circular cylinder in cross flow

Re_D	C	m
1–40	0.75	0.4
40–1000	0.51	0.5
1×10^3–2×10^5	0.26	0.6
2×10^5–1×10^6	0.076	0.7

Zhukauskas correlation

The correlation proposed by Zhukauskas (1972) is given by

$$\boxed{Nu = C \, Re_D^m \, Pr^n (Pr_\infty / Pr_w)^{1/4}} \tag{4.4-7}$$

where

$$n = \begin{cases} 0.37 & \text{if } Pr \leqslant 10 \\ 0.36 & \text{if } Pr > 10 \end{cases}$$

and the values of C and m are given in Table 4.3. All properties except Pr_w should be evaluated at T_∞ in Eq. (4.4-7).

Churchill-Bernstein correlation

Churchill and Bernstein (1977) proposed a single comprehensive equation that covers the entire range of Re_D for which data are available, as well as for a wide range of Pr. This equation is in the form

$$\boxed{Nu = 0.3 + \frac{0.62 \, Re_D^{1/2} \, Pr^{1/3}}{[1 + (0.4/Pr)^{2/3}]^{1/4}} \left[1 + \left(\frac{Re_D}{282,000} \right)^{5/8} \right]^{4/5}} \tag{4.4-8}$$

where all properties are evaluated at the film temperature. Equation (4.4-8) is recommended when

$$Re_D \, Pr > 0.2$$

4.4.2.1 *Calculation of the heat transfer rate*   Once the average heat transfer coefficient is estimated by using correlations, the rate of heat transferred is calculated as

$$\boxed{\dot{Q} = (\pi D L)\langle h \rangle |T_w - T_\infty|} \tag{4.4-9}$$

Example 4.10 Assume that a person can be approximated as a cylinder of 0.3 m diameter and 1.8 m height with a surface temperature of 30 °C. Calculate the rate of heat loss from the body while this person is subjected to a 4 m/s wind with a temperature of −10 °C.

Solution

Physical properties

The film temperature is $(30 - 10)/2 = 10\,°\text{C}$

For air at $-10\,°\text{C}$ (263 K): $\begin{cases} \mu = 16.7 \times 10^{-6}\ \text{kg/m·s} \\ \nu = 12.44 \times 10^{-6}\ \text{m}^2/\text{s} \\ k = 23.28 \times 10^{-3}\ \text{W/m·K} \\ \text{Pr} = 0.72 \end{cases}$

For air at $10\,°\text{C}$ (280 K): $\begin{cases} \nu = 14.18 \times 10^{-6}\ \text{m}^2/\text{s} \\ k = 24.86 \times 10^{-3}\ \text{W/m·K} \\ \text{Pr} = 0.714 \end{cases}$

For air at $30\,°\text{C}$ (303 K): $\begin{cases} \mu = 18.64 \times 10^{-6}\ \text{kg/m·s} \\ \text{Pr} = 0.71 \end{cases}$

Assumption

1. Steady-state conditions prevail.

Analysis

The rate of heat loss from the body can be calculated from Eq. (4.4-9):

$$\dot{Q} = (\pi D L) \langle h \rangle (T_w - T_\infty) \tag{1}$$

Determination of $\langle h \rangle$ in Eq. (1) requires the Reynolds number to be known. The Reynolds numbers at T_∞ and T_f are

$$\text{at} \quad T_\infty = -10\,°\text{C} \quad \text{Re}_D = \frac{D\,v_\infty}{\nu} = \frac{(0.3)(4)}{12.44 \times 10^{-6}} = 9.65 \times 10^4$$

$$\text{at} \quad T_f = 10\,°\text{C} \quad \text{Re}_D = \frac{D\,v_\infty}{\nu} = \frac{(0.3)(4)}{14.18 \times 10^{-6}} = 8.46 \times 10^4$$

Whitaker correlation

The use of Eq. (4.4-6) gives the Nusselt number as

$$\text{Nu} = \left(0.4\,\text{Re}_D^{1/2} + 0.06\,\text{Re}_D^{2/3}\right)\text{Pr}^{0.4}(\mu_\infty/\mu_w)^{1/4}$$

$$= \left[0.4\,(9.65 \times 10^4)^{1/2} + 0.06\,(9.65 \times 10^4)^{2/3}\right](0.72)^{0.4}\left(\frac{16.7 \times 10^{-6}}{18.64 \times 10^{-6}}\right)^{1/4}$$

$$= 214$$

Hence, the average heat transfer coefficient is

$$\langle h \rangle = \text{Nu}\left(\frac{k}{D}\right) = (214)\left(\frac{23.28 \times 10^{-3}}{0.3}\right) = 16.6\ \text{W/m}^2\text{·K}$$

Substitution of this result into Eq. (1) gives the rate of heat loss as

$$\dot{Q} = (\pi \times 0.3 \times 1.8)(16.6)\big[30 - (-10)\big] = 1126 \text{ W}$$

Zhukauskas correlation

For $Re_D = 9.65 \times 10^4$ and $Pr < 10$, $n = 0.37$, and from Table 4.3 the constants are $C = 0.26$ and $m = 0.6$. Hence, the use of Eq. (4.4-7) gives

$$Nu = 0.26 \, Re_D^{0.6} \, Pr^{0.37} (Pr_\infty / Pr_w)^{1/4}$$

$$= 0.26(9.65 \times 10^4)^{0.6}(0.72)^{0.37}\left(\frac{0.72}{0.71}\right)^{1/4} = 226$$

Therefore, the average heat transfer coefficient and the rate of heat loss from the body are

$$\langle h \rangle = Nu\left(\frac{k}{D}\right) = (226)\left(\frac{23.28 \times 10^{-3}}{0.3}\right) = 17.5 \text{ W/m}^2\cdot\text{K}$$

$$\dot{Q} = (\pi \times 0.3 \times 1.8)(17.5)\big[30 - (-10)\big] = 1188 \text{ W}$$

Churchill-Bernstein correlation

The use of Eq. (4.4-8) gives

$$Nu = 0.3 + \frac{0.62 \, Re_D^{1/2} \, Pr^{1/3}}{[1 + (0.4/Pr)^{2/3}]^{1/4}}\left[1 + \left(\frac{Re_D}{282{,}000}\right)^{5/8}\right]^{4/5}$$

$$= 0.3 + \frac{0.62(8.46 \times 10^4)^{1/2}(0.714)^{1/3}}{[1 + (0.4/0.714)^{2/3}]^{1/4}}\left[1 + \left(\frac{8.46 \times 10^4}{282{,}000}\right)^{5/8}\right]^{4/5} = 193$$

The average heat transfer coefficient and the rate of heat loss from the body are

$$\langle h \rangle = Nu\left(\frac{k}{D}\right) = (193)\left(\frac{24.86 \times 10^{-3}}{0.3}\right) = 16 \text{ W/m}^2\cdot\text{K}$$

$$\dot{Q} = (\pi \times 0.3 \times 1.8)(16)\big[30 - (-10)\big] = 1086 \text{ W}$$

Comment: The rate of heat loss predicted by the Zhukauskas correlation is 9% greater than that calculated using the Churchill-Bernstein correlation. It is important to note that no two correlations will give exactly the same result.

4.4.3   Mass Transfer Correlations

Bedingfield and Drew (1950) proposed the following correlation for cross- and parallel-flow of gases to the cylinder in which mass transfer to or from the ends of the cylinder is not considered:

$$\boxed{Sh = 0.281 \, Re_D^{1/2} \, Sc^{0.44}} \tag{4.4-10}$$

Equation (4.4-10) is valid for

$$400 \leqslant \mathrm{Re}_D \leqslant 25{,}000 \qquad 0.6 \leqslant \mathrm{Sc} \leqslant 2.6$$

For liquids the correlation obtained by Linton and Sherwood (1950) may be used:

$$\boxed{\mathrm{Sh} = 0.281\,\mathrm{Re}_D^{0.6}\,\mathrm{Sc}^{1/3}} \qquad (4.4\text{-}11)$$

Equation (4.4-11) is valid for

$$400 \leqslant \mathrm{Re}_D \leqslant 25{,}000 \qquad \mathrm{Sc} \leqslant 3000$$

4.4.3.1 *Calculation of the mass transfer rate* Once the average mass transfer coefficient is estimated by using correlations, the rate of mass of species $\mathcal{A}$ transferred is calculated as

$$\boxed{\dot{m}_A = (\pi DL)\langle k_c \rangle |c_{A_w} - c_{A_\infty}| \mathcal{M}_A} \qquad (4.4\text{-}12)$$

where $\mathcal{M}_A$ is the molecular weight of species $\mathcal{A}$.

Example 4.11 A cylindrical pipe of 5 cm outside diameter is covered with a thin layer of ethanol. Air at 30 °C flows normal to the pipe with a velocity of 3 m/s. Determine the average mass transfer coefficient.

Solution

Physical properties

Diffusion coefficient of ethanol ($\mathcal{A}$) in air ($\mathcal{B}$) at 30 °C (303 K) is

$$(\mathcal{D}_{AB})_{303} = (\mathcal{D}_{AB})_{313}\left(\frac{303}{313}\right)^{3/2} = (1.45 \times 10^{-5})\left(\frac{303}{313}\right)^{3/2} = 1.38 \times 10^{-5}\ \mathrm{m^2/s}$$

For air at 30 °C (303 K): $\nu = 16 \times 10^{-6}\ \mathrm{m^2/s}$

The Schmidt number is

$$\mathrm{Sc} = \frac{\nu}{\mathcal{D}_{AB}} = \frac{16 \times 10^{-6}}{1.38 \times 10^{-5}} = 1.16$$

Assumptions

1. Steady-state conditions prevail.
2. Isothermal system.

Analysis

The Reynolds number is

$$\mathrm{Re}_D = \frac{D v_\infty}{\nu} = \frac{(5 \times 10^{-2})(3)}{16 \times 10^{-6}} = 9375$$

The use of the correlation proposed by Bedingfield and Drew, Eq. (4.4-10), gives

$$\text{Sh} = 0.281 \, \text{Re}_D^{1/2} \, \text{Sc}^{0.44} = 0.281(9375)^{1/2}(1.16)^{0.44} = 29$$

Therefore, the average mass transfer coefficient is

$$\langle k_c \rangle = \text{Sh}\left(\frac{\mathcal{D}_{AB}}{D}\right) = (29)\left(\frac{1.38 \times 10^{-5}}{5 \times 10^{-2}}\right) = 8 \times 10^{-3} \text{ m/s}$$

4.5 FLOW IN CIRCULAR PIPES

The rate of work done, $\dot{W}$, to pump a fluid can be determined from the expression

$$\dot{W} = \dot{m} \, \widehat{W} = \dot{m}\left(\int \widehat{V} \, dP\right) \tag{4.5-1}$$

where $\dot{m}$ and $\widehat{V}$ are the mass flow rate and the specific volume of the fluid, respectively. Note that the term in parentheses on the right-hand side of Eq. (4.5-1) is known as the *shaft work* in thermodynamics[3]. For an incompressible fluid, i.e., $\widehat{V} = 1/\rho = $ constant, Eq. (4.5-1) simplifies to

$$\dot{W} = Q|\Delta P| \tag{4.5-2}$$

where Q is the volumetric flow rate of the fluid. Combination of Eq. (4.5-2) with Eq. (3.1-11) gives

$$F_D \langle v \rangle = Q|\Delta P| \tag{4.5-3}$$

or,

$$\left[(\pi DL)\left(\frac{1}{2}\rho\langle v \rangle^2\right)f\right]\langle v \rangle = Q|\Delta P| \tag{4.5-4}$$

Expressing the average velocity in terms of the volumetric flow rate

$$\langle v \rangle = \frac{Q}{\pi D^2/4} \tag{4.5-5}$$

reduces Eq. (4.5-4) to

$$\boxed{|\Delta P| = \frac{32\rho L f Q^2}{\pi^2 D^5}} \tag{4.5-6}$$

Engineering problems associated with pipe flow are classified as follows:

- Determine the pressure drop, $|\Delta P|$, or the pump size, $\dot{W}$; given the volumetric flow rate, Q, the pipe diameter, D, and the physical properties of the fluid, ρ and μ.
- Determine the volumetric flow rate, Q; given the pressure drop, $|\Delta P|$, the pipe diameter, D, and the physical properties of the fluid, ρ and μ.
- Determine the pipe diameter, D; given the volumetric flow rate, Q, the pressure drop, $|\Delta P|$, and the physical properties of the fluid, ρ and μ.

---

[3]Work done on the system is considered positive.

4.5.1 Friction Factor Correlations

4.5.1.1 *Laminar flow correlation*   For laminar flow in a circular pipe, i.e., $\text{Re} = D\langle v \rangle \rho / \mu <$ 2100, the solution of the equations of change gives[4]

$$\boxed{f = \frac{16}{\text{Re}}} \tag{4.5-7}$$

The friction factor f appearing in Eqs. (4.5-6) and (4.5-7) is also called the *Fanning friction factor*. However, this is not the only definition for f available in the literature. Another commonly used definition for f is the *Darcy friction factor*, f_D, which is four times larger than the Fanning friction factor, i.e., $f_D = 4f$. Therefore, for laminar flow

$$f_D = \frac{64}{\text{Re}} \tag{4.5-8}$$

4.5.1.2 *Turbulent flow correlation*   Since no theoretical solution exists for turbulent flow, the friction factor is usually determined from the *Moody chart* (1944) in which it is expressed as a function of the Reynolds number, Re, and the relative pipe wall roughness, ε/D. Moody prepared this chart by using the equation proposed by Colebrook (1938)

$$\boxed{\frac{1}{\sqrt{f}} = -4\log\left(\frac{\varepsilon/D}{3.7065} + \frac{1.2613}{\text{Re}\sqrt{f}}\right)} \tag{4.5-9}$$

where ε is the surface roughness of the pipe wall in meters.

4.5.1.3 *Solutions to the engineering problems*

I. Laminar flow

For flow in a pipe, the Reynolds number is defined by

$$\text{Re} = \frac{D\langle v \rangle \rho}{\mu} = \frac{4\rho Q}{\pi \mu D} \tag{4.5-10}$$

Substitution of Eq. (4.5-10) into Eq. (4.5-7) yields

$$f = \frac{4\pi \mu D}{\rho Q} \tag{4.5-11}$$

■ **Calculate $|\Delta P|$ or $\dot{W}$; given Q and D**

Substitution of Eq. (4.5-11) into Eq. (4.5-6) gives

$$\boxed{|\Delta P| = \frac{128\mu L Q}{\pi D^4}} \tag{4.5-12}$$

---

[4]See Section 9.1.3.1 in Chapter 9.

The pump size can be calculated from Eq. (4.5-2) as

$$\dot{W} = \frac{128\mu L Q^2}{\pi D^4} \qquad (4.5\text{-}13)$$

■ **Calculate Q; given $|\Delta P|$ and D**

Rearrangement of Eq. (4.5-12) gives

$$Q = \frac{\pi D^4 |\Delta P|}{128\mu L} \qquad (4.5\text{-}14)$$

■ **Calculate D; given Q and $|\Delta P|$**

Rearrangement of Eq. (4.5-12) gives

$$D = \left(\frac{128\mu L Q}{\pi |\Delta P|}\right)^{1/4} \qquad (4.5\text{-}15)$$

II. Turbulent flow

■ **Calculate $|\Delta P|$ or $\dot{W}$; given Q and D**

For the given values of Q and D, the Reynolds number can be determined using Eq. (4.5-10). However, when the values of Re and ε/D are known, determination of f from Eq. (4.5-9) requires an iterative procedure since f appears on both sides of the equation. To avoid iterative solutions, efforts have been directed to express the friction factor, f, as an explicit function of the Reynolds number, Re, and the relative pipe wall roughness, ε/D.

Gregory and Fogarasi (1985) compared the predictions of the twelve explicit relations with Eq. (4.5-9) and recommended the use of the correlation proposed by Chen (1979):

$$\frac{1}{\sqrt{f}} = -4\log\left(\frac{\varepsilon/D}{3.7065} - \frac{5.0452}{\text{Re}}\log A\right) \qquad (4.5\text{-}16)$$

where

$$A = \left(\frac{\varepsilon/D}{2.5497}\right)^{1.1098} + \left(\frac{7.1490}{\text{Re}}\right)^{0.8981} \qquad (4.5\text{-}17)$$

Thus, in order to calculate the pressure drop using Eq. (4.5-16), the following procedure should be followed through which an iterative solution is avoided:

a) Calculate the Reynolds number from Eq. (4.5-10),
b) Substitute Re into Eq. (4.5-16) and determine f,
c) Use Eq. (4.5-6) to find the pressure drop. Finally, the pump size can be determined by using Eq. (4.5-2).

Example 4.12 What is the required pressure drop per unit length in order to pump water at a volumetric flow rate of $0.03 \text{ m}^3/\text{s}$ at $20\,^\circ\text{C}$ through a commercial steel pipe ($\varepsilon = 4.6 \times 10^{-5}$ m) 20 cm in diameter?

Solution

Physical properties

For water at $20\,^\circ\text{C}$ (293 K): $\begin{cases} \rho = 999 \text{ kg/m}^3 \\ \mu = 1001 \times 10^{-6} \text{ kg/m·s} \end{cases}$

Analysis

The Reynolds number is determined from Eq. (4.5-10) as

$$\text{Re} = \frac{4\rho\,Q}{\pi\mu D} = \frac{(4)(999)(0.03)}{\pi(1001 \times 10^{-6})(0.2)} = 191 \times 10^3$$

Substitution of this value into Eqs. (4.5-17) and (4.5-16) gives

$$A = \left(\frac{\varepsilon/D}{2.5497}\right)^{1.1098} + \left(\frac{7.1490}{\text{Re}}\right)^{0.8981}$$

$$= \left[\frac{(4.6 \times 10^{-5}/0.2)}{2.5497}\right]^{1.1098} + \left(\frac{7.1490}{191 \times 10^3}\right)^{0.8981} = 1.38 \times 10^{-4}$$

$$\frac{1}{\sqrt{f}} = -4\log\left(\frac{\varepsilon/D}{3.7065} - \frac{5.0452}{\text{Re}}\log A\right)$$

$$= -4\log\left[\frac{(4.6 \times 10^{-5}/0.2)}{3.7065} - \frac{5.0452}{191 \times 10^3}\log(1.38 \times 10^{-4})\right] = 15.14$$

Hence, the friction factor is

$$f = 4.36 \times 10^{-3}$$

Thus, Eq. (4.5-6) gives the pressure drop per unit pipe length as

$$\frac{|\Delta P|}{L} = \frac{32\rho f Q^2}{\pi^2 D^5} = \frac{(32)(999)(4.36 \times 10^{-3})(0.03)^2}{\pi^2(0.2)^5} = 40 \text{ Pa/m}$$

■ **Calculate** Q; **given** $|\Delta P|$ **and** D

In this case, rearrangement of Eq. (4.5-6) gives

$$f = \left(\frac{Y}{Q}\right)^2 \tag{4.5-18}$$

where Y is defined by

$$Y = \sqrt{\frac{\pi^2 D^5 |\Delta P|}{32\rho L}} \tag{4.5-19}$$

Substitution of Eqs. (4.5-10) and (4.5-18) into Eq. (4.5-9) yields

$$Q = -4Y \log\left(\frac{\varepsilon/D}{3.7065} + \frac{\mu D}{\rho Y}\right) \qquad (4.5\text{-}20)$$

Thus, the procedure to calculate the volumetric flow rate becomes:

a) Calculate Y from Eq. (4.5-19),
b) Substitute Y into Eq. (4.5-20) and determine the volumetric flow rate.

Example 4.13 What is the volumetric flow rate of water in m^3/s at $20\,^\circ C$ that can be delivered through a commercial steel pipe ($\varepsilon = 4.6 \times 10^{-5}$ m) 20 cm in diameter when the pressure drop per unit length of the pipe is 40 Pa/m?

Solution

Physical properties

For water at $20\,^\circ C$ (293 K): $\begin{cases} \rho = 999 \text{ kg/m}^3 \\ \mu = 1001 \times 10^{-6} \text{ kg/m·s} \end{cases}$

Analysis

Substitution of the given values into Eq. (4.5-19) yields

$$Y = \sqrt{\frac{\pi^2 D^5 |\Delta P|}{32\rho L}} = \sqrt{\frac{\pi^2 (0.2)^5 (40)}{(32)(999)}} = 1.99 \times 10^{-3}$$

Hence, Eq. (4.5-20) gives the volumetric flow rate as

$$Q = -4Y \log\left(\frac{\varepsilon/D}{3.7065} + \frac{\mu D}{\rho Y}\right)$$

$$= -(4)(1.99 \times 10^{-3}) \log\left[\frac{(4.6 \times 10^{-5}/0.2)}{3.7065} + \frac{(1001 \times 10^{-6})(0.2)}{(999)(1.99 \times 10^{-3})}\right] = 0.03 \text{ m}^3/\text{s}$$

■ **Calculate D; given Q and $|\Delta P|$**

Swamee and Jain (1976) and Cheng and Turton (1990) presented explicit equations to solve problems of this type. These equations, however, are unnecessarily complex. A simpler equation can be obtained by using the procedure suggested by Tosun and Akşahin (1993) as follows. Equation (4.5-6) can be rearranged in the form

$$f = (DN)^5 \qquad (4.5\text{-}21)$$

where N is defined by

$$N = \left(\frac{\pi^2 |\Delta P|}{32\rho L Q^2}\right)^{1/5} \qquad (4.5\text{-}22)$$

For turbulent flow, the value of f varies between 0.00025 and 0.01925. Using an average value of 0.01 for f gives a relationship between D and N as

$$D = \frac{0.4}{N} \tag{4.5-23}$$

Substitution of Eq. (4.5-21) into the left-hand side of Eq. (4.5-9), and substitution of Eqs. (4.5-10), (4.5-23), and $f = 0.01$ into the right-hand side of Eq. (4.5-9) give

$$\boxed{D = \frac{0.574}{N} \left(\left\{ \log \left[\varepsilon N + 5.875 \left(\frac{\mu}{\rho Q N} \right) \right] - 0.171 \right\}^2 \right)^{-1/5}} \tag{4.5-24}$$

The procedure to calculate the pipe diameter becomes:

a) Calculate N from Eq. (4.5-22),
b) Substitute N into Eq. (4.5-24) and determine the pipe diameter.

Example 4.14 Water at 20 °C is to be pumped through a commercial steel pipe ($\varepsilon = 4.6 \times 10^{-5}$ m) at a volumetric flow rate of 0.03 m³/s. Determine the diameter of the pipe if the allowable pressure drop per unit length of pipe is 40 Pa/m.

Solution

Physical properties

For water at 20 °C (293 K): $\begin{cases} \rho = 999 \text{ kg/m}^3 \\ \mu = 1001 \times 10^{-6} \text{ kg/m·s} \end{cases}$

Analysis

Equation (4.5-22) gives

$$N = \left(\frac{\pi^2 |\Delta P|}{32 \rho L Q^2} \right)^{1/5} = \left[\frac{\pi^2 (40)}{(32)(999)(0.03)^2} \right]^{1/5} = 1.69$$

Hence, Eq. (4.5-24) gives the pipe diameter as

$$D = \frac{0.574}{N} \left(\left\{ \log \left[\varepsilon N + 5.875 \left(\frac{\mu}{\rho Q N} \right) \right] - 0.171 \right\}^2 \right)^{-1/5}$$

$$= \frac{0.574}{1.69} \left(\left\{ \log \left[(4.6 \times 10^{-5})(1.69) + \frac{(5.875)(1001 \times 10^{-6})}{(999)(0.03)(1.69)} \right] - 0.171 \right\}^2 \right)^{-1/5}$$

$$= 0.2 \text{ m}$$

4.5.2 Heat Transfer Correlations

For heat transfer in circular pipes, various correlations have been suggested depending on the flow conditions, i.e., laminar or turbulent.

4.5.2.1 *Laminar flow correlation* For laminar flow heat transfer in a circular tube with constant wall temperature, Sieder and Tate (1936) proposed the following correlation:

$$\mathrm{Nu} = 1.86 \left[\mathrm{Re}\,\mathrm{Pr}(D/L)\right]^{1/3} (\mu/\mu_w)^{0.14}$$

(4.5-25)

in which all properties except μ_w are evaluated at the mean bulk temperature. Equation (4.5-25) is valid for

$$13 \leqslant \mathrm{Re} \leqslant 2030 \qquad 0.48 \leqslant \mathrm{Pr} \leqslant 16{,}700 \qquad 0.0044 \leqslant \mu/\mu_w \leqslant 9.75$$

The analytical solution[5] to this problem is only possible for very long tubes, i.e., $L/D \to \infty$. In this case the Nusselt number remains constant at 3.66.

4.5.2.2 *Turbulent flow correlations* The following correlations approximate the physical situation quite well for the cases of constant wall temperature and constant wall heat flux:

Dittus-Boelter correlation

Dittus and Boelter (1930) proposed the following correlation in which all physical properties are evaluated at the mean bulk temperature:

$$\mathrm{Nu} = 0.023\,\mathrm{Re}^{4/5}\,\mathrm{Pr}^{n}$$

(4.5-26)

where

$$n = \begin{cases} 0.4 & \text{for heating} \\ 0.3 & \text{for cooling} \end{cases}$$

The Dittus-Boelter correlation is valid when

$$0.7 \leqslant \mathrm{Pr} \leqslant 160 \qquad \mathrm{Re} \geqslant 10{,}000 \qquad L/D \geqslant 10$$

Sieder-Tate correlation

The correlation proposed by Sieder and Tate (1936) is

$$\mathrm{Nu} = 0.027\,\mathrm{Re}^{4/5}\,\mathrm{Pr}^{1/3} (\mu/\mu_w)^{0.14}$$

(4.5-27)

in which all properties except μ_w are evaluated at the mean bulk temperature. Equation (4.5-27) is valid for

$$0.7 \leqslant \mathrm{Pr} \leqslant 16{,}700 \qquad \mathrm{Re} \geqslant 10{,}000 \qquad L/D \geqslant 10$$

Whitaker correlation

The equation proposed by Whitaker (1972) is

$$\mathrm{Nu} = 0.015\,\mathrm{Re}^{0.83}\,\mathrm{Pr}^{0.42} (\mu/\mu_w)^{0.14}$$

(4.5-28)

---

[5] See Section 9.3.1.2 in Chapter 9.

in which the Prandtl number dependence is based on the work of Friend and Metzner (1958), and the functional dependence of μ/μ_w is from Sieder and Tate (1936). All physical properties except μ_w are evaluated at the mean bulk temperature. The Whitaker correlation is valid for

$$2300 \leqslant \text{Re} \leqslant 1 \times 10^5 \qquad 0.48 \leqslant \text{Pr} \leqslant 592 \qquad 0.44 \leqslant \mu/\mu_w \leqslant 2.5$$

4.5.2.3   *Calculation of the heat transfer rate*   Once the average heat transfer coefficient is calculated from correlations by using Eqs. (4.5-25)–(4.5-28), then the rate of energy transferred is calculated as

$$\boxed{\dot{Q} = (\pi D L)\langle h \rangle \Delta T_{LM}} \tag{4.5-29}$$

where ΔT_{LM}, *logarithmic mean temperature difference*, is defined by

$$\Delta T_{LM} = \frac{(T_w - T_b)_{in} - (T_w - T_b)_{out}}{\ln\left[\dfrac{(T_w - T_b)_{in}}{(T_w - T_b)_{out}}\right]} \tag{4.5-30}$$

The derivation of Eq. (4.5-29) is given in Section 9.3.1 in Chapter 9.

Example 4.15 Steam condensing on the outer surface of a thin-walled circular tube of 65 mm diameter maintains a uniform surface temperature of 100 °C. Oil flows through the tube at an average velocity of 1 m/s. Determine the length of the tube in order to increase oil temperature from 40 °C to 60 °C. Physical properties of the oil are as follows:

$$\text{At } 50\,^\circ\text{C:} \begin{cases} \mu = 12.4 \times 10^{-3} \text{ kg/m·s} \\ \nu = 4.28 \times 10^{-5} \text{ m}^2/\text{s} \\ \text{Pr} = 143 \end{cases}$$

At 100 °C: $\mu = 9.3 \times 10^{-3}$ kg/m·s.

Solution

Assumptions

1. Steady-state conditions prevail.
2. Physical properties remain constant.
3. Changes in kinetic and potential energies are negligible.

Analysis

System: Oil in the pipe

The inventory rate equation for mass becomes

$$\text{Rate of mass in} = \text{Rate of mass out} = \dot{m} = \rho \langle v \rangle (\pi D^2/4) \tag{1}$$

On the other hand, the inventory rate equation for energy reduces to

$$\text{Rate of energy in} = \text{Rate of energy out} \tag{2}$$

The terms in Eq. (2) are expressed by

$$\text{Rate of energy in} = \dot{m}\,\widehat{C}_P\,(T_{b_{in}} - T_{ref}) + \pi\,DL\langle h\rangle \Delta T_{LM} \tag{3}$$

$$\text{Rate of energy out} = \dot{m}\,\widehat{C}_P\,(T_{b_{out}} - T_{ref}) \tag{4}$$

Since the wall temperature is constant, the expression for ΔT_{LM}, Eq. (4.5-30), becomes

$$\Delta T_{LM} = \frac{T_{b_{out}} - T_{b_{in}}}{\ln\left(\dfrac{T_w - T_{b_{in}}}{T_w - T_{b_{out}}}\right)} \tag{5}$$

Substitution of Eqs. (1), (3), (4) and (5) into Eq. (2) gives

$$\frac{L}{D} = \frac{1}{4}\,\frac{\langle v\rangle \rho\,\widehat{C}_P}{\langle h\rangle}\,\ln\left(\frac{T_w - T_{b_{in}}}{T_w - T_{b_{out}}}\right) \tag{6}$$

Noting that $\text{St}_H = \langle h\rangle/(\langle v\rangle \rho\,\widehat{C}_P) = \text{Nu}/(\text{Re}\,\text{Pr})$, Eq. (6) becomes

$$\frac{L}{D} = \frac{1}{4}\,\frac{1}{\text{St}_H}\,\ln\left(\frac{T_w - T_{b_{in}}}{T_w - T_{b_{out}}}\right) = \frac{1}{4}\,\frac{\text{Re}\,\text{Pr}}{\text{Nu}}\,\ln\left(\frac{T_w - T_{b_{in}}}{T_w - T_{b_{out}}}\right) \tag{7}$$

To determine Nu (or $\langle h\rangle$), first the Reynolds number must be calculated. The mean bulk temperature is $(40 + 60)/2 = 50\,°C$ and the Reynolds number is

$$\text{Re} = \frac{D\langle v\rangle}{\nu} = \frac{(65 \times 10^{-3})(1)}{4.28 \times 10^{-5}} = 1519 \quad \Rightarrow \quad \text{Laminar flow}$$

Since the flow is laminar, Eq. (4.5-25) must be used, i.e.,

$$\text{Nu} = 1.86\big[\text{Re}\,\text{Pr}(D/L)\big]^{1/3}(\mu/\mu_w)^{0.14} \tag{8}$$

Substitution of Eq. (8) into Eq. (7) yields

$$\frac{L}{D} = \text{Re}\,\text{Pr}\left[\frac{(\mu/\mu_w)^{-0.14}}{(4)(1.86)}\,\ln\left(\frac{T_w - T_{b_{in}}}{T_w - T_{b_{out}}}\right)\right]^{3/2}$$

$$= (1519)(143)\left[\frac{(12.4 \times 10^{-3}/9.3 \times 10^{-3})^{-0.14}}{(4)(1.86)}\,\ln\left(\frac{100 - 40}{100 - 60}\right)\right]^{3/2} = 2602$$

The tube length is then

$$L = (2602)(65 \times 10^{-3}) = 169 \text{ m}$$

Example 4.16 Air at $20\,°C$ enters a circular pipe of 1.5 cm internal diameter with a velocity of 50 m/s. Steam condenses on the outside of the pipe so as to keep the surface temperature of the pipe at $150\,°C$.

a) Calculate the length of the pipe required to increase air temperature to $90\,°C$.
b) Discuss the effect of surface roughness on the length of the pipe.

Solution

Physical properties

The mean bulk temperature is $(20 + 90)/2 = 55\,°C$

For air at $20\,°C$ (293 K): $\rho = 1.2047\ \text{kg/m}^3$

For air at $55\,°C$ (328 K): $\begin{cases} \mu = 19.8 \times 10^{-6}\ \text{kg/m·s} \\ \nu = 18.39 \times 10^{-6}\ \text{m}^2/\text{s} \\ \text{Pr} = 0.707 \end{cases}$

For air at $150\,°C$ (423 K): $\mu = 23.86 \times 10^{-6}\ \text{kg/m·s}$.

Analysis

a) System: Air in the pipe

The inventory rate equation for mass reduces to

$$\text{Rate of mass of air in} = \text{Rate of mass of air out} = \dot{m} \tag{1}$$

Note that for compressible fluids like air both density and average velocity depend on temperature and pressure. Therefore, using the inlet conditions

$$\dot{m} = (\pi D^2/4)\big(\rho \langle v \rangle\big)_{inlet} = \left[\frac{\pi(0.015)^2}{4} \right](1.2047)(50) = 1.06 \times 10^{-2}\ \text{kg/s}$$

In problems dealing with the flow of compressible fluids, it is customary to define mass velocity, G, as

$$G = \frac{\dot{m}}{A} = \rho \langle v \rangle \tag{2}$$

The advantage of using G is the fact that it remains constant for steady flow of compressible fluids through ducts of uniform cross-section. In this case

$$G = (1.2047)(50) = 60.24\ \text{kg/m}^2\text{·s}$$

The inventory rate equation for energy is written as

$$\text{Rate of energy in} = \text{Rate of energy out} \tag{3}$$

Equations (3)–(5) of Example 4.15 are also applicable to this problem. Therefore, we get

$$\frac{L}{D} = \frac{1}{4} \frac{\text{Re}\,\text{Pr}}{\text{Nu}} \ln\left(\frac{T_w - T_{b_{in}}}{T_w - T_{b_{out}}} \right) \tag{4}$$

The Nusselt number in Eq. (4) can be determined only if the Reynolds number is known. The Reynolds number is calculated as

$$\text{Re} = \frac{DG}{\mu} = \frac{(0.015)(60.24)}{19.80 \times 10^{-6}} = 45{,}636 \quad \Rightarrow \quad \text{Turbulent flow}$$

The value of L depends on the correlations as follows:

Dittus-Boelter correlation

Substitution of Eq. (4.5-26) into Eq. (4) gives

$$\frac{L}{D} = \frac{\mathrm{Re}^{0.2}\,\mathrm{Pr}^{0.6}}{0.092}\ln\left(\frac{T_w - T_{b_{in}}}{T_w - T_{b_{out}}}\right) = \frac{(45{,}636)^{0.2}(0.707)^{0.6}}{0.092}\ln\left(\frac{150-20}{150-90}\right) = 58.3$$

Therefore, the required length is

$$L = (58.3)(1.5) = 87 \text{ cm}$$

Sieder-Tate correlation

Substitution of Eq. (4.5-27) into Eq. (4) gives

$$\frac{L}{D} = \frac{\mathrm{Re}^{0.2}\,\mathrm{Pr}^{2/3}(\mu/\mu_w)^{-0.14}}{0.108}\ln\left(\frac{T_w - T_{b_{in}}}{T_w - T_{b_{out}}}\right)$$

$$= \frac{(45{,}636)^{0.2}(0.707)^{2/3}}{0.108}\left(\frac{19.80 \times 10^{-6}}{23.86 \times 10^{-6}}\right)^{-0.14}\ln\left(\frac{150-20}{150-90}\right) = 49.9$$

Therefore, the required length is

$$L = (49.9)(1.5) = 75 \text{ cm}$$

Whitaker correlation

Substitution of Eq. (4.5-28) into Eq. (4) gives

$$\frac{L}{D} = \frac{\mathrm{Re}^{0.17}\,\mathrm{Pr}^{0.58}(\mu/\mu_w)^{-0.14}}{0.06}\ln\left(\frac{T_w - T_{b_{in}}}{T_w - T_{b_{out}}}\right)$$

$$= \frac{(45{,}636)^{0.17}(0.707)^{0.58}}{0.06}\left(\frac{19.80 \times 10^{-6}}{23.86 \times 10^{-6}}\right)^{-0.14}\ln\left(\frac{150-20}{150-90}\right) = 67$$

Therefore, the required length is

$$L = (67)(1.5) = 101 \text{ cm}$$

b) Note that Eq. (4) is also expressed in the form

$$\frac{L}{D} = \frac{1}{4}\frac{1}{\mathrm{St_H}}\ln\left(\frac{T_w - T_{b_{in}}}{T_w - T_{b_{out}}}\right) \tag{5}$$

The use of the Chilton-Colburn analogy, i.e., $f/2 = \mathrm{St_H}\,\mathrm{Pr}^{2/3}$, reduces Eq. (5) to

$$\frac{L}{D} = \frac{1}{2}\frac{\mathrm{Pr}^{2/3}}{f}\ln\left(\frac{T_w - T_{b_{in}}}{T_w - T_{b_{out}}}\right) = \frac{1}{2}\frac{(0.707)^{2/3}}{f}\ln\left(\frac{150-20}{150-90}\right) = \frac{0.3068}{f} \tag{6}$$

The friction factor can be calculated from the Chen correlation, Eq. (4.5-16)

$$\frac{1}{\sqrt{f}} = -4\log\left(\frac{\varepsilon/D}{3.7065} - \frac{5.0452}{\mathrm{Re}}\log A\right)$$

where

$$A = \left(\frac{\varepsilon/D}{2.5497}\right)^{1.1098} + \left(\frac{7.1490}{\text{Re}}\right)^{0.8981}$$

For various values of ε/D, the calculated values of f, L/D and L are given as follows:

ε/D	f	L/D	L (cm)
0	0.0053	57.9	86.9
0.001	0.0061	50.3	75.5
0.002	0.0067	45.8	68.7
0.003	0.0072	42.6	63.9
0.004	0.0077	39.8	59.7

Comment: The increase in surface roughness increases the friction factor and hence power consumption. On the other hand, the increase in surface roughness causes an increase in the heat transfer coefficient with a concomitant decrease in pipe length.

4.5.3   Mass Transfer Correlations

Mass transfer in cylindrical tubes is encountered in a variety of operations, such as wetted wall columns, reverse osmosis, and cross-flow ultrafiltration. As in the case of heat transfer, mass transfer correlations depend on whether the flow is laminar or turbulent.

4.5.3.1   *Laminar flow correlation*   For laminar flow mass transfer in a circular tube with a constant wall concentration, an expression analogous to Eq. (4.5-25) is given by

$$\boxed{\text{Sh} = 1.86\left[\text{Re}\,\text{Sc}(D/L)\right]^{1/3}}$$
(4.5-31)

Equation (4.5-31) is valid for

$$\left[\text{Re}\,\text{Sc}(D/L)\right]^{1/3} \geqslant 2$$

4.5.3.2   *Turbulent flow correlations*

Gilliland-Sherwood correlation

Gilliland and Sherwood (1934) correlated the experimental results obtained from wetted wall columns in the form

$$\boxed{\text{Sh} = 0.023\,\text{Re}^{0.83}\,\text{Sc}^{0.44}}$$
(4.5-32)

which is valid for

$$2000 \leqslant \text{Re} \leqslant 35{,}000 \qquad 0.6 \leqslant \text{Sc} \leqslant 2.5$$

Linton-Sherwood correlation

The correlation proposed by Linton and Sherwood (1950) is given by

$$\boxed{\mathrm{Sh} = 0.023\,\mathrm{Re}^{0.83}\,\mathrm{Sc}^{1/3}}$$

(4.5-33)

Equation (4.5-33) is valid for

$$2000 \leqslant \mathrm{Re} \leqslant 70{,}000 \qquad 0.6 \leqslant \mathrm{Sc} \leqslant 2500$$

4.5.3.3 *Calculation of the mass transfer rate*   Once the average mass transfer coefficient is calculated from correlations given by Eqs. (4.5-31)–(4.5-33), then the rate of mass of species $\mathcal{A}$ transferred is calculated as

$$\boxed{\dot{m}_A = (\pi\,DL)\langle k_c\rangle (\Delta c_A)_{LM}\,\mathcal{M}_A}$$

(4.5-34)

where $\mathcal{M}_A$ is the molecular weight of species $\mathcal{A}$, and $(\Delta c_A)_{LM}$, *logarithmic mean concentration difference*, is defined by

$$(\Delta c_A)_{LM} = \frac{(c_{A_w} - c_{A_b})_{in} - (c_{A_w} - c_{A_b})_{out}}{\ln\!\left[\dfrac{(c_{A_w} - c_{A_b})_{in}}{(c_{A_w} - c_{A_b})_{out}}\right]}$$

(4.5-35)

The derivation of Eq. (4.5-34) is given in Section 9.5.1 in Chapter 9.

Example 4.17 A smooth tube with an internal diameter of 2.5 cm is cast from solid naphthalene. Pure air enters the tube at an average velocity of 9 m/s. If the average air pressure is 1 atm and the temperature is 40 °C, estimate the tube length required for the average concentration of naphthalene vapor in the air to reach 25% of the saturation value.

Solution

Physical properties

Diffusion coefficient of naphthalene ($\mathcal{A}$) in air ($\mathcal{B}$) at 40 °C (313 K) is

$$(\mathcal{D}_{AB})_{313} = (\mathcal{D}_{AB})_{300}\left(\frac{313}{300}\right)^{3/2} = (0.62 \times 10^{-5})\left(\frac{313}{300}\right)^{3/2} = 6.61 \times 10^{-6}\ \mathrm{m^2/s}$$

For air at 40 °C (313 K): $\nu = 16.95 \times 10^{-6}\ \mathrm{m^2/s}$

The Schmidt number is

$$\mathrm{Sc} = \frac{\nu}{\mathcal{D}_{AB}} = \frac{16.95 \times 10^{-6}}{6.61 \times 10^{-6}} = 2.56$$

Assumptions

1. Steady-state conditions prevail.
2. The system is isothermal.

Analysis

System: Air in the naphthalene tube

If naphthalene is designated as species $\mathcal{A}$, then the rate equation for the conservation of species $\mathcal{A}$ becomes

$$\text{Rate of moles of } \mathcal{A} \text{ in} = \text{Rate of moles of } \mathcal{A} \text{ out} \tag{1}$$

The terms in Eq. (1) are expressed by

$$\text{Rate of moles of } \mathcal{A} \text{ in} = \pi D L \langle k_c \rangle (\Delta c_A)_{LM} \tag{2}$$

$$\text{Rate of moles of } \mathcal{A} \text{ out} = \mathcal{Q}(c_{A_b})_{out} = (\pi D^2 / 4) \langle v \rangle (c_{A_b})_{out} \tag{3}$$

Since the concentration at the wall is constant, the expression for $(\Delta c_A)_{LM}$, Eq. (4.5-35), becomes

$$(\Delta c_A)_{LM} = \frac{(c_{A_b})_{out}}{\ln\left[\dfrac{c_{A_w}}{c_{A_w} - (c_{A_b})_{out}}\right]} \tag{4}$$

Substitution of Eqs. (2)–(4) into Eq. (1) gives

$$\frac{L}{D} = -\frac{1}{4}\frac{\langle v \rangle}{\langle k_c \rangle} \ln\left[1 - \frac{(c_{A_b})_{out}}{c_{A_w}}\right] = -\frac{1}{4}\frac{\langle v \rangle}{\langle k_c \rangle} \ln(1 - 0.25) = 0.072\left(\frac{\langle v \rangle}{k_c}\right) \tag{5}$$

Note that Eq. (5) can also be expressed in the form

$$\frac{L}{D} = 0.072\left(\frac{1}{\text{St}_M}\right) = 0.072\left(\frac{\text{Re Sc}}{\text{Sh}}\right) \tag{6}$$

The value of L depends on the correlations as follows:

Chilton-Colburn analogy

Substitution of Eq. (3.5-13) into Eq. (6) gives

$$\frac{L}{D} = 0.072\,\frac{2}{f}\,\text{Sc}^{2/3} \tag{7}$$

The Reynolds number is

$$\text{Re} = \frac{D\langle v \rangle}{\nu} = \frac{(2.5 \times 10^{-2})(9)}{16.95 \times 10^{-6}} = 13{,}274 \quad \Rightarrow \quad \text{Turbulent flow}$$

The friction factor can be calculated from the Chen correlation, Eq. (4.5-16). Taking $\varepsilon/D \approx 0$,

$$A = \left(\frac{\varepsilon/D}{2.5497}\right)^{1.1098} + \left(\frac{7.1490}{\text{Re}}\right)^{0.8981} = \left(\frac{7.1490}{13{,}274}\right)^{0.8981} = 1.16 \times 10^{-3}$$

$$\frac{1}{\sqrt{f}} = -4\log\left[-\frac{5.0452}{13{,}274}\log(1.16 \times 10^{-3})\right] \quad \Rightarrow \quad f = 0.0072$$

Hence Eq. (7) becomes

$$\frac{L}{D} = \frac{(0.072)(2)(2.56)^{2/3}}{0.0072} = 37.4$$

The required length is then

$$L = (37.4)(2.5) = 93.5 \text{ cm}$$

Linton-Sherwood correlation

Substitution of Eq. (4.5-33) into Eq. (6) gives

$$\frac{L}{D} = 3.13 \, \text{Re}^{0.17} \, \text{Sc}^{2/3} = 3.13(13,274)^{0.17}(2.56)^{2/3} = 29.4$$

The tube length is

$$L = (29.4)(2.5) = 73.5 \text{ cm}$$

4.5.4   Flow in Non-Circular Ducts

The correlations given for the friction factor, heat transfer coefficient, and mass transfer coefficient are only valid for ducts of circular cross-section. These correlations can be used for flow in non-circular ducts by introducing the concept of *hydraulic equivalent diameter*, D_h, defined by

$$D_h = 4\left(\frac{\text{Flow area}}{\text{Wetted perimeter}}\right) \tag{4.5-36}$$

The Reynolds number based on the hydraulic equivalent diameter is

$$\text{Re}_h = \frac{D_h \langle v \rangle \rho}{\mu} \tag{4.5-37}$$

so that the friction factor, based on the hydraulic equivalent diameter, is related to Re_h in the form

$$f_h = \Omega\left(\frac{16}{\text{Re}_h}\right) \tag{4.5-38}$$

where Ω depends on the geometry of the system. Since $\Omega = 1$ only for a circular pipe, the use of the hydraulic equivalent diameter is not recommended for laminar flow (Bird *et al.*, 2002; Fahien, 1983). The hydraulic equivalent diameter for various geometries is shown in Table 4.4.

Example 4.18 Water flows at an average velocity of 5 m/s through a duct of equilateral triangular cross-section with one side, a, being equal to 2 cm. Electric wires are wrapped around the outer surface of the duct to provide a constant wall heat flux of 100 W/cm^2. If the inlet water temperature is 25 °C and the duct length is 1.5 m, calculate:

a) The power required to pump water through the duct,
b) The exit water temperature,
c) The average heat transfer coefficient.

Table 4.4. The hydraulic equivalent diameter for various geometries

Geometry	D_h
b	$2b$
b, a (rectangle)	$\dfrac{2ab}{a+b}$
$60°$, a (triangle)	$\dfrac{a}{\sqrt{3}}$
D_i, D_o (annulus)	$D_o - D_i$

Solution

Physical properties

For water at 25 °C (298 K):
$$\begin{cases} \rho = 997 \text{ kg/m}^3 \\ \mu = 892 \times 10^{-6} \text{ kg/m·s} \\ \widehat{C}_P = 4180 \text{ J/kg·K} \end{cases}$$

Assumptions

1. Steady-state conditions prevail.
2. Changes in kinetic and potential energies are negligible.
3. Variations in ρ and $\widehat{C}_P$ with temperature are negligible.

Analysis

System: Water in the duct

a) The power required is calculated from Eq. (3.1-11)

$$\dot{W} = F_D \langle v \rangle = \left[(3aL) \left(\frac{1}{2} \rho \langle v \rangle^2 \right) f \right] \langle v \rangle \tag{1}$$

The friction factor in Eq. (1) can be calculated from the modified form of the Chen correlation, Eq. (4.5-16)

$$\frac{1}{\sqrt{f}} = -4 \log \left(\frac{\varepsilon/D}{3.7065} - \frac{5.0452}{\mathrm{Re}_h} \log A \right) \tag{2}$$

where

$$A = \left(\frac{\varepsilon/D}{2.5497} \right)^{1.1098} + \left(\frac{7.1490}{\mathrm{Re}_h} \right)^{0.8981} \tag{3}$$

The hydraulic equivalent diameter and the Reynolds number are

$$D_h = \frac{a}{\sqrt{3}} = \frac{2}{\sqrt{3}} = 1.155 \text{ cm}$$

$$\mathrm{Re}_h = \frac{D_h \langle v \rangle \rho}{\mu} = \frac{(1.155 \times 10^{-2})(5)(997)}{892 \times 10^{-6}} = 64{,}548 \quad \Rightarrow \quad \text{Turbulent flow}$$

Substitution of these values into Eqs. (3) and (2) and taking $\varepsilon/D \approx 0$ give

$$A = \left(\frac{7.1490}{\mathrm{Re}_h} \right)^{0.8981} = \left(\frac{7.1490}{64{,}548} \right)^{0.8981} = 2.8 \times 10^{-4}$$

$$\frac{1}{\sqrt{f}} = -4 \log \left[-\frac{5.0452}{64{,}548} \log(2.8 \times 10^{-4}) \right] \quad \Rightarrow \quad f = 0.0049$$

Hence, the power required is calculated from Eq. (1) as

$$\dot{W} = \left\{ (3)(2 \times 10^{-2})(1.5) \left[\frac{1}{2}(997)(5)^2 \right] (0.0049) \right\} (5) = 27.5 \text{ W}$$

b) The inventory rate equation for mass is

$$\text{Rate of mass in} = \text{Rate of mass out} = \dot{m} = \rho \langle v \rangle \left(\frac{\sqrt{3} a^2}{4} \right) \tag{4}$$

$$\dot{m} = (997)(5) \left[\frac{\sqrt{3}(2 \times 10^{-2})^2}{4} \right] = 0.863 \text{ kg/s}$$

The inventory rate equation for energy reduces to

$$\text{Rate of energy in} = \text{Rate of energy out} \tag{5}$$

The terms in Eq. (5) are expressed by

$$\text{Rate of energy in} = \dot{m}\,\widehat{C}_P\,(T_{b_{in}} - T_{ref}) + \dot{Q}_w \tag{6}$$

$$\text{Rate of energy out} = \dot{m}\,\widehat{C}_P\,(T_{b_{out}} - T_{ref}) \tag{7}$$

where $\dot{Q}_w$ is the rate of heat transfer to water from the lateral surfaces of the duct. Substitution of Eqs. (6) and (7) into Eq. (5) gives

$$T_{b_{out}} = T_{b_{in}} + \frac{\dot{Q}_w}{\dot{m}\,\widehat{C}_P} = 25 + \frac{(3)(2)(150)(100)}{(0.863)(4180)} = 50\,^\circ\text{C}$$

c) The mean bulk temperature is $(25 + 50)/2 = 37.5\,^\circ\text{C}$. At this temperature

$$k = 628 \times 10^{-3}\,\text{W/m·K} \qquad \text{and} \qquad \text{Pr} = 4.62$$

The use of the Dittus-Boelter correlation, Eq. (4.5-26), gives

$$\text{Nu} = 0.023\,\text{Re}_P^{4/5}\,\text{Pr}^{0.4} = 0.023(64{,}548)^{4/5}(4.62)^{0.4} = 299$$

Therefore, the average heat transfer coefficient is

$$\langle h \rangle = \text{Nu}\left(\frac{k}{D_h}\right) = (299)\left(\frac{628 \times 10^{-3}}{1.155 \times 10^{-2}}\right) = 16{,}257\,\text{W/m}^2\text{·K}$$

4.6 FLOW IN PACKED BEDS

The chemical and energy industries deal predominantly with multiphase and multicomponent systems in which considerable attention is devoted to increasing the interfacial contact between the phases to enhance property transfers and chemical reactions at these extended surface interfaces. As a result, packed beds are extensively used in the chemical process industries. Some examples are gas absorption, catalytic reactors, and deep bed filtration.

4.6.1 Friction Factor Correlations

The friction factor for packed beds, f_{pb}, is defined by

$$f_{pb} = \frac{\epsilon^3}{1 - \epsilon}\,\frac{D_P|\Delta P|}{\rho v_o^2 L} \tag{4.6-1}$$

where ϵ is the *porosity* (or *void volume fraction*), D_P is the particle diameter, and v_o is the *superficial velocity*. The superficial velocity is obtained by dividing the volumetric flow rate by the total cross-sectional area of the bed. Note that the actual flow area is a fraction of the total cross-sectional area.

Example 4.19 Water flows through a concentric annulus at a volumetric flow rate of $5\,\text{m}^3/\text{min}$. The diameters of the inner and the outer pipes are 30 cm and 50 cm, respectively. Calculate the superficial velocity.

Solution

If the inner and outer pipe diameters are designated by D_i and D_o, respectively, the superficial velocity, v_o, is defined by

$$v_o = \frac{Q}{\pi D_o^2/4} = \frac{5}{\pi (0.5)^2/4} = 25.5 \text{ m/min}$$

The actual average velocity, $\langle v \rangle_{act}$, in the annulus is

$$\langle v \rangle_{act} = \frac{Q}{\pi (D_o^2 - D_i^2)/4} = \frac{5}{\pi [(0.5)^2 - (0.3)^2]/4} = 40 \text{ m/min}$$

Comment: The superficial velocity is always lower than the actual average velocity by a factor of porosity, which is equal to $[1 - (D_i/D_o)^2]$ in this example.

For packed beds, the Reynolds number is defined by

$$\text{Re}_{pb} = \frac{D_P v_o \rho}{\mu} \frac{1}{1 - \epsilon} \tag{4.6-2}$$

For laminar flow, the relationship between the friction factor and the Reynolds number is given by

$$\boxed{f_{pb} = \frac{150}{\text{Re}_{pb}}} \qquad \text{Re}_{pb} < 10 \tag{4.6-3}$$

which is known as the *Kozeny-Carman equation*.

In the case of turbulent flow, i.e., $\text{Re}_{pb} > 1000$, the relationship between Re_{pb} and f_{pb} is given by the *Burke-Plummer equation* in the form

$$\boxed{f_{pb} = 1.75} \qquad \text{Re}_{pb} > 1000 \tag{4.6-4}$$

The so-called *Ergun equation* (1952) is simply the summation of the Kozeny-Carman and the Burke-Plummer equations

$$\boxed{f_{pb} = \frac{150}{\text{Re}_{pb}} + 1.75} \tag{4.6-5}$$

Example 4.20 A column of 0.8 m^2 cross-section and 30 m height is packed with spherical particles of diameter 6 mm. A fluid with $\rho = 1.2$ kg/m^3 and $\mu = 1.8 \times 10^{-5}$ kg/m·s flows through the bed at a mass flow rate of 0.65 kg/s. If the pressure drop is measured as 3200 Pa, calculate the porosity of the bed:

a) Analytically,
b) Numerically.

Solution

Assumption

1. The system is isothermal.

Analysis

The superficial velocity through the packed bed is

$$v_o = \frac{0.65}{(1.2)(0.8)} = 0.677 \text{ m/s}$$

Substitution of the values into Eqs. (4.6-1) and (4.6-2) gives the friction factor and the Reynolds number as a function of porosity in the form

$$f_{pb} = \frac{\epsilon^3}{1-\epsilon} \frac{D_P |\Delta P|}{\rho v_o^2 L} = \frac{\epsilon^3}{1-\epsilon} \left[\frac{(6 \times 10^{-3})(3200)}{(1.2)(0.677)^2 (30)} \right] = 1.164 \left(\frac{\epsilon^3}{1-\epsilon} \right) \qquad (1)$$

$$\text{Re}_{pb} = \frac{D_P v_o \rho}{\mu} \frac{1}{1-\epsilon} = \left[\frac{(6 \times 10^{-3})(0.677)(1.2)}{1.8 \times 10^{-5}} \right] \frac{1}{1-\epsilon} = 270.8 \left(\frac{1}{1-\epsilon} \right) \qquad (2)$$

Substitution of Eqs. (1) and (2) into Eq. (4.6-5) gives

$$\epsilon^3 - 0.476 \epsilon^2 + 2.455 \epsilon - 1.979 = 0 \qquad (3)$$

a) Equation (3) can be solved analytically by using the procedure described in Section A.7.1.2 in Appendix A. In order to calculate the discriminant, the terms M and N must be calculated from Eqs. (A.7-5) and (A.7-6), respectively:

$$M = \frac{(3)(2.455) - (0.476)^2}{9} = 0.793$$

$$N = \frac{-(9)(0.476)(2.455) + (27)(1.979) + (2)(0.476)^3}{54} = 0.799$$

Therefore, the discriminant is

$$\Delta = M^3 + N^2 = (0.793)^3 + (0.799)^2 = 1.137$$

Since $\Delta > 0$, Eq. (3) has only one real root as given by Eq. (A.7-7). The terms S and T in this equation are calculated as

$$S = \left(N + \sqrt{\Delta} \right)^{1/3} = \left(0.799 + \sqrt{1.137} \right)^{1/3} = 1.231$$

$$T = \left(N - \sqrt{\Delta} \right)^{1/3} = \left(0.799 - \sqrt{1.137} \right)^{1/3} = -0.644$$

Hence the average porosity of the bed is

$$\epsilon = 1.231 - 0.644 + \frac{0.476}{3} = 0.746$$

b) Equation (3) is rearranged as

$$F(\epsilon) = \epsilon^3 - 0.476\,\epsilon^2 + 2.455\,\epsilon - 1.979 = 0 \tag{4}$$

From Eq. (A.7-25) the iteration scheme is

$$\epsilon_k = \epsilon_{k-1} - \frac{0.02\,\epsilon_{k-1}\,F(\epsilon_{k-1})}{F(1.01\,\epsilon_{k-1}) - F(0.99\,\epsilon_{k-1})} \tag{5}$$

Assuming a starting value of $\epsilon_o = 0.7$, the iterations are given in the table below:

k	ϵ_k
0	0.7
1	0.746
2	0.745
3	0.745

4.6.2 Heat Transfer Correlation

Whitaker (1972) proposed the following correlation for heat transfer in packed beds:

$$\boxed{\mathrm{Nu}_{pb} = \left(0.4\,\mathrm{Re}_{pb}^{1/2} + 0.2\,\mathrm{Re}_{pb}^{2/3}\right)\mathrm{Pr}^{0.4}} \tag{4.6-6}$$

The Nusselt number in Eq. (4.6-6) is defined by

$$\mathrm{Nu}_{pb} = \frac{\langle h \rangle D_P}{k}\,\frac{\epsilon}{1-\epsilon} \tag{4.6-7}$$

Equation (4.6-6) is valid when

$$3.7 \leqslant \mathrm{Re}_{pb} \leqslant 8000 \qquad 0.34 \leqslant \epsilon \leqslant 0.74 \qquad \mathrm{Pr} \approx 0.7$$

All properties in Eq. (4.6-6) are evaluated at the average fluid temperature in the bed.

4.6.2.1 *Calculation of the heat transfer rate*   Once the average heat transfer coefficient is determined, the rate of heat transfer is calculated from

$$\boxed{\dot{Q} = a_v V \langle h \rangle \Delta T_{LM}} \tag{4.6-8}$$

where V is the total volume of the packed bed and a_v is the packing surface area per unit volume defined by

$$a_v = \frac{6(1-\epsilon)}{D_P} \tag{4.6-9}$$

4.6.3 Mass Transfer Correlation

Dwivedi and Upadhyay (1977) proposed a single correlation for both gases and liquids in packed and fluidized beds in terms of the j-factor as

$$\epsilon j_{M_{pb}} = \frac{0.765}{(\text{Re}_{pb}^*)^{0.82}} + \frac{0.365}{(\text{Re}_{pb}^*)^{0.386}} \qquad (4.6\text{-}10)$$

which is valid for $0.01 \leqslant \text{Re}_{pb}^* \leqslant 15{,}000$. The terms $j_{M_{pb}}$ and Re_{pb}^* in Eq. (4.6-10) are defined by

$$j_{M_{pb}} = \left(\frac{\langle k_c \rangle}{v_o}\right) \text{Sc}^{2/3} \qquad (4.6\text{-}11)$$

and

$$\text{Re}_{pb}^* = \frac{D_P v_o \rho}{\mu} \qquad (4.6\text{-}12)$$

4.6.3.1 *Calculation of the mass transfer rate*   Once the average mass transfer coefficient is determined, the rate of mass transfer of species $\mathcal{A}$, $\dot{m}_A$, is given by

$$\dot{m}_A = a_v V \langle k_c \rangle (\Delta c_A)_{LM} \mathcal{M}_A \qquad (4.6\text{-}13)$$

Example 4.21 Instead of using a naphthalene pipe as in Example 4.17, it is suggested to form a packed bed of porosity 0.45 in a pipe, 2.5 cm in internal diameter, by using naphthalene spheres 5 mm in diameter. Pure air at $40\,°\text{C}$ flows at a superficial velocity of 9 m/s through the bed. Determine the length of the packed bed required for the average concentration of naphthalene vapor in the air to reach 25% of the saturation value.

Solution

Physical properties

Diffusion coefficient of naphthalene ($\mathcal{A}$) in air ($\mathcal{B}$) at $40\,°\text{C}$ (313 K) is

$$(\mathcal{D}_{AB})_{313} = (\mathcal{D}_{AB})_{300}\left(\frac{313}{300}\right)^{3/2} = (0.62 \times 10^{-5})\left(\frac{313}{300}\right)^{3/2} = 6.61 \times 10^{-6}\ \text{m}^2/\text{s}$$

For air at $40\,°\text{C}$ (313 K): $\nu = 16.95 \times 10^{-6}\ \text{m}^2/\text{s}$

The Schmidt number is

$$\text{Sc} = \frac{\nu}{\mathcal{D}_{AB}} = \frac{16.95 \times 10^{-6}}{6.61 \times 10^{-6}} = 2.56$$

Assumptions

1. Steady-state conditions prevail.
2. The system is isothermal.
3. The diameter of the naphthalene spheres does not change appreciably.

Analysis

System: Air in the packed bed

Under steady conditions, the conservation statement for naphthalene, species $\mathcal{A}$, becomes

$$\text{Rate of moles of } \mathcal{A} \text{ in} = \text{Rate of moles of } \mathcal{A} \text{ out} \tag{1}$$

The terms in Eq. (1) are expressed by

$$\text{Rate of moles of } \mathcal{A} \text{ in} = a_v V \langle k_c \rangle (\Delta c_A)_{LM} \tag{2}$$

$$\text{Rate of moles of } \mathcal{A} \text{ out} = \mathcal{Q}(c_{A_b})_{out} = (\pi D^2/4)\, v_o (c_{A_b})_{out} \tag{3}$$

Since the concentration at the surface of the naphthalene spheres is constant, the expression for $(\Delta c_A)_{LM}$, Eq. (4.5-35), becomes

$$(\Delta c_A)_{LM} = \frac{(c_{A_b})_{out}}{\ln\left[\dfrac{c_{A_w}}{c_{A_w} - (c_{A_b})_{out}}\right]} \tag{4}$$

Substitution of Eqs. (2)–(4) into Eq. (1) and noting that $V = (\pi D^2/4)L$ give

$$L = -\frac{v_o}{\langle k_c \rangle a_v} \ln\left[1 - \frac{(c_{A_b})_{out}}{c_{A_w}}\right] \tag{5}$$

Note that for a circular pipe, i.e., $a_v = 4/D$, the above equation reduces to Eq. (5) in Example 4.17.

The interfacial area per unit volume, a_v, is calculated from Eq. (4.6-9) as

$$a_v = \frac{6(1 - \epsilon)}{D_P} = \frac{6(1 - 0.45)}{0.005} = 660 \text{ m}^{-1}$$

To determine the average mass transfer coefficient from Eq. (4.6-10), first it is necessary to calculate the Reynolds number

$$\text{Re}_{pb}^* = \frac{D_P v_o}{\nu} = \frac{(0.005)(9)}{16.95 \times 10^{-6}} = 2655$$

Substitution of this value into Eq. (4.6-10) gives

$$\epsilon j_{M_{pb}} = \frac{0.765}{(\text{Re}_{pb}^*)^{0.82}} + \frac{0.365}{(\text{Re}_{pb}^*)^{0.386}} = \frac{0.765}{(2655)^{0.82}} + \frac{0.365}{(2655)^{0.386}} = 0.0186$$

in which $\epsilon j_{M_{pb}}$ is given by Eq. (4.6-11). Therefore, the average mass transfer coefficient is

$$\langle k_c \rangle = 0.0186 \frac{v_o}{\epsilon \, \text{Sc}^{2/3}} = \frac{(0.0186)(9)}{(0.45)(2.56)^{2/3}} = 0.2 \text{ m/s}$$

The length of the bed is calculated from Eq. (5) as

$$L = -\frac{9}{(0.2)(660)} \ln(1 - 0.25) = 0.02 \text{ m}$$

Comment: The use of a packed bed increases the mass transfer area between air and solid naphthalene. This in turn causes a drastic decrease in the length of the equipment.

NOTATION

A	area, m^2
a_v	packing surface area per unit volume, 1/m
$\widehat{C}_P$	heat capacity at constant pressure, kJ/kg·K
c_i	concentration of species i, kmol/m^3
D	diameter, m
D_h	hydraulic equivalent diameter, m
D_P	particle diameter, m
$\mathcal{D}_{AB}$	diffusion coefficient for system $\mathcal{A}$-$\mathcal{B}$, m^2/s
F_D	drag force, N
f	friction factor
G	mass velocity, kg/m^2·s
g	acceleration of gravity, m/s^2
j_H	Chilton-Colburn j-factor for heat transfer
j_M	Chilton-Colburn j-factor for mass transfer
k	thermal conductivity, W/m·K
k_c	mass transfer coefficient, m/s
L	length, m
M	mass, kg
$\dot{m}$	mass flow rate, kg/s
$\mathcal{M}$	molecular weight, kg/kmol
$\dot{n}$	molar flow rate, kmol/s
P	pressure, Pa
$\dot{Q}$	heat transfer rate, W
Q	volumetric flow rate, m^3/s
q	heat flux, W/m^2
$\mathcal{R}$	gas constant, J/mol·K
T	temperature, °C or K
t	time, s
V	volume, m^3
v	velocity, m/s
v_o	superficial velocity, m/s
v_t	terminal velocity, m/s
W	work, J; width, m
$\dot{W}$	rate of work, W
x	rectangular coordinate, m
Δ	difference
ϵ	porosity
ε	surface roughness of the pipe, m
μ	viscosity, kg/m·s
ν	kinematic viscosity, m^2/s
ρ	density, kg/m^3

Overlines

$\sim$ per mole
$\frown$ per unit mass

Bracket

$\langle a \rangle$ average value of a

Superscript

sat   saturation

Subscripts

A, B	species in binary systems
b	bulk
c	transition from laminar to turbulent
ch	characteristic
f	film
i	species in multicomponent systems
in	inlet
LM	log-mean
out	outlet
pb	packed bed
w	wall or surface
∞	free-stream

Dimensionless Numbers

Ar	Archimedes number
Pr	Prandtl number
Nu	Nusselt number
Re	Reynolds number
Re_D	Reynolds number based on the diameter
Re_h	Reynolds number based on the hydraulic equivalent diameter
Re_L	Reynolds number based on the length
Re_x	Reynolds number based on the distance x
Sc	Schmidt number
Sh	Sherwood number
St_H	Stanton number for heat transfer
St_M	Stanton number for mass transfer

REFERENCES

Bedingfield, C.H. and T.B. Drew, 1950, Analogy between heat transfer and mass transfer. A psychometric study, Ind. Eng. Chem. 42, 1164.

Bird, R.B., W.E. Stewart and E.N. Lightfoot, 2002, Transport Phenomena, 2nd Ed., Wiley, New York.

Blausius, H., 1908, Grenzschleten in Flussigkeiten mit kleiner Reibung, Z. Angew. Math. Phys. 56, 1.

Chen, N.H., 1979, An explicit equation for friction factor in pipe, Ind. Eng. Chem. Fund. 18 (3), 296.

Cheng, X.X. and R. Turton, 1990, How to calculate pipe size without iteration, Chem. Eng. 97 (Nov.), 187.

Churchill, S.W., 1977, Friction factor equation spans all fluid flow regimes, Chem. Eng. 84 (Nov. 7), 91.

Churchill, S.W. and M. Bernstein, 1977, A correlating equation for forced convection from gases and liquids to a circular cylinder in cross flow, J. Heat Transfer, 99, 300.

Colebrook, C.F., 1938–9, Turbulent flow in pipes with particular reference to the transition region between the smooth and rough pipe laws, J. Inst. Civil Eng. 11, 133.

Dittus, F.W. and L.M.K. Boelter, 1930, University of California Publications on Engineering, Vol. 2, p. 443, Berkeley.

Dwivedi, P.N. and S.N. Upadhyay, 1977, Particle-fluid mass transfer in fixed and fluidized beds, Ind. Eng. Chem. Process Des. Dev. 16, 157.

Ergun, S., 1952, Fluid flow through packed columns, Chem. Eng. Prog. 48, 89.

Fahien, R.W., 1983, Fundamentals of Transport Phenomena, McGraw-Hill, New York.

Friend, W.L. and A.B. Metzner, 1958, Turbulent heat transfer inside tubes and the analogy among heat, mass, and momentum transfer, AIChE Journal 4, 393.

Frossling, N., 1938, Beitr. Geophys. 52, 170.

Gilliland, E.R. and T.K. Sherwood, 1934, Diffusion of vapors into air streams, Ind. Eng. Chem. 26, 516.

Gregory, G.A. and M. Fogarasi, 1985, Alternate to standard friction factor equation, Oil Gas J. 83, 120.

Lapple, C.E. and C.B. Shepherd, 1940, Calculation of particle trajectories, Ind. Eng. Chem. 32, 605.

Linton, W.H. and T.K. Sherwood, 1950, Mass transfer from solid shapes to water in streamline and turbulent flow, Chem. Eng. Prog. 46, 258.

Moody, L.F., 1944, Friction factors for pipe flow, Trans. ASME 66, 671.

Ranz, W.E. and W.R. Marshall, 1952, Evaporation from drops – Part I, Chem. Eng. Prog. 48, 141.

Sieder, E.N. and G.E. Tate, 1936, Heat transfer and pressure drop of liquids in tubes, Ind. Eng. Chem. 28, 1429.

Steinberger, R.L. and R.E. Treybal, 1960, Mass transfer from a solid sphere to a flowing liquid stream, AIChE Journal 6, 227.

Swamee, P.K. and A.K. Jain, 1976, Explicit equations for pipe-flow problems, J. Hydr. Div. ASCE 102, 657.

Tosun, İ. and İ. Akşahin, 1992, Explicit expressions for the friction factor, Unpublished report, Middle East Technical University.

Tosun, İ. and İ. Akşahin, 1993, Calculate critical piping parameters, Chem. Eng. 100 (March), 165. For corrections see also Chem. Eng. 100 (July), 8.

Turton, R. and N.N. Clark, 1987, An explicit relationship to predict spherical particle terminal velocity, Powder Technology 53, 127.

Turton, R. and O. Levenspiel, 1986, A short note on the drag correlation for spheres, Powder Technology 47, 83.

Whitaker, S., 1972, Forced convection heat transfer correlations for flow in pipes, past flat plates, single cylinders, single spheres, and for flow in packed beds and tube bundles, AIChE Journal 18, 361.

Zhukauskas, A., 1972, Advances in Heat Transfer, Vol. 8: Heat Transfer from Tubes in Cross Flow, Eds. J.P. Hartnett and T.F. Irvine, Jr., Academic Press, New York.

SUGGESTED REFERENCES FOR FURTHER STUDY

Brodkey, R.S. and H.C. Hershey, 1988, Transport Phenomena: A Unified Approach, McGraw-Hill, New York.

Hines, A.L. and R.N. Maddox, 1985, Mass Transfer – Fundamentals and Applications, Prentice-Hall, Englewood Cliffs, New Jersey.

Incropera, F.P. and D.P. DeWitt, 2002, Fundamentals of Heat and Mass Transfer, 5th Ed., Wiley, New York.

Middleman, S., 1998, An Introduction to Mass and Heat Transfer – Principles of Analysis and Design, Wiley, New York.

Skelland, A.H.P., 1974, Diffusional Mass Transfer, Wiley, New York.

Whitaker, S., 1976, Elementary Heat Transfer Analysis, Pergamon Press, New York.

PROBLEMS

4.1 A flat plate of length 2 m and width 30 cm is to be placed parallel to an air stream at a temperature of 25 °C. Which side of the plate, i.e., length or width, should be in the direction of flow so as to minimize the drag force if:

a) The velocity of air is 7 m/s,
b) The velocity of air is 30 m/s.

(**Answer:** a) Length   b) Width)

4.2 Air at atmospheric pressure and 200 °C flows at 8 m/s over a flat plate 150 cm long in the direction of flow and 70 cm wide.

a) Estimate the rate of cooling of the plate so as to keep the surface temperature at 30 °C.
b) Calculate the drag force exerted on the plate.

(**Answer:** a) 1589 W   b) 0.058 N)

4.3 Water at 15 °C flows at 0.15 m/s over a flat plate 1 m long in the direction of flow and 0.3 m wide. If energy is transferred from the top and bottom surfaces of the plate to the flowing stream at a steady rate of 3500 W, determine the temperature of the plate surface.

(**Answer:** 35 °C)

4.4 Fins are used to increase the area available for heat transfer between metal walls and poorly conducting fluids such as gases. A simple rectangular fin is shown below.

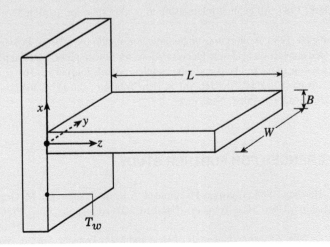

If one assumes,

- $T = T(z)$ only,
- No heat is lost from the end or from the edges,
- The average heat transfer coefficient, $\langle h \rangle$, is constant and uniform over the entire surface of the fin,
- The thermal conductivity of the fin, k, is constant,
- The temperature of the medium surrounding the fin, T_∞, is uniform,
- The wall temperature, T_w, is constant,

the resulting steady-state temperature distribution is given by

$$\frac{T - T_\infty}{T_w - T_\infty} = \frac{\cosh\left[\Lambda\left(1 - \dfrac{z}{L}\right)\right]}{\cosh \Lambda}$$

where

$$\Lambda = \sqrt{\frac{2\langle h \rangle L^2}{k B}}$$

If the rate of heat loss from the fin is 478 W, determine the average heat transfer coefficient for the following conditions: $T_\infty = 175\,°C$; $T_w = 260\,°C$; $k = 105$ W/m·K; $L = 4$ cm; $W = 30$ cm; $B = 5$ mm.

(**Answer:** 400 W/m^2·K)

4.5 Consider the rectangular fin given in Problem 4.4. One of the problems of practical interest is the determination of the optimum values of B and L to maximize the heat transfer rate from the fin for a fixed volume, V, and W. Show that the optimum dimensions are given by

$$B_{opt} \simeq \left(\frac{\langle h \rangle V^2}{k W^2}\right)^{1/3} \quad \text{and} \quad L_{opt} \simeq \left(\frac{k V}{\langle h \rangle W}\right)^{1/3}$$

4.6 Consider the rectangular fin given in Problem 4.4. If a laminar flow region exists over the plate, show that the optimum value of W for the maximum heat transfer rate from the fin for a fixed volume, V, and thickness, B, is given by

$$W_{opt} = 1.2\, V^{4/5} B^{-6/5} \left[\left(\frac{k_f}{k}\right) \mathrm{Pr}^{1/3} \sqrt{\frac{v_\infty}{\nu}}\right]^{2/5}$$

where k_f is the thermal conductivity of the fluid.

4.7 A thin aluminum fin ($k = 205$ W/m·K) of length $L = 20$ cm has two ends attached to two parallel walls with temperatures $T_o = 100\,°C$ and $T_L = 90\,°C$ as shown in the figure below. The fin loses heat by convection to the ambient air at $T_\infty = 30\,°C$ with an average heat transfer coefficient of $\langle h \rangle = 120$ W/m^2·K through the top and bottom surfaces (heat loss from the edges may be considered negligible).

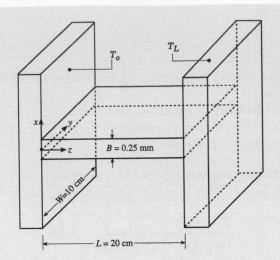

One of your friends assumes that there is no internal generation of energy within the fin and determines the steady-state temperature distribution within the fin as

$$\frac{T - T_\infty}{T_o - T_\infty} = e^{Nz} - 2\Omega \sinh Nz$$

in which N and Ω are defined as

$$N = \sqrt{\frac{2\langle h \rangle}{kB}} \quad \text{and} \quad \Omega = \frac{e^{NL} - \left(\dfrac{T_L - T_\infty}{T_o - T_\infty}\right)}{2 \sinh NL}$$

a) Show that there is indeed no internal generation of energy within the fin.
b) Determine the location and the value of the minimum temperature within the fin.

(**Answer:** $z = 0.1$ cm, $T = 30.14\,°C$)

4.8 Rework Example 4.8 by using the Ranz-Marshall correlation, Eq. (4.3-33), the Frossling correlation, Eq. (4.3-34), and the modified Frossling correlation, Eq. (4.3-35). Why do the resulting Sherwood numbers differ significantly from 541?

4.9 In an experiment carried out at $20\,°C$, a glass sphere of density 2620 kg/m^3 falls through carbon tetrachloride ($\rho = 1590$ kg/m^3 and $\mu = 9.58 \times 10^{-4}$ kg/m·s) with a terminal velocity of 65 cm/s. Determine the diameter of the sphere.

(**Answer:** 21 mm)

4.10 A CO_2 bubble is rising in a glass of beer 20 cm tall. Estimate the time required for a bubble 5 mm in diameter to reach the top if the properties of CO_2 and beer can be taken as equal to those of air and water, respectively.

(**Answer:** 0.54 s)

4.11 Show that the use of the Dittus-Boelter correlation, Eq. (4.5-26), together with the Chilton-Colburn analogy, Eq. (3.5-12), yields

$$f \simeq 0.046\,\mathrm{Re}^{-0.2}$$

which is a good power-law approximation for the friction factor in smooth circular pipes. Calculate f for $\mathrm{Re} = 10^5$, 10^6 and 10^7 using this approximate equation and compare the values with those obtained by using the Chen correlation, Eq. (4.5-16).

4.12 For laminar flow of an incompressible Newtonian fluid in a circular pipe, Eq. (4.5-12) indicates that the pressure drop is proportional to the volumetric flow rate. For fully turbulent flow show that the pressure drop in a pipe is proportional to the square of the volumetric flow rate.

4.13 Determine the power to pump a fluid at a volumetric flow rate of 1.1×10^{-3} m^3/s through a 3 cm diameter horizontal smooth pipe 10 m long. The physical properties of the fluid are given as $\rho = 935$ kg/m^3 and $\mu = 1.92 \times 10^{-3}$ kg/m·s.

(**Answer:** 10.4 W)

4.14 The purpose of blood pressure in the human body is to push blood to the tissues of the organism so that they can perform their functions. Each time the heart beats, it pumps out blood into the arteries. The blood pressure reaches its maximum value, i.e., systolic pressure, when the heart contracts to pump the blood. In between beats, the heart is at rest and the blood pressure falls to a minimum value, diastolic pressure. An average healthy person has systolic and diastolic pressures of 120 and 80 mmHg, respectively. The human body has about 5.6 L of blood. If it takes 20 s for blood to circulate throughout the body, estimate the power output of the heart.

(**Answer:** 3.73 W)

4.15 Water is in isothermal turbulent flow at 20 °C through a horizontal pipe of circular cross-section with 10 cm inside diameter. The following experimental values of velocity are measured as a function of radial distance r:

r (cm)	0.5	1.5	2.5	3.5	4.5
v_z (m/s)	0.394	0.380	0.362	0.337	0.288

The velocity distribution is proposed in the form

$$v_z = v_{\max}\left(1 - \frac{r}{R}\right)^{1/n}$$

where $v_{\max}$ is the maximum velocity and R is the radius of the pipe. Calculate the pressure drop per unit length of the pipe.

(**Answer:** 12.3 Pa/m)

4.16   In Example 4.15, the length to diameter ratio is expressed as

$$\frac{L}{D} = \frac{1}{4}\frac{1}{St_H}\ln\left(\frac{T_w - T_{b_{in}}}{T_w - T_{b_{out}}}\right)$$

Use the Chilton-Colburn analogy, i.e.,

$$\frac{f}{2} = St_H\,Pr^{2/3}$$

and evaluate the value of L/D. Is it a realistic value? Why/why not?

4.17   Water at $10\,°C$ enters a circular pipe of internal diameter 2.5 cm with an average velocity of 1.2 m/s. Steam condenses on the outside of the pipe so as to keep the surface temperature of the pipe at $82\,°C$. If the length of the pipe is 5 m, determine the outlet temperature of water.

(**Answer:** $51\,°C$)

4.18   Dry air at 1 atm pressure and $50\,°C$ enters a circular pipe of 12 cm internal diameter with an average velocity of 10 cm/s. The inner surface of the pipe is coated with a thin absorbent material soaked with water at $20\,°C$. If the length of the pipe is 6 m, calculate the amount of water vapor carried out of the pipe per hour.

(**Answer:** 0.067 kg/h)

4.19   A column with an internal diameter of 50 cm and a height of 2 m is packed with spherical particles 3 mm in diameter so as to form a packed bed with $\epsilon = 0.45$. Estimate the power required to pump a Newtonian liquid ($\mu = 70 \times 10^{-3}$ kg/m·s; $\rho = 1200$ kg/m^3) through the packed bed at a mass flow rate of 1.2 kg/s.

(**Answer:** 39.6 W)

4.20   The drag force, F_D, is defined as the interfacial transfer of momentum from the fluid to the solid. In Chapter 3, power, $\dot{W}$, is given by Eq. (3.1-11) as

$$\dot{W} = F_D v_{ch} \tag{1}$$

For flow in conduits, power is also expressed by Eq. (4.5-2) in the form

$$\dot{W} = Q|\Delta P| \tag{2}$$

a) For flow in a circular pipe, the characteristic velocity is taken as the average velocity. For this case, use Eqs. (1) and (2) to show that

$$F_D = A|\Delta P| \tag{3}$$

where A is the cross-sectional area of the pipe.
b) For flow through packed beds, the characteristic velocity is taken as the actual average velocity or interstitial velocity, i.e.,

$$v_{ch} = \frac{v_o}{\epsilon} \tag{4}$$

in which v_o is the superficial velocity and ϵ is the porosity of the bed. Show that

$$F_D = \epsilon A |\Delta P| \tag{5}$$

where A is the cross-sectional area of the packed bed.

c) In fluidization, the drag force on each particle should support its effective weight, i.e., weight minus buoyancy. Show that the drag force is given by

$$F_D = AL(1 - \epsilon)(\rho_P - \rho)g\epsilon \tag{6}$$

where L is the length of the bed, and ρ and ρ_P are the densities of the fluid and solid particle, respectively. Note that in the calculation of the buoyancy force the volume occupied by solid particles should be multiplied by the density of suspension, i.e., $\epsilon\rho + (1-\epsilon)\rho_P$, instead of by ρ.

Combine Eqs. (5) and (6) to get

$$\frac{|\Delta P|}{L} = g(1 - \epsilon)(\rho_P - \rho) \tag{7}$$

which is a well-known equation in fluidization.

4.21 A 15×90 m lawn is covered by a layer of ice 0.15 mm thick at $-4\,°C$. The wind at a temperature of $0\,°C$ with 15% relative humidity blows in the direction of the short side of the lawn. If the wind velocity is 10 m/s, estimate the time required for the ice layer to disappear by sublimation under steady conditions. The vapor pressure and the density of ice at $-4\,°C$ are 3.28 mmHg and 917 kg/m^3, respectively.

(**Answer:** 41 min)

5

RATE OF GENERATION IN MOMENTUM, ENERGY, AND MASS TRANSPORT

In Chapter 1, the generation rate per unit volume is designated by $\Re$. Integration of this quantity over the volume of the system gives the generation rate in the conservation statement. In this chapter, explicit expressions for $\Re$ will be developed for the cases of momentum, energy, and mass transport.

5.1 RATE OF GENERATION IN MOMENTUM TRANSPORT

In general, forces acting on a particle can be classified as surface forces and body forces. *Surface forces*, such as normal stresses (pressure) and tangential stresses, act by direct contact on a surface. *Body forces*, however, act at a distance on a volume. Gravitational, electrical and electromagnetic forces are examples of body forces.

For solid bodies Newton's second law of motion states that

$$\begin{pmatrix} \text{Summation of forces} \\ \text{acting on a system} \end{pmatrix} = \begin{pmatrix} \text{Time rate of change of} \\ \text{momentum of a system} \end{pmatrix} \qquad (5.1\text{-}1)$$

in which forces acting on a system include both surface and body forces. Equation (5.1-1) can be extended to fluid particles by considering the rate of flow of momentum into and out of the volume element, i.e.,

$$\begin{pmatrix} \text{Rate of} \\ \text{momentum in} \end{pmatrix} - \begin{pmatrix} \text{Rate of} \\ \text{momentum out} \end{pmatrix} + \begin{pmatrix} \text{Summation of forces} \\ \text{acting on a system} \end{pmatrix}$$
$$= \begin{pmatrix} \text{Time rate of change of} \\ \text{momentum of a system} \end{pmatrix} \qquad (5.1\text{-}2)$$

On the other hand, for a given system, the inventory rate equation for momentum can be expressed as

$$\begin{pmatrix} \text{Rate of} \\ \text{momentum in} \end{pmatrix} - \begin{pmatrix} \text{Rate of} \\ \text{momentum out} \end{pmatrix} + \begin{pmatrix} \text{Rate of momentum} \\ \text{generation} \end{pmatrix}$$
$$= \begin{pmatrix} \text{Rate of momentum} \\ \text{accumulation} \end{pmatrix} \qquad (5.1\text{-}3)$$

Comparison of Eqs. (5.1-2) and (5.1-3) indicates that

$$\begin{pmatrix} \text{Rate of momentum} \\ \text{generation} \end{pmatrix} = \begin{pmatrix} \text{Summation of forces} \\ \text{acting on a system} \end{pmatrix} \qquad (5.1\text{-}4)$$

in which the forces acting on a system are the pressure force (surface force) and the gravitational force (body force).

5.1.1 Momentum Generation as a Result of Gravitational Force

Consider a basketball player holding a ball in his/her hands. When (s)he drops the ball, it starts to accelerate as a result of gravitational force. According to Eq. (5.1-4), the rate of momentum generation is given by

$$\text{Rate of momentum generation} = Mg \tag{5.1-5}$$

where M is the mass of the ball and g is the gravitational acceleration. Therefore, the rate of momentum generation per unit volume, $\Re$, is given by

$$\boxed{\Re = \rho g} \tag{5.1-6}$$

5.1.2 Momentum Generation as a Result of Pressure Force

Consider the steady flow of an incompressible fluid in a pipe as shown in Figure 5.1. The rate of mechanical energy required to pump the fluid is given by Eq. (4.5-3) as

$$\dot{W} = F_D \langle v \rangle = \mathcal{Q} |\Delta P| \tag{5.1-7}$$

Since the volumetric flow rate, $\mathcal{Q}$, is the product of average velocity, $\langle v \rangle$, with the cross-sectional area, A, Eq. (5.1-7) reduces to

$$A |\Delta P| - F_D = 0 \tag{5.1-8}$$

For the system whose boundaries are indicated by a dotted line in Figure 5.1, the conservation of mass states that

$$\dot{m}_{in} = \dot{m}_{out} \tag{5.1-9}$$

or,

$$\left(\rho \langle v \rangle A \right)_{in} = \left(\rho \langle v \rangle A \right)_{out} \quad \Rightarrow \quad \langle v \rangle_{in} = \langle v \rangle_{out} \tag{5.1-10}$$

On the other hand, the conservation statement for momentum, Eq. (5.1-3), takes the form

$$\begin{pmatrix} \text{Rate of} \\ \text{momentum in} \end{pmatrix} - \begin{pmatrix} \text{Rate of} \\ \text{momentum out} \end{pmatrix} + \begin{pmatrix} \text{Rate of momentum} \\ \text{generation} \end{pmatrix} = 0 \tag{5.1-11}$$

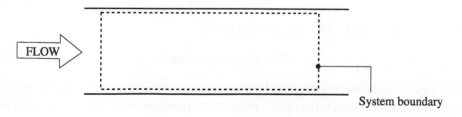

Figure 5.1. Flow through a pipe.

and can be expressed as

$$\left(\dot{m}\langle v\rangle\right)_{in} - \left[\left(\dot{m}\langle v\rangle\right)_{out} + F_D\right] + \Re(AL) = 0 \tag{5.1-12}$$

where $\Re$ is the rate of momentum generation per unit volume. Note that the rate of momentum transfer from the fluid to the pipe wall manifests itself as a drag force. The use of Eqs. (5.1-9) and (5.1-10) simplifies Eq. (5.1-12) to

$$\Re(AL) - F_D = 0 \tag{5.1-13}$$

Comparison of Eqs. (5.1-8) and (5.1-13) indicates that the rate of momentum generation per unit volume is equal to the pressure gradient, i.e.,

$$\boxed{\Re = \frac{|\Delta P|}{L}} \tag{5.1-14}$$

5.1.3 Modified Pressure

Equations (5.1-6) and (5.1-14) indicate that the presence of pressure and/or gravity forces can be interpreted as a source of momentum. In fluid mechanics, it is customary to combine these two forces in a single term and express the rate of momentum generation per unit volume as

$$\Re = \frac{|\Delta \mathcal{P}|}{L} \tag{5.1-15}$$

where $\mathcal{P}$ is the *modified pressure*[1] defined by

$$\mathcal{P} = P + \rho g h \tag{5.1-16}$$

in which h is the distance measured in the direction opposite to gravity from any chosen reference plane.

5.1.3.1 *Physical interpretation of the modified pressure* Consider a stagnant liquid in a storage tank open to the atmosphere. Let z be the distance measured from the surface of the liquid in the direction of gravity. The hydrostatic pressure distribution within the fluid is given by

$$P = P_{atm} + \rho g z \tag{5.1-17}$$

For this case the modified pressure is defined as

$$\mathcal{P} = P - \rho g z \tag{5.1-18}$$

Substitution of Eq. (5.1-18) into Eq. (5.1-17) gives

$$\mathcal{P} = P_{atm} = \text{constant} \tag{5.1-19}$$

The simplicity of defining the modified pressure comes from the fact that it is always constant under static conditions, whereas the hydrostatic pressure varies as a function of position.

---

[1]The term $\mathcal{P}$ is also called equivalent pressure, dynamic pressure, and piezometric pressure.

Table 5.1. Pressure difference in flow through a pipe with different orientation

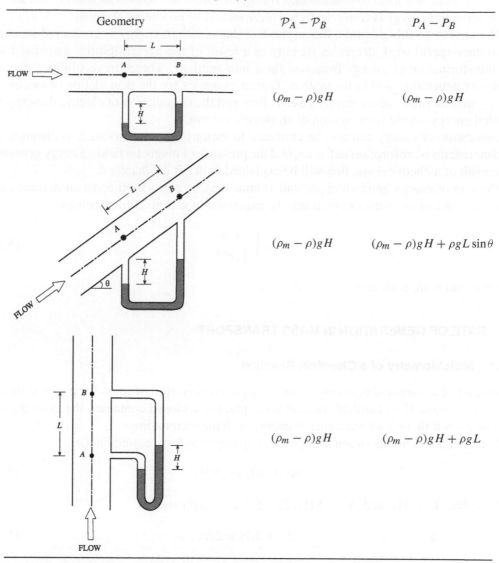

Geometry	$\mathcal{P}_A - \mathcal{P}_B$	$P_A - P_B$
	$(\rho_m - \rho)gH$	$(\rho_m - \rho)gH$
	$(\rho_m - \rho)gH$	$(\rho_m - \rho)gH + \rho g L \sin\theta$
	$(\rho_m - \rho)gH$	$(\rho_m - \rho)gH + \rho g L$

Suppose that you measure a pressure difference over a length L of a pipe. It is difficult to estimate whether this pressure difference comes from a flow situation or hydrostatic distribution. However, any variation in $\mathcal{P}$ implies a flow. Another distinct advantage of defining modified pressure is that the difference in $\mathcal{P}$ is independent of the orientation of the pipe as shown in Table 5.1.

5.2 RATE OF GENERATION IN ENERGY TRANSPORT

Let us consider the following paradox: "*One of the most important problems that the world faces today is the energy shortage. According to the first law of thermodynamics, energy is converted from one form to another and transferred from one system to another but its total is conserved. If energy is conserved, then there should be no energy shortage."*

The answer to this dilemma lies in the fact that although energy is conserved its ability to produce useful work decreases steadily as a result of the irreversibilities associated with the transformation of energy from one form into another[2]. These irreversibilities give rise to energy generation within the system. Typical examples are the degradation of mechanical energy into thermal energy during viscous flow and the degradation of electrical energy into thermal energy during transmission of an electric current.

Generation of energy can also be attributed to various other factors such as chemical and nuclear reactions, absorption radiation, and the presence of magnetic fields. Energy generation as a result of a chemical reaction will be explained in detail in Chapter 6.

The rate of energy generation per unit volume may be considered constant in most cases. If it is dependent on temperature, it may be expressed in various forms such as

$$\Re = \begin{cases} a + bT \\ \Re_o \, e^{aT} \end{cases} \tag{5.2-1}$$

where a and b are constants.

5.3   RATE OF GENERATION IN MASS TRANSPORT

5.3.1   Stoichiometry of a Chemical Reaction

Balancing of a chemical equation is based on the conservation of mass for a closed thermodynamic system. If a chemical reaction takes place in a closed container, the mass does not change even if there is an exchange of energy with the surroundings.

Consider a reaction between nitrogen and hydrogen to form ammonia, i.e.,

$$N_2 + 3H_2 = 2NH_3 \tag{5.3-1}$$

If $A_1 = N_2$, $A_2 = H_2$, and $A_3 = NH_3$, Eq. (5.3-1) is expressed as

$$A_1 + 3A_2 = 2A_3 \tag{5.3-2}$$

It is convenient to write all the chemical species on one side of the equation and give a positive sign to the species regarded as the products of the reaction. Thus,

$$2A_3 - A_1 - 3A_2 = 0 \tag{5.3-3}$$

or,

$$\sum_{i=1}^{s} \alpha_i A_i = 0 \tag{5.3-4}$$

where α_i is the stoichiometric coefficient of the ith chemical species (positive if the species is a product, negative if the species is a reactant), s is the total number of species in the reaction,

---

[2]Note that 1000 J at 100 °C is much more valuable than 1000 J at 20 °C.

and A_i is the chemical symbol for the ith chemical species, representing the molecular weight of the species.

Each chemical species, A_i, is the sum of the chemical elements, E_j, such that

$$A_i = \sum_{j=1}^{t} \beta_{ji} E_j \qquad (5.3\text{-}5)$$

where β_{ji} represents the number of chemical elements E_j in the chemical species A_i, and t is the total number of chemical elements. Substitution of Eq. (5.3-5) into Eq. (5.3-4) gives

$$\sum_{i=1}^{s} \alpha_i \left(\sum_{j=1}^{t} \beta_{ji} E_j \right) = \sum_{j=1}^{t} \left(\sum_{i=1}^{s} \alpha_i \beta_{ji} \right) E_j = 0 \qquad (5.3\text{-}6)$$

Since all the E_j are linearly independent[3], then

$$\boxed{\sum_{i=1}^{s} \alpha_i \beta_{ji} = 0} \qquad j = 1, 2, \ldots, t \qquad (5.3\text{-}7)$$

Equation (5.3-7) is used to balance chemical equations.

Example 5.1 Consider the reaction between N_2 and H_2 to form NH_3

$$\alpha_1 N_2 + \alpha_2 H_2 + \alpha_3 NH_3 = 0$$

Show how one can apply Eq. (5.3-7) to balance this equation.

Solution

If $A_1 = N_2$, $A_2 = H_2$ and $A_3 = NH_3$, the above equation can be expressed as

$$\alpha_1 A_1 + \alpha_2 A_2 + \alpha_3 A_3 = 0 \qquad (1)$$

If we let $E_1 = N$ $(j = 1)$ and $E_2 = H$ $(j = 2)$, then Eq. (5.3-7) becomes

$$\alpha_1 \beta_{11} + \alpha_2 \beta_{12} + \alpha_3 \beta_{13} = 0 \qquad \text{for} \quad j = 1 \qquad (2)$$

$$\alpha_1 \beta_{21} + \alpha_2 \beta_{22} + \alpha_3 \beta_{23} = 0 \qquad \text{for} \quad j = 2 \qquad (3)$$

---

[3]The expression

$$\sum_{i=1}^{n} \alpha_i x_i = \alpha_1 x_1 + \alpha_2 x_2 + \cdots + \alpha_n x_n$$

where $\{\alpha_1, \alpha_2, \ldots, \alpha_n\}$ is a set of scalars, is called a *linear combination* of the elements of the set $S = \{x_1, x_2, \ldots, x_n\}$. The elements of the set S are said to be *linearly dependent* if there exists a set of scalars $\{\alpha_1, \alpha_2, \ldots, \alpha_n\}$ with elements α_i **not** all equal to zero, such that the linear combination $\sum_{i=1}^{n} \alpha_i x_i = 0$ holds. If $\sum_{i=1}^{n} \alpha_i x_i = 0$ holds for all $\alpha_i = 0$, then the set S is *linearly independent*.

or,

$$\alpha_1(2) + \alpha_2(0) + \alpha_3(1) = 0 \tag{4}$$

$$\alpha_1(0) + \alpha_2(2) + \alpha_3(3) = 0 \tag{5}$$

Solutions of Eqs. (4) and (5) give

$$\alpha_1 = -\frac{1}{2}\alpha_3 \qquad \alpha_2 = -\frac{3}{2}\alpha_3 \tag{6}$$

If we take $\alpha_3 = 2$, then $\alpha_1 = -1$ and $\alpha_2 = -3$. Hence, the reaction becomes

$$N_2 + 3H_2 = 2NH_3$$

Comment: Stoichiometric coefficients have units. For example, in the above equation the stoichiometric coefficient of H_2 indicates that there are 3 moles of H_2 per mole of N_2.

5.3.2 The Law of Combining Proportions

Stoichiometric coefficients have the units of moles of i per mole of basis species, where the basis species is arbitrarily chosen. The law of combining proportions states that

$$\frac{\text{moles of } i \text{ reacted}}{(\text{moles of } i/\text{mole of basis species})} = \text{moles of basis species} \tag{5.3-8}$$

or,

$$\frac{n_i - n_{i_o}}{\alpha_i} = \varepsilon \tag{5.3-9}$$

where ε is called the *molar extent* of the reaction[4]. Rearrangement of Eq. (5.3-9) gives

$$\boxed{n_i = n_{i_o} + \alpha_i \, \varepsilon} \tag{5.3-10}$$

Note that once ε has been determined, the number of moles of any chemical species participating in the reaction can be determined by using Eq. (5.3-10).

The molar extent of the reaction should not be confused with the *fractional conversion variable*, X, which can only take values between 0 and 1. The molar extent of the reaction is an extensive property measured in moles and its value can be greater than unity.

It is also important to note that the fractional conversion may be different for each reacting species, i.e.,

$$\boxed{X_i = \frac{n_{i_o} - n_i}{n_{i_o}}} \tag{5.3-11}$$

On the other hand, molar extent is unique for a given reaction. Comparison of Eqs. (5.3-10) and (5.3-11) indicates that

$$\boxed{\varepsilon = \frac{n_{i_o}}{(-\alpha_i)} X} \tag{5.3-12}$$

---

[4]The term ε has been given various names in the literature, such as degree of advancement, reaction of coordinate, degree of reaction, and progress variable.

The total number of moles, n_T, of a reacting mixture at any instant can be calculated by the summation of Eq. (5.3-10) over all species, i.e.,

$$n_T = n_{T_o} + \overline{\alpha}\,\varepsilon \qquad (5.3\text{-}13)$$

where n_{T_o} is the initial total number of moles and $\overline{\alpha} = \sum_i \alpha_i$.

Example 5.2 A system containing 1 mol A_1, 2 mol A_2, and 7 mol A_3 undergoes the following reaction

$$A_1(g) + A_2(g) + 3/2A_3(g) \rightarrow A_4(g) + 3A_5(g)$$

Determine the limiting reactant and fractional conversion with respect to each reactant if the reaction goes to completion.

Solution

Since $n_i \geqslant 0$, it is possible to conclude from Eq. (5.3-10) that the limiting reactant has the least positive value of $n_{i_o}/(-\alpha_i)$. The values given in the following table indicate that the limiting reactant is A_1.

Species	$n_{i_o}/(-\alpha_i)$
A_1	1
A_2	2
A_3	4.67

Note that the least positive value of $n_{i_o}/(-\alpha_i)$ is also the greatest possible value of ε. Since the reaction goes to completion, species A_1 will be completely depleted and $\varepsilon = 1$. Using Eq. (5.3-12), fractional conversion values are given as follows:

Species	X
A_1	1
A_2	0.50
A_3	0.21

Example 5.3 A system containing 3 mol A_1 and 4 mol A_2 undergoes the following reaction

$$2A_1(g) + 3A_2(g) \rightarrow A_3(g) + 2A_4(g)$$

Calculate the mole fractions of each species if $\varepsilon = 1.1$. What is the fractional conversion based on the limiting reactant?

Solution

Using Eq. (5.3-10), the number of moles of each species is expressed as

$$n_1 = 3 - 2\varepsilon = 3 - (2)(1.1) = 0.8 \text{ mol}$$
$$n_2 = 4 - 3\varepsilon = 4 - (3)(1.1) = 0.7 \text{ mol}$$
$$n_3 = \varepsilon = 1.1 \text{ mol}$$
$$n_4 = 2\varepsilon = (2)(1.1) = 2.2 \text{ mol}$$

Therefore, the total number of moles is 4.8 and the mole fraction of each species is

$$x_1 = \frac{0.8}{4.8} = 0.167$$

$$x_2 = \frac{0.7}{4.8} = 0.146$$

$$x_3 = \frac{1.1}{4.8} = 0.229$$

$$x_4 = \frac{2.2}{4.8} = 0.458$$

The fractional conversion, X, based on the limiting reactant A_2 is

$$X = \frac{4 - 0.7}{4} = 0.825$$

The molar concentration of the ith species, c_i, is defined by

$$c_i = \frac{n_i}{V} \tag{5.3-14}$$

Therefore, division of Eq. (5.3-10) by the volume V gives

$$\frac{n_i}{V} = \frac{n_{i_o}}{V} + \alpha_i \frac{\varepsilon}{V} \tag{5.3-15}$$

or,

$$\boxed{c_i = c_{i_o} + \alpha_i \xi} \tag{5.3-16}$$

where c_{i_o} is the initial molar concentration of the ith species and ξ is the *intensive extent* of the reaction in moles per unit volume. Note that ξ is related to conversion, X, by

$$\boxed{\xi = \frac{c_{i_o}}{(-\alpha_i)} X_i} \tag{5.3-17}$$

The total molar concentration, c, of a reacting mixture at any instant can be calculated by the summation of Eq. (5.3-16) over all species, i.e.,

$$\boxed{c = c_o + \overline{\alpha} \xi} \tag{5.3-18}$$

where c_o is the initial total molar concentration.

When more than one reaction takes place in a reactor, Eq. (5.3-10) takes the form

$$\boxed{n_{ij} = n_{ij_o} + \alpha_{ij} \varepsilon_j} \tag{5.3-19}$$

where

n_{ij} = number of moles of the ith species in the jth reaction

n_{ij_o} = initial number of moles of the ith species in the jth reaction

α_{ij} = stoichiometric coefficient of the ith species in the jth reaction

ε_j = extent of the jth reaction

Summation of Eq. (5.3-19) over all reactions taking place in a reactor gives

$$\sum_j n_{ij} = \sum_j n_{ij_o} + \sum_j \alpha_{ij}\varepsilon_j \qquad (5.3\text{-}20)$$

or,

$$\boxed{n_i = n_{i_o} + \sum_j \alpha_{ij}\varepsilon_j} \qquad (5.3\text{-}21)$$

Example 5.4 The following two reactions occur simultaneously in a batch reactor:

$$C_2H_6 = C_2H_4 + H_2$$

$$C_2H_6 + H_2 = 2CH_4$$

A mixture of 85 mol% C_2H_6 and 15% inerts is fed into a reactor and the reactions proceed until 25% C_2H_4 and 5% CH_4 are formed. Determine the percentage of each species in a reacting mixture.

Solution

Basis: 1 mol of a reacting mixture

Let ε_1 and ε_2 be the extents of the first and second reactions, respectively. Then the number of moles of each species can be expressed as

$$n_{C_2H_6} = 0.85 - \varepsilon_1 - \varepsilon_2$$

$$n_{C_2H_4} = \varepsilon_1$$

$$n_{H_2} = \varepsilon_1 - \varepsilon_2$$

$$n_{CH_4} = 2\varepsilon_2$$

$$n_{inert} = 0.15$$

The total number of moles, n_T, is

$$n_T = 1 + \varepsilon_1$$

The mole fractions of C_2H_4 and CH_4 are given in the problem statement. These values are used to determine the extent of the reactions as

$$x_{C_2H_4} = \frac{\varepsilon_1}{1 + \varepsilon_1} = 0.25 \quad \Rightarrow \quad \varepsilon_1 = 0.333$$

$$x_{CH_4} = \frac{2\varepsilon_2}{1 + \varepsilon_1} = 0.05 \quad \Rightarrow \quad \varepsilon_2 = 0.033$$

Therefore, the mole fractions of C_2H_6, H_2, and the inerts are

$$x_{C_2H_6} = \frac{0.85 - \varepsilon_1 - \varepsilon_2}{1 + \varepsilon_1} = \frac{0.85 - 0.333 - 0.033}{1 + 0.333} = 0.363$$

$$x_{H_2} = \frac{\varepsilon_1 - \varepsilon_2}{1 + \varepsilon_1} = \frac{0.333 - 0.033}{1 + 0.333} = 0.225$$

$$x_{inert} = \frac{0.15}{1 + 0.333} = 0.112$$

5.3.3 Rate of Reaction

The rate of a chemical reaction, r, is defined by

$$\boxed{r = \frac{1}{V}\frac{d\varepsilon}{dt}} \tag{5.3-22}$$

where V is the volume physically occupied by the reacting fluid. Since both V and $d\varepsilon/dt$ are positive, the reaction rate is intrinsically positive. Note that the reaction rate has the units of moles reacted per unit time per unit volume of the reaction mixture. The reaction rate expression, r, has the following characteristics:

- It is an intensive property,
- It is independent of the reactor type,
- It is independent of a process.

Changes in the molar extent of the reaction can be related to the changes in the number of moles of species i by differentiating Eq. (5.3-10). The result is

$$d\varepsilon = \frac{1}{\alpha_i}dn_i \tag{5.3-23}$$

Substitution of Eq. (5.3-23) into Eq. (5.3-22) gives

$$\boxed{r = \frac{1}{\alpha_i}\frac{1}{V}\frac{dn_i}{dt}} \tag{5.3-24}$$

If the rate of generation of species i per unit volume, $\Re_i$, is defined by

$$\Re_i = \frac{1}{V}\frac{dn_i}{dt} \tag{5.3-25}$$

then

$$\boxed{\Re_i = \alpha_i r} \tag{5.3-26}$$

Therefore, $\Re_i$ is negative if i appears as a reactant; $\Re_i$ is positive if i is a product.

Example 5.5 For the reaction

$$3A \rightarrow B + C$$

express the reaction rate in terms of the time rate of change of species A, B, and C.

Solution

Application of Eq. (5.3-24) gives the rate as

$$r = -\frac{1}{3}\frac{1}{V}\frac{dn_A}{dt} = \frac{1}{V}\frac{dn_B}{dt} = \frac{1}{V}\frac{dn_C}{dt} \tag{1}$$

If V is constant, then Eq. (1) reduces to

$$r = -\frac{1}{3}\frac{dc_A}{dt} = \frac{dc_B}{dt} = \frac{dc_C}{dt} \tag{2}$$

Comment: The rate of reaction is equal to the time derivative of a concentration only when the volume of the reacting mixture is constant.

In the case of several reactions, $\Re_i$ is defined by

$$\boxed{\Re_i = \sum_j \alpha_{ij} r_j} \tag{5.3-27}$$

where r_j is the rate of the jth reaction.

The reaction rate is a function of temperature and concentration and is assumed to be the product of two functions; one is dependent only on the temperature and the other is dependent only on the concentration, i.e.,

$$r(T, c_i) = k(T) f(c_i) \tag{5.3-28}$$

The function $k(T)$ is called the *rate constant* and its dependence on the temperature is given by

$$k(T) = A T^m e^{-\mathcal{E}/\mathcal{R}T} \tag{5.3-29}$$

where A is a constant, $\mathcal{E}$ is the activation energy, $\mathcal{R}$ is the gas constant, and T is the absolute temperature. The power of temperature, m, is given by

$$m = \begin{cases} 0 & \text{from the Arrhenius relation} \\ 1/2 & \text{from the kinetic theory of gases} \\ 1 & \text{from statistical mechanics} \end{cases} \tag{5.3-30}$$

In engineering practice the *Arrhenius relation*, i.e.,

$$k(T) = A e^{-\mathcal{E}/\mathcal{R}T} \tag{5.3-31}$$

is generally considered valid[5], and the rate constant can be determined by running the same reaction at different temperatures. The data from these experiments are found to be linear on a semi-log plot of k versus $1/T$.

The function $f(c_i)$ depends on the concentration of all the species in the chemical reaction. Since the reaction rate is usually largest at the start of the reaction and eventually decreases

---

[5]Deviations from the Arrhenius relationship are discussed by Maheshwari and Akella (1988).

to reach a zero-rate at equilibrium, the function $f(c_i)$ is taken to be a power function of the concentration of the reactants.

If $f(c_i)$ were a power function of the products of the reaction, the reaction rate would increase rather than decrease with time. These reactions are called *autocatalytic*.

For normal decreasing rate reactions

$$f(c_i) = \prod_i c_i^{\gamma_i} \tag{5.3-32}$$

where c_i is the concentration of a reactant. Thus, the constitutive equation for the reaction rate is

$$\boxed{r = k \prod_i c_i^{\gamma_i}} \tag{5.3-33}$$

The *order of a reaction*, n, refers to the powers to which the concentrations are raised, i.e.,

$$n = \sum_i \gamma_i \tag{5.3-34}$$

It should be pointed out that there is not necessarily a connection between the order and the stoichiometry of the reaction.

NOTATION

A	area, m^2
c	concentration, $kmol/m^3$
$\mathcal{E}$	activation energy, $kJ/kmol$
F_D	drag force, N
g	acceleration of gravity, m/s^2
h	elevation, m
k	reaction rate constant
L	length, m
M	mass, kg
$\dot{m}$	mass flow rate, kg/s
n	number of moles, kmol
n_{ij}	number of moles of the ith species in the jth reaction
P	pressure, Pa
$\mathcal{P}$	modified pressure, Pa
Q	volumetric flow rate, m^3/s
r	rate of a chemical reaction, $kmol/m^3 \cdot s$
$\mathfrak{R}$	rate of generation (momentum, energy, mass) per unit volume
T	temperature, °C or K
t	time, s
V	volume, m^3
v	velocity, m/s
$\dot{W}$	rate of work, W

X fractional conversion

x_i mole fraction of species i

z rectangular coordinate, m

α_i stoichiometric coefficient of species i

α_{ij} stoichiometric coefficient of the ith species in the jth reaction

$\overline{\alpha}$ $\sum_i \alpha_i$

Δ difference

ε molar extent of a reaction, kmol

ξ intensive extent of a reaction, kmol/m^3

ρ density, kg/m^3

ρ_m density of manometer fluid, kg/m^3

Bracket

$\langle a \rangle$ average value of a

Subscripts

atm atmospheric

in inlet

o initial

out out

T total

REFERENCE

Maheshwari, M. and L. Akella, 1988, Calculation of pre-exponential term in kinetic rate expression, Chem. Eng. Ed. (Summer), 150.

SUGGESTED REFERENCES FOR FURTHER STUDY

Aris, R., 1969, Elementary Chemical Reactor Analysis, Prentice-Hall, Englewood Cliffs, New Jersey.

Fahien, R.W., 1983, Fundamentals of Transport Phenomena, McGraw-Hill, New York.

Sandler, S.I., 2006, Chemical, Biochemical, and Engineering Thermodynamics, 4th Ed., Wiley, New York.

Smith, J.C., H.C. Van Ness and M.M. Abbott, 2001, Introduction to Chemical Engineering Thermodynamics, 6th Ed., McGraw-Hill, New York.

6

STEADY-STATE MACROSCOPIC BALANCES

The use of correlations in the determination of momentum, energy and mass transfer from one phase to another under steady-state conditions is covered in Chapter 4. Although some examples in Chapter 4 make use of steady-state macroscopic balances, systematic treatment of these balances for the conservation of chemical species, mass, and energy is not presented. The basic steps in the development of steady-state macroscopic balances are as follows:

- *Define your system:* A system is any region that occupies a volume and has a boundary.
- *If possible, draw a simple sketch:* A simple sketch helps in the understanding of the physical picture.
- *List the assumptions:* Simplify the complicated problem to a mathematically tractable form by making reasonable assumptions.
- *Write down the inventory rate equation for each of the basic concepts relevant to the problem at hand:* Since the accumulation term vanishes for steady-state cases, macroscopic inventory rate equations reduce to algebraic equations. Note that in order to have a mathematically determinate system the number of independent inventory rate equations must be equal to the number of dependent variables.
- *Use engineering correlations to evaluate the transfer coefficients:* In macroscopic modeling, empirical equations that represent transfer phenomena from one phase to another contain transfer coefficients, such as the heat transfer coefficient in Newton's law of cooling. These coefficients can be evaluated by using the engineering correlations given in Chapter 4.
- *Solve the algebraic equations.*

6.1 CONSERVATION OF CHEMICAL SPECIES

The inventory rate equation given by Eq. (1.1-1) holds for every conserved quantity φ. Therefore, the conservation statement for the mass of the ith chemical species under steady conditions is given by

$$\begin{pmatrix} \text{Rate of mass} \\ \text{of } i \text{ in} \end{pmatrix} - \begin{pmatrix} \text{Rate of mass} \\ \text{of } i \text{ out} \end{pmatrix} + \begin{pmatrix} \text{Rate of generation} \\ \text{of mass } i \end{pmatrix} = 0 \qquad (6.1-1)$$

The mass of i may enter or leave the system by two means: (i) by inlet or outlet streams, (ii) by exchange of mass between the system and its surroundings through the boundaries of the system, i.e., interphase mass transfer.

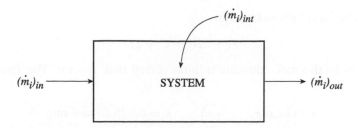

Figure 6.1. Steady-state flow system with fixed boundaries.

For a system with a single inlet and a single outlet stream as shown in Figure 6.1, Eq. (6.1-1) can be expressed as

$$(\dot{m}_i)_{in} - (\dot{m}_i)_{out} \pm (\dot{m}_i)_{int} + \left(\sum_j \alpha_{ij} r_j \right) \mathcal{M}_i V_{sys} = 0 \tag{6.1-2}$$

in which the molar rate of generation of species i per unit volume, $\mathfrak{R}_i$, is expressed by Eq. (5.3-27). The terms $(\dot{m}_i)_{in}$ and $(\dot{m}_i)_{out}$ represent the inlet and outlet mass flow rates of species i, respectively, and $\mathcal{M}_i$ is the molecular weight of species i. The interphase mass transfer rate, $(\dot{m}_i)_{int}$, is expressed as

$$(\dot{m}_i)_{int} = A_M \langle k_c \rangle (\Delta c_i)_{ch} \mathcal{M}_i \tag{6.1-3}$$

where $(\Delta c_i)_{ch}$ is the characteristic concentration difference. Note that $(\dot{m}_i)_{int}$ is considered *positive* when mass is added to the system.

As stated in Section 2.4.1, the mass flow rate of species i, $\dot{m}_i$, is given by

$$\dot{m}_i = \rho_i \langle v \rangle A = \rho_i \mathcal{Q} \tag{6.1-4}$$

Therefore, Eq. (6.1-2) takes the form

$$(\mathcal{Q}\rho_i)_{in} - (\mathcal{Q}\rho_i)_{out} \pm A_M \langle k_c \rangle (\Delta c_i)_{ch} \mathcal{M}_i + \left(\sum_j \alpha_{ij} r_j \right) \mathcal{M}_i V_{sys} = 0 \tag{6.1-5}$$

Sometimes it is more convenient to work on a molar basis. Division of Eqs. (6.1-2) and (6.1-5) by the molecular weight of species i, $\mathcal{M}_i$, gives

$$(\dot{n}_i)_{in} - (\dot{n}_i)_{out} \pm (\dot{n}_i)_{int} + \left(\sum_j \alpha_{ij} r_j \right) V_{sys} = 0 \tag{6.1-6}$$

and

$$(\mathcal{Q}c_i)_{in} - (\mathcal{Q}c_i)_{out} \pm A_M \langle k_c \rangle (\Delta c_i)_{ch} + \left(\sum_j \alpha_{ij} r_j \right) V_{sys} = 0 \tag{6.1-7}$$

where $\dot{n}_i$ and c_i are the molar flow rate and molar concentration of species i, respectively.

Example 6.1 The liquid phase reaction

$$A + 2B \rightarrow C + 2D$$

takes place in an isothermal, constant-volume stirred tank reactor. The rate of reaction is expressed by

$$r = kc_A c_B \qquad with \qquad k = 0.025 \text{ L/mol·min}$$

The feed stream consists of equal concentrations of species A and B at a value of 1 mol/L. Determine the residence time required to achieve 60% conversion of species B under steady conditions.

Solution

Assumption

1. As a result of perfect mixing, concentrations of species within the reactor are uniform, i.e., $(c_i)_{out} = (c_i)_{sys}$.

Analysis

System: Contents of the reactor

Since the reactor volume is constant, the inlet and outlet volumetric flow rates are the same and equal to Q. Therefore, the inventory rate equation for conservation of species B, Eq. (6.1-7), becomes

$$Q(c_B)_{in} - Q(c_B)_{sys} - \left[2k(c_A)_{sys}(c_B)_{sys}\right]V_{sys} = 0 \tag{1}$$

where $(c_A)_{sys}$ and $(c_B)_{sys}$ represent the molar concentration of species A and B in the reactor, respectively. Dropping the subscript "sys" and defining the residence time, τ, as $\tau = V/Q$ reduces Eq. (1) to

$$(c_B)_{in} - c_B - (2kc_A c_B)\tau = 0 \tag{2}$$

or,

$$\tau = \frac{(c_B)_{in} - c_B}{2kc_A c_B} \tag{3}$$

Using Eq. (5.3-17), the extent of the reaction can be calculated as

$$\xi = \frac{(c_B)_{in}}{(-\alpha_B)} X_B = \frac{(1)(0.6)}{2} = 0.3 \text{ mol/L} \tag{4}$$

Therefore, the concentrations of species A and B in the reactor are

$$c_A = (c_A)_{in} + \alpha_A \xi = 1 - 0.3 = 0.7 \text{ mol/L} \tag{5}$$

$$c_B = (c_B)_{in} + \alpha_B \xi = 1 - (2)(0.3) = 0.4 \text{ mol/L} \tag{6}$$

Substitution of the numerical values into Eq. (3) gives

$$\tau = \frac{1 - 0.4}{(2)(0.025)(0.7)(0.4)} = 42.9 \text{ min}$$

6.2   CONSERVATION OF MASS

Summation of Eq. (6.1-2) over all species gives the total mass balance in the form

$$\boxed{\dot{m}_{in} - \dot{m}_{out} \pm \dot{m}_{int} = 0}$$
(6.2-1)

Note that the term

$$\sum_i \left(\sum_j \alpha_{ij} r_j \right) \mathcal{M}_i = 0$$
(6.2-2)

since mass is conserved. Equation (6.2-2) implies that the rate of production of mass for the entire system is zero. However, if chemical reactions take place within the system, an individual species may be produced.

On the other hand, summation of Eq. (6.1-6) over all species gives the total mole balance as

$$\boxed{\dot{n}_{in} - \dot{n}_{out} \pm \dot{n}_{int} + \left[\sum_i \left(\sum_j \alpha_{ij} r_j \right) \right] V_{sys} = 0}$$
(6.2-3)

In this case the generation term is not zero because moles are not conserved.

Example 6.2 A liquid phase irreversible reaction

$$A \rightarrow B$$

takes place in a series of four continuous stirred tank reactors as shown in the figure below.

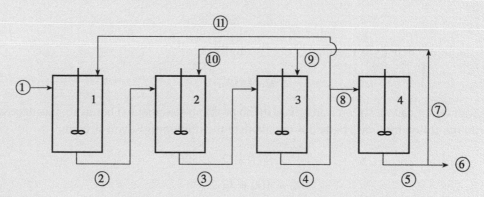

The rate of reaction is given by

$$r = kc_A \qquad \text{with} \qquad k = 3 \times 10^5 \exp\left(-\frac{4200}{T} \right)$$

in which k is in h^{-1} and T is in degrees Kelvin. The temperature and the volume of each reactor are given as follows:

Reactor	Temperature	Volume
No	(°C)	(L)
1	35	800
2	45	1000
3	70	1200
4	60	900

Determine the concentration of species A in each reactor if the feed to the first reactor contains 1.5 mol/L of A and the volumetric flow rates of the streams are given as follows:

Stream	Volumetric Flow Rate
No	(L/h)
1	500
7	200
9	50
11	100

Solution

Assumptions

1. Steady-state conditions prevail.
2. Concentrations of species within the reactor are uniform as a result of perfect mixing.
3. Liquid density remains constant.

Analysis

Conservation of total mass, Eq. (6.2-1), reduces to

$$\dot{m}_{in} = \dot{m}_{out} \tag{1}$$

Since the liquid density is constant, Eq. (1) simplifies to

$$Q_{in} = Q_{out} \tag{2}$$

Only four out of eleven streams are given in the problem statement. Therefore, it is necessary to write the following mass balances to calculate the remaining seven streams:

$$Q_1 = Q_6 = 500$$

$$500 + 100 = Q_2$$

$$Q_2 + Q_{10} = Q_3$$

$$Q_3 + 50 = Q_4$$

$$Q_8 = Q_5$$

$$Q_5 = Q_6 + 200$$

$$200 = 50 + Q_{10}$$

Simultaneous solution of the above equations gives the volumetric flow rate of each stream as:

Stream No	Volumetric Flow Rate (L/h)
1	500
2	600
3	750
4	800
5	700
6	500
7	200
8	700
9	50
10	150
11	100

For each reactor, the reaction rate constant is

$$k = 3 \times 10^5 \exp\left[-\frac{4200}{(35 + 273)}\right] = 0.359 \text{ h}^{-1} \qquad \text{for reactor \# 1}$$

$$k = 3 \times 10^5 \exp\left[-\frac{4200}{(45 + 273)}\right] = 0.551 \text{ h}^{-1} \qquad \text{for reactor \# 2}$$

$$k = 3 \times 10^5 \exp\left[-\frac{4200}{(70 + 273)}\right] = 1.443 \text{ h}^{-1} \qquad \text{for reactor \# 3}$$

$$k = 3 \times 10^5 \exp\left[-\frac{4200}{(60 + 273)}\right] = 0.999 \text{ h}^{-1} \qquad \text{for reactor \# 4}$$

For each reactor, the conservation statement for species $\mathcal{A}$, Eq. (6.1-7), can be written in the form

$$(500)(1.5) + 100c_{A_3} - 600c_{A_1} - (0.359c_{A_1})(800) = 0$$

$$600c_{A_1} + 150c_{A_4} - 750c_{A_2} - (0.551c_{A_2})(1000) = 0$$

$$750c_{A_2} + 50c_{A_4} - 800c_{A_3} - (1.443c_{A_3})(1200) = 0$$

$$700c_{A_3} - 700c_{A_4} - (0.999c_{A_4})(900) = 0$$

Simplification gives

$$8.872c_{A_1} - c_{A_3} = 7.5$$

$$4c_{A_1} - 8.673c_{A_2} + c_{A_4} = 0$$

$$15c_{A_2} - 50.632c_{A_3} + c_{A_4} = 0$$

$$c_{A_3} - 2.284c_{A_4} = 0$$

The above equations are written in matrix notation[1] as

$$
\begin{bmatrix}
8.872 & 0 & -1 & 0 \\
4 & -8.673 & 0 & 1 \\
0 & 15 & -50.632 & 1 \\
0 & 0 & 1 & -2.284
\end{bmatrix}
\begin{bmatrix}
c_{A_1} \\
c_{A_2} \\
c_{A_3} \\
c_{A_4}
\end{bmatrix}
=
\begin{bmatrix}
7.5 \\
0 \\
0 \\
0
\end{bmatrix}
$$

Therefore, the solution is

$$
\begin{bmatrix}
c_{A_1} \\
c_{A_2} \\
c_{A_3} \\
c_{A_4}
\end{bmatrix}
=
\begin{bmatrix}
8.872 & 0 & -1 & 0 \\
4 & -8.673 & 0 & 1 \\
0 & 15 & -50.632 & 1 \\
0 & 0 & 1 & -2.284
\end{bmatrix}^{-1}
\begin{bmatrix}
7.5 \\
0 \\
0 \\
0
\end{bmatrix}
$$

$$
=
\begin{bmatrix}
0.115 & -0.004 & -0.002 & -0.003 \\
0.054 & -0.119 & -0.002 & -0.053 \\
0.016 & -0.036 & -0.021 & -0.025 \\
0.007 & -0.016 & -0.009 & -0.449
\end{bmatrix}
\begin{bmatrix}
7.5 \\
0 \\
0 \\
0
\end{bmatrix}
$$

The multiplication gives the concentrations in each reactor as

$$
\begin{bmatrix}
c_{A_1} \\
c_{A_2} \\
c_{A_3} \\
c_{A_4}
\end{bmatrix}
=
\begin{bmatrix}
0.859 \\
0.402 \\
0.120 \\
0.053
\end{bmatrix}
$$

6.3 CONSERVATION OF ENERGY

The conservation statement for *total* energy under steady conditions takes the form

$$
\begin{pmatrix} \text{Rate of} \\ \text{energy in} \end{pmatrix} - \begin{pmatrix} \text{Rate of} \\ \text{energy out} \end{pmatrix} + \begin{pmatrix} \text{Rate of energy} \\ \text{generation} \end{pmatrix} = 0 \tag{6.3-1}
$$

The first law of thermodynamics states that total energy can be neither created nor destroyed. Therefore, the rate of generation term in Eq. (6.3-1) equals zero.

Energy may enter or leave the system by two means: (*i*) by inlet and/or outlet streams, (*ii*) by exchange of energy between the system and its surroundings through the boundaries of the system in the form of heat and work.

For a system with a single inlet and a single outlet stream as shown in Figure 6.2, Eq. (6.3-1) can be expressed as

$$
(\dot{E}_{in} + \dot{Q}_{int} + \dot{W}) - \dot{E}_{out} = 0 \tag{6.3-2}
$$

where the interphase heat transfer rate, $\dot{Q}_{int}$, is expressed as

$$
\dot{Q}_{int} = A_H \langle h \rangle (\Delta T)_{ch} \tag{6.3-3}
$$

---

[1] Matrix operations are given in Section A.9 in Appendix A.

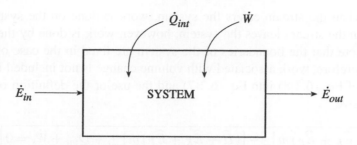

Figure 6.2. Steady-state flow system with fixed boundaries interchanging energy in the form of heat and work with the surroundings.

in which $(\Delta T)_{ch}$ is the characteristic temperature difference. Note that $\dot{Q}_{int}$ is considered positive when energy is added to the system. Similarly, $\dot{W}$ is also considered positive when work is done on the system.

As stated in Section 2.4.2, the rate of energy entering or leaving the system, $\dot{E}$, is expressed as

$$\dot{E} = \widehat{E}\dot{m} \qquad (6.3\text{-}4)$$

Therefore, Eq. (6.3-2) becomes

$$(\widehat{E}\dot{m})_{in} - (\widehat{E}\dot{m})_{out} + \dot{Q}_{int} + \dot{W} = 0 \qquad (6.3\text{-}5)$$

To determine the total energy per unit mass, $\widehat{E}$, consider an astronaut on the space shuttle *Atlantis*. When the astronaut looks at the earth, (s)he sees that the earth has an external kinetic energy due to its rotation and its motion around the sun. The earth also has an internal kinetic energy as a result of all the objects, i.e., people, cars, planes, etc., moving on its surface that the astronaut cannot see. A physical object is usually composed of smaller objects, each of which can have a variety of internal and external energies. The sum of the internal and external energies of the smaller objects is usually apparent as internal energy of the larger objects.

The above discussion indicates that the total energy of any system is expressed as the sum of its internal and external energies. Kinetic and potential energies constitute the external energy, while the energy associated with the translational, rotational, and vibrational motion of molecules and atoms is considered the internal energy. Therefore, total energy per unit mass can be expressed as

$$\widehat{E} = \widehat{U} + \widehat{E}_K + \widehat{E}_P \qquad (6.3\text{-}6)$$

where $\widehat{U}$, $\widehat{E}_K$, and $\widehat{E}_P$ represent internal, kinetic, and potential energies per unit mass, respectively. Substitution of Eq. (6.3-6) into Eq. (6.3-5) gives

$$\left[\left(\widehat{U} + \widehat{E}_K + \widehat{E}_P\right)\dot{m}\right]_{in} - \left[\left(\widehat{U} + \widehat{E}_K + \widehat{E}_P\right)\dot{m}\right]_{out} + \dot{Q}_{int} + \dot{W} = 0 \qquad (6.3\text{-}7)$$

The rate of work done on the system by the surroundings is given by

$$\dot{W} = \underbrace{\dot{W}_s}_{\text{Shaft work}} + \underbrace{(P\widehat{V}\dot{m})_{in} - (P\widehat{V}\dot{m})_{out}}_{\text{Flow work}} \qquad (6.3\text{-}8)$$

In Figure 6.2, when the stream enters the system, work is done on the system by the surroundings. When the stream leaves the system, however, work is done by the system on the surroundings. Note that the boundaries of the system are fixed in the case of a steady-state flow system. Therefore, work associated with volume change is not included in Eq. (6.3-8).

Substitution of Eq. (6.3-8) into Eq. (6.3-7) and the use of the definition of enthalpy, i.e., $\widehat{H} = \widehat{U} + P\widehat{V}$, give

$$\boxed{\left[\left(\widehat{H} + \widehat{E}_K + \widehat{E}_P\right)\dot{m}\right]_{in} - \left[\left(\widehat{H} + \widehat{E}_K + \widehat{E}_P\right)\dot{m}\right]_{out} + \dot{Q}_{int} + \dot{W}_s = 0} \tag{6.3-9}$$

which is known as the *steady-state energy equation*.

The kinetic and potential energy terms in Eq. (6.3-9) are expressed in the form

$$\widehat{E}_K = \frac{1}{2}v^2 \tag{6.3-10}$$

and

$$\widehat{E}_P = gh \tag{6.3-11}$$

where g is the acceleration of gravity and h is the elevation with respect to a reference plane.

Enthalpy, on the other hand, depends on temperature and pressure. Change in enthalpy is expressed by

$$d\widehat{H} = \widehat{C}_P\,dT + \widehat{V}(1 - \beta T)\,dP \tag{6.3-12}$$

where β is the *coefficient of volume expansion* and is defined by

$$\beta = \frac{1}{\widehat{V}}\left(\frac{\partial \widehat{V}}{\partial T}\right)_P = -\frac{1}{\rho}\left(\frac{\partial \rho}{\partial T}\right)_P \tag{6.3-13}$$

Note that

$$\beta = \begin{cases} 0 & \text{for an incompressible fluid} \\ 1/T & \text{for an ideal gas} \end{cases} \tag{6.3-14}$$

When the changes in the kinetic and potential energies between the inlet and outlet of the system are negligible, Eq. (6.3-9) reduces to

$$\boxed{(\widehat{H}\dot{m})_{in} - (\widehat{H}\dot{m})_{out} + \dot{Q}_{int} + \dot{W}_s = 0} \tag{6.3-15}$$

In terms of molar quantities, Eqs. (6.3-9) and (6.3-15) are written as

$$\boxed{\left[\left(\widetilde{H} + \widetilde{E}_K + \widetilde{E}_P\right)\dot{n}\right]_{in} - \left[\left(\widetilde{H} + \widetilde{E}_K + \widetilde{E}_P\right)\dot{n}\right]_{out} + \dot{Q}_{int} + \dot{W}_s = 0} \tag{6.3-16}$$

and

$$\boxed{(\widetilde{H}\dot{n})_{in} - (\widetilde{H}\dot{n})_{out} + \dot{Q}_{int} + \dot{W}_s = 0} \tag{6.3-17}$$

6.3.1 Energy Equation Without a Chemical Reaction

In the case of no chemical reaction, Eqs. (6.3-9) and (6.3-16) are used to determine energy interactions. If kinetic and potential energy changes are negligible, then these equations reduce to Eqs. (6.3-15) and (6.3-17), respectively. The use of the energy equation requires the enthalpy change to be known or calculated. For some substances, such as steam and ammonia, enthalpy values are either tabulated or given in the form of a graph as a function of temperature and pressure. In that case enthalpy changes can be determined easily. If enthalpy values are not tabulated, then the determination of enthalpy depending on the values of temperature and pressure in a given process is given below.

6.3.1.1 *Constant pressure and no phase change*   Since $dP = 0$, integration of Eq. (6.3-12) gives

$$\widehat{H} = \int_{T_{ref}}^{T} \widehat{C}_P \, dT \tag{6.3-18}$$

in which $\widehat{H}$ is taken as zero at T_{ref}. Substitution of Eq. (6.3-18) into Eq. (6.3-15) gives

$$\dot{m}_{in}\left(\int_{T_{ref}}^{T_{in}} \widehat{C}_P \, dT\right) - \dot{m}_{out}\left(\int_{T_{ref}}^{T_{out}} \widehat{C}_P \, dT\right) + \dot{Q}_{int} + \dot{W}_s = 0 \tag{6.3-19}$$

If $\widehat{C}_P$ is independent of temperature, Eq. (6.3-19) reduces to

$$\dot{m}_{in}\widehat{C}_P(T_{in} - T_{ref}) - \dot{m}_{out}\widehat{C}_P(T_{out} - T_{ref}) + \dot{Q}_{int} + \dot{W}_s = 0 \tag{6.3-20}$$

Example 6.3 It is required to cool a gas composed of 75 mole % N_2, 15% CO_2, and 10% O_2 from 800 °C to 350 °C. Determine the cooling duty of the heat exchanger if the heat capacity expressions are in the form

$$\widetilde{C}_P (\text{J/mol·K}) = a + bT + cT^2 + dT^3 \qquad T [=] K$$

where the coefficients a, b, c, and d are given by

Species	a	$b \times 10^2$	$c \times 10^5$	$d \times 10^5$
N_2	28.882	−0.1570	0.8075	−2.8706
O_2	25.460	1.5192	−0.7150	1.3108
CO_2	21.489	5.9768	−3.4987	7.4643

Solution

Assumptions

1. Ideal gas behavior.
2. Changes in kinetic and potential energies are negligible.
3. Pressure drop in the heat exchanger is negligible.

Analysis

System: Gas stream in the heat exchanger

Since $\dot{n}_{int} = 0$ and there is no chemical reaction, Eq. (6.2-3) reduces to

$$\dot{n}_{in} = \dot{n}_{out} = \dot{n} \tag{1}$$

Therefore, Eq. (6.3-19) becomes

$$\dot{Q}_{int} = \dot{n}\left(\int_{T_{ref}}^{T_{out}} \widetilde{C}_P\, dT - \int_{T_{ref}}^{T_{in}} \widetilde{C}_P\, dT\right) = \dot{n}\left(\int_{T_{in}}^{T_{out}} \widetilde{C}_P\, dT\right) \tag{2}$$

or,

$$\widetilde{Q}_{int} = \int_{T_{in}}^{T_{out}} \widetilde{C}_P\, dT \tag{3}$$

where $\widetilde{Q}_{int} = \dot{Q}_{int}/\dot{n}$, $T_{in} = 1073$ K, and $T_{out} = 623$ K.

The molar heat capacity of the gas stream, $\widetilde{C}_P$, can be calculated by multiplying the mole fraction of each component by the respective heat capacity and adding them together, i.e.,

$$\widetilde{C}_P = \sum_{i=1}^{3} x_i (a_i + b_i T + c_i T^2 + d_i T^3)$$

$$= 27.431 + 0.931 \times 10^{-2} T + 0.009 \times 10^{-5} T^2 - 0.902 \times 10^{-9} T^3 \tag{4}$$

Substitution of Eq. (4) into Eq. (3) and integration give

$$\widetilde{Q}_{int} = -15{,}662 \text{ J/mol}$$

The minus sign indicates that heat must be removed from the gas stream.

6.3.1.2 *Constant pressure with phase change* When we start heating a substance at constant pressure, a typical variation in temperature as a function of time is given in Figure 6.3.

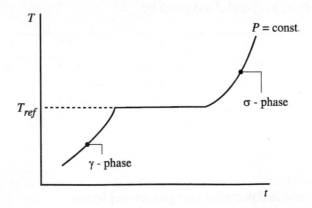

Figure 6.3. Temperature-time relationship as the substance transforms from the γ-phase to the σ-phase.

Let T_{ref} be the temperature at which phase change from the γ-phase to the σ-phase, or vice versa, takes place. If we choose the γ-phase enthalpy as zero at the reference temperature, then enthalpies of the σ- and γ-phases at any given temperature T are given as

$$
\widehat{H} =
\begin{cases}
\displaystyle\int_{T_{ref}}^{T} (\widetilde{C}_P)_\sigma \, dT & \sigma\text{-phase} \\[4mm]
\displaystyle -\widehat{\lambda} - \int_{T}^{T_{ref}} (\widetilde{C}_P)_\gamma \, dT & \gamma\text{-phase}
\end{cases}
\tag{6.3-21}
$$

where $\widehat{\lambda} = \widehat{H}_\sigma - \widehat{H}_\gamma$ at the reference temperature.

Example 6.4 One way of cooling a can of cola on a hot summer day is to wrap a piece of wet cloth around the can and expose it to a gentle breeze. Calculate the steady-state temperature of the can if the air temperature is 35 °C.

Solution

Assumptions

1. Steady-state conditions prevail.
2. Ideal gas behavior.

Analysis

System: Wet cloth and the cola can

The inventory rate equation for energy becomes

$$
\text{Rate of energy in} = \text{Rate of energy out}
\tag{1}
$$

Let the steady-state temperature of the cloth and that of cola be T_w. The rate of energy entering the system is given by

$$
\text{Rate of energy in} = A_H \langle h \rangle (T_\infty - T_w)
\tag{2}
$$

in which A_H and T_∞ represent the heat transfer area and air temperature, respectively. On the other hand, the rate of energy leaving the system is expressed in the form

$$
\text{Rate of energy out} = \dot{n}_A \left[\widetilde{\lambda}_A + (\widetilde{C}_P)_A (T_\infty - T_w) \right]
\tag{3}
$$

where $\dot{n}_A$ represents the rate of moles of water, i.e., species $\mathcal{A}$, evaporated and is given by

$$
\dot{n}_A = A_M \langle k_c \rangle \left(c_{A_w} - c_{A_\infty} \right)
\tag{4}
$$

in which A_M represents the mass transfer area. Substitution of Eqs. (2), (3) and (4) into Eq. (1) and using

$$
A_H = A_M \qquad c_{A_\infty} \simeq 0 \qquad \widetilde{\lambda}_A \gg (\widetilde{C}_P)_A (T_\infty - T_w)
$$

give

$$
T_\infty - T_w = c_{A_w} \widetilde{\lambda}_A \left(\frac{\langle k_c \rangle}{\langle h \rangle} \right)
\tag{5}
$$

The ratio $\langle k_c \rangle / \langle h \rangle$ can be estimated by the use of the Chilton-Colburn analogy, i.e., $j_H = j_M$, as

$$\frac{\text{St}_H}{\text{St}_M} = \left(\frac{\text{Sc}}{\text{Pr}}\right)^{2/3} \quad \Rightarrow \quad \frac{\langle k_c \rangle}{\langle h \rangle} = \frac{1}{\rho \widehat{C}_P} \left(\frac{\text{Pr}}{\text{Sc}}\right)^{2/3} \tag{6}$$

The use of Eq. (6) in Eq. (5) yields

$$T_\infty - T_w = \frac{c_{A_w} \widetilde{\lambda}_A}{(\rho \widehat{C}_P)_B} \left(\frac{\text{Pr}}{\text{Sc}}\right)_B^{2/3} \tag{7}$$

where the properties ρ, $\widehat{C}_P$, Pr, and Sc belong to air, species $\mathcal{B}$. The concentration of species $\mathcal{A}$ at the interface, c_{A_w}, is given by

$$c_{A_w} = \frac{P_A^{sat}}{\mathcal{R} T_w} \tag{8}$$

It should be remembered that the quantities c_{A_w} and $\widetilde{\lambda}_A$ must be evaluated at T_w, whereas ρ_B, $\widehat{C}_{P_B}$, Pr_B, and Sc_B must be evaluated at $T_f = (T_w + T_\infty)/2$. Since T_w is unknown, a trial-and-error procedure will be used in order to determine T_w as follows:

Step 1: Assume $T_w = 15\,°\text{C}$

Step 2: Determine the physical properties:

For water at $15\,°\text{C}$ (288 K): $\begin{cases} P_A^{sat} = 0.01703 \text{ bar} \\ \widetilde{\lambda}_A = 2466 \times 18 = 44{,}388 \text{ kJ/kmol} \end{cases}$

The saturation concentration is

$$c_{A_w} = \frac{P_A^{sat}}{\mathcal{R} T_w} = \frac{0.01703}{(8.314 \times 10^{-2})(15 + 273)} = 7.11 \times 10^{-4} \text{ kmol/m}^3$$

The film temperature is $T_f = (35 + 15)/2 = 25\,°\text{C}$.

For air at $25\,°\text{C}$ (298 K): $\begin{cases} \rho = 1.1845 \text{ kg/m}^3 \\ \nu = 15.54 \times 10^{-6} \text{ m}^2/\text{s} \\ \widehat{C}_P = 1.005 \text{ kJ/kg·K} \\ \text{Pr} = 0.712 \end{cases}$

The diffusion coefficient of water in air is

$$\mathcal{D}_{AB} = (2.88 \times 10^{-5}) \left(\frac{298}{313}\right)^{3/2} = 2.68 \times 10^{-5} \text{ m}^2/\text{s}$$

The Schmidt number is

$$\text{Sc} = \frac{\nu}{\mathcal{D}_{AB}} = \frac{15.54 \times 10^{-6}}{2.68 \times 10^{-5}} = 0.58$$

Step 3: Substitute the values into Eq. (7) and check whether the right- and left-hand sides are equal to each other:

$$T_\infty - T_w = 35 - 15 = 20$$

$$\frac{c_{A_w} \widetilde{\lambda}_A}{(\rho \widehat{C}_P)_B} \left(\frac{\mathrm{Pr}}{\mathrm{Sc}}\right)_B^{2/3} = \frac{(7.11 \times 10^{-4})(44{,}388)}{(1.1845)(1.005)} \left(\frac{0.712}{0.58}\right)^{2/3} = 30.4$$

Since the left- and right-hand sides of Eq. (7) are quite different from each other, another value of T_w should be assumed.

Assume $T_w = 11\,°C$

For water at $11\,°C$ (284 K): $\begin{cases} P_A^{sat} = 0.01308 \text{ bar} \\ \widetilde{\lambda}_A = 2475.4 \times 18 = 44{,}557 \text{ kJ/kmol} \end{cases}$

The saturation concentration is

$$c_{A_w} = \frac{P_A^{sat}}{\mathcal{R} T_w} = \frac{0.01308}{(8.314 \times 10^{-2})(11 + 273)} = 5.54 \times 10^{-4} \text{ kmol/m}^3$$

The film temperature is $T_f = (35 + 11)/2 = 23\,°C$.

For air at $23\,°C$ (296 K): $\begin{cases} \rho = 1.1926 \text{ kg/m}^3 \\ \nu = 15.36 \times 10^{-6} \text{ m}^2/\text{s} \\ \widehat{C}_P = 1.005 \text{ kJ/kg·K} \\ \mathrm{Pr} = 0.713 \end{cases}$

The diffusion coefficient of water in air is

$$\mathcal{D}_{AB} = (2.88 \times 10^{-5}) \left(\frac{296}{313}\right)^{3/2} = 2.65 \times 10^{-5} \text{ m}^2/\text{s}$$

The Schmidt number is

$$\mathrm{Sc} = \frac{\nu}{\mathcal{D}_{AB}} = \frac{15.36 \times 10^{-6}}{2.65 \times 10^{-5}} = 0.58$$

The left- and right-hand sides of Eq. (7) now become

$$T_\infty - T_w = 35 - 11 = 24$$

$$\frac{c_{A_w} \widetilde{\lambda}_A}{(\rho \widehat{C}_P)_B} \left(\frac{\mathrm{Pr}}{\mathrm{Sc}}\right)_B^{2/3} = \frac{(5.54 \times 10^{-4})(44{,}557)}{(1.1926)(1.005)} \left(\frac{0.713}{0.58}\right)^{2/3} = 23.6$$

Therefore, the steady-state temperature is $11\,°C$.

Comment: Whenever a gas flows over a liquid, the temperature of the liquid decreases as a result of evaporation. This process is known as evaporative cooling. The resulting steady-state temperature, on the other hand, is called the wet-bulb temperature.

6.3.1.3 *Variable pressure and no phase change* Enthalpy of an ideal gas is dependent only on temperature and is expressed by Eq. (6.3-18). Therefore, in problems involving ideal gases, variation in pressure has no effect on the enthalpy change. In the case of incompressible fluids, Eq. (6.3-12) reduces to

$$\widehat{H} = \int_{T_{ref}}^{T} \widehat{C}_P \, dT + \widehat{V}(P - P_{ref}) \qquad (6.3-22)$$

in which the enthalpy is taken as zero at the reference temperature and pressure. At low and moderate pressures, the second term on the right-hand side of Eq. (6.3-22) is usually considered negligible.

Example 6.5 A certain process requires a steady supply of compressed air at 600 kPa and 50 °C at the rate of 0.2 kg/s. For this purpose, air at ambient conditions of 100 kPa and 20 °C is first compressed to 600 kPa in an adiabatic compressor, and then it is fed to a heat exchanger where it is cooled to 50 °C at constant pressure. As cooling medium, water is used and it enters the heat exchanger at 15 °C and leaves at 40 °C. Determine the mass flow rate of water if the rate of work done on the compressor is 44 kJ/s.

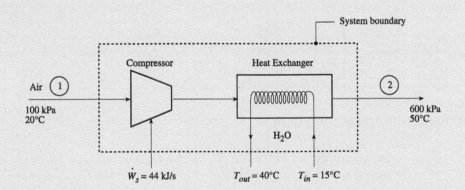

Solution

Assumptions

1. Steady-state conditions prevail.
2. Changes in kinetic and potential energies are negligible.
3. There is no heat loss from the heat exchanger to the surroundings.
4. Heat capacities of air and water remain essentially constant at the values of 1 kJ/kg·K and 4.178 kJ/kg·K, respectively.

Analysis

System: Compressor and heat exchanger

Conservation of total mass, Eq. (6.2-1), reduces to

$$\dot{m}_1 = \dot{m}_2 = \dot{m} \tag{1}$$

Therefore, Eq. (6.3-15) becomes

$$\dot{m}_{air}(\widehat{H}_1 - \widehat{H}_2)_{air} - \dot{Q}_{int} + \dot{W}_s = 0 \tag{2}$$

in which the enthalpy change of the air and the interphase heat transfer rate are given by

$$(\widehat{H}_1 - \widehat{H}_2)_{air} = (\widehat{C}_P)_{air}(T_1 - T_2)_{air} \tag{3}$$

$$\dot{Q}_{int} = (\dot{m}\widehat{C}_P)_{H_2O}(T_{out} - T_{in})_{H_2O} \tag{4}$$

Substitution of Eqs. (3) and (4) into Eq. (2) and rearrangement give

$$\dot{m}_{H_2O} = \frac{(\dot{m}\widehat{C}_P)_{air}(T_1 - T_2)_{air} + \dot{W}_s}{(\widehat{C}_P)_{H_2O}(T_{out} - T_{in})_{H_2O}} = \frac{(0.2)(1)(20 - 50) + 44}{(4.178)(40 - 15)} = 0.364 \text{ kg/s} \tag{5}$$

Comment: The definition of a system plays a crucial role in the solution of the problem. Note that there is no need to find out the temperature or pressure at the exit of the compressor. If, however, one chooses the compressor and heat exchanger as two separate systems, then the pressure and temperature at the exit of the compressor must be calculated.

6.3.2 Energy Equation with a Chemical Reaction

6.3.2.1 *Thermochemistry* Thermochemistry deals with the changes in energy in chemical reactions. The difference between the enthalpy of one mole of a pure compound and the total enthalpy of the elements of which it is composed is called the *heat of formation*, $\Delta \widetilde{H}_f$, of the compound. The *standard heat of formation*, $\Delta \widetilde{H}_f^o$, is the heat of formation when both the compound and its elements are at standard conditions as shown in Figure 6.4. The superscript ^{o} implies the *standard state*. Since enthalpy is a state function, it is immaterial whether or not the reaction could take place at standard conditions.

The standard state is usually taken as the stable form of the element or compound at the temperature of interest, T, and under 1 atm (1.013 bar). Therefore, the word *standard* refers not to any particular temperature, but to unit pressure of 1 atm. The elements in their standard states are taken as the reference state and are assigned an enthalpy of zero. The standard heat of formation of many compounds is usually tabulated at 25 °C and can readily be found in Perry's Chemical Engineers' Handbook (1997) and thermodynamics textbooks. For example, the standard heat of formation of ethyl benzene, C_8H_{10}, in the gaseous state is 29,790 J/mol at 298 K. Consider the formation of ethyl benzene from its elements by the reaction

$$8C(s) + 5H_2(g) = C_8H_{10}(g)$$

The standard heat of formation is given by

$$\left(\Delta \widetilde{H}_f^o\right)_{C_8H_{10}} = \widetilde{H}_{C_8H_{10}}^o - 8\widetilde{H}_C^o - 5\widetilde{H}_{H_2}^o = 29,790 \text{ J/mol}$$

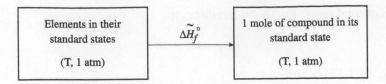

Figure 6.4. Calculation of the standard heat of formation, $\Delta\tilde{H}_f^o$.

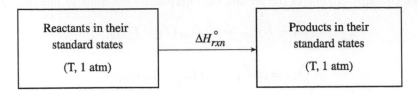

Figure 6.5. Calculation of the standard heat of reaction, ΔH_{rxn}^o.

Since $\tilde{H}_C^o = \tilde{H}_{H_2}^o = 0$, it follows that

$$\left(\Delta\tilde{H}_f^o\right)_{C_8H_{10}} = \tilde{H}_{C_8H_{10}}^o = 29{,}790 \text{ J/mol}$$

It is possible to generalize this result in the form

$$\boxed{\left(\Delta\tilde{H}_f^o\right)_i = \tilde{H}_i^o} \tag{6.3-23}$$

The standard heat of formation of a substance is just the standard heat of reaction in which one mole of it is formed from elementary species. Therefore, the *standard heat of reaction*, ΔH_{rxn}^o, is the difference between the total enthalpy of the pure product mixture and that of the pure reactant mixture at standard conditions as shown in Figure 6.5.

The standard heat of reaction can be calculated as

$$\Delta H_{rxn}^o = \sum_i \alpha_i^o \tilde{H}_i^o \tag{6.3-24}$$

Substitution of Eq. (6.3-23) into Eq. (6.3-24) gives

$$\Delta H_{rxn}^o = \sum_i \alpha_i \left(\Delta\tilde{H}_f^o\right)_i \tag{6.3-25}$$

Note that the standard heat of formation of an element is zero.

If heat is evolved in the reaction, the reaction is called *exothermic*. If heat is absorbed, the reaction is called *endothermic*. Therefore,

$$\Delta H_{rxn}^o \begin{cases} > 0 & \text{for an endothermic reaction} \\ < 0 & \text{for an exothermic reaction} \end{cases} \tag{6.3-26}$$

If the standard heat of reaction is known at 298 K, then its value at any other temperature can be found as follows: The variation of the standard heat of reaction as a function of temperature under constant pressure is given by

$$d\Delta H^o_{rxn} = \left(\frac{\partial \Delta H^o_{rxn}}{\partial T}\right)_{P=1} dT \tag{6.3-27}$$

The term $(\partial \Delta H^o_{rxn}/\partial T)_P$ can be expressed as

$$\left(\frac{\partial \Delta H^o_{rxn}}{\partial T}\right)_P = \frac{\partial}{\partial T}\left(\sum_i \alpha^o_i \widetilde{H}^o_i\right) = \sum_i \alpha_i \left(\frac{\partial \widetilde{H}^o_i}{\partial T}\right)_P = \sum_i \alpha_i \widetilde{C}^o_{P_i} = \Delta \widetilde{C}^o_P \tag{6.3-28}$$

Substitution of Eq. (6.3-28) into Eq. (6.3-27) and integration give

$$\boxed{\Delta H^o_{rxn}(T) = \Delta H^o_{rxn}(T = 298 \text{ K}) + \int_{298}^T \Delta \widetilde{C}^o_P \, dT} \tag{6.3-29}$$

6.3.2.2 *Energy balance around a continuous stirred tank reactor* An energy balance in a continuous stirred tank reactor (CSTR) with the following assumptions is a good example of the energy balance with a chemical reaction:

1. Steady-state conditions prevail.
2. Stirring does not contribute much energy to the system, i.e., $\dot{W}_s \simeq 0$.
3. Volume of the system is constant, i.e., inlet and outlet volumetric flow rates are equal.
4. As a result of perfect mixing, the temperature and concentration of the system are uniform, i.e., $c_{out} = c_{sys}$ and $T_{out} = T_{sys}$.
5. Changes in kinetic and potential energies are negligible.

Since a chemical reaction is involved in this case, it is more appropriate to work on a molar basis. Therefore, Eq. (6.3-17) simplifies to

$$(\widetilde{H}\dot{n})_{in} - (\widetilde{H}\dot{n})_{out} + \dot{Q}_{int} = 0 \tag{6.3-30}$$

Any molar quantity of a mixture, $\widetilde{\psi}$, can be expressed in terms of partial molar quantities[2], $\overline{\psi}_i$, as

$$\widetilde{\psi} = \sum_i x_i \overline{\psi}_i \tag{6.3-31}$$

Multiplication of Eq. (6.3-31) by molar flow rate, $\dot{n}$, gives

$$\widetilde{\psi}\dot{n} = \sum_i \dot{n}_i \overline{\psi}_i \tag{6.3-32}$$

Therefore, Eq. (6.3-30) is expressed as

$$\left[\sum_i \dot{n}_i \overline{H}_i(T_{in})\right]_{in} - \left[\sum_i \dot{n}_i \overline{H}_i(T)\right]_{out} + \dot{Q}_{int} = 0 \tag{6.3-33}$$

---

[2]Partial molar quantities, unlike molar quantities of pure substances, depend also on the composition of the mixture.

On the other hand, the macroscopic mole balance for species i, Eq. (6.1-6), is

$$(\dot{n}_i)_{in} - (\dot{n}_i)_{out} + V_{sys} \sum_j \alpha_{ij} r_j = 0 \qquad (6.3\text{-}34)$$

Multiplication of Eq. (6.3-34) by $\overline{H}_i(T)$ and summation over all species give

$$\left[\sum_i \dot{n}_i \overline{H}_i(T) \right]_{in} - \left[\sum_i \dot{n}_i \overline{H}_i(T) \right]_{out} - V_{sys} \sum_j r_j (-\Delta H_{rxn,j}) = 0 \qquad (6.3\text{-}35)$$

where the heat of reaction is defined by

$$\Delta H_{rxn,j} = \sum_i \alpha_{ij} \overline{H}_i(T) \qquad (6.3\text{-}36)$$

Subtraction of Eq. (6.3-35) from Eq. (6.3-33) yields

$$\left\{ \sum_i \dot{n}_i \left[\overline{H}_i(T_{in}) - \overline{H}_i(T) \right] \right\}_{in} + \dot{Q}_{int} + V_{sys} \sum_j r_j (-\Delta H_{rxn,j}) = 0 \qquad (6.3\text{-}37)$$

Dividing Eq. (6.3-37) by the volumetric flow rate, Q, gives

$$\left\{ \sum_i c_i \left[\overline{H}_i(T_{in}) - \overline{H}_i(T) \right] \right\}_{in} + \frac{\dot{Q}_{int}}{Q} + \tau \sum_j r_j (-\Delta H_{rxn,j}) = 0 \qquad (6.3\text{-}38)$$

where τ is the residence time defined by

$$\tau = \frac{V_{sys}}{Q} \qquad (6.3\text{-}39)$$

The partial molar heat capacity of species i, $\overline{C}_{P_i}$, is related to the partial molar enthalpy as

$$\overline{C}_{P_i} = \left(\frac{\partial \overline{H}_i}{\partial T} \right)_P \qquad (6.3\text{-}40)$$

If $\overline{C}_{P_i}$ is independent of temperature, then integration of Eq. (6.3-40) gives

$$\overline{H}_i(T_{in}) - \overline{H}_i(T) = \overline{C}_{P_i}(T_{in} - T) \qquad (6.3\text{-}41)$$

Substitution of Eqs. (6.3-40) and (6.3-41) into Eq. (6.3-38) yields

$$\boxed{(C_P)_{in}(T_{in} - T) + \frac{\dot{Q}_{int}}{Q} + \tau \sum_j r_j (-\Delta H_{rxn,j}) = 0} \qquad (6.3\text{-}42)$$

where

$$(C_P)_{in} = \sum_i (c_i)_{in} \overline{C}_{P_i} \qquad (6.3\text{-}43)$$

It should be noted that the reaction rate expression in Eq. (6.3-42) contains a reaction rate constant, k, expressed in the form

$$k = Ae^{-\mathcal{E}/\mathcal{R}T} \tag{6.3-44}$$

Therefore, Eq. (6.3-42) is highly nonlinear in temperature.

Once the feed composition, stoichiometry and order of the chemical reaction, heat of reaction, and reaction rate constant are known, conservation statements for chemical species and energy contain five variables, namely, inlet temperature, T_{in}; extent of reaction, ξ; reactor temperature, T; residence time, τ; and interphase heat transfer rate, $\dot{Q}_{int}$. Therefore, three variables must be known, while the remaining two can be calculated from the conservation of chemical species and energy. Among these variables, T_{in} is the variable associated with the feed, ξ and T are the variables associated with the product, and τ and $\dot{Q}_{int}$ are the variables of design.

Example 6.6 A liquid feed to a jacketed CSTR consists of 2000 mol/m^3 A and 2400 mol/m^3 B. A second-order irreversible reaction takes place as

$$A + B \rightarrow 2C$$

The rate of reaction is given by

$$r = kc_A c_B$$

where the reaction rate constant at 298 K is $k = 8.4 \times 10^{-6}$ m^3/mol·min, and the activation energy is 50,000 J/mol. The reactor operates isothermally at 65 °C. The molar heat capacity at constant pressure and the standard heat of formation of species A, B, and C at 298 K are given as follows:

Species	$\widetilde{C}_P^o$ (J/mol·K)	$\Delta\widetilde{H}_f^o$ (kJ/mol)
A	175	−60
B	130	−75
C	110	−90

a) Calculate the residence time required to obtain 80% conversion of species A.
b) What should be the volume of the reactor if species C are to be produced at a rate of 820 mol/min?
c) If the feed enters the reactor at a temperature of 25 °C, determine the rate of heat that must be removed from the reactor to maintain isothermal operation.
d) If the heat transfer coefficient is 1050 W/m^2·K and the average cooling fluid temperature is 15 °C, estimate the required heat transfer area.

Solution

Assumptions

1. As a result of perfect mixing, concentrations of the species within the reactor are uniform, i.e., $(c_i)_{out} = (c_i)_{sys}$.

2. Solution nonidealities are negligible, i.e., $\overline{C}_{P_i} = \widetilde{C}_{P_i}$; $\Delta H_{rxn} = \Delta H_{rxn}^o$
3. There is no heat loss from the reactor.

Analysis

System: Contents of the reactor

a) Since the reactor volume is constant, the inlet and outlet volumetric flow rates are the same and equal to $\mathcal{Q}$. Therefore, the inventory rate equation for conservation of species $\mathcal{A}$, Eq. (6.1-7), becomes

$$\mathcal{Q}(c_A)_{in} - \mathcal{Q}(c_A)_{sys} - \left[k(c_A)_{sys}(c_B)_{sys} \right] V_{sys} = 0 \tag{1}$$

where $(c_A)_{sys}$ and $(c_B)_{sys}$ represent the molar concentrations of species $\mathcal{A}$ and $\mathcal{B}$ in the reactor, respectively. Dropping the subscript "sys" and dividing Eq. (1) by the volumetric flow rate, $\mathcal{Q}$, gives

$$\tau = \frac{(c_A)_{in} - c_A}{k c_A c_B} \tag{2}$$

Using Eq. (5.3-17), the extent of reaction can be calculated as

$$\xi = \frac{(c_A)_{in}}{(-\alpha_A)} X_A = \frac{(2000)(0.8)}{1} = 1600 \text{ mol/m}^3 \tag{3}$$

Therefore, the concentrations of species $\mathcal{A}$, $\mathcal{B}$, and $\mathcal{C}$ in the reactor are

$$c_A = (c_A)_{in} + \alpha_A \xi = 2000 - 1600 = 400 \text{ mol/m}^3 \tag{4}$$

$$c_B = (c_B)_{in} + \alpha_B \xi = 2400 - 1600 = 800 \text{ mol/m}^3 \tag{5}$$

$$c_C = (c_C)_{in} + \alpha_C \xi = (2)(1600) = 3200 \text{ mol/m}^3 \tag{6}$$

If k_1 and k_2 represent the rate constants at temperatures of T_1 and T_2, respectively, then

$$k_2 = k_1 \exp\left[-\frac{\mathcal{E}}{\mathcal{R}} \left(\frac{1}{T_2} - \frac{1}{T_1} \right) \right] \tag{7}$$

Therefore, the reaction rate constant at 65 °C (338 K) is

$$k = 8.4 \times 10^{-6} \exp\left[-\frac{50,000}{8.314} \left(\frac{1}{338} - \frac{1}{298} \right) \right] = 9.15 \times 10^{-5} \text{ m}^3/\text{mol·min} \tag{8}$$

Substitution of numerical values into Eq. (2) gives

$$\tau = \frac{2000 - 400}{(9.15 \times 10^{-5})(400)(800)} = 54.6 \text{ min}$$

b) The reactor volume, V, is given by

$$V = \tau \mathcal{Q}$$

The volumetric flow rate can be determined from the production rate of species $\mathcal{C}$, i.e.,

$$c_C Q = 820 \quad \Rightarrow \quad Q = \frac{820}{3200} = 0.256 \text{ m}^3/\text{min}$$

Hence, the reactor volume is

$$V = (54.6)(0.256) = 14 \text{ m}^3$$

c) For this problem, Eq. (6.3-42) simplifies to

$$\dot{Q}_{int} = -Q(C_P)_{in}(T_{in} - T) - V(kc_A c_B)(-\Delta H_{rxn}^o) \tag{9}$$

The standard heat of reaction at 298 K is

$$\Delta H_{rxn}^o(298) = \sum_i \alpha_i (\Delta \tilde{H}_f^o)_i = (-1)(-60) + (-1)(-75) + (2)(-90) = -45 \text{ kJ/mol}$$

The standard heat of reaction at 338 K is given by Eq. (6.3-29)

$$\Delta H_{rxn}^o(338) = \Delta H_{rxn}^o(298 \text{ K}) + \int_{298}^{338} \Delta \tilde{C}_P^o \, dT$$

where

$$\Delta \tilde{C}_P^o = \sum_i \alpha_i \tilde{C}_{P_i}^o = (-1)(175) + (-1)(130) + (2)(110) = -85 \text{ J/mol·K}$$

Hence

$$\Delta H_{rxn}^o(338) = -45,000 + (-85)(338 - 298) = -48,400 \text{ J/mol}$$

On the other hand, the use of Eq. (6.3-43) gives

$$(C_P)_{in} = \sum_i (c_i)_{in} \tilde{C}_{P_i} = (2000)(175) + (2400)(130) = 662,000 \text{ J/m}^3\text{·K}$$

Therefore, substitution of the numerical values into Eq. (9) yields

$$\dot{Q}_{int} = -(0.256)(662,000)(25 - 65)$$

$$- (14)\big[(9.15 \times 10^{-5})(400)(800)\big](48,400) = -13 \times 10^6 \text{ J/min}$$

The minus sign indicates that the system, i.e., reactor, loses energy to the surroundings.

d) The application of Newton's law of cooling gives

$$|\dot{Q}_{int}| = A_H \langle h \rangle (T_{reactor} - T_{coolant})$$

or,

$$A_H = \frac{13 \times 10^6}{(1050)(65 - 15)(60)} = 4.1 \text{ m}^2$$

NOTATION

A	area, m^2
A_H	heat transfer area, m^2
A_M	mass transfer area, m^2
$\widehat{C}_P$	heat capacity at constant pressure, kJ/kg·K
c	concentration, $kmol/m^3$
$\mathcal{D}_{AB}$	diffusion coefficient for system $\mathcal{A}$-$\mathcal{B}$, m^2/s
E	total energy, J
E_K	kinetic energy, J
E_P	potential energy, J
$\dot{E}$	rate of energy, J/s
$\mathcal{E}$	activation energy, J/mol
g	acceleration of gravity, m/s^2
H	enthalpy, J
h	elevation, m
k	reaction rate constant
k_c	mass transfer coefficient, m/s
$\dot{m}$	mass flow rate, kg/s
$\mathcal{M}$	molecular weight, kg/kmol
$\dot{n}$	molar flow rate, kmol/s
P	pressure, Pa
$\dot{Q}$	heat transfer rate, W
$\mathcal{Q}$	volumetric flow rate, m^3/s
r	rate of a chemical reaction, $kmol/m^3 \cdot s$
$\mathcal{R}$	gas constant, J/mol·K
T	temperature, °C or K
t	time, s
U	internal energy, J
V	volume, m^3
v	velocity, m/s
$\dot{W}$	rate of work, W
$\dot{W}_s$	rate of shaft work, W
X	fractional conversion
x_i	mole fraction of species i
α_i	stoichiometric coefficient of species i
α_{ij}	stoichiometric coefficient of the ith species in the jth reaction
β	coefficient of volume expansion, Eq. (6.3-13), K^{-1}
Δ	difference
$\Delta\widetilde{H}_f$	heat of formation, J/mol
ΔH_{rxn}	heat of reaction, J
λ	latent heat of vaporization, J
μ	viscosity, kg/m·s
ν	kinematic viscosity, m^2/s
ξ	intensive extent of a reaction, $kmol/m^3$

ρ density, kg/m^3

τ residence time, s

Overlines

$\sim$ per mole

$\frown$ per unit mass

$-$ partial molar

Bracket

$\langle a \rangle$ average value of a

Superscripts

o standard state

sat   saturation

Subscripts

A, B species in binary systems

ch   characteristic

f   film

i   species in multicomponent systems

in   inlet

int   interphase

j   reaction number

out   outlet

ref   reference

sys   system

Dimensionless Numbers

Pr   Prandtl number

Sc   Schmidt number

St$_\text{H}$ Stanton number for heat transfer

St$_\text{M}$ Stanton number for mass transfer

REFERENCES

Kauschus, W., J. Demont and K. Hartmann, 1978, On the steady states of continuous stirred tank reactors, Chem. Eng. Sci. 33, 1283.

Perry, R.H., D.W. Green and J.O. Maloney, Eds., 1997, Perry's Chemical Engineers' Handbook, 7th Ed., McGraw-Hill, New York.

SUGGESTED REFERENCES FOR FURTHER STUDY

Aris, R., 1969, Elementary Chemical Reactor Analysis, Prentice-Hall, Englewood Cliffs, New Jersey.

Felder, R.M. and R.W. Rousseau, 2000, Elementary Principles of Chemical Processes, 3rd Ed., Wiley, New York.

Fogler, H.S., 1992, Elements of Chemical Reaction Engineering, 2nd Ed., Prentice-Hall, Englewood Cliffs, New Jersey.

Hill, C.G., 1977, An Introduction to Chemical Engineering Kinetics and Reactor Design, Wiley, New York.

Myers, A.L. and W.D. Seider, 1976, Introduction to Chemical Engineering and Computer Calculations, Prentice-Hall, Englewood Cliffs, New Jersey.

Sandler, S.I., 2006, Chemical, Biochemical, and Engineering Thermodynamics, 4th Ed., Wiley, New York.

PROBLEMS

6.1   Water at 20 °C is flowing at steady-state through a piping system as shown in the figure below.

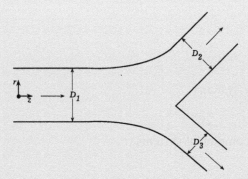

The velocity distribution (in m/s) in a pipe with $D_1 = 4$ cm is given by

$$v_z = 3\left(1 - \frac{r}{R_1}\right)^{1/7}$$

where $R_1 = D_1/2$ and r is the radial coordinate. If the volumetric flow rate of water through a pipe with $D_3 = 1$ cm is 0.072 m^3/min, calculate the volumetric flow rate of water (in cm^3/s) through a pipe with $D_2 = 2$ cm.

(**Answer:** 1880 cm^3/s)

6.2   2520 kg/h of oil is to be cooled from 180 °C to 110 °C in a countercurrent heat exchanger as shown in the figure below. Calculate the flow rate of water passing through the heat exchanger for the following cases:

a) The cooling water, which enters the heat exchanger at 15 °C, is mixed with water at 30 °C at the exit of the heat exchanger to obtain 2415 kg/h of process water at 60 °C to be used in another location in the plant.

b) The cooling water, which enters the heat exchanger at 30 °C, is mixed with water at 30 °C at the exit of the heat exchanger to obtain 2415 kg/h of process water at 60 °C to be used in another location in the plant.

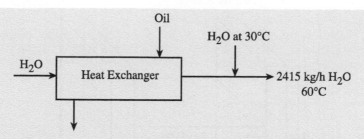

Assume that oil and water have constant heat capacities of 2.3 and 4.2 kJ/kg·K, respectively.

(**Answer:** a) 1610 kg/h)

6.3 The following parallel reactions take place in an isothermal, constant-volume CSTR:

$$A \to 2B \quad r = k_1 c_A \quad k_1 = 1.3 \text{ s}^{-1}$$
$$3A \to C \quad r = k_2 c_A \quad k_2 = 0.4 \text{ s}^{-1}$$

Pure A is fed to the reactor at a concentration of 350 mol/m^3.

a) Determine the residence time required to achieve 85% conversion of species A under steady conditions.

b) Determine the concentrations of species B and C.

(**Answer:** a) $\tau = 2.27$ s b) $c_B = 309.9$ mol/m^3, $c_C = 47.7$ mol/m^3)

6.4 Species A undergoes the following consecutive first-order reactions in the liquid phase in an isothermal, constant-volume CSTR:

$$A \xrightarrow{k_1} B \xrightarrow{k_2} C$$

where $k_1 = 1.5$ s^{-1} and $k_2 = 0.8$ s^{-1}. If the feed to the reactor consists of pure A, determine the residence time required to maximize the concentration of species B under steady conditions.

(**Answer:** 0.913 s)

6.5 An isomerization reaction

$$A \rightleftharpoons B$$

takes place in a constant-volume CSTR. The feed to the reactor consists of pure A. The rate of the reaction is given by

$$r = k_1 c_A - k_2 c_B$$

For the maximum conversion of species A at a given residence time, determine the reactor temperature.

$$\left(\textbf{Answer: } T = \frac{\mathcal{E}_2/\mathcal{R}}{\ln\{A_2\tau[(\mathcal{E}_2/\mathcal{E}_1) - 1]\}} \right)$$

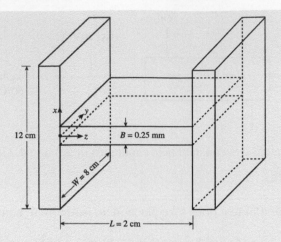

Figure 6.6. Schematic diagram for Problem 6.6.

6.6 Two electronic components ($k = 190$ W/m·K) are to be cooled by passing 0.2 m^3/s of air at 25 °C between them. To enhance the rate of heat loss, it is proposed to install equally spaced rectangular aluminum plates between the electronic components as shown in Figure 6.6.

The rate of heat loss from the electronic component on the left, i.e., $z = 0$, must be 500 W and the temperature should not exceed 80 °C, while the other component must dissipate 2 kW with a maximum allowable temperature of 90 °C. Determine the number of plates that must be placed per cm between the electronic components (use the temperature distribution given in Problem 4.7).

(**Answer:** One possible solution is 10 fins per cm)

6.7 As shown in Example 6.4, the wet-bulb temperature can be calculated from

$$T_\infty - T_w = \frac{c_{A_w} \widetilde{\lambda}_A}{(\rho \, \widehat{C}_P)_B} \left(\frac{\text{Pr}}{\text{Sc}} \right)_B^{2/3} \tag{1}$$

by a trial-and-error procedure because both c_{A_w} and $\widetilde{\lambda}_A$ must be evaluated at T_w, whereas ρ_B, $\widehat{C}_{P_B}$, Pr_B and Sc_B must be evaluated at the film temperature. In engineering applications, an approximate equation used to estimate the wet-bulb temperature is given by

$$T_w^2 - T_\infty \, T_w + \phi = 0 \tag{2}$$

where

$$\phi = \frac{P_A^{sat} \, T_\infty \mathcal{M}_A \widehat{\lambda}_A}{P_\infty \mathcal{M}_B \widehat{C}_{P_B}} \left(\frac{\text{Pr}}{\text{Sc}} \right)_B^{2/3} \tag{3}$$

Develop Eq. (2) from Eq. (1) and indicate the assumptions involved in the derivation.

6.8 An exothermic, first-order, irreversible reaction

$$A \rightarrow B$$

takes place in a constant-volume, jacketed CSTR.

a) Show that the conservation equations for chemical species A and energy take the form

$$Q[(c_A)_{in} - c_A] - kc_A V = 0 \qquad (1)$$

$$[Q(C_P)_{in} + A_H \langle h \rangle](T_m - T) + V k c_A (-\Delta H_{rxn}) = 0 \qquad (2)$$

where T_m is a weighted mean temperature defined by

$$T_m = \frac{Q(C_P)_{in} T_{in} + A_H \langle h \rangle T_c}{Q(C_P)_{in} + A_H \langle h \rangle} \qquad (3)$$

in which $\langle h \rangle$ is the average heat transfer coefficient, T_c is the cooling fluid temperature, and A_H is the heat transfer area.

b) Show that the elimination of c_A between Eqs. (1) and (2) leads to

$$[Q(C_P)_{in} + A_H \langle h \rangle](T_m - T) + \frac{kQV(c_A)_{in}}{Q + kV}(-\Delta H_{rxn}) = 0 \qquad (4)$$

c) In terms of the following dimensionless quantities

$$\theta = \frac{\mathcal{E}}{\mathcal{R}}\left(\frac{1}{T_m} - \frac{1}{T}\right) \qquad \chi = \frac{[Q(C_P)_{in} + A_H \langle h \rangle]T_m}{Q(c_A)_{in}(-\Delta H_{rxn})}$$

$$A_m = A e^{-\mathcal{E}/\mathcal{R}T_m} \qquad \beta = \frac{\mathcal{R}T_m}{\mathcal{E}}(1 + \chi) \qquad \frac{1}{\gamma} = \frac{\mathcal{R}T_m Q \chi}{\mathcal{E} V A_m}$$

show that Eq. (4) takes the form

$$e^{\theta} = \frac{\theta}{\gamma(1 - \beta\theta)} \qquad (5)$$

d) To determine the roots of Eq. (5) for given values of γ and β, it is more convenient to rearrange Eq. (5) in the form

$$F(\theta) = \ln\left[\frac{\theta}{\gamma(1 - \beta\theta)}\right] \qquad (6)$$

Examine the behavior of the function in Eq. (6) and conclude that

- At least one steady-state solution exists when $\beta \geqslant 0.25$,
- Two steady-state solutions exist when $\beta < 0.25$ and $\gamma = \gamma_{min} < \gamma_{max}$ or $\gamma_{min} < \gamma = \gamma_{max}$,
- Three steady-state solutions exist when $\beta < 0.25$ and $\gamma_{min} < \gamma < \gamma_{max}$,

where $\gamma_{\min}$ and $\gamma_{\max}$ are defined by

$$\gamma_{\min} = \left(\frac{1 + \sqrt{1 - 4\beta}}{2\beta}\right)^2 \exp\left[-\left(\frac{1 + \sqrt{1 - 4\beta}}{2\beta}\right)\right] \tag{7}$$

$$\gamma_{\max} = \left(\frac{2}{1 + \sqrt{1 - 4\beta}}\right)^2 \exp\left[-\left(\frac{2}{1 + \sqrt{1 - 4\beta}}\right)\right] \tag{8}$$

The existence of more than one steady-state solution is referred to as *multiple steady-states*. For more detailed information on this problem see Kauschus *et al.* (1978).

7

UNSTEADY-STATE MACROSCOPIC BALANCES

In this chapter we will consider unsteady-state transfer processes between phases by assuming no gradients within each phase. Since the dependent variables, such as temperature and concentration, are considered uniform within a given phase, the resulting macroscopic balances are ordinary differential equations in time.

The basic steps in the development of unsteady macroscopic balances are similar to those for steady-state balances given in Chapter 6. These can be briefly summarized as follows:

- Define your system.
- If possible, draw a simple sketch.
- List the assumptions.
- Write down the inventory rate equation for each of the basic concepts relevant to the problem at hand.
- Use engineering correlations to evaluate the transfer coefficients.
- Write down the initial conditions: the number of initial conditions must be equal to the sum of the order of differential equations written for the system.
- Solve the ordinary differential equations.

7.1 APPROXIMATIONS USED IN MODELING OF UNSTEADY-STATE PROCESSES

7.1.1 Pseudo-Steady-State Approximation

As stated in Chapter 1, the general inventory rate equation can be expressed in the form

$$\begin{pmatrix} \text{Rate of} \\ \text{input} \end{pmatrix} - \begin{pmatrix} \text{Rate of} \\ \text{output} \end{pmatrix} + \begin{pmatrix} \text{Rate of} \\ \text{generation} \end{pmatrix} = \begin{pmatrix} \text{Rate of} \\ \text{accumulation} \end{pmatrix} \qquad (7.1\text{-}1)$$

Remember that the molecular and convective fluxes constitute the input and output terms. Among the terms appearing on the left-hand side of Eq. (7.1-1), molecular transport is the slowest process. Therefore, in a given unsteady-state process, the term on the right-hand side of Eq. (7.1-1) may be considered negligible if

$$\begin{pmatrix} \text{Rate of} \\ \text{molecular transport} \end{pmatrix} \gg \begin{pmatrix} \text{Rate of} \\ \text{accumulation} \end{pmatrix} \qquad (7.1\text{-}2)$$

or,

$$(\text{Diffusivity}) \begin{pmatrix} \text{Gradient of} \\ \text{Quantity/Volume} \end{pmatrix} (\text{Area}) \gg \frac{\text{Difference in quantity}}{\text{Characteristic time}} \qquad (7.1\text{-}3)$$

Note that the "Gradient of Quantity/Volume" is expressed in the form

$$\text{Gradient of Quantity/Volume} = \frac{\text{Difference in Quantity/Volume}}{\text{Characteristic length}} \qquad (7.1\text{-}4)$$

On the other hand, volume and area are expressed in terms of characteristic length as

$$\text{Volume} = (\text{Characteristic length})^3 \qquad (7.1\text{-}5)$$

$$\text{Area} = (\text{Characteristic length})^2 \qquad (7.1\text{-}6)$$

Substitution of Eqs. (7.1-4)–(7.1-6) into Eq. (7.1-3) gives

$$\boxed{\frac{(\text{Diffusivity})(\text{Characteristic time})}{(\text{Characteristic length})^2} \gg 1} \qquad (7.1\text{-}7)$$

In the literature, the dimensionless term on the left-hand side of Eq. (7.1-7) is known as the *Fourier number* and designated by τ.

In engineering analysis, the neglect of the unsteady-state term is often referred to as the *pseudo-steady-state* (or *quasi-steady-state*) approximation. However, it should be noted that the pseudo-steady-state approximation is only valid if the constraint given by Eq. (7.1-7) is satisfied.

Example 7.1 We are testing a 2 cm thick insulating material. The density, thermal conductivity, and heat capacity of the insulating material are 255 kg/m^3, 0.07 W/m·K, and 1300 J/kg·K, respectively. If our experiments take 10 min, is it possible to assume pseudo-steady-state behavior?

Solution

For the pseudo-steady-state approximation to be valid, Eq. (7.1-7) must be satisfied, i.e.,

$$\frac{\alpha t_{ch}}{L_{ch}^2} \gg 1$$

The thermal diffusivity, α, of the insulating material is

$$\alpha = \frac{k}{\rho \widehat{C}_P} = \frac{0.07}{(255)(1300)} = 2.11 \times 10^{-7} \text{ m}^2/\text{s}$$

Hence,

$$\frac{\alpha t_{ch}}{L_{ch}^2} = \frac{(2.11 \times 10^{-7})(10)(60)}{(2 \times 10^{-2})^2} = 0.32 < 1$$

which indicates that we have an unsteady-state problem at hand.

7.1.2 No Variation of Dependent Variable Within the Phase of Interest

In engineering analysis it is customary to neglect spatial variations in either temperature or concentration within the solid. Although this approximation simplifies the mathematical problem, it is only possible under certain circumstances as will be shown in the following development.

Let us consider the transport of a quantity φ from the solid phase to the fluid phase through a solid-fluid interface. Under steady conditions without generation, the inventory rate equation, Eq. (1.1-1), for the interface takes the form

$$\begin{pmatrix} \text{Rate of transport of } \varphi \text{ from} \\ \text{the solid to the interface} \end{pmatrix} = \begin{pmatrix} \text{Rate of transport of } \varphi \text{ from} \\ \text{the interface to the fluid} \end{pmatrix} \qquad (7.1\text{-}8)$$

Since the molecular flux of φ is dominant within the solid phase, Eq. (7.1-8) reduces to

$$\begin{pmatrix} \text{Molecular flux of } \varphi \text{ from} \\ \text{the solid to the interface} \end{pmatrix} = \begin{pmatrix} \text{Flux of } \varphi \text{ from} \\ \text{the interface to the fluid} \end{pmatrix} \qquad (7.1\text{-}9)$$

or,

$$\left[\begin{pmatrix} \text{Transport} \\ \text{property} \end{pmatrix} \begin{pmatrix} \text{Gradient of} \\ \text{driving force} \end{pmatrix} \right]_{solid} = \left[\begin{pmatrix} \text{Transfer} \\ \text{coefficient} \end{pmatrix} \begin{pmatrix} \text{Difference in} \\ \text{Quantity/Volume} \end{pmatrix} \right]_{fluid}$$

$$(7.1\text{-}10)$$

The gradient of driving force is expressed in the form

$$\text{Gradient of driving force} = \frac{\text{Difference in driving force}}{\text{Characteristic length}} \qquad (7.1\text{-}11)$$

On the other hand, "Difference in Quantity/Volume" can be expressed as

$$\begin{pmatrix} \text{Difference in} \\ \text{Quantity/Volume} \end{pmatrix} = \begin{pmatrix} \dfrac{\text{Transport property}}{\text{Diffusivity}} \end{pmatrix} \begin{pmatrix} \text{Difference in} \\ \text{driving force} \end{pmatrix} \qquad (7.1\text{-}12)$$

Substitution of Eqs. (7.1-11) and (7.1-12) into the left- and right-hand sides of Eq. (7.1-10), respectively, gives

$$\text{Bi} = \frac{\begin{pmatrix} \text{Characteristic} \\ \text{length} \end{pmatrix}}{\begin{pmatrix} \text{Transport} \\ \text{property} \end{pmatrix}_{solid}} \left[\frac{\begin{pmatrix} \text{Transfer} \\ \text{coefficient} \end{pmatrix} \begin{pmatrix} \text{Transport} \\ \text{property} \end{pmatrix}}{\text{Diffusivity}} \right]_{fluid} \qquad (7.1\text{-}13)$$

in which Bi designates the *Biot number* defined by

$$\text{Bi} = \frac{(\text{Difference in driving force})_{solid}}{(\text{Difference in driving force})_{fluid}} \qquad (7.1\text{-}14)$$

Therefore, the Biot numbers for heat and mass transfer are defined as

$$\text{Bi}_\text{H} = \frac{\langle h \rangle L_{ch}}{k_{solid}} \qquad \text{and} \qquad \text{Bi}_\text{M} = \frac{\langle k_c \rangle L_{ch}}{(\mathcal{D}_{AB})_{solid}} \qquad (7.1\text{-}15)$$

It is important to distinguish the difference between the Biot and the Nusselt (or the Sherwood) numbers. The transport properties in the Biot numbers, Eq. (7.1-15), are referred to the solid, whereas the transport properties in the Nusselt and the Sherwood numbers, Eqs. (3.4-11) and (3.4-12), are referred to the fluid. Some textbooks define the characteristic length, L_{ch}, as the ratio of the volume to the surface area. In general, it should be the distance over which significant changes in temperature or concentration take place.

When the Biot number is small, one can conclude from Eq. (7.1-14) that

$$\begin{pmatrix} \text{Difference in} \\ \text{driving force} \end{pmatrix}_{solid} \ll \begin{pmatrix} \text{Difference in} \\ \text{driving force} \end{pmatrix}_{fluid} \qquad (7.1\text{-}16)$$

Therefore, dependent variables may be considered uniform within the solid phase only if $\text{Bi} \ll 1$. This approach is known as *lumped-parameter analysis*.

It is also possible to define the Biot numbers in terms of the time scales. Using the quantities given in Table 3.3, the Biot numbers are given by

$$\text{Bi}_\text{H} = \frac{\text{Conductive time scale}}{\text{Convective time scale for heat transport}} = \frac{L_{ch}^2/\alpha}{L_{ch}/(h/\rho\widehat{C}_P)} = \frac{hL_{ch}}{k} \qquad (7.1\text{-}17)$$

$$\text{Bi}_\text{M} = \frac{\text{Diffusive time scale}}{\text{Convective time scale for mass transport}} = \frac{L_{ch}^2/\mathcal{D}_{AB}}{L_{ch}/k_c} = \frac{k_c L_{ch}}{\mathcal{D}_{AB}} \qquad (7.1\text{-}18)$$

7.2 CONSERVATION OF CHEMICAL SPECIES

The conservation statement for the mass of the ith chemical species is given by

$$\begin{pmatrix} \text{Rate of mass} \\ \text{of } i \text{ in} \end{pmatrix} - \begin{pmatrix} \text{Rate of mass} \\ \text{of } i \text{ out} \end{pmatrix} + \begin{pmatrix} \text{Rate of generation} \\ \text{of mass } i \end{pmatrix} = \begin{pmatrix} \text{Rate of accumulation} \\ \text{of mass } i \end{pmatrix}$$

$$(7.2\text{-}1)$$

For a system with a single inlet and a single outlet stream as shown in Figure 7.1, Eq. (7.2-1) can be expressed as

$$\boxed{(\dot{m}_i)_{in} - (\dot{m}_i)_{out} \pm (\dot{m}_i)_{int} + V_{sys}\mathcal{M}_i \sum_j \alpha_{ij} r_j = \frac{d(m_i)_{sys}}{dt}} \qquad (7.2\text{-}2)$$

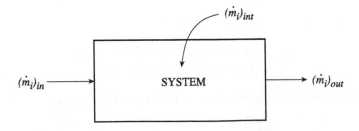

Figure 7.1. Unsteady-state flow system exchanging mass with the surroundings.

The interphase mass transfer rate, $(\dot{m}_i)_{int}$, is considered positive when mass is added to the system and is expressed by

$$(\dot{m}_i)_{int} = A_M \langle k_c \rangle (\Delta c_i)_{ch} \mathcal{M}_i \qquad (7.2\text{-}3)$$

Substitution of Eq. (7.2-3) into Eq. (7.2-2) gives

$$(\mathcal{Q}\rho_i)_{in} - (\mathcal{Q}\rho_i)_{out} \pm A_M \langle k_c \rangle (\Delta c_i)_{ch} \mathcal{M}_i + V_{sys} \mathcal{M}_i \sum_j \alpha_{ij} r_j = \frac{d(m_i)_{sys}}{dt} \qquad (7.2\text{-}4)$$

On a molar basis, Eqs. (7.2-2) and (7.2-4) take the form

$$(\dot{n}_i)_{in} - (\dot{n}_i)_{out} \pm (\dot{n}_i)_{int} + V_{sys} \sum_j \alpha_{ij} r_j = \frac{d(n_i)_{sys}}{dt} \qquad (7.2\text{-}5)$$

and

$$(\mathcal{Q}c_i)_{in} - (\mathcal{Q}c_i)_{out} \pm A_M \langle k_c \rangle (\Delta c_i)_{ch} + V_{sys} \sum_j \alpha_{ij} r_j = \frac{d(n_i)_{sys}}{dt} \qquad (7.2\text{-}6)$$

7.3 CONSERVATION OF TOTAL MASS

Summation of Eq. (7.2-2) over all species gives the total mass balance in the form

$$\dot{m}_{in} - \dot{m}_{out} \pm \dot{m}_{int} = \frac{dm_{sys}}{dt} \qquad (7.3\text{-}1)$$

Note that the term $\sum_i \alpha_{ij} \mathcal{M}_i$ is zero since mass is conserved. On the other hand, summation of Eq. (7.2-5) over all species gives the total mole balance as

$$\dot{n}_{in} - \dot{n}_{out} \pm \dot{n}_{int} + V_{sys} \sum_j \overline{\alpha}_j r_j = \frac{dn_{sys}}{dt} \qquad (7.3\text{-}2)$$

where

$$\overline{\alpha}_j = \sum_i \alpha_{ij} \qquad (7.3\text{-}3)$$

The generation term in Eq. (7.3-2) is not zero because moles are not conserved. This term vanishes only when $\overline{\alpha}_j = 0$ for all values of j.

Example 7.2 An open cylindrical tank of height H and diameter D is initially half full of a liquid. At time $t = 0$, the liquid is fed into the tank at a constant volumetric flow rate of $\mathcal{Q}_{in}$, and at the same time it is allowed to drain out through an orifice of diameter D_o at the bottom of the tank. Express the variation in the liquid height as a function of time.

Solution

Assumptions

1. Rate of evaporation from the liquid surface is negligible.
2. Liquid is incompressible.
3. Pressure distribution in the tank is hydrostatic.

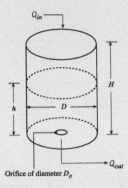

Orifice of diameter D_o

Analysis

System: Fluid in the tank

The inventory rate equation for total mass, Eq. (7.3-1), reduces to

$$\rho Q_{in} - \rho \langle v_o \rangle A_o = \frac{d(Ah\rho)}{dt} \tag{1}$$

where $\langle v_o \rangle$ is the average velocity through the orifice, i.e., the volumetric flow rate divided by the cross-sectional area; A_o and A are the cross-sectional areas of the orifice and the tank, respectively. Since ρ and A are constant, Eq. (1) becomes

$$Q_{in} - \langle v_o \rangle A_o = A \frac{dh}{dt} \tag{2}$$

In order to proceed further, $\langle v_o \rangle$ must be related to h.

For flow in a pipe of uniform cross-sectional area A, the pressure drop across an orifice is given by

$$\langle v_o \rangle = \frac{C_o}{\sqrt{1 - \beta^4}} \sqrt{\frac{2|\Delta P|}{\rho}} \tag{3}$$

where β is the ratio of the orifice diameter to the pipe diameter, $|\Delta P|$ is the pressure drop across the orifice, and C_o is the orifice coefficient. The value of C_o is generally determined from experiments and given as a function of β and the Reynolds number, Re_o, defined by

$$\mathrm{Re}_o = \frac{D_o \langle v_o \rangle \rho}{\mu} \tag{4}$$

For $\beta < 0.25$, the term $\sqrt{1 - \beta^4}$ is almost unity. On the other hand, when $\mathrm{Re}_o > 20{,}000$, experimental measurements show that $C_o \simeq 0.61$. Hence, Eq. (3) reduces to

$$\langle v_o \rangle = 0.61 \sqrt{\frac{2|\Delta P|}{\rho}} \tag{5}$$

Since the pressure in the tank is hydrostatic, $|\Delta P| \simeq \rho g h$ and Eq. (5) becomes

$$\langle v_o \rangle = 0.61 \sqrt{2gh} = 2.7 \sqrt{h} \tag{6}$$

Substitution of Eq. (6) into Eq. (2) gives the governing differential equation for the liquid height in the tank as

$$2.7 \left(\frac{A_o}{A} \right) (\Omega - \sqrt{h}) = \frac{dh}{dt} \tag{7}$$

where

$$\Omega = \frac{Q_{in}}{2.7 A_o} \tag{8}$$

Note that the system reaches steady-state when $dh/dt = 0$ at which point the liquid height, h_s, is given by

$$h_s = \Omega^2 \tag{9}$$

Now it is worthwhile to investigate two cases:

Case (i) Liquid level in the tank increases

At $t = 0$, the liquid level in the tank is $H/2$. Therefore, the liquid level increases, i.e., $dh/dt > 0$ in Eq. (7), if

$$\Omega^2 > H/2 \tag{10}$$

Rearrangement of Eq. (7) gives

$$\int_0^t dt = \frac{1}{2.7} \left(\frac{A}{A_o} \right) \int_{H/2}^h \frac{dh}{\Omega - \sqrt{h}} \tag{11}$$

Integration of Eq. (11) yields

$$t = 0.74 \left(\frac{A}{A_o} \right) \left[\sqrt{\frac{H}{2}} - \sqrt{h} + \Omega \ln \left(\frac{\Omega - \sqrt{H/2}}{\Omega - \sqrt{h}} \right) \right] \tag{12}$$

Equations (9) and (10) indicate that $h_s > H/2$. When $h_s > H$, the steady-state condition can never be achieved in the tank. The time required to fill the tank, t_f, is

$$t_f = 0.74 \left(\frac{A}{A_o} \right) \left[\sqrt{\frac{H}{2}} - \sqrt{H} + \Omega \ln \left(\frac{\Omega - \sqrt{H/2}}{\Omega - \sqrt{H}} \right) \right] \tag{13}$$

If $H/2 < h_s < H$, then the time, t_∞, required for the level of the tank to reach 99% of the steady-state value is

$$t_\infty = 0.74 \left(\frac{A}{A_o}\right) \left[\sqrt{\frac{H}{2}} - \sqrt{0.99}\,\Omega + \Omega \ln\left(\frac{\Omega - \sqrt{H/2}}{\Omega - \sqrt{0.99}\,\Omega} \right) \right] \qquad (14)$$

Case (*ii*) Liquid level in the tank decreases

The liquid level in the tank decreases, i.e., $dh/dt < 0$ in Eq. (7), if

$$\Omega^2 < H/2 \qquad (15)$$

Equation (12) is also valid for this case. Equations (9) and (15) imply that $h_s < H/2$. Since h_s cannot be negative, this further implies that it is impossible to empty the tank under these circumstances. The time required for the level of the tank to reach 99% of the steady-state value is also given by Eq. (14).

The ratio h/H is plotted versus $t/[0.74(A/A_o)\sqrt{H}]$ with $\Omega/\sqrt{H}$ as a parameter in the figure below.

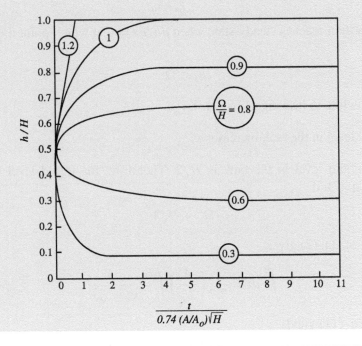

Example 7.3 A liquid phase irreversible reaction

$$A \rightarrow B$$

takes place in a CSTR of volume V_T. The reactor is initially empty. At $t = 0$, a solution of species A at concentration c_{A_o} flows into the reactor at a constant volumetric flow rate of Q_{in}. No liquid leaves the reactor until the liquid volume reaches a value of V_T. The rate of reaction is given by

$$r = kc_A$$

If the reaction takes place under isothermal conditions, express the concentration of species $\mathcal{A}$ within the reactor as a function of time.

Solution

Assumptions

1. Well-mixed system, i.e., the temperature and the concentration of the contents of the reactor are uniform.
2. The density of the reaction mixture is constant.

Analysis

System: Contents of the reactor

The problem should be considered in three parts: the filling period, the unsteady-state period, and the steady-state period.

i) The filling period

During this period, there is no outlet stream from the reactor. Hence, the conservation of total mass, Eq. (7.3-1), is given by

$$\rho \mathcal{Q}_{in} = \frac{dm_{sys}}{dt} \tag{1}$$

Since $\mathcal{Q}_{in}$ and ρ are constant, integration of Eq. (1) and the use of the initial condition, $m_{sys} = 0$ at $t = 0$, give

$$m_{sys} = \mathcal{Q}_{in}\rho t \tag{2}$$

Since $m_{sys} = \rho V_{sys}$, Eq. (2) can also be expressed as

$$V_{sys} = \mathcal{Q}_{in} t \tag{3}$$

From Eq. (3), the time required to fill the reactor, t^*, is calculated as $t^* = V_T / \mathcal{Q}_{in}$, where V_T is the volume of the reactor.

The inventory rate equation based on the moles of species $\mathcal{A}$, Eq. (7.2-6), reduces to

$$\mathcal{Q}_{in} c_{A_o} - k c_A V_{sys} = \frac{dn_A}{dt} \tag{4}$$

where V_{sys}, the volume of the reaction mixture, is dependent on time. The molar concentration can be expressed in terms of the number of moles as

$$c_A = \frac{n_A}{V_{sys}} \tag{5}$$

such that Eq. (4) can be rearranged in the form

$$\int_0^{n_A} \frac{dn_A}{\mathcal{Q}_{in} c_{A_o} - k n_A} = \int_0^t dt \tag{6}$$

Integration gives

$$n_A = \frac{Q_{in}c_{A_o}}{kt}\left[1 - \exp(-kt)\right] \qquad (7)$$

Substitution of Eq. (7) into Eq. (5) and the use of Eq. (3) give the concentration as a function of time as

$$c_A = \frac{c_{A_o}}{kt}\left[1 - \exp(-kt)\right] \qquad 0 \leqslant t \leqslant V_T/Q_{in} \qquad (8)$$

The concentration c_A^* at the instant the tank is full, i.e., at $t = t^* = V_T/Q_{in}$, is

$$c_A^* = \frac{Q_{in}c_{A_o}}{kV_T}\left[1 - \exp\left(-\frac{kV_T}{Q_{in}}\right)\right] \qquad (9)$$

ii) The unsteady-state period

Since the total volume of the reactor V_T is constant, then the inlet and outlet volumetric flow rates are the same, i.e.,

$$Q_{in} = Q_{out} = Q \qquad (10)$$

The inventory rate equation for the moles of species $\mathcal{A}$, Eq. (7.2-6), is

$$Qc_{A_o} - Qc_A - kc_AV_T = \frac{d(c_AV_T)}{dt} \qquad (11)$$

Equation (11) can be rearranged in the form

$$\frac{1}{\tau}\left[c_{A_o} - c_A(1 + k\tau)\right] = \frac{dc_A}{dt} \qquad (12)$$

where τ is the residence time defined by

$$\tau = \frac{V_T}{Q} \qquad (13)$$

Equation (12) is a separable equation and can be written in the form

$$\tau \int_{c_A^*}^{c_A} \frac{dc_A}{c_{A_o} - c_A(1 + k\tau)} = \int_{t^*}^{t} dt \qquad (14)$$

Integration of Eq. (14) gives the concentration distribution as

$$c_A = \frac{c_{A_o}}{1 + k\tau} + \left(c_A^* - \frac{c_{A_o}}{1 + k\tau}\right)\exp\left[-\frac{(1 + k\tau)(t - t^*)}{\tau}\right] \qquad (15)$$

iii) The steady-state period

The concentration in the tank reaches its steady-state value, c_{A_s}, as $t \to \infty$. In this case, the exponential term in Eq. (15) vanishes and the result is

$$c_{A_s} = \frac{c_{A_o}}{1 + k\tau} \qquad (16)$$

Note that Eq. (16) can also be obtained from Eq. (12) by letting $dc_A/dt = 0$. The time required for the concentration to reach 99% of its steady-state value, t_∞, is

$$t_\infty = t^* + \frac{\tau}{1+k\tau} \ln \left\{ 100 \left[1 - \left(\frac{1+k\tau}{k\tau} \right) [1 - \exp(-k\tau)] \right] \right\} \tag{17}$$

When $k\tau \ll 1$, i.e., a slow first-order reaction, Eq. (17) simplifies to

$$t_\infty - t^* = 4.6\tau \tag{18}$$

Example 7.4 A sphere of naphthalene, 2 cm in diameter, is suspended in air at 90 °C. Estimate the time required for the diameter of the sphere to be reduced to one-half its initial value if:

a) The air is stagnant,
b) The air is flowing past the naphthalene sphere with a velocity of 5 m/s.

Solution

Physical properties

For naphthalene (species $\mathcal{A}$) at 90 °C (363 K): $\begin{cases} \rho_A^S = 1145 \text{ kg/m}^3 \\ \mathcal{M}_A = 128 \\ P_A^{sat} = 11.7 \text{ mmHg} \end{cases}$

Diffusion coefficient of species $\mathcal{A}$ in air (species $\mathcal{B}$) is

$$(\mathcal{D}_{AB})_{363} = (0.62 \times 10^{-5}) \left(\frac{363}{300} \right)^{3/2} = 8.25 \times 10^{-6} \text{ m}^2/\text{s}$$

For air at 90 °C (363 K): $\nu = 21.95 \times 10^{-6} \text{ m}^2/\text{s}$

The Schmidt number is

$$\text{Sc} = \frac{\nu}{\mathcal{D}_{AB}} = \frac{21.95 \times 10^{-6}}{8.25 \times 10^{-6}} = 2.66$$

Assumptions

1. Pseudo-steady-state behavior.
2. Ideal gas behavior.

Analysis

System: Naphthalene sphere

The terms appearing in the conservation of species $\mathcal{A}$, Eq. (7.2-2), are

$$(\dot{m}_A)_{in} = (\dot{m}_A)_{out} = 0$$
$$(\dot{m}_A)_{int} = -\left(\pi D_P^2 \right) \langle k_c \rangle (c_{A_w} - c_{A_\infty}) \mathcal{M}_A$$
$$r = 0$$
$$(m_A)_{sys} = V_{sys} \rho_A^S = \left(\pi D_P^3/6 \right) \rho_A^S$$

Therefore, Eq. (7.2-2) reduces to

$$-\left(\pi D_P^2\right)\langle k_c\rangle(c_{A_w} - c_{A_\infty})\mathcal{M}_A = \frac{d}{dt}\left(\frac{\pi D_P^3}{6}\rho_A^S\right) \tag{1}$$

Taking $c_{A_\infty} = 0$ and rearrangement give

$$t = \frac{\rho_A^S}{2\mathcal{M}_A c_{A_w}} \int_{D_o/2}^{D_o} \frac{dD_P}{\langle k_c\rangle} \tag{2}$$

where D_o is the initial diameter of the naphthalene sphere.

The average mass transfer coefficient, $\langle k_c\rangle$, can be related to the diameter of the sphere, D_P, by using one of the mass transfer correlations given in Section 4.3.3. The use of the Ranz-Marshall correlation, Eq. (4.3-33), gives

$$\text{Sh} = 2 + 0.6\,\text{Re}_P^{1/2}\,\text{Sc}^{1/3} \tag{3}$$

a) When air is stagnant, i.e., $\text{Re}_P = 0$, Eq. (3) reduces to

$$\text{Sh} = \frac{\langle k_c\rangle D_P}{\mathcal{D}_{AB}} = 2 \quad \Rightarrow \quad \langle k_c\rangle = \frac{2\mathcal{D}_{AB}}{D_P} \tag{4}$$

Substitution of Eq. (4) into Eq. (2) and integration give

$$t = \frac{3}{32}\frac{\rho_A^S D_o^2}{\mathcal{M}_A c_{A_w}\mathcal{D}_{AB}} \tag{5}$$

The saturation concentration of naphthalene, c_{A_w}, is

$$c_{A_w} = \frac{P_A^{sat}}{\mathcal{R}T} = \frac{11.7/760}{(0.08205)(90+273)} = 5.17 \times 10^{-4}\ \text{kmol/m}^3 \tag{6}$$

Substitution of the values into Eq. (5) gives the required time as

$$t = \frac{3}{32}\frac{(1145)(0.02)^2}{(128)(5.17 \times 10^{-4})(8.25 \times 10^{-6})} = 2.59 \times 10^5\ \text{s} \simeq 3\ \text{days}$$

b) When air flows with a certain velocity, the Ranz-Marshall correlation can be expressed as

$$\frac{\langle k_c\rangle D_P}{\mathcal{D}_{AB}} = 2 + 0.6\left(\frac{D_P v_\infty}{\nu}\right)^{1/2}\text{Sc}^{1/3}$$

or,

$$\langle k_c\rangle = \frac{1}{D_P}\left(\alpha + \beta\sqrt{D_P}\right) \tag{7}$$

where the coefficients α and β are defined by

$$\alpha = 2\mathcal{D}_{AB} = 2(8.25 \times 10^{-6}) = 1.65 \times 10^{-5} \tag{8}$$

$$\beta = 0.6 \mathcal{D}_{AB}(v_\infty/v)^{1/2} \, \mathrm{Sc}^{1/3}$$

$$= (0.6)(8.25 \times 10^{-6})\left(\frac{5}{21.95 \times 10^{-6}}\right)^{1/2} (2.66)^{1/3} = 3.27 \times 10^{-3} \qquad (9)$$

Substitution of Eqs. (7)–(9) into Eq. (2) gives

$$t = \frac{1145}{(2)(128)(5.17 \times 10^{-4})} \int_{0.01}^{0.02} \left(\frac{D_P}{1.65 \times 10^{-5} + 3.27 \times 10^{-3}\sqrt{D_P}}\right) dD_P$$

Analytical evaluation of the above integral is possible and the result is

$$t = 3097 \text{ s} \simeq 52 \text{ min}$$

Verification of the pseudo-steady-state approximation

$$\frac{\mathcal{D}_{AB}t}{D_P^2} = \frac{(8.25 \times 10^{-6})(3097)}{(2 \times 10^{-2})^2} = 64 \gg 1$$

7.4 CONSERVATION OF MOMENTUM

According to Newton's second law of motion, the conservation statement for linear momentum is expressed as

$$\begin{pmatrix} \text{Time rate of change of} \\ \text{linear momentum of a body} \end{pmatrix} = \begin{pmatrix} \text{Forces acting} \\ \text{on a body} \end{pmatrix} \qquad (7.4\text{-}1)$$

In Section 4.3, we considered the balance of forces acting on a single spherical particle of diameter D_P, falling in a stagnant fluid with a constant terminal velocity v_t. In the case of an accelerating sphere, an additional force, called the *fluid inertia force*, acts besides the gravitational, buoyancy, and drag forces. This force arises from the fact that the fluid around the sphere is also accelerated from rest, resulting in a change in the momentum of the fluid. The rate of change of fluid momentum shows up as an additional force acting on the sphere, pointing in the direction opposite to the motion of the sphere. This additional force has a magnitude equal to one-half the rate of change of momentum of a sphere of liquid moving at the same velocity as the solid sphere. Therefore, Eq. (7.4-1) is written in the form

$$\begin{pmatrix} \text{Time rate of change of} \\ \text{linear momentum of a sphere} \end{pmatrix} = \begin{pmatrix} \text{Gravitational} \\ \text{force} \end{pmatrix} - \begin{pmatrix} \text{Buoyancy} \\ \text{force} \end{pmatrix}$$

$$- \begin{pmatrix} \text{Drag} \\ \text{force} \end{pmatrix} - \begin{pmatrix} \text{Fluid inertia} \\ \text{force} \end{pmatrix} \qquad (7.4\text{-}2)$$

and can be expressed as

$$\frac{\pi D_P^3}{6}\rho_P \frac{dv}{dt} = \frac{\pi D_P^3}{6}\rho_P g - \frac{\pi D_P^3}{6}\rho g - \left(\frac{\pi D_P^2}{4}\right)\left(\frac{1}{2}\rho v^2\right) f - \frac{\pi D_P^3}{12}\rho \frac{dv}{dt} \qquad (7.4\text{-}3)$$

where ρ_P and D_P represent the density and diameter of the solid sphere, respectively, and ρ is the fluid density. Simplification of Eq. (7.4-3) gives

$$D_P(\rho_P + 0.5\rho)\frac{dv}{dt} = D_P(\rho_P - \rho)g - \frac{3}{4}\rho v^2 f \tag{7.4-4}$$

The friction factor, f, is usually given as a function of the Reynolds number, Re_P, defined by

$$\mathrm{Re}_P = \frac{D_P v \rho}{\mu} \tag{7.4-5}$$

Therefore, it is much more convenient to express the velocity, v, in terms of Re_P. Thus, Eq. (7.4-4) takes the form

$$(\rho_P + 0.5\rho)\frac{D_P^2}{\mu}\frac{d\,\mathrm{Re}_P}{dt} = \mathrm{Ar} - \frac{3}{4} f \,\mathrm{Re}_P^2 \tag{7.4-6}$$

where Ar is the Archimedes number defined by Eq. (4.3-6). Note that when the particle reaches its terminal velocity, i.e., $d\,\mathrm{Re}_P/dt = 0$, Eq. (7.4-6) reduces to Eq. (4.3-4). Integration of Eq. (7.4-6) gives

$$t = \frac{(\rho_P + 0.5\rho)D_P^2}{\mu} \int_0^{\mathrm{Re}_P} \left(\mathrm{Ar} - \frac{3}{4} f \,\mathrm{Re}_P^2\right)^{-1} d\,\mathrm{Re}_P \tag{7.4-7}$$

A friction factor-Reynolds number relationship is required to carry out the integration. Substitution of the Turton-Levenspiel correlation, Eq. (4.3-10), into Eq. (7.4-7) gives

$$t = \frac{(\rho_P + 0.5\rho)D_P^2}{\mu} \int_0^{\mathrm{Re}_P} \left(\mathrm{Ar} - 18\,\mathrm{Re}_P - 3.114\,\mathrm{Re}_P^{1.657} - \frac{0.31\,\mathrm{Re}_P^2}{1 + 16{,}300\,\mathrm{Re}_P^{-1.09}}\right)^{-1} d\,\mathrm{Re}_P \tag{7.4-8}$$

Equation (7.4-8) should be evaluated numerically.

Example 7.5 Calculate the time required for a spherical lead particle, 1.5 mm in diameter, to reach 60% of its terminal velocity in air at 50 °C.

Solution

Physical properties

For air at 50 °C (323 K): $\begin{cases} \rho = 1.0928 \ \mathrm{kg/m^3} \\ \mu = 19.57 \times 10^{-6} \ \mathrm{kg/m \cdot s} \end{cases}$

For lead at 50 °C: $\rho = 11{,}307 \ \mathrm{kg/m^3}$

Analysis

When the particle reaches its terminal velocity, the value of the Reynolds number can be calculated from Eq. (4.3-12). The Archimedes number is

$$\mathrm{Ar} = \frac{D_P^3 g \rho(\rho_P - \rho)}{\mu^2} = \frac{(1.5 \times 10^{-3})^3 (9.8)(1.0928)(11{,}307)}{(19.57 \times 10^{-6})^2} = 1.067 \times 10^6$$

Substitution of this value into Eq. (4.3-12) gives the Reynolds number under steady conditions as

$$\text{Re}_P \,|_{v=v_t} = \frac{\text{Ar}}{18}(1 + 0.0579\,\text{Ar}^{0.412})^{-1.214}$$

$$= \frac{1.067 \times 10^6}{18}\left[1 + 0.0579(1.067 \times 10^6)^{0.412}\right]^{-1.214} = 1701$$

In this problem it is required to calculate the time for the particle to reach a Reynolds number of

$$\text{Re}_P = (0.6)(1701) = 1021$$

Therefore, the required time can be calculated from Eq. (7.4-8) as

$$t = \frac{(11,307)(1.5 \times 10^{-3})^2}{19.57 \times 10^{-6}}\,I \tag{1}$$

where

$$I = \int_0^{\text{Re}_P} \left(1.067 \times 10^6 - 18\,\text{Re}_P - 3.114\,\text{Re}_P^{1.657} - \frac{0.31\,\text{Re}_P^2}{1 + 16,300\,\text{Re}_P^{-1.09}}\right)^{-1} d\,\text{Re}_P$$

The value of I can be determined by using one of the numerical techniques given in Section A.8-4 in Appendix A. The use of the Gauss-Legendre quadrature is shown below. According to Eq. (A.8-13)

$$\text{Re}_P = \frac{1021}{2}(u + 1)$$

and the five-point quadrature is given by

$$I = \frac{1021}{2}\sum_{i=0}^4 w_i F(u_i) \tag{2}$$

where the function $F(u)$ is given by

$$F(u) = \frac{1}{1.067 \times 10^6 - 9189(u+1) - 95602(u+1)^{1.657} - \dfrac{80,789(u+1)^2}{1 + 18.22(u+1)^{-1.09}}}$$

The values of w_i and $F(u_i)$ are given up to three decimals in the following table:

i	u_i	w_i	$F(u_i) \times 10^6$	$w_i F(u_i) \times 10^6$
0	0.000	0.569	1.044	0.594
1	+0.538	0.479	1.187	0.569
2	−0.538	0.479	0.966	0.463
3	+0.906	0.237	1.348	0.319
4	−0.906	0.237	0.940	0.223
			$\sum_{i=0}^4 w_i F(u_i) = 2.17 \times 10^{-6}$	

Therefore, the value of I can be calculated from Eq. (2) as

$$I = \frac{1021}{2}(2.17 \times 10^{-6}) = 1.11 \times 10^{-3}$$

Substitution of this value into Eq. (1) gives

$$t = \frac{(11,307)(1.5 \times 10^{-3})^2(1.11 \times 10^{-3})}{19.57 \times 10^{-6}} = 1.44 \text{ s}$$

7.5 CONSERVATION OF ENERGY

The conservation statement for total energy under unsteady-state conditions is given by

$$\begin{pmatrix} \text{Rate of} \\ \text{energy in} \end{pmatrix} - \begin{pmatrix} \text{Rate of} \\ \text{energy out} \end{pmatrix} = \begin{pmatrix} \text{Rate of energy} \\ \text{accumulation} \end{pmatrix} \qquad (7.5\text{-}1)$$

For a system shown in Figure 7.2, following the discussion explained in Section 6.3, Eq. (7.5-1) is written as

$$\left[(\widehat{U} + \widehat{E}_K + \widehat{E}_P)\dot{m}\right]_{in} - \left[(\widehat{U} + \widehat{E}_K + \widehat{E}_P)\dot{m}\right]_{out} + \dot{Q}_{int} + \dot{W}$$

$$= \frac{d}{dt}\left[(\widehat{U} + \widehat{E}_K + \widehat{E}_P)m\right]_{sys} \qquad (7.5\text{-}2)$$

Note that, contrary to the steady-state flow system, the boundaries of this system are not fixed in space. Therefore, besides shaft and flow works, work associated with the expansion or compression of the system boundaries must be included in $\dot{W}$, thus resulting in the form

$$\dot{W} = \underbrace{-P_{sys}\frac{dV_{sys}}{dt}}_{A} + \underbrace{\dot{W}_s}_{B} + \underbrace{(P\widehat{V}\dot{m})_{in} - (P\widehat{V}\dot{m})_{out}}_{C} \qquad (7.5\text{-}3)$$

where terms A, B, and C represent, respectively, work associated with the expansion or compression of the system boundaries, shaft work, and flow work.

Substitution of Eq. (7.5-3) into Eq. (7.5-2) and the use of the definition of enthalpy, i.e., $\widehat{H} = \widehat{U} + P\widehat{V}$, give

$$\left[(\widehat{H} + \widehat{E}_K + \widehat{E}_P)\dot{m}\right]_{in} - \left[(\widehat{H} + \widehat{E}_K + \widehat{E}_P)\dot{m}\right]_{out} + \dot{Q}_{int} - P_{sys}\frac{dV_{sys}}{dt} + \dot{W}_s$$

$$= \frac{d}{dt}\left[(\widehat{U} + \widehat{E}_K + \widehat{E}_P)m\right]_{sys} \qquad (7.5\text{-}4)$$

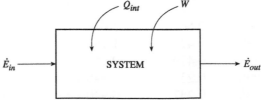

Figure 7.2. Unsteady-state flow system exchanging energy in the form of heat and work with the surroundings.

which is known as the *general energy equation*. Note that under steady conditions Eq. (7.5-4) reduces to Eq. (6.3-9). In terms of molar quantities, Eq. (7.5-4) is written as

$$\left[(\widetilde{H} + \widetilde{E}_K + \widetilde{E}_P)\dot{n}\right]_{in} - \left[(\widetilde{H} + \widetilde{E}_K + \widetilde{E}_P)\dot{n}\right]_{out} + \dot{Q}_{int} - P_{sys}\frac{dV_{sys}}{dt} + \dot{W}_s$$

$$= \frac{d}{dt}\left[(\widetilde{U} + \widetilde{E}_K + \widetilde{E}_P)n\right]_{sys} \tag{7.5-5}$$

When the changes in the kinetic and potential energies between the inlet and outlet of the system as well as within the system are negligible, Eq. (7.5-4) reduces to

$$(\widehat{H}\dot{m})_{in} - (\widehat{H}\dot{m})_{out} + \dot{Q}_{int} - P_{sys}\frac{dV_{sys}}{dt} + \dot{W}_s = \frac{d}{dt}(\widehat{U}m)_{sys} \tag{7.5-6}$$

The accumulation term in Eq. (7.5-6) can be expressed in terms of enthalpy as

$$\frac{d}{dt}(\widehat{U}m)_{sys} = \frac{d}{dt}\left[(\widehat{H} - P\widehat{V})m\right]_{sys} = \frac{d}{dt}(\widehat{H}m)_{sys} - P_{sys}\frac{dV_{sys}}{dt} - V_{sys}\frac{dP_{sys}}{dt} \tag{7.5-7}$$

Substitution of Eq. (7.5-7) into Eq. (7.5-6) gives

$$\boxed{(\widehat{H}\dot{m})_{in} - (\widehat{H}\dot{m})_{out} + \dot{Q}_{int} + V_{sys}\frac{dP_{sys}}{dt} + \dot{W}_s = \frac{d}{dt}(\widehat{H}m)_{sys}} \tag{7.5-8}$$

On a molar basis, Eq. (7.5-8) can be expressed as

$$\boxed{(\widetilde{H}\dot{n})_{in} - (\widetilde{H}\dot{n})_{out} + \dot{Q}_{int} + V_{sys}\frac{dP_{sys}}{dt} + \dot{W}_s = \frac{d}{dt}(\widetilde{H}n)_{sys}} \tag{7.5-9}$$

Example 7.6 Air at atmospheric pressure and 25 °C is flowing at a velocity of 5 m/s over a copper sphere, 1.5 cm in diameter. The sphere is initially at a temperature of 50 °C. How long will it take to cool the sphere to 30 °C? How much heat is transferred from the sphere to the air?

Solution

Physical properties

For air at 25 °C (298 K): $\begin{cases} \mu = 18.41 \times 10^{-6} \text{ kg/m·s} \\ \nu = 15.54 \times 10^{-6} \text{ m}^2\text{/s} \\ k = 25.96 \times 10^{-3} \text{ W/m·K} \\ Pr = 0.712 \end{cases}$

For air at 40 °C (313 K): $\mu = 19.11 \times 10^{-6}$ kg/m·s

For copper at 40 °C (313 K): $\begin{cases} \rho = 8924 \text{ kg/m}^3 \\ \widehat{C}_P = 387 \text{ J/kg·K} \\ k = 397 \text{ W/m·K} \end{cases}$

Assumptions

1. No temperature gradients exist within the sphere, i.e., $Bi_H \ll 1$.
2. The average heat transfer coefficient on the surface of the sphere is constant.
3. The physical properties of copper are independent of temperature.
4. Pseudo-steady-state behavior.

Analysis

System: Copper sphere

For the problem at hand, the terms in Eq. (7.5-8) are

$$\dot{m}_{in} = \dot{m}_{out} = 0$$

$$\dot{W}_s = 0$$

$$\dot{Q}_{int} = -\left(\pi D_P^2\right)\langle h\rangle(T - T_\infty)$$

$$\frac{dP_{sys}}{dt} = 0$$

$$m_{sys} = \left(\pi D_P^3/6\right)\rho_{Cu}$$

$$\widehat{H}_{sys} = (\widehat{C}_P)_{Cu}(T - T_{ref})$$

where T is the copper sphere temperature at any instant and T_∞ is the air temperature. Therefore, Eq. (7.5-8) becomes

$$-\pi D_P^2\langle h\rangle(T - T_\infty) = \left(\frac{\pi D_P^3}{6}\right)(\rho \widehat{C}_P)_{Cu}\frac{dT}{dt} \tag{1}$$

Integration of Eq. (1) with the initial condition that $T = T_i$ at $t = 0$ gives

$$t = \frac{D_P}{6\langle h\rangle}(\rho \widehat{C}_P)_{Cu}\ln\left(\frac{T_i - T_\infty}{T - T_\infty}\right) \tag{2}$$

To determine the average heat transfer coefficient, $\langle h\rangle$, first it is necessary to calculate the Reynolds number:

$$Re_P = \frac{D_P v_\infty}{\nu} = \frac{(0.015)(5)}{15.54 \times 10^{-6}} = 4826$$

The use of the Whitaker correlation, Eq. (4.3-30), gives

$$Nu = 2 + \left(0.4 Re_P^{1/2} + 0.06 Re_P^{2/3}\right) Pr^{0.4}(\mu_\infty/\mu_w)^{1/4}$$

$$= 2 + \left[0.4(4826)^{1/2} + 0.06(4826)^{2/3}\right](0.712)^{0.4}\left(\frac{18.41 \times 10^{-6}}{19.11 \times 10^{-6}}\right)^{1/4} = 40.9$$

The average heat transfer coefficient is

$$\langle h\rangle = Nu\left(\frac{k}{D_P}\right) = (40.9)\left(\frac{25.96 \times 10^{-3}}{0.015}\right) = 71 \text{ W/m}^2\cdot\text{K}$$

Therefore, the time required for cooling is

$$t = \frac{(0.015)(8924)(387)}{(6)(71)} \ln\left(\frac{50-25}{30-25}\right) = 196 \text{ s}$$

The amount of energy transferred from the sphere to the air can be calculated from

$$Q_{int} = \int_0^t \dot{Q}_{int}\, dt = \pi D_P^2 \langle h \rangle \int_0^t (T - T_\infty)\, dt \tag{3}$$

Substitution of Eq. (2) into Eq. (3) and integration yield

$$Q_{int} = \left(\frac{\pi D_P^3}{6}\right)(\rho \widehat{C}_P)_{Cu}(T_i - T_\infty)\left\{1 - \exp\left[-\frac{6\langle h\rangle t}{D_P(\rho \widehat{C}_P)_{Cu}}\right]\right\} \tag{4}$$

Note that from Eq. (2)

$$\exp\left[-\frac{6\langle h\rangle t}{D_P(\rho \widehat{C}_P)_{Cu}}\right] = \frac{T - T_\infty}{T_i - T_\infty} \tag{5}$$

Substitution of Eq. (5) into Eq. (4) gives

$$Q_{int} = \left(\frac{\pi D_P^3}{6}\right)(\rho \widehat{C}_P)_{Cu}(T_i - T) = \left[\frac{\pi (0.015)^3}{6}\right][(8924)(387)](50 - 30) = 122 \text{ J} \tag{6}$$

Verification of assumptions

- Assumption # 1

$$\text{Bi}_H = \frac{\langle h\rangle(D_P/2)}{k_{Cu}} = \frac{(71)(0.015/2)}{397} = 1.34 \times 10^{-3} \ll 1$$

- Assumption # 4

$$\frac{\alpha t}{D_P^2} = \left[\frac{397}{(8924)(387)}\right]\frac{(196)}{(0.015)^2} = 100 \gg 1$$

Comment: Note that Eq. (6) can be simply obtained from the first law of thermodynamics written for a closed system. Considering the copper sphere as a system,

$$\Delta U = Q_{int} + W \quad \Rightarrow \quad Q_{int} = \Delta U = m\widehat{C}_V \Delta T \simeq m\widehat{C}_P \Delta T$$

Example 7.7 A solid sphere at a uniform temperature of T_1 is suddenly immersed in a well-stirred fluid of temperature T_o in an insulated tank ($T_1 > T_o$).

a) Determine the temperatures of the sphere and the fluid as a function of time.
b) Determine the steady-state temperatures of the sphere and the fluid.

Solution

Assumptions

1. The physical properties of the sphere and the fluid are independent of temperature.
2. The average heat transfer coefficient on the surface of the sphere is constant.

3. The sphere and the fluid have uniform but unequal temperatures at any instant, i.e., $Bi_H \ll 1$ and mixing is perfect.

Analysis

a) Since the fluid and the sphere are at different temperatures at a given instant, it is necessary to write two differential equations: one for the fluid, and one for the sphere.

System: Solid sphere

The terms in Eq. (7.5-8) are

$$\dot{m}_{in} = \dot{m}_{out} = 0$$

$$\dot{W}_s = 0$$

$$\dot{Q}_{int} = -\left(\pi D_P^2\right)\langle h \rangle (T_s - T_f)$$

$$\frac{dP_{sys}}{dt} = 0$$

$$m_{sys} = \left(\pi D_P^3/6\right)\rho_s$$

$$\widehat{H}_{sys} = \widehat{C}_{P_s}(T_s - T_{ref})$$

where D_P is the diameter of the sphere, and subscripts s and f stand for the sphere and the fluid, respectively. Therefore, Eq. (7.5-8) becomes

$$-\phi_s(T_s - T_f) = \frac{dT_s}{dt} \tag{1}$$

where

$$\phi_s = \frac{6\langle h \rangle}{D_P \widehat{C}_{P_s} \rho_s} \tag{2}$$

System: Fluid in the tank

The terms in Eq. (7.5-8) are

$$\dot{m}_{in} = \dot{m}_{out} = 0$$

$$\dot{W}_s = 0$$

$$\dot{Q}_{int} = \left(\pi D_P^2\right)\langle h \rangle (T_s - T_f)$$

$$\frac{dP_{sys}}{dt} = 0$$

$$m_{sys} = m_f$$

$$\widehat{H}_{sys} = \widehat{C}_{P_f}(T_f - T_{ref})$$

Hence, Eq. (7.5-8) reduces to

$$\phi_f(T_s - T_f) = \frac{dT_f}{dt} \tag{3}$$

where

$$\phi_f = \frac{\langle h \rangle \pi D_P^2}{m_f \widehat{C}_{P_f}} \tag{4}$$

From Eq. (1), the fluid temperature, T_f, is given in terms of the sphere temperature, T_s, as

$$T_f = T_s + \frac{1}{\phi_s}\frac{dT_s}{dt} \tag{5}$$

Substitution of Eq. (5) into Eq. (3) gives

$$\frac{d^2 T_s}{dt^2} + \phi\frac{dT_s}{dt} = 0 \tag{6}$$

where

$$\phi = \phi_f + \phi_s \tag{7}$$

Two initial conditions are necessary to solve this second-order ordinary differential equation. One of the initial conditions is

$$\text{at } t = 0 \ \ T_s = T_1 \tag{8}$$

The other initial condition can be obtained from Eq. (5) as

$$\text{at } t = 0 \ \ \frac{dT_s}{dt} = \phi_s(T_o - T_1) \tag{9}$$

The solution of Eq. (6) subject to the initial conditions defined by Eqs. (8) and (9) is

$$T_s = T_1 - \frac{\phi_s}{\phi}(T_1 - T_o)\big[1 - \exp(-\phi t)\big] \tag{10}$$

The use of Eq. (10) in Eq. (5) gives the fluid temperature in the form

$$T_f = T_1 - \frac{T_1 - T_o}{\phi}\big[\phi_s + \phi_f \exp(-\phi t)\big] \tag{11}$$

b) Under steady conditions, i.e., $t \to \infty$, Eqs. (10) and (11) reduce to

$$T_s = T_f = T_\infty = \frac{\phi_f T_1 + \phi_s T_o}{\phi} \tag{12}$$

Comment: Note that the final steady-state temperature, T_∞, can be simply obtained by the application of the first law of thermodynamics. Taking the sphere and the fluid together as a system, we get

$$\Delta U = \frac{\pi D_P^3}{6}\rho_s \widehat{C}_{P_s}(T_\infty - T_1) + m_f \widehat{C}_{P_f}(T_\infty - T_o) = 0 \tag{13}$$

Noting that

$$\frac{\phi_f}{\phi_s} = \frac{\pi D_P^3}{6} \frac{\rho_s \widehat{C}_{P_s}}{m_f \widehat{C}_{P_f}} \tag{14}$$

Equation (13) reduces to

$$\frac{\phi_f}{\phi_s}(T_\infty - T_1) + (T_\infty - T_o) = 0 \tag{15}$$

Solution of Eq. (15) results in Eq. (12).

Example 7.8 A spherical steel tank of volume 0.5 m^3 initially contains air at 7 bar and 50 °C. A relief valve is opened and air is allowed to escape at a constant flow rate of 12 mol/min.

a) If the tank is well insulated, estimate the temperature and pressure of air within the tank after 5 minutes.
b) If heating coils are placed in the tank to maintain the air temperature at 50 °C, estimate the pressure of air and the amount of heat transferred after 5 minutes.

Air may be assumed to be an ideal gas with a constant $\widetilde{C}_P$ of 29 J/mol·K.

Solution

a) System: Contents of the tank

Assumptions

1. Properties of the tank contents are uniform, i.e., $\widetilde{H}_{out} = \widetilde{H}_{sys}$.
2. Heat transfer between the system and its surroundings is almost zero. Note that the insulation around the tank does not necessarily imply that $\widetilde{Q}_{int} = 0$. Since the tank wall is in the surroundings, there will be heat transfer between the tank wall and the air remaining in the tank during the evacuation process. Heat transfer may be considered negligible when (*i*) the mass of the wall is small, (*ii*) the process takes place rapidly (remember that heat transfer is a slow process).

Analysis

Since $\dot{n}_{in} = \dot{n}_{int} = 0$ and there is no chemical reaction, Eq. (7.3-2) reduces to

$$-\dot{n}_{out} = \frac{dn_{sys}}{dt} \quad \Rightarrow \quad -12 = \frac{dn_{sys}}{dt} \tag{1}$$

Integration of Eq. (1) yields

$$n_{sys} = n_o - 12t \tag{2}$$

where n_o is the number of moles of air initially present in the tank, i.e.,

$$n_o = \frac{P_o V}{\mathcal{R} T_o} = \frac{(7)(0.5)}{(8.314 \times 10^{-5})(50 + 273)} = 130.3 \text{ mol}$$

On the other hand, the inventory rate equation for energy, Eq. (7.5-5), takes the form

$$-\tilde{H}_{out}\,\dot{n}_{out} = \frac{d(n\tilde{U})_{sys}}{dt} = n_{sys}\frac{d\tilde{U}_{sys}}{dt} + \tilde{U}_{sys}\frac{dn_{sys}}{dt} \tag{3}$$

Substitution of Eqs. (1) and (2) into Eq. (3) gives

$$-12(\tilde{H}_{out} - \tilde{U}_{sys}) = (n_o - 12t)\frac{d\tilde{U}_{sys}}{dt} \tag{4}$$

Since $\tilde{H} = \tilde{U} + P\tilde{V} = \tilde{U} + \mathcal{R}T$, the use of the first assumption enables us to express the left-hand side of Eq. (4) as

$$\tilde{H}_{out} - \tilde{U}_{sys} = \tilde{H}_{sys} - \tilde{U}_{sys} = (\tilde{U}_{sys} + \mathcal{R}T_{sys}) - \tilde{U}_{sys} = \mathcal{R}T_{sys} \tag{5}$$

On the other hand, the right-hand side of Eq. (4) is expressed in terms of temperature as

$$\frac{d\tilde{U}_{sys}}{dt} = \tilde{C}_V\frac{dT_{sys}}{dt} \tag{6}$$

Hence, substitution of Eqs. (5) and (6) into Eq. (4) gives

$$-12\mathcal{R}T_{sys} = (n_o - 12t)\tilde{C}_V\frac{dT_{sys}}{dt} \tag{7}$$

For an ideal gas

$$\tilde{C}_P = \tilde{C}_V + \mathcal{R} \quad \Rightarrow \quad \frac{\tilde{C}_V}{\mathcal{R}} = \gamma - 1 \tag{8}$$

where

$$\gamma = \frac{\tilde{C}_P}{\tilde{C}_V} = \frac{29}{29 - 8.314} = 1.4 \tag{9}$$

Note that Eq. (7) is a separable equation. Substitution of Eq. (8) into Eq. (7) and rearrangement yield

$$-12(\gamma - 1)\int_0^t \frac{dt}{n_o - 12t} = \int_{T_o}^{T_{sys}} \frac{dT_{sys}}{T_{sys}} \tag{10}$$

Integration gives

$$T_{sys} = T_o\left(\frac{n_o - 12t}{n_o}\right)^{\gamma - 1} \tag{11}$$

The variation of pressure as a function of time can be estimated by using the ideal gas law, i.e.,

$$P_{sys} = \frac{n_{sys}\mathcal{R}T_{sys}}{V} \tag{12}$$

Substitution of Eqs. (2) and (11) into Eq. (12) gives

$$P_{sys} = \frac{\mathcal{R}T_o}{V}(n_o - 12t)\left(\frac{n_o - 12t}{n_o}\right)^{\gamma-1} \tag{13}$$

Since $\mathcal{R}T_o/V = P_o/n_o$, Eq. (13) reduces to

$$P_{sys} = P_o\left(\frac{n_o - 12t}{n_o}\right)^{\gamma} \tag{14}$$

Substitution of the numerical values into Eqs. (11) and (14) gives T_{sys} and P_{sys}, respectively, after 5 minutes as

$$T_{sys} = (50 + 273)\left[\frac{130.3 - (12)(5)}{130.3}\right]^{1.4-1} = 252.4 \text{ K}$$

$$P_{sys} = 7\left[\frac{130.3 - (12)(5)}{130.3}\right]^{1.4} = 2.95 \text{ bar}$$

Comment: Note that Eq. (11) can be rearranged in the form

$$\frac{T_{sys}}{T_o} = \left(\frac{n_{sys}}{n_o}\right)^{\gamma-1} \tag{15}$$

The use of the ideal gas law to express the number of moles gives

$$\frac{T_{sys}}{T_o} = \left(\frac{P_{sys}}{P_o}\right)^{\gamma-1}\left(\frac{T_o}{T_{sys}}\right)^{\gamma-1} \quad \Rightarrow \quad \frac{T_{sys}}{T_o} = \left(\frac{P_{sys}}{P_o}\right)^{(\gamma-1)/\gamma} \tag{16}$$

which is a well-known equation for a closed system undergoing a reversible adiabatic (or isentropic) process. Therefore, the gas remaining in the tank at the end of 5 min undergoes reversible adiabatic expansion throughout the process.

b) System: Contents of the tank

Assumption

1. Properties of the tank contents are uniform, i.e., $\widetilde{H}_{out} = \widetilde{H}_{sys}$.

Analysis

Equation (7.3-2) becomes

$$-\dot{n}_{out} = \frac{dn_{sys}}{dt} \quad \Rightarrow \quad -12 = \frac{dn_{sys}}{dt} \tag{17}$$

Integration of Eq. (17) yields

$$n_{sys} = n_o - 12t \tag{18}$$

where n_o is the number of moles of air initially present in the tank, i.e.,

$$n_o = \frac{P_o V}{\mathcal{R} T_o} = \frac{(7)(0.5)}{(8.314 \times 10^{-5})(50 + 273)} = 130.3 \text{ mol}$$

In this case, the process is isothermal and, as a result, the pressure of the system can be directly calculated from the ideal gas law, i.e.,

$$P_{sys} = \left(\frac{\mathcal{R} T_{sys}}{V}\right) n_{sys} \tag{19}$$

The use of Eq. (18) in Eq. (19) results in

$$P_{sys} = \left(\frac{\mathcal{R} T_{sys}}{V}\right)(n_o - 12t) = P_o - 12\left(\frac{\mathcal{R} T_{sys}}{V}\right)t \tag{20}$$

Substitution of the numerical values gives

$$P = 7 - \frac{(12)(8.314 \times 10^{-5})(50 + 273)(5)}{0.5} = 3.78 \text{ bar}$$

The amount of heat supplied by the heating coils is determined from the inventory rate equation for energy, Eq. (7.5-5). Simplification of this equation gives

$$-\widetilde{H}_{out} \dot{n}_{out} + \dot{Q}_{int} = \frac{d(n\widetilde{U})_{sys}}{dt} = \widetilde{U}_{sys} \frac{dn_{sys}}{dt} \tag{21}$$

Since the process is isothermal, $\widetilde{U}_{sys}$ remains constant. Substituting Eq. (17) into Eq. (21) and using the fact that $\widetilde{H}_{out} = \widetilde{H}_{sys}$ yield

$$\dot{Q}_{int} = 12(\widetilde{H}_{sys} - \widetilde{U}_{sys}) = 12\mathcal{R} T_{sys} = (12)(8.314)(50 + 273) = 32{,}225 \text{ J/min}$$

Therefore, the amount of heat transferred is

$$Q_{int} = \dot{Q}_{int} t = (32{,}225)(5) = 161{,}125 \text{ J}$$

7.5.1 Unsteady-State Energy Balance Around a Continuous Stirred Tank Reactor

An unsteady-state energy balance in a continuous stirred tank reactor (CSTR) follows the same line as the steady-state case given in Section 6.3.2.2. Using the same assumptions, the resulting energy balance becomes

$$\left[\sum_i \dot{n}_i \overline{H}_i(T_{in})\right]_{in} - \left[\sum_i \dot{n}_i \overline{H}_i(T)\right]_{out} + \dot{Q}_{int} = \frac{d}{dt}\left[\sum_i n_i \overline{H}_i(T)\right]_{sys} \tag{7.5-10}$$

On the other hand, the macroscopic mole balance for species i, Eq. (7.2-5), is

$$(\dot{n}_i)_{in} - (\dot{n}_i)_{out} + V_{sys} \sum_j \alpha_{ij} r_j = \frac{d(n_i)_{sys}}{dt} \tag{7.5-11}$$

Multiplication of Eq. (7.5-11) by $\overline{H}_i(T)$ and summation over all species give

$$\left[\sum_i \dot{n}_i \overline{H}_i(T)\right]_{in} - \left[\sum_i \dot{n}_i \overline{H}_i(T)\right]_{out} - V_{sys}\sum_j r_j(-\Delta H_{rxn,j}) = \left[\sum_i \overline{H}_i(T)\frac{dn_i}{dt}\right]_{sys}$$

$$(7.5\text{-}12)$$

Subtraction of Eq. (7.5-12) from Eq. (7.5-10) yields

$$\sum_i (\dot{n}_i)_{in}\left[\overline{H}_i(T_{in}) - \overline{H}_i(T)\right] + \dot{Q}_{int} + V_{sys}\sum_j r_j(-\Delta H_{rxn,j}) = \left[\sum_i n_i(T)\frac{d\overline{H}_i(T)}{dt}\right]_{sys}$$

$$(7.5\text{-}13)$$

Dividing Eq. (7.5-13) by the volumetric flow rate, $\mathcal{Q}$, gives

$$\sum_i (c_i)_{in}\left[\overline{H}_i(T_{in}) - \overline{H}_i(T)\right] + \frac{\dot{Q}_{int}}{\mathcal{Q}} + \tau\sum_j r_j(-\Delta H_{rxn,j}) = \tau\left[\sum_i c_i(T)\frac{d\overline{H}_i(T)}{dt}\right]_{sys}$$

$$(7.5\text{-}14)$$

where τ is the residence time. Expressing the partial molar enthalpy of species i in terms of the partial molar heat capacity by Eq. (6.3-41) gives

$$\boxed{(C_P)_{in}(T_{in} - T) + \frac{\dot{Q}_{int}}{\mathcal{Q}} + \tau\sum_j r_j(-\Delta H_{rxn,j}) = \tau(C_P)_{sys}\frac{dT}{dt}} \qquad (7.5\text{-}15)$$

where

$$(C_P)_{in} = \sum_i (c_i)_{in}\overline{C}_{P_i} \qquad (7.5\text{-}16)$$

$$(C_P)_{sys} = \sum_i (c_i)_{sys}\overline{C}_{P_i} \qquad (7.5\text{-}17)$$

Note that Eq. (7.5-15) reduces to Eq. (6.3-42) under steady conditions. On the other hand, for a batch reactor, i.e., no inlet or outlet streams, Eq. (7.5-15) takes the form

$$\boxed{\dot{Q}_{int} + V_{sys}\sum_j r_j(-\Delta H_{rxn,j}) = V_{sys}(C_P)_{sys}\frac{dT}{dt}} \qquad (7.5\text{-}18)$$

It is important to note that Eqs. (7.5-15) and (7.5-18) are valid for systems in which pressure remains constant.

Example 7.9 The reaction described in Example 6.6 is to be carried out in a batch reactor that operates adiabatically. The reactor is initially charged with 2000 moles of species $\mathcal{A}$ and 2400 moles of species $\mathcal{B}$ at a temperature of 25 °C. Determine the time required for 80% conversion of $\mathcal{A}$ if the reactor volume is 1 m³.

Solution

System: Contents of the reactor

The conservation statement for species $\mathcal{A}$, Eq. (7.2-5), is

$$-kc_A c_B V = \frac{d\dot{n}_A}{dt} \qquad (1)$$

or,

$$-kn_A n_B = V \frac{dn_A}{dt} \qquad (2)$$

The number of moles of species $\mathcal{A}$ and $\mathcal{B}$ in terms of the molar extent of the reaction, ε, is given by

$$n_A = n_{A_o} + \alpha_A \varepsilon = 2000 - \varepsilon \qquad (3)$$

$$n_B = n_{B_o} + \alpha_B \varepsilon = 2400 - \varepsilon \qquad (4)$$

The molar extent of the reaction can be calculated from Eq. (5.3-12) as

$$\varepsilon = \frac{n_{A_o}}{(-\alpha_A)} X_A = \frac{(2000)(0.8)}{1} = 1600 \text{ mol} \qquad (5)$$

Substitution of Eqs. (3) and (4) into Eq. (2) and rearrangement give

$$t = V \int_0^{1600} \frac{d\varepsilon}{k(2000 - \varepsilon)(2400 - \varepsilon)} \qquad (6)$$

Note that Eq. (6) cannot be integrated directly since the reaction rate constant, k, is dependent on ε via temperature.

The energy equation must be used to determine the variation of temperature as a function of the molar extent of the reaction. For an adiabatic reactor, i.e., $\dot{Q}_{int} = 0$, Eq. (7.5-18) reduces to

$$r(-\Delta H_{rxn}^o) = (C_P)_{sys} \frac{dT}{dt} \qquad (7)$$

Substitution of Eqs. (5.3-22) and (7.5-17) into Eq. (7) yields

$$(-\Delta H_{rxn}^o) \frac{d\varepsilon}{dt} = \left[\left(\sum_i n_{i_o} \tilde{C}_{P_i} \right) + \Delta \tilde{C}_P^o \varepsilon \right] \frac{dT}{dt} \qquad (8)$$

In this problem

$$\Delta \tilde{C}_P^o = -85 \text{ J/mol·K} \qquad (9)$$

$$\sum_i n_{i_o} \tilde{C}_{P_i} = (2000)(175) + (2400)(130) = 662{,}000 \qquad (10)$$

$$\Delta H_{rxn}^o = -45{,}000 - 85(T - 298) \qquad (11)$$

Substitution of Eqs. (9)–(11) into Eq. (8) and rearrangement give

$$\int_0^\varepsilon \frac{d\varepsilon}{662,000 - 85\varepsilon} = \int_{298}^T \frac{dT}{45,000 + 85(T - 298)} \tag{12}$$

Integration gives

$$T = 298 + \frac{45,000\varepsilon}{662,000 - 85\varepsilon} \tag{13}$$

Now it is possible to evaluate Eq. (6) numerically. The use of Simpson's rule with $n = 8$, i.e., $\Delta\varepsilon = 200$, gives

ε (mol/m^3)	T (K)	$\left[k(2000 - \varepsilon)(2400 - \varepsilon)\right]^{-1} \times 10^4$
0	298	248
200	312	121.9
400	326.7	63.3
600	342.2	34.9
800	358.6	20.5
1000	376	12.9
1200	394.4	8.9
1400	414	6.9
1600	434.9	6.5

The application of Eq. (A.8-12) in Appendix A reduces Eq. (6) to

$$t = \frac{200}{3}\left[248 + 4(121.9 + 34.9 + 12.9 + 6.9) + 2(63.3 + 20.5 + 8.9) + 6.5\right] \times 10^{-4}$$

$$= 7.64 \text{ min} \tag{14}$$

7.6 DESIGN OF A SPRAY TOWER FOR THE GRANULATION OF MELT

The purpose of this section is to apply the concepts covered in this chapter to a practical design problem. A typical tower for melt granulation is shown in Figure 7.3. The dimensions of the tower must be determined such that the largest melt particles solidify before striking the walls or the floor of the tower. Mathematical modeling of this tower can be accomplished by considering the unsteady-state macroscopic energy balances for the melt particles in conjunction with their settling velocities. This enables one to determine the cooling time and thus the dimensions of the tower.

It should be remembered that mathematical modeling is a highly interactive process. It is customary to build the initial model as simple as possible by making assumptions. Experience gained in working through this simplified model gives a feeling for the problem and builds confidence. The process is repeated several times, each time relaxing one of the assumptions and thus making the model more realistic. In the design procedure presented below, the following assumptions are made:

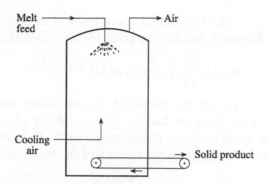

Figure 7.3. Schematic diagram of a spray cooling tower.

1. The particle falls at a constant terminal velocity.
2. Energy losses from the tower are negligible.
3. Particles do not shrink or expand during solidification, i.e., solid and melt densities are almost the same.
4. The temperature of the melt particle is uniform at any instant, i.e., Bi ≪ 1.
5. The physical properties are independent of temperature.
6. Solid particles at the bottom of the tower are at temperature T_s, the solidification temperature.

7.6.1   Determination of Tower Diameter

The mass flow rate of air can be calculated from the energy balance around the tower:

$$\begin{pmatrix} \text{Rate of energy} \\ \text{gained by air} \end{pmatrix} = \begin{pmatrix} \text{Rate of energy lost} \\ \text{by the melt particles} \end{pmatrix} \tag{7.6-1}$$

or,

$$\dot{m}_a \langle \widehat{C}_{P_a} \rangle \left[(T_a)_{out} - (T_a)_{in} \right] = \dot{m}_m \left\{ \widehat{C}_{P_m} \left[(T_m)_{in} - T_s \right] + \widehat{\lambda} \right\} \tag{7.6-2}$$

where the subscripts a and m stand for the air and the melt particle, respectively, and $\widehat{\lambda}$ is the latent heat of fusion per unit mass.

Once the air mass flow rate, $\dot{m}_a$, is calculated from Eq. (7.6-2), the diameter of the tower is calculated as

$$\dot{m}_a = \left(\frac{\pi D^2}{4} \right) v_a \rho_a \quad \Rightarrow \quad D = \sqrt{\frac{4 \dot{m}_a}{\pi \rho_a v_a}} \tag{7.6-3}$$

7.6.2   Determination of Tower Height

Tower height, H, is determined from

$$H = v_t t \tag{7.6-4}$$

The terminal velocity of the falling particle, v_t, is determined by using the formulas given in Section 4.3. The required cooling time, t, is determined from the unsteady-state energy balance around the melt particle.

7.6.2.1 *Terminal velocity*   The Turton-Clark correlation is an explicit relationship between the Archimedes and the Reynolds numbers as given by Eq. (4.3-12), i.e.,

$$\mathrm{Re}_P = \frac{\mathrm{Ar}}{18}(1 + 0.0579\,\mathrm{Ar}^{0.412})^{-1.214} \tag{7.6-5}$$

The Archimedes number, Ar, can be calculated directly when the particle diameter and the physical properties of the fluid are known. The use of Eq. (7.6-5) then determines the Reynolds number. In this case, however, the definition of the Reynolds number involves the relative velocity, v_r, rather than the terminal velocity of the melt particle, i.e.,

$$\mathrm{Re}_P = \frac{D_P v_r \rho_a}{\mu_a} \tag{7.6-6}$$

Since the air and the melt particle flow in countercurrent direction to each other, the relative velocity, v_r, is

$$v_r = v_t + v_a \tag{7.6-7}$$

7.6.2.2 *Cooling time*   The total cooling time consists of two parts: the cooling period during which the melt temperature decreases from the temperature at the inlet to T_s, and the solidification period during which the temperature of the melt remains at T_s.

i) **Cooling period:** Considering the melt particle as a system, the terms appearing in Eq. (7.5-8) become

$$\dot{m}_{in} = \dot{m}_{out} = 0$$

$$\dot{W}_s = 0$$

$$\dot{Q}_{int} = -\left(\pi D_P^2\right)\langle h \rangle \left(T_m - \langle T_a \rangle\right)$$

$$\frac{dP_{sys}}{dt} = 0$$

$$m_{sys} = \left(\pi D_P^3/6\right)\rho_m$$

$$\widehat{H}_{sys} = \widehat{C}_{P_m}(T_m - T_{ref})$$

where $\langle T_a \rangle$ is the average air temperature, i.e., $[(T_a)_{in} + (T_a)_{out}]/2$. Hence, Eq. (7.5-8) takes the form

$$-6\langle h \rangle \left(T_m - \langle T_a \rangle\right) = D_P \rho_m \widehat{C}_{P_m} \frac{dT_m}{dt} \tag{7.6-8}$$

Equation (7.6-8) is a separable equation and rearrangement yields

$$\int_0^{t_1} dt = -\frac{D_P \rho_m \widehat{C}_{P_m}}{6\langle h \rangle} \int_{(T_m)_{in}}^{T_s} \frac{dT_m}{T_m - \langle T_a \rangle} \tag{7.6-9}$$

Integration of Eq. (7.6-9) gives the cooling time, t_1, as

$$t_1 = \frac{D_P \rho_m \widehat{C}_{P_m}}{6\langle h \rangle} \ln\left[\frac{(T_m)_{in} - \langle T_a \rangle}{T_s - \langle T_a \rangle}\right] \tag{7.6-10}$$

The average heat transfer coefficient, $\langle h \rangle$ in Eq. (7.6-10) can be calculated from the Whitaker correlation, Eq. (4.3-30), i.e.,

$$\mathrm{Nu} = 2 + \left(0.4\,\mathrm{Re}_P^{1/2} + 0.06\,\mathrm{Re}_P^{2/3}\right)\mathrm{Pr}^{0.4}(\mu_\infty/\mu_w)^{1/4} \qquad (7.6\text{-}11)$$

ii) **Solidification period:** During the solidification process, solid and liquid phases coexist and temperature remains constant at T_s. Considering the particle as a system, the terms appearing in Eq. (7.5-8) become

$$\dot{m}_{in} = \dot{m}_{out} = 0$$

$$\dot{W}_s = 0$$

$$\dot{Q}_{int} = -\left(\pi D_P^2\right)\langle h \rangle\left(T_s - \langle T_a \rangle\right)$$

$$\frac{dP_{sys}}{dt} = 0$$

$$m_{sys} = m_l + m_s$$

$$T_{ref} = T_s \Rightarrow \begin{cases} \widehat{H}_l = 0 \\ \widehat{H}_s = -\widehat{\lambda} \end{cases} \Rightarrow (m\widehat{H})_{sys} = m_l\widehat{H}_l + m_s\widehat{H}_s = -\widehat{\lambda}\,m_s$$

where m_l and m_s represent the liquid and solidified portions of the particle, respectively. Therefore, Eq. (7.5-8) reduces to

$$\pi D_P^2 \langle h \rangle\left(T_s - \langle T_a \rangle\right) = \widehat{\lambda}\,\frac{dm_s}{dt} \qquad (7.6\text{-}12)$$

Integration of Eq. (7.6-12) gives the time required for solidification, t_2, as

$$\boxed{t_2 = \frac{\widehat{\lambda}\rho_m D_P}{6\,\langle h \rangle\left(T_s - \langle T_a \rangle\right)}} \qquad (7.6\text{-}13)$$

Therefore, the total time, t, in Eq. (7.6-4) is

$$t = t_1 + t_2 \qquad (7.6\text{-}14)$$

Example 7.10 Determine the dimensions of the spray cooling tower for the following conditions:

Production rate $= 3000$ kg/h $\qquad D_P = 2$ mm $\qquad \rho_m = 1700$ kg/m^3

$(T_a)_{in} = 10\,^\circ$C $\qquad (T_a)_{out} = 20\,^\circ$C $\qquad (T_m)_{in} = 110\,^\circ$C $\qquad T_s = 70\,^\circ$C

$v_a = 2$ m/s $\qquad \widehat{\lambda} = 186$ kJ/kg $\qquad \widehat{C}_{P_m} = 1.46$ kJ/kg·K

Solution

Physical properties

The average air temperature is $(10 + 20)/2 = 15\,°C$.

For air at $15\,°C$ (288 K):
$$
\begin{cases}
\rho = 1.2 \text{ kg/m}^3 \\
\mu = 17.93 \times 10^{-6} \text{ kg/m·s} \\
k = 25.22 \times 10^{-3} \text{ W/m·K} \\
\widehat{C}_P = 1.004 \text{ kJ/kg·K} \\
\text{Pr} = 0.714
\end{cases}
$$

Analysis

The mass flow rate of air, $\dot{m}_a$, is calculated from Eq. (7.6-2) as

$$
\dot{m}_a = \frac{\dot{m}_m\{\widehat{C}_{P_m}[(T_m)_{in} - T_s] + \widehat{\lambda}\}}{\langle\widehat{C}_{P_a}\rangle[(T_a)_{out} - (T_a)_{in}]}
$$

$$
= \frac{(3000)[(1.46)(110 - 70) + 186]}{(1.004)(20 - 10)} = 73{,}028 \text{ kg/h}
$$

The use of Eq. (7.6-3) gives the tower diameter as

$$
D = \sqrt{\frac{4\dot{m}_a}{\pi \rho_a v_a}} = \sqrt{\frac{(4)(73{,}028)}{\pi(1.2)(2)(3600)}} = 3.3 \text{ m}
$$

The use of Eq. (4.3-6) gives the Archimedes number as

$$
\text{Ar} = \frac{D_P^3 g \rho_a (\rho_m - \rho_a)}{\mu_a^2} = \frac{(2 \times 10^{-3})^3(9.8)(1.2)(1700 - 1.2)}{(17.93 \times 10^{-6})^2} = 4.97 \times 10^5
$$

Hence, the Reynolds number and the relative velocity are

$$
\text{Re}_P = \frac{\text{Ar}}{18}(1 + 0.0579\,\text{Ar}^{0.412})^{-1.214}
$$

$$
= \frac{4.97 \times 10^5}{18}\left[1 + 0.0579(4.97 \times 10^5)^{0.412}\right]^{-1.214} = 1134
$$

$$
v_r = \frac{\mu_a \text{Re}_P}{\rho_a D_P} = \frac{(17.93 \times 10^{-6})(1134)}{(1.2)(2 \times 10^{-3})} = 8.5 \text{ m/s}
$$

Therefore, the terminal velocity of the particle is

$$
v_t = v_r - v_a = 8.5 - 2 = 6.5 \text{ m/s}
$$

The use of the Whitaker correlation, Eq. (7.6-11), with $\mu_\infty/\mu_w \approx 1$, gives

$$
\text{Nu} = 2 + \left(0.4\,\text{Re}_P^{1/2} + 0.06\,\text{Re}_P^{2/3}\right)\text{Pr}^{0.4}(\mu_\infty/\mu_w)^{1/4}
$$

$$
= 2 + \left[0.4(1134)^{1/2} + 0.06(1134)^{2/3}\right](0.714)^{0.4} = 19.5
$$

Hence, the average heat transfer coefficient is

$$\langle h \rangle = \mathrm{Nu}\left(\frac{k}{D_P}\right) = (19.5)\left(\frac{25.22 \times 10^{-3}}{2 \times 10^{-3}}\right) = 246 \ \mathrm{W/m^2 \cdot K}$$

The time required for cooling and solidification can be calculated from Eqs. (7.6-10) and (7.6-13), respectively:

$$t_1 = \frac{D_P \rho_m \widehat{C}_{P_m}}{6 \langle h \rangle} \ln\left[\frac{(T_m)_{in} - \langle T_a \rangle}{T_s - \langle T_a \rangle}\right] = \frac{(2 \times 10^{-3})(1700)(1460)}{(6)(246)} \ln\left(\frac{110 - 15}{70 - 15}\right) = 1.8 \ \mathrm{s}$$

$$t_2 = \frac{\widehat{\lambda} \rho_m D_P}{6 \langle h \rangle (T_s - \langle T_a \rangle)} = \frac{(186{,}000)(1700)(2 \times 10^{-3})}{(6)(246)(70 - 15)} = 7.8 \ \mathrm{s}$$

Therefore, the tower height is

$$H = (6.5)(1.8 + 7.8) = 62.4 \ \mathrm{m}$$

NOTATION

A	area, $\mathrm{m^2}$
A_M	mass transfer area, $\mathrm{m^2}$
$\widehat{C}_V$	heat capacity at constant volume, $\mathrm{kJ/kg \cdot K}$
$\widehat{C}_P$	heat capacity at constant pressure, $\mathrm{kJ/kg \cdot K}$
c	concentration, $\mathrm{kmol/m^3}$
D_P	particle diameter, m
$\mathcal{D}_{AB}$	diffusion coefficient for system $\mathcal{A}$-$\mathcal{B}$, $\mathrm{m^2/s}$
E_K	kinetic energy, J
E_P	potential energy, J
$\dot{E}$	rate of energy, J/s
$\mathcal{E}$	activation energy, J/mol
f	friction factor
g	acceleration of gravity, $\mathrm{m/s^2}$
H	enthalpy, J
h	elevation, m; heat transfer coefficient, $\mathrm{W/m^2 \cdot K}$
k	thermal conductivity, $\mathrm{W/m \cdot K}$
k_c	mass transfer coefficient, m/s
L	length, m
$\dot{m}$	mass flow rate, kg/s
$\mathcal{M}$	molecular weight, kg/kmol
$\dot{n}$	molar flow rate, kmol/s
P	pressure, Pa
$\dot{Q}$	heat transfer rate, W
$\mathcal{Q}$	volumetric flow rate, $\mathrm{m^3/s}$
r	rate of a chemical reaction, $\mathrm{kmol/m^3 \cdot s}$

$\mathcal{R}$ gas constant, J/mol·K
T temperature, °C or K
t time, s
U internal energy, J
V volume, m^3
v velocity, m/s
$\dot{W}$ rate of work, W
$\dot{W}_s$ rate of shaft work, W
X fractional conversion
x_i mole fraction of species i

α thermal diffusivity, m^2/s
α_{ij} stoichiometric coefficient of the ith species in the jth reaction
γ $\widetilde{C}_P/\widetilde{C}_V$
Δ difference
ΔH_{rxn} heat of reaction, J
ε molar extent of a reaction, kmol
λ latent heat, J
v kinematic viscosity (or momentum diffusivity), m^2/s
ρ density, kg/m^3
τ residence time, s

Overlines

$\sim$ per mole
$\wedge$ per unit mass
$-$ partial molar

Bracket

$\langle a \rangle$ average value of a

Superscripts

o standard state
S solid
sat saturation

Subscripts

A, B species in binary systems
a air
ch characteristic
i species in multicomponent systems
in inlet
int interphase
j reaction number

m	melt
out	outlet
P	particle
sys	system
w	surface or wall
∞	free-stream

Dimensionless Numbers

Ar	Archimedes number
Bi_H	Biot number for heat transfer
Bi_M	Biot number for mass transfer
Pr	Prandtl number
Re	Reynolds number
Sc	Schmidt number
Sh	Sherwood number

REFERENCES

Bondy Lippa, 1983, Heat transfer in agitated vessels, Chem. Eng. 90, 62.

Chapra, S.C. and R.P. Canale, 1991, Long-term phenomenological model of phosphorous and oxygen for stratified lakes, Wat. Res. 25, 707.

Eubank, P.T., M.G. Johnson, J.C. Holste and K.R. Hall, 1985, Thermal errors in metering pumps, Ind. Eng. Chem. Fund. 24, 392.

Foster, T.C., Time required to empty a vessel, 1981, Chem. Eng. 88 (May 4), 105.

Horwitz, B.A., 1981, How to cool your beer more quickly, Chem. Eng. 88 (Aug. 10), 97.

Kinsley, G.R., 2001, Properly purge and inert storage vessels, Chem. Eng. Prog. 97(2), 57.

Tosun, İ. and İ. Akşahin, 1993, Predict heating and cooling times accurately, Chem. Eng. 100 (Nov.), 183.

SUGGESTED REFERENCES FOR FURTHER STUDY

Bird, R.B., W.E. Stewart and E.N. Lightfoot, 2002, Transport Phenomena, 2nd Ed., Wiley, New York.

Churchill, S.W., 1974, The Interpretation and Use of Rate Data: The Rate Concept, Scripta Publishing Co., Washington, D.C.

Felder, R.M. and R.W. Rousseau, 2000, Elementary Principles of Chemical Processes, 3rd Ed., Wiley, New York.

Incropera, F.P. and D.P. DeWitt, 2002, Fundamentals of Heat and Mass Transfer, 5th Ed., Wiley, New York.

Whitwell, J.C. and R.K. Toner, 1969, Conservation of Mass and Energy, McGraw-Hill, New York.

PROBLEMS

7.1 Purging is the addition of an inert gas, such as nitrogen or carbon dioxide, to a piece of process equipment that contains flammable vapors or gases to provide the space nonignitable for a certain period of time. One of the purging methods is *sweep-through purging* (Kinsley, 2001) in which a purge gas is introduced into a vessel at one opening and the mixed gas is withdrawn at another opening and vented either to the atmosphere or to an air-pollution control device.

A 80 m^3 tank is initially charged with air at atmospheric pressure. Determine the volume of nitrogen that must be swept through it in order to reduce the oxygen concentration to 1% by volume.

(**Answer:** 243.6 m^3)

7.2 Two perfectly stirred tanks with capacities of 1.5 and 0.75 m^3 are connected in such a way that the effluent from the first passes into the second. Both tanks are initially filled with salt solution 0.5 kg/L in concentration. If pure water is fed into the first tank at a rate of 75 L/min, determine the salt concentration in the second tank after 10 minutes?

(**Answer:** 0.423 kg/L)

7.3 Two vertical tanks placed on a platform are connected by a horizontal pipe 5 cm in diameter as shown in Figure 7.4. Each tank is 2 m deep and 1 m in diameter. At first, the valve on the pipe is closed and one tank is full while the other is empty. When the valve is opened, the average velocity through the pipe is given by

$$\langle v \rangle = 2\sqrt{h}$$

where $\langle v \rangle$ is the average velocity in m/s and h is the difference between the levels in the two tanks in meters. Calculate the time for the levels in the two tanks to become equal.

(**Answer:** 4.7 min)

7.4 **a)** A stream containing 10% species $\mathcal{A}$ by weight starts to flow at a rate of 2 kg/min into a tank originally holding 300 kg of pure $\mathcal{B}$. Simultaneously, a valve at the bottom of the tank is opened and the tank contents are also withdrawn at a rate of 2 kg/min. Considering perfect mixing within the tank, determine the time required for the exit stream to contain 5% species $\mathcal{A}$ by weight.

b) Consider the problem in part (a). As a result of the malfunctioning of the exit valve, tank contents are withdrawn at a rate of 2.5 kg/min instead of 2 kg/min. How long does it take for the exit stream to contain 5% species $\mathcal{A}$ in this case?

(**Answer:** a) 104 min b) 95.5 min)

7.5 The following levels were measured for the flow system shown in Figure 7.5. The cross-sectional area of each tank is 1.5 m^2.

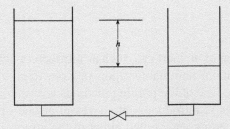

Figure 7.4. Schematic diagram for Problem 7.3.

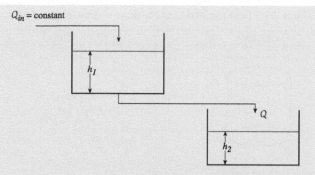

Figure 7.5 Schematic diagram for Problem 7.5.

t	h_1	h_2
(min)	(cm)	(cm)
0	50	30
1	58	35
2	67	40
3	74	46
4	82	51
5	89	58
6	96	64

a) Determine the value of Q_{in}.

b) If the flow rate of the stream leaving the first tank, Q, is given as

$$Q = \beta \sqrt{h_1}$$

determine the value of β.

(**Answer:** a) 0.2 m^3/min b) 0.1 m$^{5/2}$/min)

7.6 Time required to empty a vessel is given for four common tank geometries by Foster (1981) as shown in Table 7.1. In each case, the liquid leaves the tank through an orifice of cross-sectional area A_o. The orifice coefficient is C_o. Assume that the pressure in each tank is atmospheric. Verify the formulas in Table 7.1.

7.7 For steady flow of an incompressible fluid through a control volume whose boundaries are stationary in space, show that Eq. (6.3-9) reduces to

$$\frac{\Delta P}{\rho} + \frac{\Delta \langle v \rangle^2}{2} + g \, \Delta h + \left(\Delta \widehat{U} - \widehat{Q}_{int} \right) = \widehat{W}_s \tag{1}$$

where Δ represents the difference between the outlet and inlet values.

a) Using the thermodynamic relations

$$d\widehat{U} = T \, d\widehat{S} - P \, d\widehat{V} \tag{2}$$

Table 7.1   Time required to empty tanks of different geometries

Geometry	Time

$$t = \frac{\pi D^2 \sqrt{h}}{\sqrt{8g}\, C_o A_o}$$

$$t = \sqrt{\frac{2}{g}}\, \frac{\pi h^{5/2} \tan^2 \theta}{5 C_o A_o}$$

$$t = \sqrt{\frac{8}{g}}\, \frac{L[D^{3/2} - (D-h)^{3/2}]}{3\, C_o A_o}$$

$$t = \sqrt{\frac{2}{g}}\, \frac{\pi h^{3/2}(D - 0.6h)}{3 C_o A_o}$$

and

$$d\widehat{S} = \frac{d\widehat{Q}_{int}}{T} + d\widehat{S}_{gen} \tag{3}$$

show that

$$d\widehat{E}_v = T\, d\widehat{S}_{gen} = d\widehat{U} - d\widehat{Q}_{int} \tag{4}$$

where $\widehat{E}_v$, the friction loss per unit mass, represents the irreversible degradation of mechanical energy into thermal energy, and $\widehat{S}_{gen}$ is the entropy generation per unit mass.

b) Substitute Eq. (4) into Eq. (1) to obtain the engineering Bernoulli equation (or macroscopic mechanical energy equation) for an incompressible fluid as

$$\frac{\Delta P}{\rho} + \frac{\Delta \langle v \rangle^2}{2} + g\, \Delta h + \widehat{E}_v - \widehat{W}_s = 0 \tag{5}$$

c) To estimate the friction loss for flow in a pipe, consider steady flow of an incompressible fluid in a horizontal pipe of circular cross-section. Simplify Eq. (5) for this case to get

$$\widehat{E}_v = \frac{|\Delta P|}{\rho} \tag{6}$$

Compare Eq. (6) with Eq. (4.5-6) and show that the friction loss per unit mass, $\widehat{E}_v$, for pipe flow is given by

$$\widehat{E}_v = \frac{2fL\langle v\rangle^2}{D} \tag{7}$$

7.8 A cylindrical tank, 5 m in diameter, discharges through a mild steel pipe system ($\varepsilon = 4.6 \times 10^{-5}$ m) connected to the tank base as shown in the figure below. The drain pipe system has an equivalent length of 100 m and a diameter of 23 cm. The tank is initially filled with water to an elevation of H with respect to the reference plane.

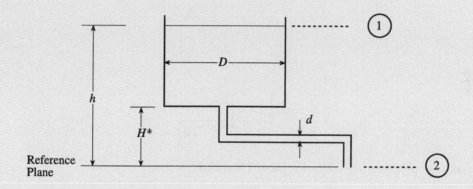

a) Apply the Bernoulli equation, Eq. (5) in Problem 7.7, to the region between planes "1" and "2" and show that

$$\langle v_2\rangle^2 = \frac{2gh}{1 + \dfrac{4fL_{eq}}{d}}$$

where L_{eq} is the equivalent length of the drain pipe.

b) Consider the tank as a system and show that the application of the unsteady-state macroscopic mass balance gives

$$dt = -\left(\frac{D}{d}\right)^2 \sqrt{\frac{1}{2g}\left(1 + \frac{4fL_{eq}}{d}\right)} \frac{dh}{\sqrt{h}} \tag{2}$$

Analytical integration of Eq. (2) is possible only if the friction factor f is constant.

c) At any instant, note that the pressure drop in the drain pipe system is equal to $\rho g(h - H^*)$. Use Eqs. (4.5-18)–(4.5-20) to determine f as a function of liquid height in the tank. Take $H^* = 1$ m, $H = 4$ m, and the final value of h as 1.5 m.

d) If f remains almost constant, then show that the integration of Eq. (2) yields

$$t = \left(\frac{D}{d}\right)^2 \sqrt{\frac{2}{g}\left(1 + \frac{4fL_{eq}}{d}\right)} (\sqrt{H} - \sqrt{h}) \qquad (3)$$

Calculate the time required for h to drop from 4 m to 1.5 m.

e) Plot the variations of $\langle v_2 \rangle$ and h as a function of time on the same plot. Show that dh/dt is negligible at all times in comparison with the liquid velocity through the drain pipe system.

(**Answer:** c) 0.0039 d) 7.7 min)

7.9 Consider the draining of a spherical tank of diameter D with associated drain piping as shown in the figure below. The tank is initially filled with water to an elevation of H with respect to the reference plane.

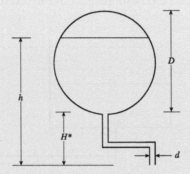

a) Repeat the procedure given in Problem 7.8 and show that

$$t = \frac{4}{d^2}\sqrt{\frac{2}{g}\left(1 + \frac{4fL_{eq}}{d}\right)}\left[\sqrt{h}\left(\frac{h^2}{5} - \frac{2}{3}X_1 h + X_2\right) - \sqrt{H}\left(\frac{H^2}{5} - \frac{2}{3}X_1 H + X_2\right)\right]$$

where

$$X_1 = H^* + R \qquad \text{and} \qquad X_2 = X_1^2 - R^2$$

b) A spherical tank, 4 m in diameter, discharges through a mild steel pipe system ($\varepsilon = 4.6 \times 10^{-5}$ m) with an equivalent length of 100 m and a diameter of 23 cm. Determine the time to drain the tank if $H^* = 1$ m and $H = 4.5$ m.

(**Answer:** b) 4.9 min)

7.10 Suspended particles in agitated vessels are frequently encountered in the chemical process industries. Some examples are mixer-settler extractors, catalytic slurry reactors, and crystallizers. The design of such equipment requires the mass transfer coefficient to be known. For this purpose, solid particles (species $\mathcal{A}$) with a known external surface area, A_o, and total mass, M_o, are added to an agitated liquid of volume V and the concentration of species $\mathcal{A}$ is recorded as a function of time.

a) Consider the liquid as a system and show that the unsteady-state macroscopic mass balance for species $\mathcal{A}$ is

$$\langle k_c \rangle A_o \left(\frac{M}{M_o} \right)^{2/3} (c_A^{sat} - c_A) = V \frac{dc_A}{dt} \tag{1}$$

where M is the total mass of solid particles at any instant and c_A^{sat} is the equilibrium solubility. Rearrange Eq. (1) in the form

$$\langle k_c \rangle = -\frac{V}{A_o (M/M_o)^{2/3}} \frac{d \ln(c_A^{sat} - c_A)}{dt} \tag{2}$$

and show how one can obtain the average mass transfer coefficient from the experimental data.

b) Another way of calculating the mass transfer coefficient is to choose experimental conditions so that only a small fraction of the initial solids is dissolved during a run. Under these circumstances, show that the average mass transfer coefficient can be calculated from the following expression:

$$\langle k_c \rangle = \frac{V}{\langle A \rangle t} \ln \left(\frac{c_A^{sat}}{c_A^{sat} - c_A} \right) \tag{3}$$

where $\langle A \rangle$ is the average surface area of the particles. Indicate the assumptions involved in the derivation of Eq. (3).

7.11   Consider Problem 7.10 in which the average mass transfer coefficient of suspended particles is known. Estimate the time required for the dissolution of solid particles as follows:

a) Write down the total mass balance for species $\mathcal{A}$ and relate the mass of the particles, M, to the concentration of species $\mathcal{A}$, c_A, as

$$\frac{M}{M_o} = 1 - \left(\frac{V}{M_o} \right) c_A \tag{1}$$

b) Substitute Eq. (1) into Eq. (1) in Problem 7.10 to get

$$dt = \alpha \frac{d\theta}{[1 - (1 + \beta^3) \theta]^{2/3} (1 - \theta)} \tag{2}$$

where

$$\theta = \frac{c_A}{c_A^{sat}} \qquad \alpha = \frac{V}{\langle k_c \rangle A_o} \qquad \beta^3 = \frac{V c_A^{sat}}{M_o} - 1 \tag{3}$$

c) Show that the integration of Eq. (2) leads to

$$t = \frac{1}{6\beta^2} \ln \left[\left(\frac{u + \beta}{1 + \beta} \right)^2 \left(\frac{1 - \beta + \beta^2}{u^2 - u\beta + \beta^2} \right) \right] + \frac{1}{\sqrt{3}\beta^2} \tan^{-1} \left\{ \frac{\sqrt{3}(u - 1)}{2\beta - 1 + u[(2/\beta) - 1]} \right\} \tag{4}$$

where

$$u^3 = 1 - (1 + \beta^3)\theta \tag{5}$$

7.12 Rework Example 7.3 if the rate of reaction is given by

$$r = kc_A^2 \tag{1}$$

a) For the filling period show that the governing differential equation is given by

$$t\frac{dc_A}{dt} + ktc_A^2 + c_A = c_{A_o} \tag{2}$$

Using the substitution

$$c_A = \frac{1}{ku}\frac{du}{dt} \tag{3}$$

show that Eq. (1) reduces to

$$\frac{d}{dt}\left(t\frac{du}{dt}\right) - c_{A_o}ku = 0 \tag{4}$$

Solve Eq. (4) and obtain the solution as

$$c_A = \sqrt{\frac{c_{A_o}}{kt}}\frac{I_1(2\sqrt{c_{A_o}kt})}{I_o(2\sqrt{c_{A_o}kt})} \tag{5}$$

Note that Eq. (2) indicates that $c_A = c_{A_o}$ at $t = 0$. Obtain the same result from Eq. (5).

b) Show that the governing differential equation for the unsteady-state period is given in the form

$$\frac{dc_A}{dt} + kc_A^2 + \frac{c_A}{\tau} = \frac{c_{A_o}}{\tau} \tag{6}$$

where τ is the residence time. Using

$$c_A = c_{A_s} + \frac{1}{z} \tag{7}$$

show that Eq. (6) reduces to

$$\frac{dz}{dt} - \beta z = k \tag{8}$$

where

$$\beta = 2kc_{A_s} + \frac{1}{\tau} \tag{9}$$

Note that c_{A_s} in Eq. (7) represents the steady-state concentration satisfying the equation

$$kc_{A_s}^2 + \frac{c_{A_s}}{\tau} = \frac{c_{A_o}}{\tau} \tag{10}$$

Solve Eq. (8) and obtain

$$c_A = c_{A_s} + \frac{1}{[(c_A^* - c_{A_s})^{-1} + (k/\beta)]\exp[\beta(t - t^*)] - (k/\beta)} \tag{11}$$

where c_A^* and t^* represent the concentration and time at the end of the filling period, respectively.

7.13 For creeping flow, i.e., $\mathrm{Re} \ll 1$, a relationship between the friction factor and the Reynolds number is given by Stokes' law, Eq. (4.3-7).

a) Substitute Eq. (4.3-7) into Eq. (7.4-7) and show that

$$v = \frac{(\rho_P - \rho)g D_P^2}{18\mu}\left\{1 - \exp\left[-\frac{18\mu t}{(\rho_P + 0.5\rho)D_P^2}\right]\right\} \tag{1}$$

b) Show that the time required for the sphere to reach 99% of its terminal velocity, t_∞, is given by

$$t_\infty = \frac{D_P^2}{3.9\mu}(\rho_P + 0.5\rho) \tag{2}$$

and investigate the cases in which the initial acceleration period is negligible.

c) Show that the distance traveled by the particle during unsteady-state fall is given by

$$s = t v_t - v_t \frac{(\rho_P - \rho)D_P^2}{18\mu}\left\{1 - \exp\left[-\frac{18\mu t}{(\rho_P + 0.5\rho)D_P^2}\right]\right\} \tag{3}$$

where v_t is the terminal velocity of the falling particle and is defined by

$$v_t = \frac{(\rho_P - \rho)g D_P^2}{18\mu} \tag{4}$$

7.14 When Newton's law is applicable, the friction factor is constant and is given by Eq. (4.3-9).

a) Substitute Eq. (4.3-9) into Eq. (7.4-7) and show that

$$\frac{v}{v_t} = \frac{1 - \exp(-\gamma t)}{1 + \exp(-\gamma t)} \tag{1}$$

where the terminal velocity, v_t, and γ are given by

$$v_t = 1.74\sqrt{\frac{(\rho_P - \rho)g D_P}{\rho}} \quad \text{and} \quad \gamma^{-1} = 1.51\left(\frac{\rho_P + 0.5\rho}{\rho}\right)\frac{D_P}{v_t} \tag{2}$$

b) Show that the distance traveled is

$$s = t v_t + \frac{2v_t}{\gamma}\ln\left[\frac{1 + \exp(-\gamma t)}{2}\right] \tag{3}$$

7.15 Consider the two-dimensional motion of a spherical particle in a fluid. When the horizontal component of velocity is very large compared to the vertical component, the process can be modeled as one-dimensional motion in the absence of a gravitational field. Using the unsteady-state momentum balance, show that

$$t = \frac{4\rho_P D_P^2}{3\mu} \int_{Re_P}^{Re_{P_o}} \frac{d\,Re_P}{f\,Re_P^2} \tag{1}$$

where Re_{P_o} is the value of the Reynolds number at $t = 0$.

a) When Stokes' law is applicable, show that the distance traveled by the particle is given by

$$s = \frac{v_o \rho_P D_P^2}{18\mu}\left[1 - \exp\left(-\frac{18\mu t}{\rho_P D_P^2}\right)\right] \tag{2}$$

where v_o is the value of velocity at $t = 0$.

b) When Newton's law is applicable, show that the distance traveled by the particle is given by

$$s = \frac{3.03\rho_P D_P}{\rho} \ln\left(1 + \frac{\rho v_o t}{3.03\rho_P D_P}\right) \tag{3}$$

7.16 Coming home with a friend to have a cold beer after work, you find out that you had left the beer on the kitchen counter. As a result of the sunlight coming through the kitchen window, it was too warm to drink.

  One way of cooling the beer is obviously to put it in the freezer. However, your friend insists that placing a can of beer in a pot in the kitchen sink, and letting cold water run over it into the pot and then into the sink shortens the cooling time. He claims that the overall heat transfer coefficient for this process is much greater than that for a can of beer sitting idly in the freezer in still air. He supports this idea with the following data reported by Horwitz (1981):

	Freezer	Tap Water
Cooling medium temperature (°C)	−21	13
Initial temperature of beer (°C)	29	29
Final temperature of beer (°C)	15	15
Time elapsed (min)	21.1	8.6

Surface area of can	$= 0.03\ m^2$
Quantity of beer in can	$= 0.355\ kg$
Heat capacity of beer	$= 4.2\ kJ/kg \cdot K$

a) Do you think that your friend is right? Show your work by calculating the heat transfer coefficient in each case. Ignore the cost and availability of water.

b) Calculate the time required to cool the beer from 29 °C to 4 °C in the freezer.

c) Suppose that you first cool the beer to 15 °C in the running water and then place the beer in the freezer. Calculate the time required to cool the beer from 29 °C to 4 °C in this case.

(**Answer:** a) $\langle h \rangle$ (freezer) $= 12.9$ W/m²·K, $\langle h \rangle$ (tap water) $= 200$ W/m²·K b) 44.5 min c) 32 min)

7.17 M kg of a liquid is to be heated from T_1 to T_2 in a well-stirred, jacketed tank by steam condensing at T_s in the jacket. The heat transfer area, A; the heat capacity of tank contents per unit mass, $\widehat{C}_P$; and the overall heat transfer coefficient, U, are known. Show that the required heating time is given by

$$t = \frac{M\widehat{C}_P}{UA} \ln\left(\frac{T_s - T_1}{T_s - T_2}\right) \tag{1}$$

Indicate the assumptions involved in the derivation of Eq. (1).

7.18 In Problem 7.17, assume that hot water, with a constant mass flow rate $\dot{m}$ and inlet temperature T_{in}, is used as a heating medium instead of steam.

a) Show that the outlet temperature of hot water, T_{out}, is given by

$$T_{out} = T + \frac{T_{in} - T}{\Omega} \tag{1}$$

where

$$\Omega = \exp\left(\frac{UA}{\dot{m}C}\right) \tag{2}$$

in which T is the temperature of the tank contents at any instant and C is the heat capacity of hot water.

b) Write down the unsteady-state energy balance and show that the time required to increase the temperature of the tank contents from T_1 to T_2 is given by

$$t = \frac{M\widehat{C}_P}{\dot{m}C}\left(\frac{\Omega}{\Omega - 1}\right)\ln\left(\frac{T_{in} - T_1}{T_{in} - T_2}\right) \tag{3}$$

c) Bondy and Lippa (1983) argued that when the difference between the outlet and inlet jacket temperatures is less than 10% of the ΔT_{LM} between the average temperature of the jacket and the temperature of the tank contents, Eq. (1) in Problem 7.17 can be used instead of Eq. (3) by replacing T_s with the average jacket temperature. Do you agree? For more information on this problem see Tosun and Akşahin (1993).

7.19 600 kg of a liquid is to be heated from 15 °C to 150 °C in a well-stirred, jacketed tank by steam condensing at 170 °C in the jacket. The heat transfer surface area of the jacket is 4.5 m² and the heat capacity of the liquid is 1850 J/kg·K. The overall heat transfer coefficient, U, varies with temperature as follows:

T (°C)	15	30	60	90	120	150
U (W/m²·K)	390	465	568	625	664	680

a) Calculate the required heating time.
b) Correlate the data in terms of the expression

$$U = A - \frac{B}{T}$$

where T is in degrees Kelvin, and calculate the required heating time.

(**Answer:** a) 11.7 min b) 13.7 min)

7.20 500 kg of a liquid is to be heated from 15 °C to 150 °C in a well-stirred, jacketed tank by steam condensing at 170 °C in the jacket. The heat transfer surface area of the jacket is 4.5 m² and the heat capacity of the liquid is 1850 J/kg·K. Calculate the average overall heat transfer coefficient if the variation of liquid temperature as a function of time is recorded as follows:

t (min)	0	2	4	6	8	10	12
T (°C)	15	59	90	112	129	140	150

(**Answer:** 564 W/m²·K)

7.21 A copper sphere ($k = 353$ W/m·K, $\rho = 8924$ kg/m³, $\widehat{C}_P = 387$ J/kg·K) of diameter 10 cm is placed in an evacuated enclosure with the enclosure walls at a very low temperature. It is heated uniformly throughout the volume by an electrical resistance heater at a rate of 1000 W.

a) Calculate the steady-state temperature of the sphere if the emissivity of the surface is 0.85.
b) If the heater is turned off, calculate the time required for the sphere to cool to 600 K by radiation alone.

Hint: First calculate the Biot number at the steady-state temperature to check the applicability of the lumped-parameter analysis. The heat transfer coefficient can be estimated with the help of Eq. (3.2-13).

(**Answer:** a) 901.5 K b) 1300 s)

7.22 A thermocouple is a sensor for measuring temperature. Its principle is based on the fact that an electric current flows in a closed circuit formed by two dissimilar metals if the two junctions are at different temperatures. The voltage produced by the flow of an electric current is converted to temperature. The measuring junction (or hot junction) is exposed to the medium whose temperature is to be measured and the reference junction (or cold junction) is connected to the measuring instrument.

The tip of the measuring junction may be approximated as a sphere. Its temperature must be the same as that of the medium in which it is placed. In other words, the sphere must reach thermal equilibrium with the medium. In practical applications, however, it takes time for the thermocouple to record the changes in the temperature of the medium. The so-called *response time of a thermocouple* is defined as the time required for a thermocouple to record 63% of the applied temperature difference.

a) Show that the response time of a thermocouple is given by

$$t = \frac{D \rho \widehat{C}_P}{6 \langle h \rangle}$$

where ρ and $\widehat{C}_P$ are the density and heat capacity of the thermocouple material, respectively, and D is the tip diameter.

b) Calculate the response time for the following values:

$$D = 1 \text{ cm} \qquad \langle h \rangle = 230 \text{ W/m}^2 \cdot \text{K} \qquad \widehat{C}_P = 1050 \text{ J/kg} \cdot \text{K} \qquad \rho = 1900 \text{ kg/m}^3$$

(Answer: b) 14.5 s)

7.23 A copper slab ($k = 401$ W/m·K, $\alpha = 117 \times 10^{-6}$ m^2/s) of thickness 2 cm is initially at a temperature of 25 °C. At $t = 0$, one side of the slab starts receiving a net heat flux of 5000 W/m^2, while the other side dissipates heat to the surrounding fluid at a temperature of 25 °C with an average heat transfer coefficient of 80 W/m^2·K.

a) How long does it take for the slab temperature to reach 70 °C?
b) Calculate the steady-state temperature.

(Answer: a) 1091 s b) 87.5 °C)

7.24 An insulated rigid tank of volume 0.1 m^3 is connected to a large pipeline carrying air at 10 bar and 120 °C. The valve between the pipeline and the tank is opened and air is admitted to the tank at a constant mass flow rate. The pressure in the tank is recorded as a function of time as follows:

t (min)	5	10	15	20	25	30
P (bar)	1.6	2.1	2.7	3.3	3.9	4.4

If the tank initially contains air at 1 bar and 20 °C, determine the mass flow rate of air entering the tank. Air may be assumed to be an ideal gas with a constant $\widetilde{C}_P$ of 29 J/mol·K.

(Answer: 7.25 g/min)

7.25 An insulated rigid tank of volume 0.2 m^3 is connected to a large pipeline carrying nitrogen at 10 bar and 70 °C. The valve between the pipeline and the tank is opened and nitrogen is admitted to the tank at a constant mass flow rate of 4 g/s. Simultaneously, nitrogen is withdrawn from the tank, also at a constant mass flow rate of 4 g/s. Calculate the temperature and pressure within the tank after 1 minute if the tank initially contains nitrogen at 2 bar and 35 °C. Nitrogen may be assumed to be an ideal gas with a constant $\widetilde{C}_P$ of 30 J/mol·K.

(**Answer:** 326.8 K, 2.12 bar)

7.26 A rigid tank of volume 0.2 m³ initially contains air at 2 bar and 35 °C. On one side it is connected to an air supply line at 10 bar and 70 °C, and on the other side it is connected to an empty rigid tank of 0.8 m³ as shown in the figure below. Both tanks are insulated and initially both valves are closed. The valve between the pipeline and the tank is opened and air starts to flow into the tank at a constant flow rate of 10 mol/min. Simultaneously, the valve between the tanks is also opened so as to provide a constant flow rate of 6 mol/min to the larger tank. Determine the temperature and pressure of air in the larger tank after 2 minutes. Air may be assumed to be an ideal gas with a constant $\widetilde{C}_P$ of 29 J/mol·K.

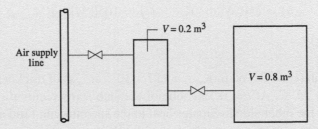

(**Answer:** 482.3 K, 0.6 bar)

7.27 Metering pumps provide a constant liquid mass flow rate for a wide variety of scientific, industrial, and medical applications. A typical pump consists of a cylinder fitted with a piston as shown in the figure below. The piston is generally located on the end of a long screw, which itself is driven at a constant velocity by a synchronous electric motor.

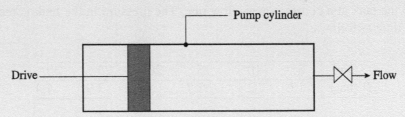

a) Assume that the manufacturer has calibrated the pump at some reference temperature, T_{ref}. Write down the unsteady-state mass balance and show that the reference mass flow rate, $\dot{m}_{ref}$, delivered by the pump is given by

$$\dot{m}_{ref} = -\rho_{ref}\frac{dV_{ref}}{dt} \tag{1}$$

where ρ_{ref} and V_{ref} are the density and the volume of the liquid in the pump cylinder at the reference temperature, respectively. Integrate Eq. (1) and show that the variation in the liquid volume as a function of time is given by

$$V_{ref} = V_{ref}^o - \left(\frac{\dot{m}_{ref}}{\rho_{ref}} \right) t \tag{2}$$

where V_{ref}^o is the volume of the cylinder at $t = 0$.

b) If the pump operates at a temperature different from the reference temperature, show that the mass flow rate provided by the pump is given by

$$\dot{m} = -\frac{d}{dt}(\rho V) \tag{3}$$

where ρ and V are the density and the volume of the pump liquid at temperature T, respectively. Expand ρ and V in a Taylor series in T about the reference temperature T_{ref} and show that

$$\rho V \simeq \rho_{ref} V_{ref} - \rho_{ref} V_{ref} (\beta_C - \beta_L)(T - T_{ref}) \tag{4}$$

where β, the coefficient of volume expansion, is defined by

$$\beta = \frac{1}{\widehat{V}} \left(\frac{\partial \widehat{V}}{\partial T} \right)_P = -\frac{1}{\rho} \left(\frac{\partial \rho}{\partial T} \right)_P \tag{5}$$

in which the subscripts L and C represent the liquid and the cylinder, respectively. Indicate the assumptions involved in the derivation of Eq. (4).

c) Show that the substitution of Eq. (4) into Eq. (3) and making use of Eqs. (1) and (2) give the fractional error in mass flow rate as

$$\frac{\dot{m} - \dot{m}_{ref}}{\dot{m}_{ref}} = -(\beta_L - \beta_C)(T - T_{ref}) + \left(\frac{V_{ref}^o}{R_o} - t \right)(\beta_L - \beta_C) \frac{dT}{dt} \tag{6}$$

where

$$R_o = -\frac{dV_{ref}}{dt} \tag{7}$$

Note that the first and the second terms on the right-hand side of Eq. (6) represent, respectively, the steady-state and the unsteady-state contributions to the error term.

d) Assume that at any instant the temperature of the pump liquid is uniform and equal to that of the surrounding fluid, i.e., the cylinder wall is diathermal, and determine the fractional error in mass flow rate for the following cases:

- The temperature of the fluid surrounding the pump, T_f, is constant. Take $\beta_C = 4 \times 10^{-5}$ K^{-1}, $\beta_L = 1.1 \times 10^{-3}$ K^{-1}, and $T_f - T_{ref} = 5$ K.
- The temperature of the surrounding fluid changes at a constant rate of 1 K/h. Take $V_{ref}^o = 500$ cm^3 and $R_o = 25$ cm^3/h.
- The surrounding fluid temperature varies periodically with time, i.e.,

$$T_f = T_{ref} + A \sin \omega t \tag{8}$$

Take $A = 1\,°C$ and $\omega = 8\pi$ h^{-1}.

e) Now assume that the liquid temperature within the pump is uniform but different from the surrounding fluid temperature as a result of a finite rate of heat transfer. If the temperature of the surrounding fluid changes as

$$T_f = T_\infty + (T_{ref} - T_\infty)e^{-\tau t} \qquad T_\infty < T_{ref} \tag{9}$$

where T_∞ is the asymptotic temperature and τ is the time constant, show that the fractional error in mass flow rate is given by

$$f = \frac{\phi}{\phi - \tau}(e^{-\tau t} - e^{-\phi t}) + e^{-\phi t} - 1 + \left(\frac{V_{ref}^o}{R_o} - t\right)\frac{\phi \tau}{\phi - \tau}(e^{-\tau t} - e^{-\phi t}) \tag{10}$$

The terms f and ϕ are defined as

$$f = -\frac{\dot{m} - \dot{m}_{ref}}{\dot{m}_{ref}}\frac{1}{(\beta_L - \beta_C)(T_{ref} - T_\infty)} \tag{11}$$

$$\phi = \frac{UA}{\rho V \widehat{C}_P} \tag{12}$$

where A is the surface area of the liquid being pumped, U is the overall heat transfer coefficient, and $\widehat{C}_P$ is the heat capacity of the pump liquid.

f) Show that the time, t^*, at which the fractional error function f achieves its maximum absolute value is given by

$$t^* = \frac{\ln(\phi/\tau)}{\phi - \tau} \tag{13}$$

This problem was studied in detail by Eubank et al. (1985).

7.28 A spherical salt, 5 cm in diameter, is suspended in a large, well-mixed tank containing a pure solvent at 25 °C. If the percent decrease in the mass of the sphere is found to be 5% in 12 minutes, calculate the average mass transfer coefficient. The solubility of salt in the solvent is 180 kg/m^3 and the density of the salt is 2500 kg/m^3.

(Answer: 8.2×10^{-6} m/s)

7.29 The phosphorous content of lakes not only depends on the external loading rate but also on the interactions between the sediments and the overlying waters. The model shown in Figure 7.6 is proposed by Chapra and Canale (1991) in which the sediment layer gains phosphorous by settling and loses phosphorous by recycle and burial.

a) Show that the governing equations for the phosphorous concentrations in the lake, P_1, and in the sediment layer, P_2, are given as

$$\dot{m}_{in} - Q_{out}P_1 - v_s A_2 P_1 + A_2\langle k_c\rangle_r P_2 = V_1\frac{dP_1}{dt} \tag{1}$$

$$v_s A_2 P_1 - A_2\langle k_c\rangle_r P_2 - A_2\langle k_c\rangle_b P_2 = V_2\frac{dP_2}{dt} \tag{2}$$

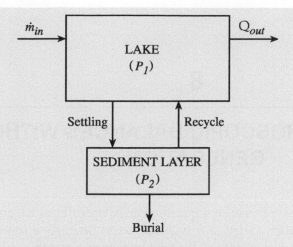

Figure 7.6 Schematic diagram for Problem 7.29.

where

$\dot{m}_{in}$ = loading rate = 2000 kg/year
Q_{out} = outflow volumetric flow rate = 80×10^6 m³/year
v_s = settling velocity of phosphorous = 40 m/year
A_2 = surface area of the sediment layer = 4.8×10^6 m²
$\langle k_c \rangle_r$ = recycle mass transfer coefficient = 2.5×10^{-2} m/year
$\langle k_c \rangle_b$ = burial mass transfer coefficient = 1×10^{-3} m/year
V_1 = volume of the lake = 53×10^6 m³
V_2 = volume of the sediment layer = 4.8×10^5 m³

b) Determine the variation of P_1 in mg/m³ as a function of time if the initial concentrations are given as $P_1 = 60$ mg/m³ and $P_2 = 500{,}000$ mg/m³.

(**Answer:** b) $P_1 = 22.9 - 165.4e^{-5.311t} + 202.5e^{-0.081t}$)

8

STEADY MICROSCOPIC BALANCES WITHOUT GENERATION

In the previous chapters, we have considered macroscopic balances in which quantities such as temperature and concentration varied only with respect to time. As a result, the inventory rate equations are written by considering the total volume as a system, and the resulting governing equations turn out to be ordinary differential equations in time. If the dependent variables such as velocity, temperature, and concentration change as a function of both position and time, then the inventory rate equations for the basic concepts are written over a differential volume element taken within the volume of the system. The resulting equations at the microscopic level are called *equations of change*.

In this chapter, we will consider steady-state microscopic balances without internal generation, and the resulting governing equations will be either ordinary or partial differential equations in position. It should be noted that the treatment for heat and mass transport is different from that for momentum transport. The main reasons for this are: (*i*) momentum is a vector quantity, while heat and mass are scalar, (*ii*) in heat and mass transport the velocity appears only in the convective flux term, while in momentum transfer it appears both in the molecular and in convective flux terms.

8.1 MOMENTUM TRANSPORT

Momentum per unit mass, by definition, is the fluid velocity, and changes in velocity can result in momentum transport. For *fully developed flow*[1] through conduits, velocity variations take place in the direction perpendicular to the flow since no-slip boundary conditions must be satisfied at the boundaries of the conduit. This results in the transfer of momentum perpendicular to the flow direction.

The inventory rate equation for momentum at the microscopic level is called the *equation of motion*. It is a vector equation with three components. For steady transfer of momentum without generation, the conservation statement for momentum reduces to

$$\text{(Rate of momentum in)} - \text{(Rate of momentum out)} = 0 \tag{8.1-1}$$

When there is no generation of momentum, it is implied that both pressure and gravity terms are zero. Hence, flow can only be generated by the movement of surfaces enclosing the fluid, and the resulting flow is called *Couette flow*. We will restrict our analysis to cases in which the following assumptions hold:

---

[1] *Fully developed flow* means there is no variation in velocity in the axial direction. In this way, the flow development regions near the entrance and exit are not taken into consideration, i.e., *end effects* are neglected.

1. Incompressible Newtonian fluid,
2. One-dimensional[2], fully developed laminar flow,
3. Constant physical properties.

The last assumption comes from the fact that temperature rise as a result of viscous dissipation during fluid motion, i.e., irreversible degradation of mechanical energy into thermal energy, is very small and cannot be detected by ordinary measuring devices in most cases. Hence, for all practical purposes the flow is assumed to be isothermal.

8.1.1 Plane Couette Flow

Consider a Newtonian fluid between two parallel plates separated by a distance B as shown in Figure 8.1. The lower plate is moved in the positive z-direction with a constant velocity of V, while the upper plate is held stationary.

The first step in the translation of Eq. (8.1-1) into mathematical terms is to postulate the functional forms of the nonzero velocity components. This can be done by making reasonable assumptions and examining the boundary conditions. For the problem at hand, the simplification of the velocity components is shown in Figure 8.2.

Since $v_z = v_z(x)$ and $v_x = v_y = 0$, Table C.1 in Appendix C indicates that the only nonzero shear-stress component is τ_{xz}. Therefore, the components of the total momentum flux are expressed as

$$\pi_{xz} = \tau_{xz} + (\rho v_z)v_x = \tau_{xz} = -\mu \frac{dv_z}{dx} \tag{8.1-2}$$

$$\pi_{yz} = \tau_{yz} + (\rho v_z)v_y = 0 \tag{8.1-3}$$

$$\pi_{zz} = \tau_{zz} + (\rho v_z)v_z = \rho v_z^2 \tag{8.1-4}$$

For a rectangular differential volume element of thickness Δx, length Δz, and width W, as shown in Figure 8.1, Eq. (8.1-1) is expressed as

$$(\pi_{zz}|_z W \Delta x + \pi_{xz}|_x W \Delta z) - (\pi_{zz}|_{z+\Delta z} W \Delta x + \pi_{xz}|_{x+\Delta x} W \Delta z) = 0 \tag{8.1-5}$$

Following the notation introduced by Bird *et al.* (2002), "in" and "out" directions for the fluxes are taken in the direction of positive x- and z-axes. Dividing Eq. (8.1-5) by $W \Delta x \Delta z$

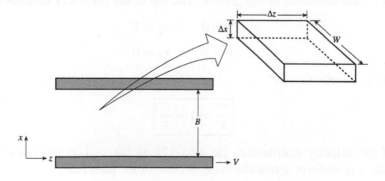

Figure 8.1. Couette flow between two parallel plates.

---

[2]*One-dimensional flow* indicates that there is only one nonzero velocity component.

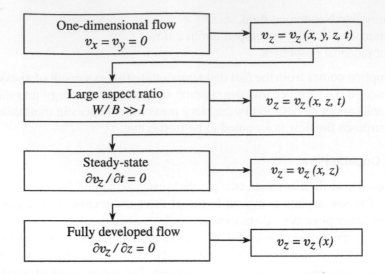

Figure 8.2. Simplification of the velocity components for Couette flow between two parallel plates.

and taking the limit as $\Delta x \to 0$ and $\Delta z \to 0$ give

$$\lim_{\Delta z \to 0} \frac{\pi_{zz}|_z - \pi_{zz}|_{z+\Delta z}}{\Delta z} + \lim_{\Delta x \to 0} \frac{\pi_{xz}|_x - \pi_{xz}|_{x+\Delta x}}{\Delta x} = 0 \qquad (8.1\text{-}6)$$

or,

$$\frac{\partial \pi_{zz}}{\partial z} + \frac{d\pi_{xz}}{dx} = 0 \qquad (8.1\text{-}7)$$

Substitution of Eqs. (8.1-2) and (8.1-4) into Eq. (8.1-7) and noting that $\partial v_z / \partial z = 0$ yield

$$\frac{d}{dx}\left(\frac{dv_z}{dx}\right) = 0 \qquad (8.1\text{-}8)$$

The solution of Eq. (8.1-8) is

$$v_z = C_1 x + C_2 \qquad (8.1\text{-}9)$$

where C_1 and C_2 are constants of integration. The use of the boundary conditions

$$\text{at} \quad x = 0 \qquad v_z = V \qquad (8.1\text{-}10)$$

$$\text{at} \quad x = B \qquad v_z = 0 \qquad (8.1\text{-}11)$$

gives the velocity distribution as

$$\boxed{\frac{v_z}{V} = 1 - \frac{x}{B}} \qquad (8.1\text{-}12)$$

The use of the velocity distribution, Eq. (8.1-12), in Eq. (8.1-2) indicates that the shear stress distribution is uniform across the cross-section of the plate, i.e.,

$$\boxed{\tau_{xz} = \frac{\mu V}{B}} \qquad (8.1\text{-}13)$$

The volumetric flow rate can be determined by integrating the velocity distribution over the cross-sectional area, i.e.,

$$Q = \int_0^W \int_0^B v_z \, dx \, dy \qquad (8.1\text{-}14)$$

Substitution of Eq. (8.1-12) into Eq. (8.1-14) gives the volumetric flow rate in the form

$$Q = \frac{WBV}{2} \qquad (8.1\text{-}15)$$

Dividing the volumetric flow rate by the flow area gives the average velocity as

$$\langle v_z \rangle = \frac{Q}{WB} = \frac{V}{2} \qquad (8.1\text{-}16)$$

8.1.2 Annular Couette Flow

Consider a Newtonian fluid in a concentric annulus as shown in Figure 8.3. The inner circular rod moves in the positive z-direction with a constant velocity of V.

For the problem at hand, the simplification of the velocity components is shown in Figure 8.4. Since $v_z = v_z(r)$ and $v_r = v_\theta = 0$, Table C.2 in Appendix C indicates that the only nonzero shear-stress component is τ_{rz}. Therefore, the components of the total momentum flux

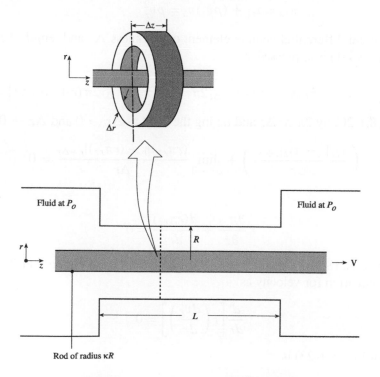

Figure 8.3. Couette flow in a concentric annulus.

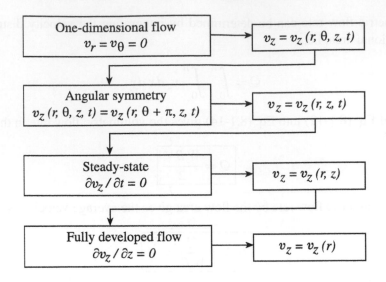

Figure 8.4. Simplification of the velocity components for Couette flow in a concentric annulus.

are given by

$$\pi_{rz} = \tau_{rz} + (\rho v_z)v_r = \tau_{rz} = -\mu \frac{dv_z}{dr} \tag{8.1-17}$$

$$\pi_{\theta z} = \tau_{\theta z} + (\rho v_z)v_\theta = 0 \tag{8.1-18}$$

$$\pi_{zz} = \tau_{zz} + (\rho v_z)v_z = \rho v_z^2 \tag{8.1-19}$$

For a cylindrical differential volume element of thickness Δr and length Δz, as shown in Figure 8.3, Eq. (8.1-1) is expressed as

$$(\pi_{zz}|_z 2\pi r \Delta r + \pi_{rz}|_r 2\pi r \Delta z) - \left[\pi_{zz}|_{z+\Delta z} 2\pi r \Delta r + \pi_{rz}|_{r+\Delta r} 2\pi (r + \Delta r)\Delta z\right] = 0 \tag{8.1-20}$$

Dividing Eq. (8.1-20) by $2\pi \Delta r \Delta z$ and taking the limit as $\Delta r \to 0$ and $\Delta z \to 0$ give

$$\lim_{\Delta z \to 0} r\left(\frac{\pi_{zz}|_z - \pi_{zz}|_{z+\Delta z}}{\Delta z}\right) + \lim_{\Delta r \to 0} \frac{(r\pi_{rz})|_r - (r\pi_{rz})|_{r+\Delta r}}{\Delta r} = 0 \tag{8.1-21}$$

or,

$$r\frac{\partial \pi_{zz}}{\partial z} + \frac{d(r\pi_{rz})}{dr} = 0 \tag{8.1-22}$$

Substitution of Eqs. (8.1-17) and (8.1-19) into Eq. (8.1-22) and noting that $\partial v_z/\partial z = 0$ give the governing equation for velocity as

$$\frac{d}{dr}\left[r\left(\frac{dv_z}{dr}\right)\right] = 0 \tag{8.1-23}$$

The solution of Eq. (8.1-23) is

$$v_z = C_1 \ln r + C_2 \tag{8.1-24}$$

where C_1 and C_2 are integration constants. The use of the boundary conditions

$$\text{at} \quad r = R \qquad v_z = 0 \tag{8.1-25}$$

$$\text{at} \quad r = \kappa R \qquad v_z = V \tag{8.1-26}$$

gives the velocity distribution as

$$\boxed{\frac{v_z}{V} = \frac{\ln(r/R)}{\ln \kappa}} \tag{8.1-27}$$

The use of the velocity distribution, Eq. (8.1-27), in Eq. (8.1-17) gives the shear stress distribution as

$$\boxed{\tau_{rz} = -\left(\frac{\mu V}{\ln \kappa}\right)\frac{1}{r}} \tag{8.1-28}$$

The volumetric flow rate is obtained by integrating the velocity distribution over the annular cross-sectional area, i.e.,

$$Q = \int_0^{2\pi} \int_{\kappa R}^{R} v_z r \, dr \, d\theta \tag{8.1-29}$$

Substitution of Eq. (8.1-27) into Eq. (8.1-29) and integration give

$$\boxed{Q = \frac{\pi R^2 V}{2}\left[\frac{1 - \kappa^2}{\ln(1/\kappa)} - 2\kappa^2\right]} \tag{8.1-30}$$

Dividing the volumetric flow rate by the flow area gives the average velocity as

$$\boxed{\langle v_z \rangle = \frac{Q}{\pi R^2 (1 - \kappa^2)} = \frac{V}{2}\left[\frac{1}{\ln(1/\kappa)} - \frac{2\kappa^2}{1 - \kappa^2}\right]} \tag{8.1-31}$$

The drag force acting on the rod is

$$F_D = -\tau_{rz}|_{r=\kappa R} 2\pi \kappa R L \tag{8.1-32}$$

The use of Eq. (8.1-28) in Eq. (8.1-32) gives

$$\boxed{F_D = \frac{2\pi \mu L V}{\ln \kappa}} \tag{8.1-33}$$

8.1.2.1 *Investigation of the limiting case*   Once the solution to a given problem is obtained, it is always advisable to investigate the limiting cases if possible, and to compare the results with the known solutions. If the results match, this does not necessarily mean that the solution is correct; however, the chances of it being correct are fairly high.

When the ratio of the radius of the inner pipe to that of the outer pipe is close to unity, i.e., $\kappa \to 1$, a concentric annulus may be considered a thin-plane slit and its curvature can be neglected. Approximation of a concentric annulus as a parallel plate requires the width, W, and the length, L, of the plate to be defined as

$$W = \pi R(1 + \kappa) \qquad (8.1\text{-}34)$$

$$B = R(1 - \kappa) \qquad (8.1\text{-}35)$$

Therefore, the product WB is equal to

$$WB = \pi R^2 (1 - \kappa^2) \quad \Longrightarrow \quad \pi R^2 = \frac{WB}{1 - \kappa^2} \qquad (8.1\text{-}36)$$

so that Eq. (8.1-30) becomes

$$Q = \frac{WBV}{2} \lim_{\kappa \to 1} \left[-\frac{1}{\ln \kappa} - 2\left(\frac{\kappa^2}{1 - \kappa^2} \right) \right] \qquad (8.1\text{-}37)$$

Substitution of $\psi = 1 - \kappa$ into Eq. (8.1-37) gives

$$Q = \frac{WBV}{2} \lim_{\psi \to 0} \left\{ -\frac{1}{\ln(1 - \psi)} - 2\left[\frac{(1 - \psi)^2}{1 - (1 - \psi)^2} \right] \right\} \qquad (8.1\text{-}38)$$

The Taylor series expansion of the term $\ln(1 - \psi)$ is

$$\ln(1 - \psi) = -\psi - \frac{1}{2} \psi^2 - \frac{1}{3} \psi^3 - \cdots \qquad -1 < \psi \leqslant 1 \qquad (8.1\text{-}39)$$

Using Eq. (8.1-39) in Eq. (8.1-38) and carrying out the divisions yield

$$Q = \frac{WBV}{2} \lim_{\psi \to 0} \left[\frac{1}{\psi} - \frac{1}{2} - \frac{\psi}{12} + \cdots \quad -2\left(\frac{1}{2\psi} - \frac{3}{4} - \frac{3\psi}{8} - \cdots \right) \right] \qquad (8.1\text{-}40)$$

or,

$$Q = \frac{WBV}{2} \lim_{\psi \to 0} \left(1 + \frac{2}{3} \psi + \cdots \right) = \frac{WBV}{2} \qquad (8.1\text{-}41)$$

which is equivalent to Eq. (8.1-15).

8.2 ENERGY TRANSPORT WITHOUT CONVECTION

The inventory rate equation for energy at the microscopic level is called the *equation of energy*. For a steady transfer of energy without generation, the conservation statement for energy reduces to

$$\text{(Rate of energy in)} = \text{(Rate of energy out)} \qquad (8.2\text{-}1)$$

The rate of energy entering and leaving the system is determined from the energy flux. As stated in Chapter 2, the total energy flux is the sum of the molecular and convective fluxes. In this case, we will restrict our analysis to cases in which convective energy flux is either zero or negligible compared with the molecular flux. This implies transfer of energy by conduction in solids and stationary liquids.

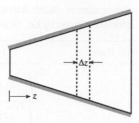

Figure 8.5. Conduction through a slightly tapered slab.

8.2.1 Conduction in Rectangular Coordinates

Consider the transfer of energy by conduction through a slightly tapered slab as shown in Figure 8.5. If the taper angle is small and the lateral surface is insulated, energy transport can be considered one-dimensional in the z-direction[3], i.e., $T = T(z)$.

Under these circumstances, Table C.4 in Appendix C indicates that the only nonzero energy flux component is e_z, and it is given by

$$e_z = q_z = -k \frac{dT}{dz} \tag{8.2-2}$$

The negative sign in Eq. (8.2-2) implies that the positive z-direction is in the direction of decreasing temperature. In a given problem, if the value of the heat flux is negative, it is implied that the flux is in the negative z-direction.

For a differential volume element of thickness Δz, as shown in Figure 8.5, Eq. (8.2-1) is expressed as

$$(Aq_z)|_z - (Aq_z)|_{z+\Delta z} = 0 \tag{8.2-3}$$

Dividing each term by Δz and taking the limit as $\Delta z \to 0$ give

$$\lim_{\Delta z \to 0} \frac{(Aq_z)|_z - (Aq_z)|_{z+\Delta z}}{\Delta z} = 0 \tag{8.2-4}$$

or,

$$\frac{d(Aq_z)}{dz} = 0 \tag{8.2-5}$$

Since flux times area gives the heat transfer rate, $\dot{Q}$, it is possible to conclude from Eq. (8.2-5) that

$$A q_z = \text{constant} = \dot{Q} \tag{8.2-6}$$

in which the area A is perpendicular to the direction of energy flux. Substitution of Eq. (8.2-2) into Eq. (8.2-6) and integration give

$$\int_0^T k(T)\,dT = -\dot{Q} \int_0^z \frac{dz}{A(z)} + C \tag{8.2-7}$$

---

[3]The z-direction in the rectangular and cylindrical coordinate systems are equivalent to each other.

Table 8.1. Heat transfer rate and temperature distribution for one-dimensional conduction in a plane wall for the boundary conditions given by Eq. (8.2-8)

Constants	Heat Transfer Rate		Temperature Distribution	
None	$\dfrac{\displaystyle\int_{T_L}^{T_o} k(T)\,dT}{\displaystyle\int_0^L \dfrac{dz}{A(z)}}$	(A)	$\dfrac{\displaystyle\int_T^{T_o} k(T)\,dT}{\displaystyle\int_{T_L}^{T_o} k(T)\,dT} = \dfrac{\displaystyle\int_0^z \dfrac{dz}{A(z)}}{\displaystyle\int_0^L \dfrac{dz}{A(z)}}$	(E)
k	$\dfrac{k(T_o - T_L)}{\displaystyle\int_0^L \dfrac{dz}{A(z)}}$	(B)	$\dfrac{T_o - T}{T_o - T_L} = \dfrac{\displaystyle\int_0^z \dfrac{dz}{A(z)}}{\displaystyle\int_0^L \dfrac{dz}{A(z)}}$	(F)
A	$\dfrac{A\displaystyle\int_{T_L}^{T_o} k(T)\,dT}{L}$	(C)	$\dfrac{\displaystyle\int_T^{T_o} k(T)\,dT}{\displaystyle\int_{T_L}^{T_o} k(T)\,dT} = \dfrac{z}{L}$	(G)
k A	$\dfrac{k(T_o - T_L)A}{L}$	(D)	$\dfrac{T_o - T}{T_o - T_L} = \dfrac{z}{L}$	(H)

where C is an integration constant. The determination of $\dot{Q}$ and C requires two boundary conditions.

When the surface temperatures are specified as

$$\text{at} \quad z = 0 \qquad T = T_o \tag{8.2-8a}$$

$$\text{at} \quad z = L \qquad T = T_L \tag{8.2-8b}$$

the resulting temperature distribution as a function of position and the heat transfer rate are given as in Table 8.1.

On the other hand, if one surface is exposed to a constant heat flux while the other is maintained at a constant temperature, i.e.,

$$\text{at} \quad z = 0 \qquad -k\frac{dT}{dz} = q_o \tag{8.2-9a}$$

$$\text{at} \quad z = L \qquad T = T_L \tag{8.2-9b}$$

the resulting temperature distribution as a function of position and the heat transfer rate are given as in Table 8.2.

The boundary conditions given by Eqs. (8.2-8) and (8.2-9) are not the only boundary conditions encountered in energy transport. For different boundary conditions, temperature distribution and heat transfer rate can be obtained from Eq. (8.2-7).

Table 8.2. Heat transfer rate and temperature distribution for one-dimensional conduction in a plane wall for the boundary conditions given by Eq. (8.2-9)

Constants	Heat Transfer Rate		Temperature Distribution	
None	$A\|_{z=0}q_o$	(A)	$\displaystyle\int_{T_L}^{T} k(T)\,dT = A\|_{z=0}q_o \int_{z}^{L} \frac{dz}{A(z)}$	(E)
k	$A\|_{z=0}q_o$	(B)	$\displaystyle T - T_L = \frac{A\|_{z=0}q_o}{k} \int_{z}^{L} \frac{dz}{A(z)}$	(F)
A	Aq_o	(C)	$\displaystyle\int_{T_L}^{T} k(T)\,dT = q_o L\left(1 - \frac{z}{L}\right)$	(G)
$\dfrac{k}{A}$	Aq_o	(D)	$\displaystyle T - T_L = \frac{q_o L}{k}\left(1 - \frac{z}{L}\right)$	(H)

Example 8.1 Consider a solid cone of circular cross-section as shown in Figure 8.6. The diameter at $z = 0$ is 8 cm and the diameter at $z = L$ is 10 cm. Calculate the steady rate of heat transfer if the lateral surface is well insulated and the thermal conductivity of the solid material as a function of temperature is given by

$$k(T) = 400 - 0.07T$$

where k is in W/m·K and T is in degrees Celsius.

Solution

The diameter increases linearly in the z-direction, i.e.,

$$D(z) = 0.05z + 0.08$$

Therefore, the cross-sectional area perpendicular to the direction of heat flux is given as a function of position in the form

$$A(z) = \frac{\pi D^2}{4} = \frac{\pi}{4}(0.05z + 0.08)^2$$

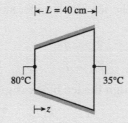

Figure 8.6. Conduction through a solid cone.

The use of Eq. (A) in Table 8.1 with $T_o = 80\,^\circ$C, $T_L = 35\,^\circ$C, and $L = 0.4$ m gives the heat transfer rate as

$$\dot{Q} = \frac{\displaystyle\int_{35}^{80} (400 - 0.07T)\,dT}{\displaystyle\int_{0}^{0.4} \frac{dz}{(\pi/4)(0.05z + 0.08)^2}} = 280 \text{ W}$$

Example 8.2 Consider the problem given in Example 2.2. Determine the temperature distribution within the slab.

Solution

With $T_L = 35\,^\circ$C, $q_o = 100{,}000$ W/m^2, $k = 398$ W/m·K, and $L = 0.04$ m, Eq. (H) in Table 8.2 gives the temperature distribution as

$$T - 35 = \frac{(100{,}000)(0.04)}{398}\left(1 - \frac{y}{0.04}\right) \quad \Rightarrow \quad T = 45.1 - 251.3y$$

Example 8.3 In rivers ice begins to form when the water cools to $0\,^\circ$C and continues to lose heat to the atmosphere. The presence of ice on rivers not only causes transportation problems but also floods when it melts. Once the ice cover is formed, its thickening depends on the rate of heat transferred from the water, through the ice cover, to the cold atmosphere. As an engineer, you are asked to estimate the increase in the thickness of the ice block as a function of time.

Solution

Assumptions

1. Pseudo-steady-state behavior.
2. River temperature is close to $0\,^\circ$C and the heat transferred from water to ice is negligible. This assumption implies that the major cause of ice thickening is the conduction of heat through the ice.

Analysis

System: Ice block

Since the density of ice is less than that of water, it floats on the river as shown in Figure 8.7. The temperatures T_m and T_s represent the melting temperature ($0\,^\circ$C) and the top surface temperature, respectively.

The temperature distribution in the ice block under steady conditions can be determined from Eq. (H) in Table 8.1 as

$$\frac{T_m - T}{T_m - T_s} = \frac{z}{L} \tag{1}$$

Therefore, the steady heat flux through the ice block is given by

$$q_z = -k\frac{dT}{dz} = \frac{k(T_m - T_s)}{L} \tag{2}$$

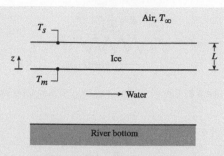

Figure 8.7. Ice block on a river.

For the ice block, the macroscopic inventory rate equation for energy is

$$-\text{Rate of energy out} = \text{Rate of energy accumulation} \qquad (3)$$

If the enthalpy of liquid water at T_m is taken as zero, then the enthalpy of solid ice is

$$\widehat{H}_{ice} = -\widehat{\lambda} - \underbrace{\int_T^{T_m} \widehat{C}_P \, dT}_{\text{Negligible}} \qquad (4)$$

Therefore, Eq. (3) is expressed as

$$-q_z A = \frac{d}{dt}\left[AL\rho(-\widehat{\lambda})\right] \qquad (5)$$

For the unsteady-state problem at hand, the pseudo-steady-state assumption implies that Eq. (2) holds at any given instant, i.e.,

$$q_z(t) = \frac{k(T_m - T_s)}{L(t)} \qquad (6)$$

Substitution of Eq. (6) into Eq. (5) and rearrangement give

$$\int_0^L L \, dL = \frac{k}{\rho \widehat{\lambda}} \int_0^t (T_m - T_s) \, dt \qquad (7)$$

Integration yields the thickness of the ice block in the form

$$L = \left[\frac{2k}{\rho \widehat{\lambda}} \int_0^t (T_m - T_s) \, dt\right]^{1/2} \qquad (8)$$

8.2.1.1 *Electrical circuit analogy* Using the analogy with Ohm's law, i.e., current = voltage/resistance, it is customary in the literature to express the rate equations in the form

$$\text{Rate} = \frac{\text{Driving force}}{\text{Resistance}} \qquad (8.2\text{-}10)$$

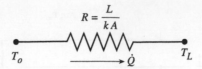

Figure 8.8. Electrical circuit analog of the plane wall.

Note that Eq. (D) in Table 8.1 is expressed as

$$\dot{Q} = \frac{T_o - T_L}{\dfrac{L}{kA}} \tag{8.2-11}$$

Comparison of Eq. (8.2-11) with Eq. (8.2-10) indicates that

$$\text{Driving force} = T_o - T_L \tag{8.2-12}$$

$$\boxed{\text{Resistance} = \frac{L}{kA} = \frac{\text{Thickness}}{(\text{Transport property})(\text{Area})}} \tag{8.2-13}$$

Hence, the electric circuit analog of the plane wall can be represented as shown in Figure 8.8. Note that the electrical circuit analogy holds only if the thermal conductivity is constant.

Example 8.4 Heat is conducted through a composite plane wall consisting of two different materials, A and B, as shown in Figure 8.9 ($T_1 > T_2$). Develop an expression to calculate the heat transfer rate under steady conditions.

Solution

Assumptions

1. Thermal conductivities of materials A and B, i.e., k_A and k_B, are constant.
2. Conduction takes place only in the z-direction.

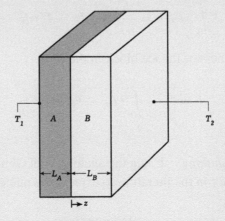

Figure 8.9. Composite plane wall in series arrangement.

Analysis

Since the area is constant, the governing equation for temperature in slab A can be easily obtained from Eq. (8.2-5) as

$$\frac{dq_z^A}{dz} = 0 \quad \Rightarrow \quad \frac{d^2 T_A}{dz^2} = 0 \tag{1}$$

The solution of Eq. (1) gives

$$T_A = C_1 z + C_2 \tag{2}$$

Similarly, the governing equation for temperature in slab B is given by

$$\frac{dq_z^B}{dz} = 0 \quad \Rightarrow \quad \frac{d^2 T_B}{dz^2} = 0 \tag{3}$$

The solution of Eq. (3) yields

$$T_B = C_3 z + C_4 \tag{4}$$

Evaluation of the constants C_1, C_2, C_3, and C_4 requires four boundary conditions. They are expressed as

$$\text{at} \quad z = -L_A \qquad T_A = T_1 \tag{5}$$

$$\text{at} \quad z = L_B \qquad T_B = T_2 \tag{6}$$

$$\text{at} \quad z = 0 \qquad T_A = T_B \tag{7}$$

$$\text{at} \quad z = 0 \qquad k_A \frac{dT_A}{dz} = k_B \frac{dT_B}{dz} \tag{8}$$

The boundary condition defined by Eq. (7) represents the condition of thermal equilibrium at the A-B interface. On the other hand, Eq. (8) indicates that the heat fluxes are continuous, i.e., equal to each other, at the A-B interface.

Application of the boundary conditions leads to the following temperature distributions within slabs A and B

$$T_A = T_1 - \frac{T_1 - T_2}{k_A} \left(\frac{z + L_A}{\dfrac{L_A}{k_A} + \dfrac{L_B}{k_B}} \right) \tag{9}$$

$$T_B = T_2 - \frac{T_1 - T_2}{k_B} \left(\frac{z - L_B}{\dfrac{L_A}{k_A} + \dfrac{L_B}{k_B}} \right) \tag{10}$$

Thus, the heat fluxes through slabs A and B are calculated as

$$q_z^A = -k_A \frac{dT_A}{dz} = \frac{T_1 - T_2}{\frac{L_A}{k_A} + \frac{L_B}{k_B}} \tag{11}$$

$$q_z^B = -k_B \frac{dT_B}{dz} = \frac{T_1 - T_2}{\frac{L_A}{k_A} + \frac{L_B}{k_B}} \tag{12}$$

The heat transfer rate through the composite plane wall is given by

$$\dot{Q} = q_z^A A = q_z^B A = \frac{T_1 - T_2}{\frac{L_A}{k_A A} + \frac{L_B}{k_B A}} \tag{13}$$

Note that Eq. (13) can be expressed in the form

$$\dot{Q} = \frac{T_1 - T_2}{R_{eq}} \tag{14}$$

where the equivalent resistance, R_{eq}, is defined by

$$R_{eq} = \frac{L_A}{k_A A} + \frac{L_B}{k_B A} = R_A + R_B \tag{15}$$

The resulting electrical circuit analog of Eq. (14), shown in Figure 8.10, indicates that the resistances are in series arrangement.

Example 8.5 Heat is conducted through a plane wall consisting of material A on the top and material B on the bottom as shown in Figure 8.11 ($T_1 > T_2$). Develop an expression to calculate the heat transfer rate under steady conditions.

Solution

Assumptions

1. Thermal conductivities of materials A and B, i.e., k_A and k_B, are constant.
2. Conduction takes place only in the z-direction.

Analysis

Since the area is constant, the governing equation for temperature in slab A is obtained from Eq. (8.2-5) as

$$\frac{dq_z^A}{dz} = 0 \quad \Rightarrow \quad \frac{d^2 T_A}{dz^2} = 0 \tag{1}$$

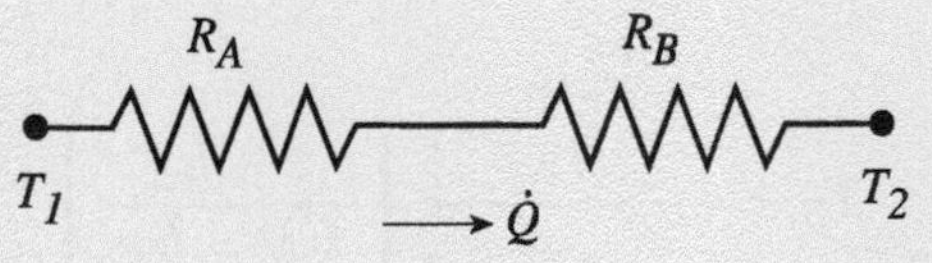

Figure 8.10. Electrical circuit analog of a plane wall in series arrangement.

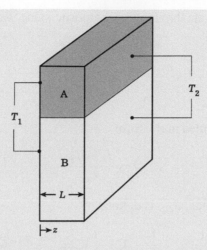

Figure 8.11. Composite plane wall in parallel arrangement.

The solution of Eq. (1) gives

$$T_A = C_1 z + C_2 \tag{2}$$

Similarly, the governing equation for temperature in slab B is given by

$$\frac{dq_z^B}{dz} = 0 \quad \Rightarrow \quad \frac{d^2 T_B}{dz^2} = 0 \tag{3}$$

The solution of Eq. (3) yields

$$T_B = C_3 z + C_4 \tag{4}$$

The boundary conditions are given as

$$\text{at} \quad z = 0 \qquad T_A = T_B = T_1 \tag{5}$$

$$\text{at} \quad z = L \qquad T_A = T_B = T_2 \tag{6}$$

Evaluation of the constants leads to the following temperature distributions

$$T_A = T_B = T_1 - \left(\frac{T_1 - T_2}{L}\right) z \tag{7}$$

The heat fluxes through slabs A and B are given by

$$q_z^A = -k_A \frac{dT_A}{dz} = k_A \left(\frac{T_1 - T_2}{L}\right) \tag{8}$$

$$q_z^B = -k_B \frac{dT_B}{dz} = k_B \left(\frac{T_1 - T_2}{L}\right) \tag{9}$$

Therefore, the heat transfer rate through the composite plane wall is given by

$$\dot{Q} = q_z^A A_A + q_z^B A_B = (T_1 - T_2)\left(\frac{1}{\dfrac{L}{k_A A_A}} + \frac{1}{\dfrac{L}{k_B A_B}}\right) \tag{10}$$

Note that Eq. (10) is represented in the form

$$\dot{Q} = \frac{T_1 - T_2}{R_{eq}} \tag{11}$$

where the equivalent resistance, R_{eq}, is defined by

$$\frac{1}{R_{eq}} = \frac{1}{\dfrac{L}{k_A A_A}} + \frac{1}{\dfrac{L}{k_B A_B}} = \frac{1}{R_A} + \frac{1}{R_B} \tag{12}$$

The resulting electrical circuit analog of Eq. (11), shown in Figure 8.12, indicates that the resistances are in parallel arrangement.

Example 8.6 For the composite wall shown in Figure 8.13, related thermal conductivities are given as $k_A = 35$ W/m·K, $k_B = 12$ W/m·K, $k_C = 23$ W/m·K, and $k_D = 5$ W/m·K.

a) Determine the steady-state heat transfer rate.

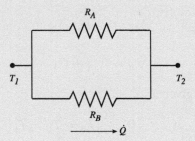

Figure 8.12. Electrical circuit analog of a plane wall in parallel arrangement.

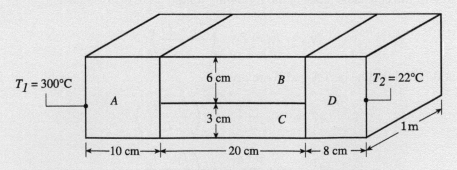

Figure 8.13. Heat conduction through a composite wall.

b) Determine the effective thermal conductivity for the composite walls. This makes it possible to consider the composite wall as a single material of thermal conductivity k_{eff}, rather than as four different materials with four different thermal conductivities.

Solution

a) An analogous electrical circuit for this case is shown below:

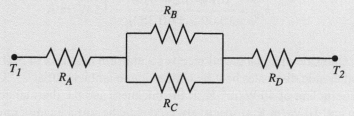

The equivalent resistance, R_o, of the two resistances in parallel is

$$R_o = \left(\frac{1}{R_B} + \frac{1}{R_C}\right)^{-1}$$

Thus, the electrical analog for the heat transfer process through the composite wall can be represented as shown below:

Using Eq. (8.2-13) the resistances are calculated as follows:

$$R_A = \frac{L_A}{k_A A} = \frac{0.1}{(35)(0.09 \times 1)} = 0.032 \text{ K/W}$$

$$R_B = \frac{L_B}{k_B A} = \frac{0.2}{(12)(0.06 \times 1)} = 0.278 \text{ K/W}$$

$$R_C = \frac{L_C}{k_C A} = \frac{0.2}{(23)(0.03 \times 1)} = 0.290 \text{ K/W}$$

$$R_D = \frac{L_D}{k_D A} = \frac{0.08}{(5)(0.09 \times 1)} = 0.178 \text{ K/W}$$

$$R_o = \left(\frac{1}{R_B} + \frac{1}{R_C}\right)^{-1} = \left(\frac{1}{0.278} + \frac{1}{0.290}\right)^{-1} = 0.142 \text{ K/W}$$

The equivalent resistance of the entire circuit is

$$R_{eq} = R_A + R_o + R_D = 0.032 + 0.142 + 0.178 = 0.352 \text{ K/W}$$

Hence, the heat transfer rate is

$$\dot{Q} = \frac{T_1 - T_2}{R_{eq}} = \frac{300 - 22}{0.352} = 790 \text{ W}$$

b) Note that

$$R_{eq} = \frac{\sum L}{k_{eff} A} \quad \Rightarrow \quad k_{eff} = \frac{\sum L}{A R_{eq}}$$

Therefore, the effective thermal conductivity is

$$k_{eff} = \frac{0.1 + 0.2 + 0.08}{(0.09 \times 1)(0.352)} = 12 \text{ W/m·K}$$

Example 8.7 One side of a plane wall receives a uniform heat flux of 500 W/m^2 due to radiation. The other side dissipates heat by convection to ambient air at 25 °C with an average heat transfer coefficient of 40 W/m^2·K. If the thickness and the thermal conductivity of the wall are 15 cm and 10 W/m·K, respectively, calculate the surface temperatures under steady conditions.

Solution

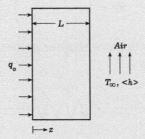

Assumption

1. Conduction takes place only in the z-direction.

Analysis

Since the area and the thermal conductivity of the wall are constant, Eq. (8.2-5) reduces to

$$\frac{dq_z}{dz} = 0 \quad \Rightarrow \quad \frac{d^2 T}{dz^2} = 0 \tag{1}$$

The solution of Eq. (1) is given by

$$T = C_1 z + C_2 \tag{2}$$

The boundary conditions are given as

$$\text{at} \quad z = 0 \qquad -k \frac{dT}{dz} = q_o \tag{3}$$

$$\text{at} \quad z = L \qquad -k \frac{dT}{dz} = \langle h \rangle (T - T_\infty) \tag{4}$$

Evaluation of the constants C_1 and C_2 leads to

$$T = T_\infty + \frac{q_o}{\langle h \rangle} + \frac{q_o L}{k} \left(1 - \frac{z}{L} \right) \tag{5}$$

Therefore, the surface temperatures can be calculated from Eq. (5) as

$$T_o = T|_{z=0} = T_\infty + q_o\left(\frac{1}{\langle h \rangle} + \frac{L}{k}\right) = 25 + \left(\frac{1}{40} + \frac{0.15}{10}\right) = 45\,°C$$

$$T_L = T|_{z=L} = T_\infty + \frac{q_o}{\langle h \rangle} = 25 + \frac{500}{40} = 37.5\,°C$$

Alternate solution: The electrical circuit analog of the system is shown in the figure below.

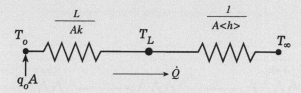

The heat transfer rate through the wall can be expressed in various forms as

$$\dot{Q} = q_o A = \frac{T_o - T_\infty}{\dfrac{L}{kA} + \dfrac{1}{\langle h \rangle A}} \tag{6a}$$

$$= \frac{T_o - T_L}{\dfrac{L}{kA}} \tag{6b}$$

$$= \frac{T_L - T_\infty}{\dfrac{1}{\langle h \rangle A}} \tag{6c}$$

Solving Eq. (6a) for T_o yields

$$T_o = T_\infty + q_o\left(\frac{L}{k} + \frac{1}{\langle h \rangle}\right) = 25 + \left(\frac{0.15}{10} + \frac{1}{40}\right) = 45\,°C$$

Solving either Eq. (6b) or Eq. (6c) for T_L gives

$$T_L = T_o - \frac{q_o L}{k} = 45 - \frac{(500)(0.15)}{10} = 37.5\,°C$$

$$T_L = T_\infty + \frac{q_o}{\langle h \rangle} = 25 + \frac{500}{40} = 37.5\,°C$$

8.2.1.2 *Transfer rate in terms of bulk fluid properties*   Consider the transfer of thermal energy from fluid A, at temperature T_A with an average heat transfer coefficient $\langle h_A \rangle$, through a solid plane wall with thermal conductivity k, to fluid B, at temperature T_B with an average heat transfer coefficient $\langle h_B \rangle$, as shown in Figure 8.14.

When the thermal conductivity and the area are constant, the heat transfer rate is calculated from Eq. (8.2-11). The use of this equation, however, requires the values of T_o and T_L to be known or measured. In common practice, it is much easier to measure the bulk fluid temperatures, T_A and T_B. It is then necessary to relate T_o and T_L to T_A and T_B.

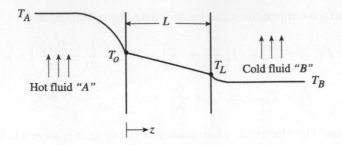

Figure 8.14. Heat transfer through a plane wall.

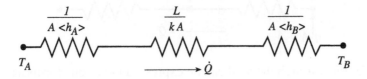

Figure 8.15. Electrical circuit analogy.

The heat transfer rates at the surfaces $z = 0$ and $z = L$ are given by Newton's law of cooling with appropriate heat transfer coefficients and expressed as

$$\dot{Q} = A\langle h_A\rangle(T_A - T_o) = A\langle h_B\rangle(T_L - T_B) \tag{8.2-14}$$

Equations (8.2-11) and (8.2-14) can be rearranged in the form

$$T_A - T_o = \dot{Q}\left(\frac{1}{A\langle h_A\rangle}\right) \tag{8.2-15}$$

$$T_o - T_L = \dot{Q}\left(\frac{L}{Ak}\right) \tag{8.2-16}$$

$$T_L - T_B = \dot{Q}\left(\frac{1}{A\langle h_B\rangle}\right) \tag{8.2-17}$$

Addition of Eqs. (8.2-15)–(8.2-17) gives

$$T_A - T_B = \dot{Q}\left(\frac{1}{A\langle h_A\rangle} + \frac{L}{Ak} + \frac{1}{A\langle h_B\rangle}\right) \tag{8.2-18}$$

or,

$$\boxed{\dot{Q} = \frac{T_A - T_B}{\dfrac{1}{A\langle h_A\rangle} + \dfrac{L}{Ak} + \dfrac{1}{A\langle h_B\rangle}}} \tag{8.2-19}$$

in which the terms in the denominator indicate that the resistances are in series. The electrical circuit analogy for this case is given in Figure 8.15.

Example 8.8 A plane wall separates hot air (A) at a temperature of 50 °C from cold air (B) at −10 °C as shown in Figure 8.16. Calculate the steady rate of heat transfer through the wall if the thermal conductivity of the wall is

a) $k = 0.7$ W/m·K
b) $k = 20$ W/m·K

Solution

Physical properties

For air at 50 °C (323 K): $\begin{cases} \nu = 17.91 \times 10^{-6} \text{ m}^2/\text{s} \\ k = 27.80 \times 10^{-3} \text{ W/m·K} \\ \text{Pr} = 0.708 \end{cases}$

For air at − 10 °C (263 K): $\begin{cases} \nu = 12.44 \times 10^{-6} \text{ m}^2/\text{s} \\ k = 23.28 \times 10^{-3} \text{ W/m·K} \\ \text{Pr} = 0.72 \end{cases}$

For air at 33.5 °C (306.5 K): $\begin{cases} \nu = 16.33 \times 10^{-6} \text{ m}^2/\text{s} \\ k = 26.59 \times 10^{-3} \text{ W/m·K} \\ \text{Pr} = 0.711 \end{cases}$

For air at 0 °C (273 K): $\begin{cases} \nu = 13.30 \times 10^{-6} \text{ m}^2/\text{s} \\ k = 24.07 \times 10^{-3} \text{ W/m·K} \\ \text{Pr} = 0.717 \end{cases}$

Analysis

The rate of heat loss can be calculated from Eq. (8.2-19), i.e.,

$$\dot{Q} = \frac{WH(T_A - T_B)}{\dfrac{1}{\langle h_A \rangle} + \dfrac{L}{k} + \dfrac{1}{\langle h_B \rangle}} \tag{1}$$

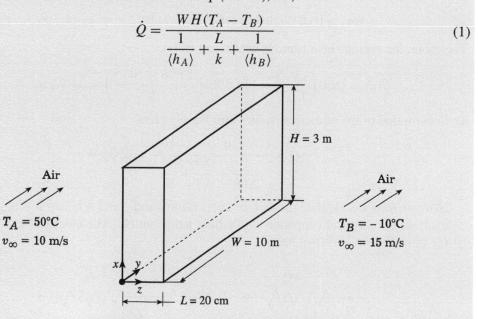

Figure 8.16. Conduction through a plane wall.

The average heat transfer coefficients, $\langle h_A \rangle$ and $\langle h_B \rangle$, can be calculated from the correlations given in Table 4.2. However, the use of these equations requires the physical properties to be evaluated at the film temperature. Since the surface temperatures of the wall cannot be determined a priori, as a first approximation, the physical properties will be evaluated at the fluid temperatures.

Left side of the wall

Note that the characteristic length in the calculation of the Reynolds number is 10 m. The Reynolds number is

$$\text{Re} = \frac{L_{ch}v_\infty}{\nu} = \frac{(10)(10)}{17.91 \times 10^{-6}} = 5.6 \times 10^6 \tag{2}$$

Since this value is between 5×10^5 and 10^8, both laminar and turbulent conditions exist on the wall. The use of Eq. (E) in Table 4.2 gives the Nusselt number as

$$\langle \text{Nu} \rangle = \left[0.037(5.6 \times 10^6)^{4/5} - 871 \right](0.708)^{1/3} = 7480 \tag{3}$$

Therefore, the average heat transfer coefficient is

$$\langle h_A \rangle = \langle \text{Nu} \rangle \left(\frac{k}{L_{ch}} \right) = (7480)\left(\frac{27.80 \times 10^{-3}}{10} \right) = 20.8 \text{ W/m}^2 \cdot \text{K} \tag{4}$$

Right side of the wall

The Reynolds number is

$$\text{Re} = \frac{L_{ch}v_\infty}{\nu} = \frac{(10)(15)}{12.44 \times 10^{-6}} = 12.1 \times 10^6 \tag{5}$$

The use of Eq. (E) in Table 4.2 gives

$$\langle \text{Nu} \rangle = \left[0.037(12.1 \times 10^6)^{4/5} - 871 \right](0.72)^{1/3} = 14{,}596 \tag{6}$$

Therefore, the average heat transfer coefficient is

$$\langle h_B \rangle = \langle \text{Nu} \rangle \left(\frac{k}{L_{ch}} \right) = (14{,}596)\left(\frac{23.28 \times 10^{-3}}{10} \right) = 34 \text{ W/m}^2 \cdot \text{K} \tag{7}$$

a) Substitution of the numerical values into Eq. (1) gives

$$\dot{Q} = \frac{(10)(3)[50 - (-10)]}{\dfrac{1}{20.8} + \dfrac{0.2}{0.7} + \dfrac{1}{34}} = 4956 \text{ W} \tag{8}$$

Now we have to calculate the surface temperatures and check whether it is appropriate to evaluate the physical properties at the fluid temperatures. The electrical circuit analogy for this problem is shown below:

The surface temperatures T_1 and T_2 can be calculated as

$$T_1 = T_A - \frac{\dot{Q}}{A \langle h_A \rangle} = 50 - \frac{4956}{(30)(20.8)} \simeq 42\,°C \tag{9}$$

$$T_2 = T_B + \frac{\dot{Q}}{A \langle h_B \rangle} = -10 + \frac{4956}{(30)(34)} \simeq -5\,°C \tag{10}$$

Therefore, the film temperatures at the left and right sides of the wall are $(42 + 50)/2 = 46\,°C$ and $(-10 - 5)/2 = -7.5\,°C$, respectively. Since these temperatures are not very different from the fluid temperatures, the heat transfer rate can be considered equal to 4956 W.

b) For $k = 20$ W/m·K, the use of Eq. (1) gives

$$\dot{Q} = \frac{(10)(3)\,[50 - (-10)]}{\dfrac{1}{20.8} + \dfrac{0.2}{20} + \dfrac{1}{34}} = 20{,}574 \text{ W} \tag{11}$$

The surface temperatures T_1 and T_2 can be calculated as

$$T_1 = T_A - \frac{\dot{Q}}{A \langle h_A \rangle} = 50 - \frac{20{,}574}{(30)(20.8)} \simeq 17\,°C \tag{12}$$

$$T_2 = T_B + \frac{\dot{Q}}{A \langle h_B \rangle} = -10 + \frac{20{,}574}{(30)(34)} \simeq 10\,°C \tag{13}$$

In this case, the film temperatures at the left and right sides are $(17 + 50)/2 = 33.5\,°C$ and $(-10 + 10)/2 = 0\,°C$, respectively. Since these values are different from the fluid temperatures, it is necessary to recalculate the average heat transfer coefficients.

Left side of the wall

Using the physical properties evaluated at $33.5\,°C$, the Reynolds number becomes

$$\text{Re} = \frac{L_{ch} v_\infty}{\nu} = \frac{(10)(10)}{16.33 \times 10^{-6}} = 6.1 \times 10^6 \tag{14}$$

The Nusselt number is

$$\langle \text{Nu} \rangle = \left[0.037(6.1 \times 10^6)^{4/5} - 871 \right](0.711)^{1/3} = 8076 \tag{15}$$

Therefore, the average heat transfer coefficient is

$$\langle h_A \rangle = \langle \text{Nu} \rangle \left(\frac{k}{L_{ch}} \right) = (8076) \left(\frac{26.59 \times 10^{-3}}{10} \right) = 21.5 \text{ W/m}^2\text{·K} \tag{16}$$

Right side of the wall

Using the physical properties evaluated at $0\,°C$, the Reynolds number becomes

$$\text{Re} = \frac{L_{ch} v_\infty}{\nu} = \frac{(10)(15)}{13.30 \times 10^{-6}} = 11.3 \times 10^6 \tag{17}$$

The use of Eq. (E) in Table 4.2 gives

$$\langle Nu \rangle = [0.037(11.3 \times 10^6)^{4/5} - 871](0.717)^{1/3} = 13{,}758 \tag{18}$$

Therefore, the average heat transfer coefficient is

$$\langle h_B \rangle = \langle Nu \rangle \left(\frac{k}{L_{ch}} \right) = (13{,}758) \left(\frac{24.07 \times 10^{-3}}{10} \right) = 33.1 \text{ W/m}^2 \cdot \text{K} \tag{19}$$

Substitution of the new values of the average heat transfer coefficients, Eqs. (16) and (19), into Eq. (1) gives the heat transfer rate as

$$\dot{Q} = \frac{(10)(3)[50 - (-10)]}{\dfrac{1}{21.5} + \dfrac{0.2}{20} + \dfrac{1}{33.1}} = 20{,}756 \text{ W} \tag{20}$$

The surface temperatures are

$$T_1 = T_A - \frac{\dot{Q}}{A \langle h_A \rangle} = 50 - \frac{20{,}756}{(30)(21.5)} \simeq 18\,°\text{C} \tag{21}$$

$$T_2 = T_B + \frac{\dot{Q}}{A \langle h_B \rangle} = -10 + \frac{20{,}756}{(30)(33.1)} \simeq 11\,°\text{C} \tag{22}$$

Since these values are almost equal to the previous ones, then the rate of heat loss is 20,756 W.

Comment: The Biot numbers, i.e., $\langle h \rangle L / k$, for this problem are calculated as follows:

	Left Side	Right Side
Part (a)	5.9	9.7
Part (b)	0.2	0.3

The physical significance of the Biot number for heat transfer, Bi_H, is given by Eq. (7.1-14). Therefore, when Bi_H is large, the temperature difference between the surface of the wall and the bulk temperature is small, and the physical properties can be calculated at the bulk fluid temperature rather than at the film temperature in engineering calculations. On the other hand, when Bi_H is small, the temperature difference between the surface of the wall and the bulk fluid temperature is large, and the physical properties must be evaluated at the film temperature. Evaluation of the physical properties at the bulk fluid temperature for small values of Bi_H may lead to erroneous results, especially if the physical properties of the fluid are strongly dependent on temperature.

8.2.2 Conduction in Cylindrical Coordinates

Consider a one-dimensional transfer of energy in the r-direction in a hollow cylindrical pipe with inner and outer radii of R_1 and R_2, respectively, as shown in Figure 8.17. Since $T =$

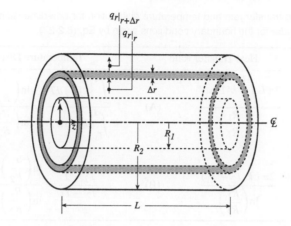

Figure 8.17. Conduction in a hollow cylindrical pipe.

$T(r)$, Table C.5 in Appendix C indicates that the only nonzero energy flux component is e_r, and it is given by

$$e_r = q_r = -k \frac{dT}{dr} \tag{8.2-20}$$

For a cylindrical differential volume element of thickness Δr, as shown in Figure 8.17, Eq. (8.2-1) is expressed in the form

$$(Aq_r)|_r - (Aq_r)|_{r+\Delta r} = 0 \tag{8.2-21}$$

Dividing Eq. (8.2-21) by Δr and taking the limit as $\Delta r \to 0$ give

$$\lim_{\Delta r \to 0} \frac{(Aq_r)|_r - (Aq_r)|_{r+\Delta r}}{\Delta r} = 0 \tag{8.2-22}$$

or,

$$\frac{d(Aq_r)}{dr} = 0 \tag{8.2-23}$$

Since flux times area gives the heat transfer rate, $\dot{Q}$, it is possible to conclude that

$$A\,q_r = \text{constant} = \dot{Q} \tag{8.2-24}$$

The area A in Eq. (8.2-24) is perpendicular to the direction of energy flux in the r-direction and is given by

$$A = 2\pi r L \tag{8.2-25}$$

Substitution of Eqs. (8.2-20) and (8.2-25) into Eq. (8.2-24) and integration give

$$\boxed{\int_0^T k(T)\,dT = -\left(\frac{\dot{Q}}{2\pi L}\right)\ln r + C} \tag{8.2-26}$$

where C is an integration constant.

Table 8.3. Heat transfer rate and temperature distribution for one-dimensional conduction in a hollow cylinder for the boundary conditions given by Eq. (8.2-27)

Constants	Heat Transfer Rate		Temperature Distribution	
None	$\dfrac{2\pi L \displaystyle\int_{T_1}^{T_2} k(T)\,dT}{\ln\left(\dfrac{R_1}{R_2}\right)}$	(A)	$\dfrac{\displaystyle\int_{T}^{T_2} k(T)\,dT}{\displaystyle\int_{T_1}^{T_2} k(T)\,dT} = \dfrac{\ln\left(\dfrac{r}{R_2}\right)}{\ln\left(\dfrac{R_1}{R_2}\right)}$	(C)
k	$\dfrac{2\pi Lk(T_2 - T_1)}{\ln\left(\dfrac{R_1}{R_2}\right)}$	(B)	$\dfrac{T_2 - T}{T_2 - T_1} = \dfrac{\ln\left(\dfrac{r}{R_2}\right)}{\ln\left(\dfrac{R_1}{R_2}\right)}$	(D)

Table 8.4. Heat transfer rate and temperature distribution for one-dimensional conduction in a hollow cylinder for the boundary conditions given by Eq. (8.2-28)

Constants	Heat Transfer Rate		Temperature Distribution	
None	$2\pi R_1 L q_1$	(A)	$\displaystyle\int_{T}^{T_2} k(T)\,dT = q_1 R_1 \ln\left(\dfrac{r}{R_2}\right)$	(C)
k	$2\pi R_1 L q_1$	(B)	$T_2 - T = \dfrac{q_1 R_1}{k} \ln\left(\dfrac{r}{R_2}\right)$	(D)

When the surface temperatures are specified as

$$\text{at} \quad r = R_1 \qquad T = T_1 \tag{8.2-27a}$$

$$\text{at} \quad r = R_2 \qquad T = T_2 \tag{8.2-27b}$$

the resulting temperature distribution as a function of radial position and the heat transfer rate are as given in Table 8.3. On the other hand, if one surface is exposed to a constant heat flux while the other is maintained at a constant temperature, i.e.,

$$\text{at} \quad r = R_1 \qquad -k\frac{dT}{dr} = q_1 \tag{8.2-28a}$$

$$\text{at} \quad r = R_2 \qquad T = T_2 \tag{8.2-28b}$$

the resulting temperature distribution as a function of radial position and the heat transfer rate are as given in Table 8.4.

8.2.2.1 *Electrical circuit analogy* Equation (B) in Table 8.3 can be expressed as

$$\dot{Q} = \frac{T_1 - T_2}{\dfrac{\ln(R_2/R_1)}{2\pi Lk}} \tag{8.2-29}$$

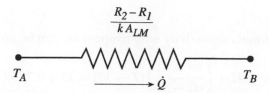

$$\frac{R_2 - R_1}{k A_{LM}}$$

T_A ———WWWW——— T_B

$\longrightarrow \dot{Q}$

Figure 8.18. Electrical circuit analog of the cylindrical wall.

Comparison of Eq. (8.2-29) with Eq. (8.2-10) indicates that the resistance is given by

$$\text{Resistance} = \frac{\ln(R_2/R_1)}{2\pi L k} \qquad (8.2\text{-}30)$$

At first, it looks as if the resistance expressions for the rectangular and the cylindrical coordinate systems are different from each other. However, the similarities between these two expressions can be shown by the following analysis.

The *logarithmic-mean area*, A_{LM}, is defined as

$$A_{LM} = \frac{A_2 - A_1}{\ln(A_2/A_1)} = \frac{2\pi L (R_2 - R_1)}{\ln(R_2/R_1)} \qquad (8.2\text{-}31)$$

Substitution of Eq. (8.2-31) into Eq. (8.2-30) gives

$$\boxed{\text{Resistance} = \frac{R_2 - R_1}{k A_{LM}}} \qquad (8.2\text{-}32)$$

Note that Eqs. (8.2-13) and (8.2-32) have the same general form:

$$\text{Resistance} = \frac{\text{Thickness}}{(\text{Transport property})(\text{Area})} \qquad (8.2\text{-}33)$$

The electrical circuit analog of the cylindrical wall can be represented as shown in Figure 8.18.

Example 8.9 Heat flows through an annular wall of inside radius $R_1 = 10$ cm and outside radius $R_2 = 15$ cm. The inside and outside surface temperatures are $60\,°\text{C}$ and $30\,°\text{C}$, respectively. The thermal conductivity of the wall is dependent on temperature as follows:

$$T = 30\,°\text{C} \qquad k = 42 \text{ W/m·K}$$
$$T = 60\,°\text{C} \qquad k = 49 \text{ W/m·K}$$

Calculate the steady rate of heat transfer if the wall has a length of 2 m.

Solution

Assumption

1. The thermal conductivity varies linearly with temperature.

Analysis

The variation in the thermal conductivity with temperature can be estimated as

$$k = 42 + \left(\frac{49 - 42}{60 - 30}\right)(T - 30) = 35 + 0.233T$$

The heat transfer rate is estimated from Eq. (A) in Table 8.3 with $R_1 = 10$ cm, $R_2 = 15$ cm, $T_1 = 60\,°C$, and $T_2 = 30\,°C$:

$$\dot{Q} = \frac{2\pi L \int_{T_2}^{T_1} k(T)\, dT}{\ln(R_2/R_1)} = \frac{2\pi(2)}{\ln(15/10)} \int_{30}^{60} (35 + 0.233T)\, dT = 42{,}291 \text{ W}$$

8.2.2.2 *Transfer rate in terms of bulk fluid properties*   The use of Eq. (8.2-29) in the calculation of the heat transfer rate requires the surface values T_1 and T_2 to be known or measured. In common practice, the bulk temperatures of the adjoining fluids to the surfaces at $R = R_1$ and $R = R_2$, i.e., T_A and T_B, are known. It is then necessary to relate T_1 and T_2 to T_A and T_B.

The heat transfer rates at the surfaces $R = R_1$ and $R = R_2$ are expressed in terms of the heat transfer coefficients by Newton's law of cooling as

$$\dot{Q} = A_1 \langle h_A \rangle (T_A - T_1) = A_2 \langle h_B \rangle (T_2 - T_B) \tag{8.2-34}$$

The surface areas A_1 and A_2 are expressed in the form

$$A_1 = 2\pi R_1 L \quad \text{and} \quad A_2 = 2\pi R_2 L \tag{8.2-35}$$

Equations (8.2-29) and (8.2-34) can be rearranged in the form

$$T_A - T_1 = \dot{Q}\left(\frac{1}{A_1 \langle h_A \rangle}\right) \tag{8.2-36}$$

$$T_1 - T_2 = \dot{Q}\left(\frac{R_2 - R_1}{A_{LM}k}\right) \tag{8.2-37}$$

$$T_2 - T_B = \dot{Q}\left(\frac{1}{A_2 \langle h_B \rangle}\right) \tag{8.2-38}$$

Addition of Eqs. (8.2-36)–(8.2-38) gives

$$T_A - T_B = \dot{Q}\left(\frac{1}{A_1 \langle h_A \rangle} + \frac{R_2 - R_1}{A_{LM}k} + \frac{1}{A_2 \langle h_B \rangle}\right) \tag{8.2-39}$$

or,

$$\dot{Q} = \frac{T_A - T_B}{\dfrac{1}{A_1 \langle h_A \rangle} + \dfrac{R_2 - R_1}{A_{LM}k} + \dfrac{1}{A_2 \langle h_B \rangle}} \tag{8.2-40}$$

in which the terms in the denominator indicate that the resistances are in series. The electrical circuit analogy for this case is given in Figure 8.19.

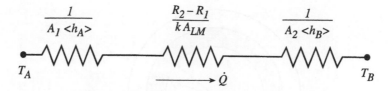

Figure 8.19. Electrical circuit analogy for Eq. (8.2-40).

In the literature, Eq. (8.2-40) is usually expressed in the form

$$Q = A_1 U_A (T_A - T_B) = A_2 U_B (T_A - T_B) \tag{8.2-41}$$

where the terms U_A and U_B are called the *overall heat transfer coefficients*. Comparison of Eq. (8.2-41) with Eq. (8.2-40) gives U_A and U_B as

$$
\begin{aligned}
U_A &= \left[\frac{1}{\langle h_A \rangle} + \frac{(R_2 - R_1)A_1}{A_{LM}k} + \frac{A_1}{\langle h_B \rangle A_2} \right]^{-1} \\
&= \left[\frac{1}{\langle h_A \rangle} + \frac{R_1 \ln(R_2/R_1)}{k} + \frac{R_1}{\langle h_B \rangle R_2} \right]^{-1}
\end{aligned}
\tag{8.2-42}
$$

and

$$
\begin{aligned}
U_B &= \left[\frac{A_2}{\langle h_A \rangle A_1} + \frac{(R_2 - R_1)A_2}{A_{LM}k} + \frac{1}{\langle h_B \rangle} \right]^{-1} \\
&= \left[\frac{R_2}{\langle h_A \rangle R_1} + \frac{R_2 \ln(R_2/R_1)}{k} + \frac{1}{\langle h_B \rangle} \right]^{-1}
\end{aligned}
\tag{8.2-43}
$$

Example 8.10 Consider a cylindrical pipe of length L with inner and outer radii of R_1 and R_2, respectively, and investigate how the rate of heat loss changes as a function of insulation thickness.

Solution

The immediate reaction of most students after reading the problem statement is "What's the point of discussing the rate of heat loss as a function of insulation thickness? Adding insulation thickness obviously decreases the rate of heat loss." This conclusion is true only for planar surfaces. In the case of curved surfaces, however, close examination of Eq. (8.2-32) indicates that while the addition of insulation increases the thickness, i.e., $R_2 - R_1$, it also increases the heat transfer area, i.e., A_{LM}. Hence, both the numerator and denominator of Eq. (8.2-32) increase when the insulation thickness increases. If the increase in the heat transfer area is greater than the increase in thickness, then resistance decreases with a concomitant increase in the rate of heat loss.

For the geometry shown in Figure 8.20, the rate of heat loss is given by

$$\dot{Q} = \frac{T_A - T_B}{\dfrac{1}{2\pi R_1 L \langle h_A \rangle} + \dfrac{\ln(R_2/R_1)}{2\pi L k_w} + \underbrace{\dfrac{\ln(R_3/R_2)}{2\pi L k_i} + \dfrac{1}{2\pi R_3 L \langle h_B \rangle}}_{X}} \tag{1}$$

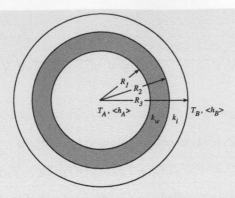

Figure 8.20. Conduction through an insulated cylindrical pipe.

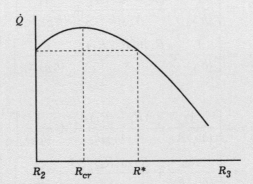

Figure 8.21. Rate of heat loss as a function of insulation thickness.

where k_w and k_i are the thermal conductivities of the wall and the insulating material, respectively. Note that the term X in the denominator of Eq. (1) is dependent on the insulation thickness. Differentiation of X with respect to R_3 gives

$$\frac{dX}{dR_3} = \frac{1}{2\pi L}\left(\frac{1}{R_3 k_i} - \frac{1}{\langle h_B \rangle R_3^2}\right) = 0 \quad \Rightarrow \quad R_3 = \frac{k_i}{\langle h_B \rangle} \tag{2}$$

To determine whether this point corresponds to a minimum or a maximum value, it is necessary to calculate the second derivative, i.e.,

$$\left.\frac{d^2 X}{dR_3^2}\right|_{R_3 = k_i/\langle h_B \rangle} = \frac{1}{2\pi L}\frac{\langle h_B \rangle^2}{k_i^3} > 0 \tag{3}$$

Therefore, at $R_3 = k_i/\langle h_B \rangle$, X has the minimum value. This implies that the rate of heat loss will reach the maximum value at $R_3 = R_{cr} = k_i/\langle h_B \rangle$, where R_{cr} is called the critical thickness of insulation. For $R_2 < R_3 \leqslant R_{cr}$, the addition of insulation causes an increase in the rate of heat loss rather than a decrease. A representative graph showing the variation in the heat transfer rate with insulation thickness is given in Figure 8.21.

Another point of interest is to determine the value of R^*, the point at which the rate of heat loss is equal to that of the bare pipe. The rate of heat loss through the bare pipe, $\dot{Q}_o$, is

$$\dot{Q}_o = \frac{T_A - T_B}{\dfrac{1}{2\pi R_1 L \langle h_A \rangle} + \dfrac{\ln(R_2/R_1)}{2\pi L k_w} + \dfrac{1}{2\pi R_2 L \langle h_B \rangle}} \tag{4}$$

On the other hand, the rate of heat loss, $\dot{Q}^*$, when $R_3 = R^*$ is

$$\dot{Q}^* = \frac{T_A - T_B}{\dfrac{1}{2\pi R_1 L \langle h_A \rangle} + \dfrac{\ln(R_2/R_1)}{2\pi L k_w} + \dfrac{\ln(R^*/R_2)}{2\pi L k_i} + \dfrac{1}{2\pi R^* L \langle h_B \rangle}} \tag{5}$$

Equating Eqs. (4) and (5) gives

$$\frac{R^*}{R_2} - \frac{\langle h_B \rangle R^*}{k_i} \ln\left(\frac{R^*}{R_2}\right) = 1 \tag{6}$$

R^* can be determined from Eq. (6) for the given values of R_2, $\langle h_B \rangle$, and k_i.

Comment: For insulating materials, the largest value of the thermal conductivity is in the order of 0.1 W/m·K. On the other hand, the smallest value of $\langle h_B \rangle$ is around 3 W/m²·K. Therefore, the maximum value of the critical radius is approximately 3.3 cm, and in most practical applications this will not pose a problem. Therefore, the critical radius of insulation is of importance only for small diameter wires or tubes.

Example 8.11 Hot fluid A flows in a pipe with inner and outer radii of R_1 and R_2, respectively. The pipe is surrounded by cold fluid B. If $R_1 = 30$ cm and $R_2 = 35$ cm, calculate the overall heat transfer coefficients and sketch the representative temperature profiles for the following cases:

a) $\langle h_A \rangle = 10$ W/m²·K; $\langle h_B \rangle = 5000$ W/m²·K; $k = 2000$ W/m·K
b) $\langle h_A \rangle = 5000$ W/m²·K; $\langle h_B \rangle = 8000$ W/m²·K; $k = 0.02$ W/m·K
c) $\langle h_A \rangle = 5000$ W/m²·K; $\langle h_B \rangle = 10$ W/m²·K; $k = 2000$ W/m·K

Solution

a) Note that the dominant resistance to heat transfer is that of fluid A. Therefore, one expects the largest temperature drop in this region. Hence Eqs. (8.2-42) and (8.2-43) give the overall heat transfer coefficients as

$$U_A = \left(\frac{1}{\langle h_A \rangle}\right)^{-1} = \langle h_A \rangle = 10 \text{ W/m}^2\text{·K}$$

$$U_B = \left(\frac{R_2}{\langle h_A \rangle R_1}\right)^{-1} = \langle h_A \rangle \left(\frac{R_1}{R_2}\right) = \frac{(10)(30)}{35} = 8.6 \text{ W/m}^2\text{·K}$$

The expected temperature profile for this case is shown below.

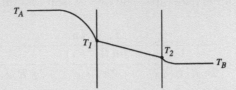

b) In this case, the dominant resistance to heat transfer is that of the pipe wall. The overall heat transfer coefficients are

$$U_A = \frac{k}{R_1 \ln(R_2/R_1)} = \frac{0.02}{(0.3) \ln(35/30)} = 0.43 \text{ W/m}^2 \cdot \text{K}$$

$$U_B = \frac{k}{R_2 \ln(R_2/R_1)} = \frac{0.02}{(0.45) \ln(35/30)} = 0.29 \text{ W/m}^2 \cdot \text{K}$$

The expected temperature profile for this case is shown below:

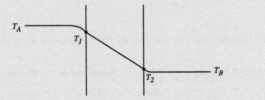

c) The dominant resistance to heat transfer is that of fluid B. Hence, the overall heat transfer coefficients are

$$U_A = \left(\frac{R_1}{\langle h_B \rangle R_2}\right)^{-1} = \langle h_B \rangle \left(\frac{R_2}{R_1}\right) = \frac{(10)(35)}{30} = 11.7 \text{ W/m}^2 \cdot \text{K}$$

$$U_B = \left(\frac{1}{\langle h_B \rangle}\right)^{-1} = \langle h_B \rangle = 10 \text{ W/m}^2 \cdot \text{K}$$

The expected temperature profile for this case is shown below:

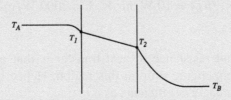

Comment: The region with the largest thermal resistance has the largest temperature drop.

8.2.3 Conduction in Spherical Coordinates

Consider one-dimensional transfer of energy in the r-direction through a hollow sphere of inner and outer radii R_1 and R_2, respectively, as shown in Figure 8.22. Since $T = T(r)$,

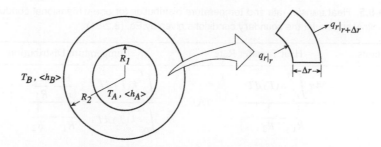

Figure 8.22. Conduction through a hollow sphere.

Table C.6 in Appendix C indicates that the only nonzero energy flux component is e_r, and it is given by

$$e_r = q_r = -k\frac{dT}{dr} \qquad (8.2\text{-}44)$$

For a spherical differential volume element of thickness Δr as shown in Figure 8.22, Eq. (8.2-1) is expressed in the form

$$(Aq_r)|_r - (Aq_r)|_{r+\Delta r} = 0 \qquad (8.2\text{-}45)$$

Dividing Eq. (8.2-45) by Δr and taking the limit as $\Delta r \to 0$ give

$$\lim_{\Delta r \to 0} \frac{(Aq_r)|_r - (Aq_r)|_{r+\Delta r}}{\Delta r} = 0 \qquad (8.2\text{-}46)$$

or,

$$\frac{d(Aq_r)}{dr} = 0 \qquad (8.2\text{-}47)$$

Since flux times area gives the heat transfer rate, $\dot{Q}$, it is possible to conclude that

$$A\,q_r = \text{constant} = \dot{Q} \qquad (8.2\text{-}48)$$

The area A in Eq. (8.2-48) is perpendicular to the direction of energy flux in the r-direction and it is given by

$$A = 4\pi r^2 \qquad (8.2\text{-}49)$$

Substitution of Eqs. (8.2-44) and (8.2-49) into Eq. (8.2-48) and integration give

$$\boxed{\int_0^T k(T)\,dT = \left(\frac{\dot{Q}}{4\pi}\right)\frac{1}{r} + C} \qquad (8.2\text{-}50)$$

where C is an integration constant.

When the surface temperatures are specified as

$$\text{at} \quad r = R_1 \qquad T = T_1 \qquad (8.2\text{-}51a)$$

$$\text{at} \quad r = R_2 \qquad T = T_2 \qquad (8.2\text{-}51b)$$

Table 8.5. Heat transfer rate and temperature distribution for one-dimensional conduction in a hollow sphere for the boundary conditions given by Eq. (8.2-51)

Constants	Heat Transfer Rate		Temperature Distribution	
None	$\dfrac{4\pi \displaystyle\int_{T_2}^{T_1} k(T)\,dT}{\dfrac{1}{R_1} - \dfrac{1}{R_2}}$	(A)	$\dfrac{\displaystyle\int_{T_2}^{T} k(T)\,dT}{\displaystyle\int_{T_2}^{T_1} k(T)\,dT} = \dfrac{\dfrac{1}{r} - \dfrac{1}{R_2}}{\dfrac{1}{R_1} - \dfrac{1}{R_2}}$	(C)
k	$\dfrac{4\pi k(T_1 - T_2)}{\dfrac{1}{R_1} - \dfrac{1}{R_2}}$	(B)	$\dfrac{T - T_2}{T_1 - T_2} = \dfrac{\dfrac{1}{r} - \dfrac{1}{R_2}}{\dfrac{1}{R_1} - \dfrac{1}{R_2}}$	(D)

Table 8.6. Heat transfer rate and temperature distribution for one-dimensional conduction in a hollow sphere for the boundary conditions given by Eq. (8.2-52)

Constants	Heat Transfer Rate		Temperature Distribution	
None	$4\pi R_1^2 q_1$	(A)	$\displaystyle\int_{T_2}^{T} k(T)\,dT = q_1 R_1^2 \left(\dfrac{1}{r} - \dfrac{1}{R_2} \right)$	(C)
k	$4\pi R_1^2 q_1$	(B)	$T - T_2 = \dfrac{q_1 R_1^2}{k} \left(\dfrac{1}{r} - \dfrac{1}{R_2} \right)$	(D)

the resulting temperature distribution as a function of radial position and the heat transfer rate are as given in Table 8.5.

On the other hand, if one surface is exposed to a constant heat flux while the other is maintained at a constant temperature, i.e.,

$$\text{at} \quad r = R_1 \qquad -k\frac{dT}{dr} = q_1 \tag{8.2-52a}$$

$$\text{at} \quad r = R_2 \qquad T = T_2 \tag{8.2-52b}$$

the resulting temperature distribution as a function of radial position and the heat transfer rate are as given in Table 8.6.

Example 8.12 A spherical metal ball of radius R is placed in an infinitely large volume of motionless fluid. The ball is maintained at a temperature of T_R, while the temperature of the fluid far from the ball is T_∞.

a) Determine the temperature distribution within the fluid.
b) Determine the rate of heat transferred to the fluid.
c) Determine the Nusselt number.
d) Calculate the heat flux at the surface of the sphere for the following values:

$$R = 2 \text{ cm} \qquad k = 0.025 \text{ W/m·K} \qquad T_R = 60\,°\text{C} \qquad T_\infty = 25\,°\text{C}$$

Solution

Assumptions

1. Steady-state conditions prevail.
2. The heat transfer from the ball to the fluid takes place only by conduction.
3. The thermal conductivity of the fluid is constant.

Analysis

a) The temperature distribution can be obtained from Eq. (D) of Table 8.5 in the form

$$\frac{T - T_\infty}{T_R - T_\infty} = \frac{R}{r} \tag{1}$$

b) The use of Eq. (B) in Table 8.5 with $T_1 = T_R$, $T_2 = T_\infty$, $R_1 = R$, and $R_2 = \infty$ gives the rate of heat transferred from the ball to the fluid as

$$\dot{Q} = \frac{4\pi k (T_R - T_\infty)}{1/R} = 4\pi R k (T_R - T_\infty) \tag{2}$$

c) The amount of heat transferred can also be calculated from Newton's law of cooling, Eq. (3.2-7), as

$$\dot{Q} = 4\pi R^2 \langle h \rangle (T_R - T_\infty) \tag{3}$$

Equating Eqs. (2) and (3) leads to

$$\frac{\langle h \rangle}{k} = \frac{1}{R} = \frac{2}{D} \tag{4}$$

Therefore, the Nusselt number is

$$\text{Nu} = \frac{\langle h \rangle D}{k} = 2 \tag{5}$$

d) One way of expressing the heat flux at the surface of the sphere is

$$q_r|_{r=R} = -k \left. \frac{dT}{dr} \right|_{r=R} \tag{6}$$

The use of Eq. (1) in Eq. (6) gives

$$q_r|_{r=R} = \frac{k(T_R - T_\infty)}{R} = \frac{(0.025)(60 - 25)}{2 \times 10^{-2}} = 43.75 \ \text{W/m}^2 \tag{7}$$

It is also possible to evaluate the heat flux at the surface of the sphere from Newton's law of cooling, i.e.,

$$q_r|_{r=R} = \langle h \rangle (T_R - T_\infty) \tag{8}$$

Since $\text{Nu} = 2$, the average heat transfer coefficient is expressed as

$$\langle h \rangle = \frac{2k}{D} = \frac{k}{R} \tag{9}$$

Substitution of Eq. (9) into Eq. (8) leads to Eq. (7).

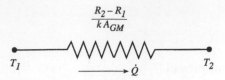

Figure 8.23. Electrical circuit analog of the spherical wall.

8.2.3.1 *Electrical circuit analogy* Equation (B) in Table 8.5 can be rearranged in the form

$$\dot{Q} = \frac{T_1 - T_2}{\dfrac{R_2 - R_1}{4\pi k R_1 R_2}} \tag{8.2-53}$$

Comparison of Eq. (8.2-53) with Eq. (8.2-10) indicates that the resistance is given by

$$\boxed{\text{Resistance} = \frac{R_2 - R_1}{4\pi k R_1 R_2}} \tag{8.2-54}$$

In order to express the resistance in the form given by Eq. (8.2-13), note that a geometric mean area, A_{GM}, is defined as

$$A_{GM} = \sqrt{A_1 A_2} = \sqrt{(4\pi R_1^2)(4\pi R_2^2)} = 4\pi R_1 R_2 \tag{8.2-55}$$

so that Eq. (8.2-54) takes the form

$$\boxed{\text{Resistance} = \frac{R_2 - R_1}{k A_{GM}} = \frac{\text{Thickness}}{(\text{Transport property})(\text{Area})}} \tag{8.2-56}$$

The electrical circuit analog of the spherical wall can be represented as shown in Figure 8.23.

8.2.3.2 *Transfer rate in terms of bulk fluid properties* The use of Eq. (8.2-53) in the calculation of the transfer rate requires the surface values T_1 and T_2 to be known or measured. In common practice, the bulk temperatures of the adjoining fluids to the surfaces at $r = R_1$ and $r = R_2$, i.e., T_A and T_B, are known. It is then necessary to relate T_1 and T_2 to T_A and T_B.

The procedure for the spherical case is similar to that for the cylindrical case and is left as an exercise for the students. If the procedure given in Section 8.2.2.2 is followed, the result is

$$\boxed{\dot{Q} = \frac{T_A - T_B}{\dfrac{1}{A_1\langle h_A \rangle} + \dfrac{R_2 - R_1}{A_{GM} k} + \dfrac{1}{A_2\langle h_B \rangle}}} \tag{8.2-57}$$

Example 8.13 Consider a spherical tank with inner and outer radii of R_1 and R_2, respectively, and investigate how the rate of heat loss varies as a function of insulation thickness.

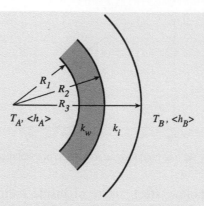

Figure 8.24. Conduction through an insulated hollow sphere.

Solution

The solution procedure for this problem is similar to that for Example 8.10. For the geometry shown in Figure 8.24, the rate of heat loss is given by

$$\dot{Q} = \frac{4\pi (T_A - T_B)}{\underbrace{\dfrac{1}{R_1^2 \langle h_A \rangle} + \dfrac{R_2 - R_1}{R_1 R_2 k_w} + \dfrac{R_3 - R_2}{R_2 R_3 k_i} + \dfrac{1}{R_3^2 \langle h_B \rangle}}_{X}} \tag{1}$$

where k_w and k_i are the thermal conductivities of the wall and the insulating material, respectively.

Differentiation of X with respect to R_3 gives

$$\frac{dX}{dR_3} = 0 \quad \Rightarrow \quad R_3 = \frac{2k_i}{\langle h_B \rangle} \tag{2}$$

To determine whether this point corresponds to a minimum or a maximum value, it is necessary to calculate the second derivative, i.e.,

$$\left. \frac{d^2 X}{dR_3^2} \right|_{R_3 = 2k_i/\langle h_B \rangle} = \frac{1}{8} \frac{\langle h_B \rangle^3}{k_i^4} > 0 \tag{3}$$

Therefore, the critical thickness of insulation for the spherical geometry is given by

$$R_{cr} = \frac{2k_i}{\langle h_B \rangle} \tag{4}$$

A representative graph showing the variation in heat transfer rate with insulation thickness is given in Figure 8.25.

Another point of interest is to determine the value of R^*, the point at which the rate of heat loss is equal to that of the bare pipe. Following the procedure given in Example 8.10, the result is

$$\left(\frac{R^*}{R_2} \right)^2 = \frac{\langle h_B \rangle R^*}{k_i} \left(\frac{R^*}{R_2} - 1 \right) + 1 \tag{5}$$

The value of R^* can be determined from Eq. (5) for the given values of R_2, $\langle h_B \rangle$, and k_i.

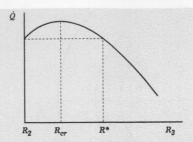

Figure 8.25. Rate of heat loss as a function of insulation thickness.

Example 8.14 Consider a hollow steel sphere of inside radius $R_1 = 10$ cm and outside radius $R_2 = 20$ cm. The inside surface is maintained at a constant temperature of $180\,°C$ and the outside surface dissipates heat to ambient temperature at $20\,°C$ by convection with an average heat transfer coefficient of 11 W/m^2·K. To reduce the rate of heat loss, it is proposed to cover the outer surface of the sphere with two types of insulating materials, X and Y, in series. Each insulating material has a thickness of 3 cm. The thermal conductivities of X and Y are 0.04 and 0.12 W/m·K, respectively. One of your friends claims that the order in which the two insulating materials are put around the sphere does not make a difference to the rate of heat loss. As an engineer, do you agree?

Solution

Physical properties

For steel: $k = 45$ W/m·K

Analysis

The rate of heat loss can be determined from Eq. (8.2-57). If the surface is first covered by X and then by Y, the rate of heat loss is

$$\dot{Q} = \cfrac{4\pi(180 - 20)}{\cfrac{0.1}{(45)(0.1)(0.2)} + \cfrac{0.03}{(0.04)(0.2)(0.23)} + \cfrac{0.03}{(0.12)(0.23)(0.26)} + \cfrac{1}{(0.26)^2(11)}}$$
$$= 91.6 \text{ W}$$

On the other hand, covering the surface first by Y and then by X gives the rate of heat loss as

$$\dot{Q} = \cfrac{4\pi(180 - 20)}{\cfrac{0.1}{(45)(0.1)(0.2)} + \cfrac{0.03}{(0.12)(0.2)(0.23)} + \cfrac{0.03}{(0.04)(0.23)(0.26)} + \cfrac{1}{(0.26)^2(11)}}$$
$$= 103.5 \text{ W}$$

Therefore, the order of the layers with different thermal conductivities does make a difference.

8.2.4 Conduction in a Fin

In the previous sections, we considered one-dimensional conduction examples. The extension of the procedure for these problems to conduction in two- or three-dimensional cases is

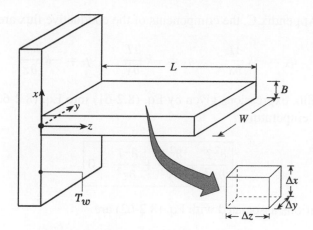

Figure 8.26. Conduction in a rectangular fin.

straightforward. The difficulty with multi-dimensional conduction problems lies in the solution of the resulting partial differential equations. An excellent book by Carslaw and Jaeger (1959) gives solutions to conduction problems with various boundary conditions.

In this section, first the governing equation for temperature distribution will be developed for three-dimensional conduction in a rectangular geometry. Then the use of area averaging[4] will be introduced to simplify the problem.

Fins are extensively used in heat transfer applications to enhance the heat transfer rate by increasing the heat transfer area. Let us consider a simple rectangular fin as shown in Figure 8.26. As engineers, we are interested in the rate of heat loss from the surfaces of the fin. This can be calculated if the temperature distribution within the fin is known. The problem will be analyzed with the following assumptions:

1. Steady-state conditions prevail.
2. The thermal conductivity of the fin is constant.
3. The average heat transfer coefficient is constant.
4. There is no heat loss from the edges or the tip of the fin.

For a rectangular volume element of thickness Δx, width Δy, and length Δz, as shown in Figure 8.26, Eq. (8.2-1) is expressed as

$$(q_x|_x \Delta y\, \Delta z + q_y|_y \Delta x\, \Delta z + q_z|_z \Delta x\, \Delta y)$$
$$- (q_x|_{x+\Delta x} \Delta y\, \Delta z + q_y|_{y+\Delta y} \Delta x\, \Delta z + q_z|_{z+\Delta z} \Delta x\, \Delta y) = 0 \qquad (8.2\text{-}58)$$

Dividing Eq. (8.2-58) by $\Delta x\, \Delta y\, \Delta z$ and taking the limit as $\Delta x \to 0$, $\Delta y \to 0$, and $\Delta z \to 0$ give

$$\lim_{\Delta x \to 0} \frac{q_x|_x - q_x|_{x+\Delta x}}{\Delta x} + \lim_{\Delta y \to 0} \frac{q_y|_y - q_y|_{y+\Delta y}}{\Delta y} + \lim_{\Delta z \to 0} \frac{q_z|_z - q_z|_{z+\Delta z}}{\Delta z} = 0 \qquad (8.2\text{-}59)$$

or,

$$\frac{\partial q_x}{\partial x} + \frac{\partial q_y}{\partial y} + \frac{\partial q_z}{\partial z} = 0 \qquad (8.2\text{-}60)$$

---

[4]The first systematic use of the area averaging technique in a textbook can be attributed to Slattery (1972).

From Table C.4 in Appendix C, the components of the conductive flux are given by

$$q_x = -k\,\frac{\partial T}{\partial x} \qquad q_y = -k\,\frac{\partial T}{\partial y} \qquad q_z = -k\,\frac{\partial T}{\partial z} \qquad (8.2\text{-}61)$$

Substitution of the flux expressions given by Eq. (8.2-61) into Eq. (8.2-60) leads to the governing equation for temperature

$$\boxed{\frac{\partial^2 T}{\partial x^2} + \frac{\partial^2 T}{\partial y^2} + \frac{\partial^2 T}{\partial z^2} = 0} \qquad (8.2\text{-}62)$$

The boundary conditions associated with Eq. (8.2-62) are

$$\text{at} \quad x = B/2 \qquad -k\,\frac{\partial T}{\partial x} = \langle h \rangle (T - T_\infty) \qquad (8.2\text{-}63)$$

$$\text{at} \quad x = -B/2 \qquad k\,\frac{\partial T}{\partial x} = \langle h \rangle (T - T_\infty) \qquad (8.2\text{-}64)$$

$$\text{at} \quad y = 0 \qquad \frac{\partial T}{\partial y} = 0 \qquad (8.2\text{-}65)$$

$$\text{at} \quad y = W \qquad \frac{\partial T}{\partial y} = 0 \qquad (8.2\text{-}66)$$

$$\text{at} \quad z = 0 \qquad T = T_w \qquad (8.2\text{-}67)$$

$$\text{at} \quad z = L \qquad \frac{\partial T}{\partial z} = 0 \qquad (8.2\text{-}68)$$

where T_∞ is the temperature of the fluid surrounding the fin.

If the measuring instrument, i.e., the temperature probe, is not sensitive enough to detect temperature variations in the x-direction, then it is necessary to change the scale of the problem to match that of the measuring device. In other words, it is necessary to average the governing equation up to the scale of the temperature measuring probe.

The area-averaged temperature is defined by

$$\langle T \rangle = \frac{\displaystyle\int_0^W \int_{-B/2}^{B/2} T\,dx\,dy}{\displaystyle\int_0^W \int_{-B/2}^{B/2} dx\,dy} = \frac{1}{WB}\int_0^W \int_{-B/2}^{B/2} T\,dx\,dy \qquad (8.2\text{-}69)$$

Note that although the local temperature, T, is dependent on x, y, and z, the area-averaged temperature, $\langle T \rangle$, depends only on z.

Area averaging is performed by integrating Eq. (8.2-62) over the cross-sectional area of the fin. The result is

$$\int_0^W \int_{-B/2}^{B/2} \frac{\partial^2 T}{\partial x^2}\,dx\,dy + \int_0^W \int_{-B/2}^{B/2} \frac{\partial^2 T}{\partial y^2}\,dx\,dy + \int_0^W \int_{-B/2}^{B/2} \frac{\partial^2 T}{\partial z^2}\,dx\,dy = 0 \qquad (8.2\text{-}70)$$

or,

$$\int_0^W \left(\frac{\partial T}{\partial x}\bigg|_{x=B/2} - \frac{\partial T}{\partial x}\bigg|_{x=-B/2}\right) dy + \int_{-B/2}^{B/2} \left(\frac{\partial T}{\partial y}\bigg|_{y=W} - \frac{\partial T}{\partial y}\bigg|_{y=0}\right) dx$$

$$+ \frac{d^2}{dz^2}\left(\int_0^W \int_{-B/2}^{B/2} T \, dx \, dy\right) = 0 \tag{8.2-71}$$

The use of the boundary conditions defined by Eqs. (8.2-63)–(8.2-66) together with the definition of the average temperature, Eq. (8.2-69), in Eq. (8.2-71) gives

$$W\left[-\frac{\langle h \rangle}{k}(T|_{x=B/2} - T_\infty) - \frac{\langle h \rangle}{k}(T|_{x=-B/2} - T_\infty)\right] + W B \frac{d^2 \langle T \rangle}{dz^2} = 0 \tag{8.2-72}$$

Since $T|_{x=B/2} = T|_{x=-B/2}$ as a result of symmetry, Eq. (8.2-72) takes the form

$$k \frac{d^2 \langle T \rangle}{dz^2} - \frac{2}{B} \langle h \rangle (T|_{x=B/2} - T_\infty) = 0 \tag{8.2-73}$$

Note that Eq. (8.2-73) contains two dependent variables, $\langle T \rangle$ and $T|_{x=B/2}$, which are at two different scales. It is generally assumed, although not expressed explicitly, that

$$\langle T \rangle \approx T|_{x=B/2} \tag{8.2-74}$$

This approximation is valid when

$$Bi_H = \frac{\langle h \rangle (B/2)}{k} \ll 1 \tag{8.2-75}$$

Substitution of Eq. (8.2-74) into Eq. (8.2-73) gives

$$\boxed{k \frac{d^2 \langle T \rangle}{dz^2} = \frac{2}{B} \langle h \rangle (\langle T \rangle - T_\infty)} \tag{8.2-76}$$

Integration of Eqs. (8.2-67) and (8.2-68) over the cross-sectional area of the fin gives the boundary conditions associated with Eq. (8.2-76) as

$$\text{at} \quad z = 0 \quad \langle T \rangle = T_w \tag{8.2-77}$$

$$\text{at} \quad z = L \quad \frac{d \langle T \rangle}{dz} = 0 \tag{8.2-78}$$

It should be kept in mind that Eqs. (8.2-62) and (8.2-76) are at two different scales. Equation (8.2-76) is obtained by averaging Eq. (8.2-62) over the cross-sectional area perpendicular to the direction of energy flux. In this way, the boundary condition, i.e., the heat transfer coefficient, is incorporated into the governing equation. The accuracy of the measurements dictates the equation to work with since the scale of the measurements should be compatible with the scale of the equation.

Table 8.7. The physical significance and the order of magnitude of the terms in Eq. (8.2-76)

Term	Physical Significance	Order of Magnitude
$k \dfrac{d^2 \langle T \rangle}{dz^2}$	Rate of conduction	$\dfrac{k(T_w - T_\infty)}{L^2}$
$\dfrac{2\langle h \rangle}{B}(\langle T \rangle - T_\infty)$	Rate of heat transfer from the fin to the surroundings	$\dfrac{2\langle h \rangle (T_w - T_\infty)}{B}$

The term $2/B$ in Eq. (8.2-76) represents the heat transfer area per unit volume of the fin, i.e.,

$$\frac{2}{B} = \frac{2LW}{BLW} = \frac{\text{Heat transfer area}}{\text{Fin volume}} \tag{8.2-79}$$

The physical significance and the order of magnitude[5] of the terms in Eq. (8.2-76) are given in Table 8.7.

Therefore, the ratio of the rate of heat transfer from the fin surface to the rate of conduction is given by

$$\frac{\text{Rate of heat transfer}}{\text{Rate of conduction}} = \frac{2\langle h \rangle (T_w - T_\infty)/B}{k(T_w - T_\infty)/L^2} = \frac{2\langle h \rangle L^2}{kB} \tag{8.2-80}$$

Before solving Eq. (8.2-76), it is convenient to express the governing equation and the boundary conditions in dimensionless form. The reason for doing this is that the inventory equations in dimensionless form represent the solution to the entire class of geometrically similar problems when they are applied to a particular geometry. Introduction of the dimensionless variables

$$\theta = \frac{\langle T \rangle - T_\infty}{T_w - T_\infty} \qquad \xi = \frac{z}{L} \qquad \Lambda = \sqrt{\frac{2\langle h \rangle L^2}{kB}} \tag{8.2-81}$$

reduces Eqs. (8.2-76)–(8.2-78) to

$$\frac{d^2\theta}{d\xi^2} = \Lambda^2 \theta \tag{8.2-82}$$

$$\text{at} \quad \xi = 0 \qquad \theta = 1 \tag{8.2-83}$$

$$\text{at} \quad \xi = 1 \qquad \frac{d\theta}{d\xi} = 0 \tag{8.2-84}$$

---

[5]The *order of magnitude* or *scale analysis* is a powerful tool for those interested in mathematical modeling. As stated by Astarita (1997), "Very often more than nine-tenths of what one can ever hope to know about a problem can be obtained from this tool, without actually solving the problem; the remaining one-tenth requires painstaking algebra and/or lots of computer time, it adds very little to our understanding of the problem, and if we have not done the first part right, all that the algebra and the computer will produce will be a lot of nonsense. Of course, when nonsense comes out of a computer people have a lot of respect for it, and that is exactly the problem." For more details on the order of magnitude analysis, see Bejan (1984), and Whitaker (1976).

The solution of Eq. (8.2-82) is

$$\theta = C_1 \sinh(\Lambda \xi) + C_2 \cosh(\Lambda \xi) \tag{8.2-85}$$

where C_1 and C_2 are constants. Application of the boundary conditions defined by Eqs. (8.2-83) and (8.2-84) leads to

$$\theta = \frac{\cosh \Lambda \cosh(\Lambda \xi) - \sinh \Lambda \sinh(\Lambda \xi)}{\cosh \Lambda} \tag{8.2-86}$$

The use of the identity

$$\cosh(x - y) = \cosh x \cosh y - \sinh x \sinh y \tag{8.2-87}$$

reduces Eq. (8.2-86) to the form

$$\boxed{\theta = \frac{\cosh [\Lambda (1 - \xi)]}{\cosh \Lambda}} \tag{8.2-88}$$

8.2.4.1 *Macroscopic equation* Integration of the governing differential equation, Eq. (8.2-76), over the volume of the system gives the macroscopic energy balance, i.e.,

$$\int_0^L \int_0^W \int_{-B/2}^{B/2} k \frac{d^2 \langle T \rangle}{dz^2} \, dx \, dy \, dz = \int_0^L \int_0^W \int_{-B/2}^{B/2} \frac{2}{B} \langle h \rangle \big(\langle T \rangle - T_\infty \big) \, dx \, dy \, dz \tag{8.2-89}$$

Evaluation of the integrations yields

$$\underbrace{BW \left(-k \frac{d \langle T \rangle}{dz} \big|_{z=0} \right)}_{\substack{\text{Rate of energy entering the} \\ \text{fin through the surface at } z = 0}} = \underbrace{2W \langle h \rangle \int_0^L \big(\langle T \rangle - T_\infty \big) \, dz}_{\substack{\text{Rate of energy loss from the top and bottom} \\ \text{surfaces of the fin to the surroundings}}} \tag{8.2-90}$$

which is the macroscopic inventory rate equation for thermal energy by considering the fin as a system. The use of Eq. (8.2-88) in Eq. (8.2-90) gives the rate of heat loss, $\dot{Q}$, from the fin as

$$\boxed{\dot{Q} = \frac{BWk(T_w - T_\infty) \Lambda \tanh \Lambda}{L}} \tag{8.2-91}$$

8.2.4.2 *Fin efficiency* The *fin efficiency*, η, is defined as the ratio of the apparent rate of heat dissipation of a fin to the ideal rate of heat dissipation if the entire fin surface were at T_w, i.e.,

$$\eta = \frac{2W \langle h \rangle \displaystyle\int_0^L \big(\langle T \rangle - T_\infty \big) \, dz}{2W \langle h \rangle (T_w - T_\infty) L} = \frac{\displaystyle\int_0^L \big(\langle T \rangle - T_\infty \big) \, dz}{(T_w - T_\infty) L} \tag{8.2-92}$$

In terms of the dimensionless quantities, Eq. (8.2-92) becomes

$$\eta = \int_0^1 \theta \, d\xi \tag{8.2-93}$$

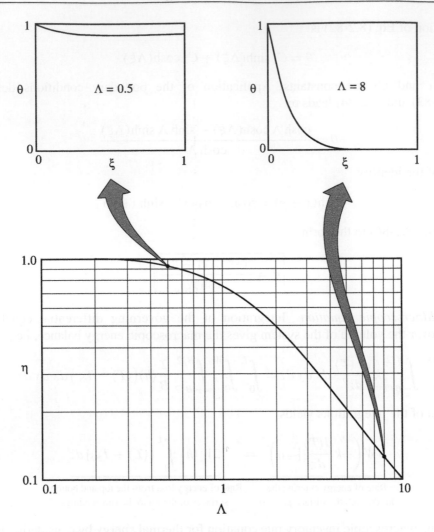

Figure 8.27. Variation in the fin efficiency, η, as a function of Λ.

Substitution of Eq. (8.2-88) into Eq. (8.2-93) gives the fin efficiency as

$$\eta = \frac{\tanh \Lambda}{\Lambda} \qquad (8.2\text{-}94)$$

The variation in the fin efficiency as a function of Λ is shown in Figure 8.27. When $\Lambda \to 0$, the rate of conduction is much larger than the rate of heat dissipation. The Taylor series expansion of η in terms of Λ gives

$$\eta = 1 - \frac{1}{3}\Lambda^2 + \frac{2}{15}\Lambda^4 - \frac{17}{315}\Lambda^6 + \cdots \qquad (8.2\text{-}95)$$

Therefore, η approaches unity as $\Lambda \to 0$, indicating that the entire fin surface is at the wall temperature.

On the other hand, large values of Λ correspond to cases in which the heat transfer rate by conduction is very slow and the rate of heat transfer from the fin surface is very rapid. Under these conditions, the fin efficiency becomes

$$\eta = \frac{1}{\Lambda} \qquad (8.2\text{-}96)$$

indicating that η approaches zero as $\Lambda \to \infty$.

Since the fin efficiency is inversely proportional to Λ, it can be improved either by increasing k and B, or by decreasing $\langle h \rangle$ and L. If the average heat transfer coefficient, $\langle h \rangle$, is increased due to an increase in the air velocity past the fin, the fin efficiency decreases. This means that the length of the fin, L, can be smaller for the larger $\langle h \rangle$ if the fin efficiency remains constant. In other words, fins are not necessary at high speeds of fluid velocity.

8.2.4.3 *Comment* In general, the governing differential equations represent the variation in the dependent variables, such as temperature and concentration, as a function of position and time. On the other hand, the transfer coefficients, which represent the interaction of the system with the surroundings, appear in the boundary conditions. If the transfer coefficients appear in the governing equations rather than in the boundary conditions, it is implied that these equations are obtained as a result of the averaging process.

Example 8.15 A plane wall of thickness 2.5 mm is made of aluminum ($k = 200$ W/m·K) and separates an air stream flowing at 40 °C from a water stream flowing at 75 °C. The average heat transfer coefficients on the air side and the water side are 20 W/m²·K and 500 W/m²·K, respectively.

a) Calculate the rate of heat transfer per m² of plane wall from the water stream to the air stream under steady conditions.

b) It is proposed to increase the rate of heat transfer by attaching aluminum fins of rectangular profile to the plane wall. To which side do we have to add fins?

c) Calculate the steady rate of heat transfer per m² of plane wall if the fins have the dimensions of $B = 1$ mm and $L = 10$ mm and are placed with a fin spacing of 125 fins/m.

Solution

Assumptions

1. Heat losses from the edges and the tip of the fin are negligible.
2. Addition of fins does not affect the heat transfer coefficient.

Analysis

a) The electrical circuit analogy of the overall system is shown below:

Therefore, the steady rate of heat transfer between water and air streams is

$$\frac{\dot{Q}}{A} = \frac{75 - 40}{\dfrac{1}{500} + \dfrac{2.5 \times 10^{-3}}{200} + \dfrac{1}{20}} = 673 \text{ W/m}^2$$

b) From the electrical circuit analogy we see that the air-side resistance is controlling the rate of heat transfer between the streams. Therefore, fins must be added to the air side, where the heat transfer coefficient is lower.

c) When fins are attached to the air side, the steady rate of heat transfer from the wall to the air stream is given by

$$\frac{\dot{Q}}{A} = A_b \langle h \rangle_{air}(T_w - T_{air}) + A_f \langle h \rangle_{air}(T_w - T_{air})\eta$$

$$= (A_b + A_f \eta)\langle h \rangle_{air}(T_w - T_{air}) \tag{1}$$

where A_b and A_f represent the area of bare wall surface and the total surface area of the fins, respectively, per m^2 of plane wall. The term T_w represents the surface temperature of the plane wall on the air side. The electrical resistance analogy for this case is represented as follows:

Therefore, the steady rate of heat transfer between the water and air streams becomes

$$\frac{\dot{Q}}{A} = \frac{T_{water} - T_{air}}{\dfrac{1}{\langle h \rangle_{water}} + \dfrac{L_{wall}}{k} + \dfrac{1}{(A_b + A_f \eta)\langle h \rangle_{air}}} \tag{2}$$

The area of bare wall surface, A_b, per m^2 of plane wall is

$$A_b = 1 - (125)(1 \times 10^{-3})(1) = 0.875 \ m^2/m^2$$

The total surface area of the fins, A_f, per m^2 of plane wall is

$$A_f = (125)\big[(2)(10 \times 10^{-3})(1)\big] = 2.5 \ m^2/m^2$$

From Eq. (8.2-81)

$$\Lambda = \sqrt{\frac{2\langle h \rangle_{air}L^2}{kB}} = \sqrt{\frac{(2)(20)(10 \times 10^{-3})^2}{(200)(1 \times 10^{-3})}} = 0.141$$

The fin efficiency, η, is given by Eq. (8.2-94)

$$\eta = \frac{\tanh \Lambda}{\Lambda} = \frac{\tanh(0.141)}{0.141} = 0.993$$

Substitution of the numerical values into Eq. (2) yields

$$\frac{\dot{Q}}{A} = \frac{75 - 40}{\dfrac{1}{500} + \dfrac{2.5 \times 10^{-3}}{200} + \dfrac{1}{\big[0.875 + (2.5)(0.993)\big](20)}} = 2070 \ W/m^2$$

indicating approximately a threefold increase in the rate of heat transfer.

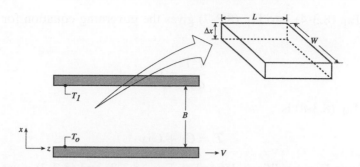

Figure 8.28. Couette flow between parallel plates.

8.3 ENERGY TRANSPORT WITH CONVECTION

Heat transfer by convection involves both the equation of motion and the equation of energy. Since we restrict the analysis to cases in which neither momentum nor energy is generated, this obviously limits the problems we might encounter.

Consider Couette flow of a Newtonian fluid between two large parallel plates under steady conditions as shown in Figure 8.28. Note that this geometry not only considers flow between parallel plates but also tangential flow between concentric cylinders. The surfaces at $x = 0$ and $x = B$ are maintained at T_o and T_1, respectively, with $T_o > T_1$. It is required to determine the temperature distribution within the fluid.

The velocity distribution for this problem is given by Eq. (8.1-12) as

$$\frac{v_z}{V} = 1 - \frac{x}{B} \tag{8.3-1}$$

On the other hand, the boundary conditions for the temperature, i.e.,

$$\text{at} \quad x = 0 \qquad T = T_o \tag{8.3-2}$$

$$\text{at} \quad x = B \qquad T = T_1 \tag{8.3-3}$$

suggest that $T = T(x)$. Therefore, Table C.4 in Appendix C indicates that the only nonzero energy flux component is e_x, and it is given by

$$e_x = q_x = -k \frac{dT}{dx} \tag{8.3-4}$$

For a rectangular volume element of thickness Δx, as shown in Figure 8.28, Eq. (8.2-1) is expressed as

$$q_x|_x WL - q_x|_{x+\Delta x} WL = 0 \tag{8.3-5}$$

Dividing Eq. (8.3-5) by $WL\,\Delta x$ and taking the limit as $\Delta x \to 0$ give

$$\lim_{\Delta x \to 0} \frac{q_x|_x - q_x|_{x+\Delta x}}{\Delta x} = 0 \tag{8.3-6}$$

or,

$$\frac{dq_x}{dx} = 0 \tag{8.3-7}$$

Substitution of Eq. (8.3-4) into Eq. (8.3-7) gives the governing equation for temperature in the form

$$\frac{d^2 T}{dx^2} = 0 \tag{8.3-8}$$

The solution of Eq. (8.3-8) is

$$T = C_1 + C_2 x \tag{8.3-9}$$

The use of the boundary conditions defined by Eqs. (8.3-2) and (8.3-3) gives the linear temperature distribution as

$$\boxed{\frac{T - T_o}{T_1 - T_o} = \frac{x}{B}} \tag{8.3-10}$$

indicating pure conduction across the fluid layer.

8.4 MASS TRANSPORT WITHOUT CONVECTION

The inventory rate equation for transfer of species A at the microscopic level is called the *equation of continuity for species* A. Under steady conditions without generation, the conservation statement for the mass of species A is given by

$$(\text{Rate of mass of } A \text{ in}) - (\text{Rate of mass of } A \text{ out}) = 0 \tag{8.4-1}$$

The rate of mass of A entering and leaving the system is determined from the mass (or molar) flux. As stated in Chapter 2, the total flux is the sum of the molecular and convective fluxes. For a one-dimensional transfer of species A in the z-direction in rectangular coordinates, the total molar flux is expressed as

$$N_{A_z} = \underbrace{-c\mathcal{D}_{AB} \frac{dx_A}{dz}}_{\text{Molecular flux}} + \underbrace{c_A v_z^*}_{\substack{\text{Convective} \\ \text{flux}}} \tag{8.4-2}$$

where v_z^* is the molar average velocity defined by Eq. (2.3-2). For a binary system composed of species A and B, the molar average velocity is given by

$$v_z^* = \frac{c_A v_{A_z} + c_B v_{B_z}}{c_A + c_B} = \frac{N_{A_z} + N_{B_z}}{c} \tag{8.4-3}$$

As we did for heat transfer, we will first consider the case of mass transfer without convection. For the transport of heat without convection, we focused our attention on conduction in solids and stationary liquids simply because energy is transferred by collisions of adjacent molecules and the migration of free electrons. In the case of mass transport, however, since species have individual velocities[6], the neglect of the convection term is not straightforward. It is

---

[6]Transport of mass by diffusion as a result of random molecular motion is called Brownian motion.

customary in the literature to neglect the convective flux in comparison with the molecular flux when mass transfer takes place in solids and stationary liquids. The reason for this can be explained as follows. Substitution of Eq. (8.4-3) into Eq. (8.4-2) gives

$$N_{A_z} = \underbrace{-c\mathcal{D}_{AB}\frac{dx_A}{dz}}_{\text{Molecular flux}} + \underbrace{x_A(N_{A_z} + N_{B_z})}_{\text{Convective flux}}$$

(8.4-4)

Since x_A is usually very small in solids and liquids, the convective flux term is considered negligible. It should be kept in mind, however, that if x_A is small, it is not necessarily implied that its gradient, i.e., dx_A/dz, is also small.

Another point of interest is the equimolar counterdiffusion in gases. The term "*equimolar counterdiffusion*" implies that for every mole of species $\mathcal{A}$ diffusing in the positive z-direction one mole of species $\mathcal{B}$ diffuses back in the negative z-direction, i.e.,

$$N_{A_z} = -N_{B_z} \quad \Rightarrow \quad c_A v_{A_z} = -c_B v_{B_z}$$

(8.4-5)

Under these circumstances, the molar average velocity, Eq. (8.4-3), becomes

$$v_z^* = \frac{N_{A_z} + (-N_{A_z})}{c} = 0$$

(8.4-6)

and the convective flux automatically drops out of Eq. (8.4-2).

8.4.1 Diffusion in Rectangular Coordinates

Consider the transfer of species $\mathcal{A}$ by diffusion through a slightly tapered slab as shown in Figure 8.29. If the taper angle is small, mass transport can be considered one-dimensional in the z-direction. Since $x_A = x_A(z)$, Table C.7 in Appendix C indicates that the only nonzero molar flux component is N_{A_z}, and it is given by

$$N_{A_z} = J_{A_z}^* = -c\mathcal{D}_{AB}\frac{dx_A}{dz}$$

(8.4-7)

The negative sign in Eq. (8.4-7) implies that the positive z-direction is in the direction of decreasing concentration. In a given problem, if the value of the mass (or molar) flux turns out to be negative, it is implied that the flux is in the negative z-direction.

Over a differential volume element of thickness Δz, as shown in Figure 8.29, Eq. (8.4-1) is written as

$$(AN_{A_z})|_z - (AN_{A_z})|_{z+\Delta z} = 0$$

(8.4-8)

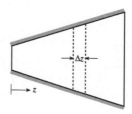

Figure 8.29. Diffusion through a slightly tapered conical duct.

Dividing Eq. (8.4-8) by Δz and taking the limit as $\Delta z \to 0$ give

$$\lim_{\Delta z \to 0} \frac{(AN_{A_z})|_z - (AN_{A_z})|_{z+\Delta z}}{\Delta z} = 0 \qquad (8.4\text{-}9)$$

or,

$$\frac{d(AN_{A_z})}{dz} = 0 \qquad (8.4\text{-}10)$$

Since flux times area gives the molar transfer rate of species $\mathcal{A}$, $\dot{n}_A$, it is possible to conclude that

$$AN_{A_z} = \text{constant} = \dot{n}_A \qquad (8.4\text{-}11)$$

in which the area A is perpendicular to the direction of mass flux. Substitution of Eq. (8.4-7) into Eq. (8.4-11) and integration give

$$\boxed{c \int_0^{x_A} \mathcal{D}_{AB}(x_A)\, dx_A = -\dot{n}_A \int_0^z \frac{dz}{A(z)} + K} \qquad (8.4\text{-}12)$$

where K is an integration constant. The determination of $\dot{n}_A$ and K requires two boundary conditions. Depending on the type of the boundary conditions used, the molar transfer rate of species $\mathcal{A}$ and the concentration distribution of species $\mathcal{A}$ as a function of position are determined from Eq. (8.4-12). When the surface concentrations are specified as

$$\text{at} \quad z = 0 \qquad x_A = x_{A_o} \qquad (8.4\text{-}13a)$$

$$\text{at} \quad z = L \qquad x_A = x_{A_L} \qquad (8.4\text{-}13b)$$

the resulting concentration distribution as a function of radial position and the molar transfer rate are given as in Table 8.8.

Table 8.8. Rate of transfer and concentration distribution for one-dimensional diffusion in rectangular coordinates for the boundary conditions given by Eq. (8.4-13)

Constants	Molar Transfer Rate		Concentration Distribution	
None	$\dfrac{c \displaystyle\int_{x_{A_L}}^{x_{A_o}} \mathcal{D}_{AB}\, dx_A}{\displaystyle\int_0^L \frac{dz}{A(z)}}$	(A)	$\dfrac{\displaystyle\int_{x_A}^{x_{A_o}} \mathcal{D}_{AB}\, dx_A}{\displaystyle\int_{x_{A_L}}^{x_{A_o}} \mathcal{D}_{AB}\, dx_A} = \dfrac{\displaystyle\int_0^z \frac{dz}{A(z)}}{\displaystyle\int_0^L \frac{dz}{A(z)}}$	(E)
$\mathcal{D}_{AB}$	$\dfrac{c\mathcal{D}_{AB}(x_{A_o} - x_{A_L})}{\displaystyle\int_0^L \frac{dz}{A(z)}}$	(B)	$\dfrac{x_{A_o} - x_A}{x_{A_o} - x_{A_L}} = \dfrac{\displaystyle\int_0^z \frac{dz}{A(z)}}{\displaystyle\int_0^L \frac{dz}{A(z)}}$	(F)
A	$\dfrac{Ac \displaystyle\int_{x_{A_L}}^{x_{A_o}} \mathcal{D}_{AB}\, dx_A}{L}$	(C)	$\dfrac{\displaystyle\int_{x_A}^{x_{A_o}} \mathcal{D}_{AB}\, dx_A}{\displaystyle\int_{x_{A_L}}^{x_{A_o}} \mathcal{D}_{AB}\, dx_A} = \dfrac{z}{L}$	(G)
$\mathcal{D}_{AB}$ A	$\dfrac{c\mathcal{D}_{AB}(x_{A_o} - x_{A_L})A}{L}$	(D)	$\dfrac{x_{A_o} - x_A}{x_{A_o} - x_{A_L}} = \dfrac{z}{L}$	(H)

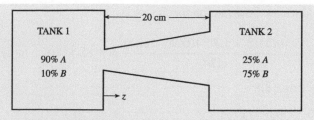

Figure 8.30. Diffusion through a conical duct.

Example 8.16 Two large tanks are connected by a truncated conical duct as shown in Figure 8.30. The diameter at $z = 0$ is 6 mm and the diameter at $z = 0.2$ m is 10 mm. Gas compositions in the tanks are given in terms of mole percentages. The pressure and temperature throughout the system are 1 atm and 25 °C, respectively, and $\mathcal{D}_{AB} = 3 \times 10^{-5}$ m^2/s.

a) Determine the initial molar flow rate of species $\mathcal{A}$ between the vessels.
b) What would be the initial molar flow rate of species $\mathcal{A}$ if the conical duct were replaced with a circular tube of 8 mm diameter?

Solution

Since the total pressure remains constant, the total number of moles in the conical duct does not change. This implies that equimolar counterdiffusion takes place within the conical duct and the molar average velocity is zero. Equation (B) in Table 8.8 gives the molar flow rate of species $\mathcal{A}$ as

$$\dot{n}_A = \frac{c\mathcal{D}_{AB}(x_{A_o} - x_{A_L})}{\displaystyle\int_0^{0.2} \frac{dz}{A(z)}} \tag{1}$$

The variation in the diameter as a function of position is represented by

$$D(z) = 0.006 + 0.02z \tag{2}$$

so that the area is

$$A(z) = \frac{\pi}{4}(0.006 + 0.02z)^2 \tag{3}$$

Substitution of Eq. (3) into Eq. (1) and integration give

$$\dot{n}_A = \frac{c\mathcal{D}_{AB}(x_{A_o} - x_{A_L})}{4244.1} \tag{4}$$

The total molar concentration is

$$c = \frac{P}{\mathcal{R}T} = \frac{101.325 \times 10^3}{(8.314 \times 10^3)(25 + 273)} = 0.041 \text{ kmol/m}^3 \tag{5}$$

Therefore, the initial molar flow rate of species $\mathcal{A}$ is

$$\dot{n}_A = \frac{(41)(3 \times 10^{-5})(0.9 - 0.25)}{4244.1} = 1.88 \times 10^{-7} \text{ mol/s} \tag{6}$$

b) From Eq. (D) in Table 8.8

$$\dot{n}_A = \frac{c\mathcal{D}_{AB}(x_{A_o} - x_{A_L})A}{L}$$

$$= \frac{(41)(3 \times 10^{-5})(0.9 - 0.25)[\pi(0.008)^2/4]}{0.2} = 2.01 \times 10^{-7} \text{ mol/s} \tag{7}$$

8.4.1.1 *Electrical circuit analogy*   The molar transfer rate of species $\mathcal{A}$ is given by Eq. (D) in Table 8.8 as

$$\dot{n}_A = \frac{c_{A_o} - c_{A_L}}{\dfrac{L}{\mathcal{D}_{AB}A}} \tag{8.4-14}$$

Comparison of Eq. (8.4-14) with Eq. (8.2-10) indicates that

$$\text{Driving force} = c_{A_o} - c_{A_L} \tag{8.4-15}$$

$$\boxed{\text{Resistance} = \frac{L}{\mathcal{D}_{AB}A} = \frac{\text{Thickness}}{(\text{Transport property})(\text{Area})}} \tag{8.4-16}$$

8.4.1.2 *Transfer rate in terms of bulk fluid properties*   Since it is much easier to measure the bulk concentrations of the adjacent solutions to the surfaces at $z = 0$ and $z = L$, it is necessary to relate the surface concentrations, x_{A_o} and x_{A_L}, to the bulk concentrations.

For energy transfer, the assumption of thermal equilibrium at a solid-fluid boundary leads to the equality of temperatures, and this condition is generally stated as "temperature is continuous at a solid-fluid boundary." In the case of mass transfer, the condition of phase equilibrium for a nonreacting multicomponent system at a solid-fluid boundary implies the equality of chemical potentials or partial molar Gibbs free energies. Therefore, concentrations at a solid-fluid boundary are not necessarily equal to each other with a resulting discontinuity in the concentration distribution. For example, consider a homogeneous membrane chemically different from the solution it is separating. In that case, the solute may be more (or less) soluble in the membrane than in the bulk solution. A typical distribution of concentration is shown in Figure 8.31. Under these conditions, a thermodynamic property $\mathcal{H}$, called the *partition coefficient*, is introduced, which relates the concentration of species in the membrane at equilibrium to the concentration in bulk solution. For the problem depicted in Figure 8.31, the partition coefficients can be defined as

$$\mathcal{H}^- = \frac{c_{A_o}}{c_{A_i}^-} \tag{8.4-17}$$

$$\mathcal{H}^+ = \frac{c_{A_L}}{c_{A_i}^+} \tag{8.4-18}$$

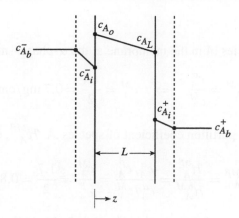

Figure 8.31. Concentration distribution across a membrane.

The molar rate of transfer of species $\mathcal{A}$ across the membrane under steady conditions can be expressed as

$$\dot{n}_A = A \langle k_c^- \rangle (c_{A_b}^- - c_{A_i}^-) = A \langle k_c^+ \rangle (c_{A_i}^+ - c_{A_b}^+) \tag{8.4-19}$$

On the other hand, the use of Eqs. (8.4-17) and (8.4-18) in Eq. (8.4-14) leads to

$$\dot{n}_A = \frac{A \mathcal{D}_{AB}(\mathcal{H}^- c_{A_i}^- - \mathcal{H}^+ c_{A_i}^+)}{L} \tag{8.4-20}$$

Equations (8.4-19)–(8.4-20) can be rearranged in the form

$$c_{A_b}^- - c_{A_i}^- = \dot{n}_A \left(\frac{1}{A \langle k_c^- \rangle} \right) \quad \| \times \mathcal{H}^- \tag{8.4-21}$$

$$\mathcal{H}^- c_{A_i}^- - \mathcal{H}^+ c_{A_i}^+ = \dot{n}_A \left(\frac{L}{A \mathcal{D}_{AB}} \right) \tag{8.4-22}$$

$$c_{A_i}^+ - c_{A_b}^+ = \dot{n}_A \left(\frac{1}{A \langle k_c^+ \rangle} \right) \quad \| \times \mathcal{H}^+ \tag{8.4-23}$$

Multiplication of Eqs. (8.4-21) and (8.4-23) by $\mathcal{H}^-$ and $\mathcal{H}^+$, respectively, and the addition of these equations with Eq. (8.4-22) give the transfer rate as

$$\dot{n}_A = \frac{c_{A_b}^- - \left(\dfrac{\mathcal{H}^+}{\mathcal{H}^-} \right) c_{A_b}^+}{\dfrac{1}{A \langle k_c^- \rangle} + \dfrac{L}{A \mathcal{D}_{AB} \mathcal{H}^-} + \left(\dfrac{\mathcal{H}^+}{\mathcal{H}^-} \right) \left(\dfrac{1}{A \langle k_c^+ \rangle} \right)} \tag{8.4-24}$$

Example 8.17 A membrane separating a liquid α-phase from a liquid β-phase is permeable to species $\mathcal{A}$. The concentrations of species $\mathcal{A}$ in the α- and β-phases are 1.4 mg/cm^3 and 1 mg/cm^3, respectively. The (α-phase/membrane) partition coefficient of species $\mathcal{A}$, $\mathcal{H}_A^{\alpha M}$, is 2 and the (α-phase/β-phase) partition coefficient of species $\mathcal{A}$, $\mathcal{H}_A^{\alpha \beta}$, is 1.7. If the average mass transfer coefficients on both sides of the membrane are very large, sketch a representative concentration distribution of species $\mathcal{A}$.

Solution

The concentration of species $\mathcal{A}$ in the membrane at the α-phase—membrane interface is

$$\mathcal{H}_A^{\alpha M} = \frac{c_A^\alpha}{c_A^M} \quad \Rightarrow \quad c_A^M = \frac{1.4}{2} = 0.7 \text{ mg/cm}^3$$

The (membrane/β-phase) partition coefficient of species $\mathcal{A}$, $\mathcal{H}_A^{M\beta}$, can be calculated as

$$\mathcal{H}_A^{M\beta} = \frac{\mathcal{H}_A^{\alpha\beta}}{\mathcal{H}_A^{\alpha M}} = \frac{c_A^\alpha/c_A^\beta}{c_A^\alpha/c_A^M} = \frac{c_A^M}{c_A^\beta} = \frac{1.7}{2} = 0.85$$

Therefore, the concentration of species $\mathcal{A}$ in the membrane at the β-phase—membrane interface is

$$c_A^M = (0.85)(1) = 0.85 \text{ mg/cm}^3$$

A representative concentration distribution of species $\mathcal{A}$ is shown in the figure below. Since the mass transfer coefficients are very large, i.e., the Biot number for mass transfer is very large, there is no variation in concentration in the α- and β-phases.

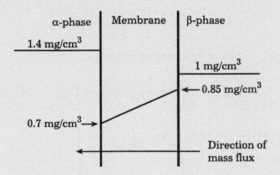

Example 8.18 Develop an expression for the transfer of species i from the concentrated α-phase to the dilute α-phase through two nonporous membranes, A and B, as shown in the figure below. Let $\mathcal{D}_A$ and $\mathcal{D}_B$ be the effective diffusion coefficients of species i in membranes A and B, respectively.

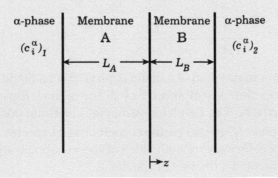

Solution

Assumption

1. Diffusion takes place only in the z-direction.

Analysis

Since the area is constant, the governing equation for the concentration of species i in membrane A can be easily obtained from Eq. (8.4-10) as

$$\frac{dN_{i_z}^A}{dz} = 0 \quad \Rightarrow \quad \frac{d^2 c_i^A}{dz^2} = 0 \tag{1}$$

The solution of Eq. (1) gives

$$c_i^A = K_1 z + K_2 \tag{2}$$

Similarly, the governing equation for the concentration of species i in membrane B is given by

$$\frac{dN_{i_z}^B}{dz} = 0 \quad \Rightarrow \quad \frac{d^2 c_i^B}{dz^2} = 0 \tag{3}$$

The solution of Eq. (3) yields

$$c_i^B = K_3 z + K_4 \tag{4}$$

Evaluation of the constants K_1, K_2, K_3, and K_4 requires four boundary conditions. They are expressed as

$$\text{at} \quad z = -L_A \qquad \mathcal{H}_i^{A\alpha} = \frac{c_i^A}{(c_i^\alpha)_1} \tag{5}$$

$$\text{at} \quad z = L_B \qquad \mathcal{H}_i^{B\alpha} = \frac{c_i^B}{(c_i^\alpha)_2} \tag{6}$$

$$\text{at} \quad z = 0 \qquad \mathcal{H}_i^{AB} = \frac{c_i^A}{c_i^B} \tag{7}$$

$$\text{at} \quad z = 0 \qquad \mathcal{D}_A \frac{dc_i^A}{dz} = \mathcal{D}_B \frac{dc_i^B}{dz} \tag{8}$$

The boundary conditions defined by Eqs. (5)–(7) assume thermodynamic equilibrium between the phases at the interfaces. On the other hand, Eq. (8) indicates that the molar fluxes are continuous, i.e., equal to each other, at the A-B interface.

Application of the boundary conditions leads to the following concentration distribution of species i within membranes A and B

$$c_i^A = (c_i^\alpha)_1 \mathcal{H}_i^{A\alpha} - \frac{1}{\mathcal{D}_A}\left[\frac{(c_i^\alpha)_1 - (c_i^\alpha)_2}{\dfrac{L_A}{\mathcal{D}_A \mathcal{H}_i^{A\alpha}} + \dfrac{L_B}{\mathcal{D}_B \mathcal{H}_i^{B\alpha}}}\right](z + L_A) \tag{9}$$

$$c_i^B = (c_i^\alpha)_2 \mathcal{H}_i^{B\alpha} - \frac{1}{\mathcal{D}_B}\left[\frac{(c_i^\alpha)_1 - (c_i^\alpha)_2}{\dfrac{L_A}{\mathcal{D}_A \mathcal{H}_i^{A\alpha}} + \dfrac{L_B}{\mathcal{D}_B \mathcal{H}_i^{B\alpha}}}\right](z - L_B) \tag{10}$$

The flux expressions are given by

$$N_{i_z}^A = -\mathcal{D}_A \frac{dc_i^A}{dz} \qquad\qquad N_{i_z}^B = -\mathcal{D}_B \frac{dc_i^B}{dz} \tag{11}$$

Thus, the molar flux of species i through membrane A is the same as that through membrane B, and is given by

$$N_{i_z}^A = N_{i_z}^B = \frac{(c_i^\alpha)_1 - (c_i^\alpha)_2}{\dfrac{L_A}{\mathcal{D}_A \mathcal{H}_i^{A\alpha}} + \dfrac{L_B}{\mathcal{D}_B \mathcal{H}_i^{B\alpha}}} \tag{12}$$

8.4.2   Diffusion in Cylindrical Coordinates

Consider one-dimensional diffusion of species $\mathcal{A}$ in the radial direction through a hollow circular pipe with inner and outer radii of R_1 and R_2, respectively, as shown in Figure 8.32.

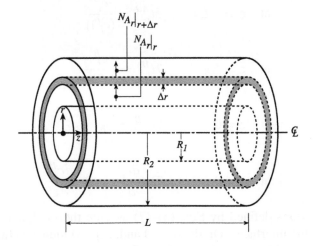

Figure 8.32. Diffusion through a hollow cylinder.

Since $x_A = x_A(r)$, Table C.8 in Appendix C indicates that the only nonzero molar flux component is N_{A_r}, and it is given by

$$N_{A_r} = J_{A_r}^* = -c\mathcal{D}_{AB}\frac{dx_A}{dr} \tag{8.4-25}$$

For a cylindrical differential volume element of thickness Δr, as shown in Figure 8.32, Eq. (8.4-1) is expressed in the form

$$(AN_{A_r})|_r - (AN_{A_r})|_{r+\Delta r} = 0 \tag{8.4-26}$$

Dividing Eq. (8.4-26) by Δr and taking the limit as $\Delta r \to 0$ give

$$\lim_{\Delta r \to 0} \frac{(AN_{A_r})|_r - (AN_{A_r})|_{r+\Delta r}}{\Delta r} = 0 \tag{8.4-27}$$

or,

$$\frac{d(AN_{A_r})}{dr} = 0 \tag{8.4-28}$$

Since flux times area gives the molar transfer rate of species $\mathcal{A}$, $\dot{n}_A$, it is possible to conclude that

$$AN_{A_r} = \text{constant} = \dot{n}_A \tag{8.4-29}$$

Note that the area A in Eq. (8.4-29) is perpendicular to the direction of mass flux, and is given by

$$A = 2\pi r L \tag{8.4-30}$$

Substitution of Eqs. (8.4-25) and (8.4-30) into Eq. (8.4-29) and integration give

$$\boxed{c\int_0^{x_A} \mathcal{D}_{AB}(x_A)\,dx_A = -\left(\frac{\dot{n}_A}{2\pi L}\right)\ln r + K} \tag{8.4-31}$$

where K is an integration constant.

When the surface concentrations are specified as

$$\text{at} \quad r = R_1 \qquad x_A = x_{A_1} \tag{8.4-32a}$$

$$\text{at} \quad r = R_2 \qquad x_A = x_{A_2} \tag{8.4-32b}$$

the resulting concentration distribution as a function of radial position and the molar transfer rate are as given in Table 8.9.

Table 8.9. Rate of transfer and concentration distribution for one-dimensional diffusion in a hollow cylinder for the boundary conditions given by Eq. (8.4-32)

Constant	Molar Transfer Rate		Concentration Distribution	
None	$\dfrac{2\pi Lc \displaystyle\int_{x_{A_1}}^{x_{A_2}} \mathcal{D}_{AB}\, dx_A}{\ln\left(\dfrac{R_1}{R_2}\right)}$	(A)	$\dfrac{\displaystyle\int_{x_A}^{x_{A_2}} \mathcal{D}_{AB}\, dx_A}{\displaystyle\int_{x_{A_1}}^{x_{A_2}} \mathcal{D}_{AB}\, dx_A} = \dfrac{\ln\left(\dfrac{r}{R_2}\right)}{\ln\left(\dfrac{R_1}{R_2}\right)}$	(C)
$\mathcal{D}_{AB}$	$\dfrac{2\pi Lc\mathcal{D}_{AB}(x_{A_2} - x_{A_1})}{\ln\left(\dfrac{R_1}{R_2}\right)}$	(B)	$\dfrac{x_{A_2} - x_A}{x_{A_2} - x_{A_1}} = \dfrac{\ln\left(\dfrac{r}{R_2}\right)}{\ln\left(\dfrac{R_1}{R_2}\right)}$	(D)

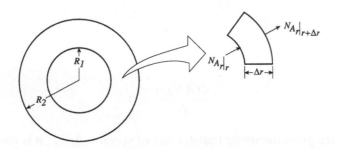

Figure 8.33. Diffusion through a hollow sphere.

8.4.3 Diffusion in Spherical Coordinates

Consider one-dimensional diffusion of species $\mathcal{A}$ in the radial direction through a hollow sphere with inner and outer radii of R_1 and R_2, respectively, as shown in Figure 8.33. Since $x_A = x_A(r)$, Table C.9 in Appendix C indicates that the only nonzero molar flux component is N_{A_r}, and it is given by

$$N_{A_r} = -c\mathcal{D}_{AB}\frac{dx_A}{dr} \tag{8.4-33}$$

For a spherical differential volume element of thickness Δr, as shown in Figure 8.33, Eq. (8.4-1) is expressed in the form

$$(AN_{A_r})|_r - (AN_{A_r})|_{r+\Delta r} = 0 \tag{8.4-34}$$

Dividing Eq. (8.4-34) by Δr and taking the limit as $\Delta r \to 0$ give

$$\lim_{\Delta r \to 0} \frac{(AN_{A_r})|_r - (AN_{A_r})|_{r+\Delta r}}{\Delta r} = 0 \tag{8.4-35}$$

or,

$$\frac{d(AN_{A_r})}{dr} = 0 \tag{8.4-36}$$

Table 8.10. Rate of transfer and concentration distribution for one-dimensional diffusion in a hollow sphere for the boundary conditions given by Eq. (8.4-40)

Constant	Molar Transfer Rate		Concentration Distribution	
None	$\dfrac{4\pi c \displaystyle\int_{x_{A_2}}^{x_{A_1}} \mathcal{D}_{AB}\, dx_A}{\dfrac{1}{R_1} - \dfrac{1}{R_2}}$	(A)	$\dfrac{\displaystyle\int_{x_{A_2}}^{x_A} \mathcal{D}_{AB}\, dx_A}{\displaystyle\int_{x_{A_2}}^{x_{A_1}} \mathcal{D}_{AB}\, dx_A} = \dfrac{\dfrac{1}{r} - \dfrac{1}{R_2}}{\dfrac{1}{R_1} - \dfrac{1}{R_2}}$	(C)
$\mathcal{D}_{AB}$	$\dfrac{4\pi c \mathcal{D}_{AB}(x_{A_1} - x_{A_2})}{\dfrac{1}{R_1} - \dfrac{1}{R_2}}$	(B)	$\dfrac{x_A - x_{A_2}}{x_{A_1} - x_{A_2}} = \dfrac{\dfrac{1}{r} - \dfrac{1}{R_2}}{\dfrac{1}{R_1} - \dfrac{1}{R_2}}$	(D)

Since flux times area gives the molar transfer rate of species A, $\dot{n}_A$, it is possible to conclude that

$$AN_{A_r} = \text{constant} = \dot{n}_A \tag{8.4-37}$$

Note that the area A in Eq. (8.4-37) is perpendicular to the direction of mass flux, and is given by

$$A = 4\pi r^2 \tag{8.4-38}$$

Substitution of Eqs. (8.4-33) and (8.4-38) into Eq. (8.4-37) and integration give

$$\boxed{c \int_0^{x_A} \mathcal{D}_{AB}(x_A)\, dx_A = \left(\frac{\dot{n}_A}{4\pi}\right)\frac{1}{r} + K} \tag{8.4-39}$$

where K is an integration constant.

When the surface concentrations are specified as

$$\text{at} \quad r = R_1 \quad\quad x_A = x_{A_1} \tag{8.4-40a}$$
$$\text{at} \quad r = R_2 \quad\quad x_A = x_{A_2} \tag{8.4-40b}$$

the resulting concentration distribution as a function of radial position and the molar transfer rate are as given in Table 8.10.

Example 8.19 Consider the transfer of species A from a spherical drop or a bubble of radius R to a stationary fluid having a concentration of c_{A_∞}.

a) Determine the concentration distribution of species A within the fluid.
b) Determine the molar rate of species A transferred to the fluid.
c) Determine the Sherwood number.

Solution

Assumptions

1. Steady-state conditions prevail.
2. The concentration at the surface of the sphere is constant at c_{A_w}.
3. Mass transfer does not affect the radius R.

Analysis

a) The concentration distribution is obtained from Eq. (D) of Table 8.10 in the form

$$\frac{c_A - c_{A\infty}}{c_{A_w} - c_{A\infty}} = \frac{R}{r} \tag{1}$$

b) The use of Eq. (B) in Table 8.10 with $c_{A_1} = c_{A_w}$, $c_{A_2} = c_{A\infty}$, $R_1 = R$, and $R_2 = \infty$ gives the molar rate of transfer of species $\mathcal{A}$ to the fluid as

$$\dot{n}_A = 4\pi \mathcal{D}_{AB} R(c_{A_w} - c_{A\infty}) \tag{2}$$

c) The molar transfer rate can also be calculated from Eq. (3.3-7) as

$$\dot{n}_A = 4\pi R^2 \langle k_c \rangle (c_{A_w} - c_{A\infty}) \tag{3}$$

Equating Eqs. (2) and (3) leads to

$$\frac{\langle k_c \rangle}{\mathcal{D}_{AB}} = \frac{1}{R} = \frac{2}{D} \tag{4}$$

Therefore, the Sherwood number is

$$\text{Sh} = \frac{\langle k_c \rangle D}{\mathcal{D}_{AB}} = 2 \tag{5}$$

Note that this problem is exactly analogous to that in Example 8.12.

8.4.4 Diffusion and Reaction in a Catalyst Pore

At first, it may seem strange to a student to see an example concerning a reaction in a catalyst pore in a chapter that deals with "steady-state microscopic balances **without generation**." In general, reactions can be classified as heterogeneous and homogeneous. A *heterogeneous reaction* occurs on the surface and is usually a catalytic reaction. A *homogeneous reaction*, on the other hand, occurs throughout a given phase. In Chapter 5, the rate of generation of species i per unit volume as a result of a chemical reaction, $\Re_i$, was given by Eq. (5.3-26) in the form

$$\Re_i = \alpha_i r \tag{8.4-41}$$

in which r represents a homogeneous reaction rate. Therefore, a homogeneous reaction rate appears in the inventory of chemical species, whereas a heterogeneous reaction rate appears in the boundary conditions.

Consider an idealized single cylindrical pore of radius R and length L in a catalyst particle as shown in Figure 8.34. The bulk gas stream has a species $\mathcal{A}$ concentration of c_{A_b}. Species $\mathcal{A}$ diffuses through the gas film and its concentration at the pore mouth, i.e., $z = 0$, is c_{A_o}. As species $\mathcal{A}$ diffuses into the catalyst pore, it undergoes a first-order irreversible reaction

$$A \rightarrow B$$

on the interior surface of the catalyst. The problem will be analyzed with the following assumptions:

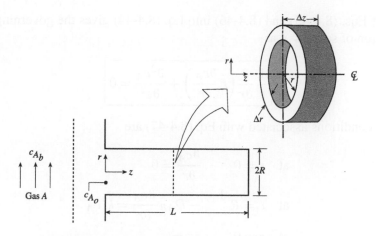

Figure 8.34. Diffusion and reaction in a cylindrical pore.

1. Steady-state conditions prevail.
2. The system is isothermal.
3. The diffusion coefficient is constant.

For a cylindrical differential volume element of thickness Δr and length Δz, as shown in Figure 8.34, Eq. (8.4-1) is expressed as

$$\left(N_{A_r}|_r 2\pi r \Delta z + N_{A_z}|_z 2\pi r \Delta r\right) - \left[N_{A_r}|_{r+\Delta r} 2\pi(r + \Delta r)\Delta z + N_{A_z}|_{z+\Delta z} 2\pi r \Delta r\right] = 0 \tag{8.4-42}$$

Dividing Eq. (8.4-42) by $2\pi \Delta r \Delta z$ and taking the limit as $\Delta r \to 0$ and $\Delta z \to 0$ give

$$\frac{1}{r} \lim_{\Delta r \to 0} \frac{(r N_{A_r})|_r - (r N_{A_r})|_{r+\Delta r}}{\Delta r} + \lim_{\Delta z \to 0} \frac{N_{A_z}|_z - N_{A_z}|_{z+\Delta z}}{\Delta z} = 0 \tag{8.4-43}$$

or,

$$\frac{1}{r} \frac{\partial}{\partial r}(r N_{A_r}) + \frac{\partial N_{A_z}}{\partial z} = 0 \tag{8.4-44}$$

Since the temperature is constant and there is no volume change due to reaction, the pressure and hence the total molar concentration, c, remain constant. Under these conditions, from Table C.8 in Appendix C, the components of the molar flux become[7]

$$N_{A_r} = -\mathcal{D}_{AB} \frac{\partial c_A}{\partial r} \tag{8.4-45}$$

$$N_{A_z} = -\mathcal{D}_{AB} \frac{\partial c_A}{\partial z} \tag{8.4-46}$$

---

[7]From the stoichiometry of the reaction, the molar average velocity is zero.

Substitution of Eqs. (8.4-45) and (8.4-46) into Eq. (8.4-44) gives the governing equation for the concentration of species $\mathcal{A}$ as

$$\boxed{\frac{1}{r}\frac{\partial}{\partial r}\left(r\frac{\partial c_A}{\partial r}\right)+\frac{\partial^2 c_A}{\partial z^2}=0}\qquad(8.4\text{-}47)$$

The boundary conditions associated with Eq. (8.4-47) are

$$\text{at}\quad r=0\qquad\frac{\partial c_A}{\partial r}=0\qquad\qquad(8.4\text{-}48)$$

$$\text{at}\quad r=R\qquad-\mathcal{D}_{AB}\frac{\partial c_A}{\partial r}=k^s c_A\qquad(8.4\text{-}49)$$

$$\text{at}\quad z=0\qquad c_A=c_{A_o}\qquad\qquad(8.4\text{-}50)$$

$$\text{at}\quad z=L\qquad\frac{\partial c_A}{\partial z}=0\qquad\qquad(8.4\text{-}51)$$

The term k^s in Eq. (8.4-49) is the first-order surface reaction rate constant and has the dimensions of m/s. In writing Eq. (8.4-51) it is implicitly assumed that no reaction takes place on the surface at $z=L$, and the term $\partial c_A/\partial z=0$ implies that there is no mass transfer through this surface.

As done in Section 8.2.4, this complicated problem will be solved by making use of the area averaging technique. The area-averaged concentration for species $\mathcal{A}$ is defined by

$$\langle c_A\rangle=\frac{\displaystyle\int_0^{2\pi}\int_0^R c_A r\,dr\,d\theta}{\displaystyle\int_0^{2\pi}\int_0^R r\,dr\,d\theta}=\frac{1}{\pi R^2}\int_0^{2\pi}\int_0^R c_A r\,dr\,d\theta\qquad(8.4\text{-}52)$$

Although the local concentration, c_A, is dependent on r and z, the area-averaged concentration, $\langle c_A\rangle$, depends only on z.

Area averaging is performed by integrating Eq. (8.4-47) over the cross-sectional area of the pore. The result is

$$\int_0^{2\pi}\int_0^R\frac{1}{r}\frac{\partial}{\partial r}\left(r\frac{\partial c_A}{\partial r}\right)r\,dr\,d\theta+\int_0^{2\pi}\int_0^R\frac{\partial^2 c_A}{\partial z^2}r\,dr\,d\theta=0\qquad(8.4\text{-}53)$$

Since the limits of the integration are constant, the order of differentiation and integration in the second term of Eq. (8.4-53) can be interchanged to obtain

$$\int_0^{2\pi}\int_0^R\frac{\partial^2 c_A}{\partial z^2}r\,dr\,d\theta=\frac{d^2}{dz^2}\left(\int_0^{2\pi}\int_0^R c_A r\,dr\,d\theta\right)=\pi R^2\frac{d^2\langle c_A\rangle}{dz^2}\qquad(8.4\text{-}54)$$

Substitution of Eq. (8.4-54) into Eq. (8.4-53) yields

$$2\pi R\left.\frac{\partial c_A}{\partial r}\right|_{r=R}+\pi R^2\frac{d^2\langle c_A\rangle}{dz^2}=0\qquad(8.4\text{-}55)$$

The use of the boundary condition given by Eq. (8.4-49) leads to

$$\mathcal{D}_{AB} \frac{d^2 \langle c_A \rangle}{dz^2} = \frac{2}{R} k^s c_A|_{r=R} \tag{8.4-56}$$

in which the dependent variables, i.e., $\langle c_A \rangle$ and $c_A|_{r=R}$, are at two different scales. It is generally assumed, although not expressed explicitly, that

$$c_A|_{r=R} \simeq \langle c_A \rangle \tag{8.4-57}$$

This approximation is valid for $\mathrm{Bi_M} = \langle k_c \rangle R / \mathcal{D}_{AB} \ll 1$. Substitution of Eq. (8.4-57) into Eq. (8.4-56) gives

$$\boxed{\mathcal{D}_{AB} \frac{d^2 \langle c_A \rangle}{dz^2} = \frac{2}{R} k^s \langle c_A \rangle} \tag{8.4-58}$$

Integration of Eqs. (8.4-50) and (8.4-51) over the cross-sectional area of the pore gives the boundary conditions associated with Eq. (8.4-58) as

$$\text{at} \quad z = 0 \qquad \langle c_A \rangle = c_{A_o} \tag{8.4-59}$$

$$\text{at} \quad z = L \qquad \frac{d \langle c_A \rangle}{dz} = 0 \tag{8.4-60}$$

Equations (8.4-47) and (8.4-58) are at two different scales. Equation (8.4-58) is obtained by averaging Eq. (8.4-47) over the cross-sectional area perpendicular to the direction of mass flux. As a result, the boundary condition, i.e., the heterogeneous reaction rate expression, appears in the conservation statement.

Note that the term $2/R$ in Eq. (8.4-58) is the catalyst surface area per unit volume, i.e.,

$$\frac{2}{R} = \frac{2\pi R L}{\pi R^2 L} = a_v = \frac{\text{Catalyst surface area}}{\text{Pore volume}} \tag{8.4-61}$$

Since the heterogeneous reaction rate expression has the units of moles/(area)(time), multiplication of this term by a_v converts the units to moles/(volume)(time).

The physical significance and the order of magnitude of the terms in Eq. (8.4-58) are given in Table 8.11. Therefore, the ratio of the rate of reaction to the rate of diffusion is given by

$$\frac{\text{Rate of reaction}}{\text{Rate of diffusion}} = \frac{2k^s c_{A_o}/R}{\mathcal{D}_{AB} c_{A_o}/L^2} = \frac{2k^s L^2}{R \mathcal{D}_{AB}} \tag{8.4-62}$$

In the literature, this ratio is often referred to as the *Thiele modulus*[8], Λ, and expressed as

$$\Lambda = \sqrt{\frac{2k^s L^2}{R \mathcal{D}_{AB}}} \tag{8.4-63}$$

---

[8]Note that the characteristic time for the surface reaction can be expressed as $(R/2)/k^s$. Therefore, the Thiele modulus can also be interpreted as the ratio of the diffusive time scale to the reaction time scale.

Table 8.11. The physical significance and the order of magnitude of the terms in Eq. (8.4-58)

Term	Physical Significance	Order of Magnitude
$\mathcal{D}_{AB} \dfrac{d^2 \langle c_A \rangle}{dz^2}$	Rate of diffusion	$\mathcal{D}_{AB} \dfrac{c_{A_o}}{L^2}$
$\dfrac{2k^s \langle c_A \rangle}{R}$	Rate of reaction	$\dfrac{2k^s c_{A_o}}{R}$

Before solving Eq. (8.4-58), it is convenient to express the governing equation and the boundary conditions in dimensionless form. Introduction of the dimensionless quantities

$$\theta = \frac{\langle c_A \rangle}{c_{A_o}} \qquad \xi = \frac{z}{L} \tag{8.4-64}$$

reduces Eqs. (8.4-58)–(8.4-60) to

$$\frac{d^2\theta}{d\xi^2} = \Lambda^2 \theta \tag{8.4-65}$$

$$\text{at} \quad \xi = 0 \qquad \theta = 1 \tag{8.4-66}$$

$$\text{at} \quad \xi = 1 \qquad \frac{d\theta}{d\xi} = 0 \tag{8.4-67}$$

Since these equations are exactly the same as those developed for the fin problem in Section 8.2.4, the solution is given by Eq. (8.2-88), i.e.,

$$\boxed{\theta = \frac{\cosh[\Lambda(1 - \xi)]}{\cosh \Lambda}} \tag{8.4-68}$$

8.4.4.1 *Macroscopic equation* Integration of the microscopic level equations over the volume of the system gives the equations at the macroscopic level. Integration of Eq. (8.4-58) over the volume of the system gives

$$\int_0^L \int_0^{2\pi} \int_0^R \mathcal{D}_{AB} \frac{d^2 \langle c_A \rangle}{dz^2} r \, dr \, d\theta \, dz = \int_0^L \int_0^{2\pi} \int_0^R \frac{2}{R} k^s \langle c_A \rangle r \, dr \, d\theta \, dz \tag{8.4-69}$$

Carrying out the integrations yields

$$\underbrace{\pi R^2 \left(-\mathcal{D}_{AB} \frac{d \langle c_A \rangle}{dz} \Big|_{z=0} \right)}_{\substack{\text{Rate of moles of species } \mathcal{A} \text{ entering} \\ \text{the pore through the surface at } z=0}} = \underbrace{2\pi R k^s \int_0^L \langle c_A \rangle \, dz}_{\substack{\text{Rate of conversion of species } \mathcal{A} \text{ to} \\ \text{species } \mathcal{B} \text{ at the catalyst surface}}} \tag{8.4-70}$$

which is the macroscopic inventory rate equation for the conservation of species $\mathcal{A}$ by considering the catalyst pore as a system. The use of Eq. (8.4-68) in Eq. (8.4-70) gives the molar rate of conversion of species $\mathcal{A}$, $\dot{n}_A$, as

$$\boxed{\dot{n}_A = \frac{\pi R^2 \mathcal{D}_{AB} c_{A_o} \Lambda \tanh \Lambda}{L}} \tag{8.4-71}$$

8.4.4.2 *Effectiveness factor*   The effectiveness factor, η, is defined as the ratio of the apparent rate of conversion to the rate if the entire internal surface were exposed to the concentration c_{A_o}, i.e.,

$$\eta = \frac{2\pi R k^s \int_0^L \langle c_A \rangle \, dz}{2\pi R k^s c_{A_o} L} = \frac{\int_0^L \langle c_A \rangle \, dz}{c_{A_o} L} \tag{8.4-72}$$

In terms of the dimensionless quantities, Eq. (8.4-72) becomes

$$\eta = \int_0^1 \theta \, d\xi \tag{8.4-73}$$

Substitution of Eq. (8.4-68) into Eq. (8.4-73) gives the effectiveness factor as

$$\boxed{\eta = \frac{\tanh \Lambda}{\Lambda}} \tag{8.4-74}$$

Note that the effectiveness factor for a first-order irreversible reaction is identical to the fin efficiency. Therefore, Figure 8.27, which shows the variation in η as a function of Λ, is also valid for this case.

When $\Lambda \to 0$, the rate of diffusion is much larger than the rate of reaction. The Taylor series expansion of η in terms of Λ gives

$$\eta = 1 - \frac{1}{3}\Lambda^2 + \frac{2}{15}\Lambda^4 - \frac{17}{315}\Lambda^6 + \cdots \tag{8.4-75}$$

Therefore, η approaches unity as $\Lambda \to 0$, indicating that the entire surface is exposed to a reactant. On the other hand, large values of Λ correspond to cases in which diffusion is very slow and the surface reaction is very rapid. Under these conditions, the effectiveness factor becomes

$$\eta = \frac{1}{\Lambda} \tag{8.4-76}$$

As $\Lambda \to \infty$, η approaches zero. This implies that a good part of the catalyst surface is starved of a reactant and hence not effective.

8.5 MASS TRANSPORT WITH CONVECTION

In the case of mass transfer, each species involved in the transfer has its own individual velocity. For a single phase system composed of the binary species A and B, the characteristic velocity for the mixture can be defined in several ways as stated in Section 2.3. If the mass transfer takes place in the z-direction, the three characteristic velocities are as given in Table 8.12.

Table 8.12. Characteristic velocities in the z-direction for a binary system

Velocity	Definition	
Mass average	$v_z = \dfrac{\rho_A v_{A_z} + \rho_B v_{B_z}}{\rho_A + \rho_B} = \dfrac{\mathcal{W}_{A_z} + \mathcal{W}_{B_z}}{\rho}$	(A)
Molar average	$v_z^* = \dfrac{c_A v_{A_z} + c_B v_{B_z}}{c_A + c_B} = \dfrac{N_{A_z} + N_{B_z}}{c}$	(B)
Volume average	$v_z^{\blacksquare} = c_A \overline{V}_A v_{A_z} + c_B \overline{V}_B v_{B_z} = \overline{V}_A N_{A_z} + \overline{V}_B N_{B_z}$	(C)

Hence, the total mass or molar flux of species $\mathcal{A}$ can be expressed as

$$\mathcal{W}_{A_z} = \underbrace{-\rho \mathcal{D}_{AB} \frac{d\omega_A}{dz}}_{\text{Molecular flux}} + \underbrace{\rho_A v_z}_{\substack{\text{Convective} \\ \text{flux}}} \tag{8.5-1}$$

$$N_{A_z} = \underbrace{-c \mathcal{D}_{AB} \frac{dx_A}{dz}}_{\text{Molecular flux}} + \underbrace{c_A v_z^*}_{\substack{\text{Convective} \\ \text{flux}}} \tag{8.5-2}$$

$$N_{A_z} = \underbrace{-\mathcal{D}_{AB} \frac{dc_A}{dz}}_{\text{Molecular flux}} + \underbrace{c_A v_z^{\blacksquare}}_{\substack{\text{Convective} \\ \text{flux}}} \tag{8.5-3}$$

The tricky part of mass transfer problems is that there is no need to have a bulk motion of the mixture as a result of external means, such as pressure drop, to have a nonzero convective flux term in Eqs. (8.5-1)–(8.5-3). Even in the case of the diffusion of species $\mathcal{A}$ through a stagnant film of $\mathcal{B}$, a nonzero convective term arises as can be seen from the following examples.

It should also be noted that, if one of the characteristic velocities is zero, this does not necessarily imply that the other characteristic velocities are also zero. For example, in Section 8.4, it was shown that the molar average velocity is zero for an equimolar counterdiffusion since $N_{A_z} = -N_{B_z}$. The mass average velocity for this case is given by

$$v_z = \frac{\mathcal{W}_{A_z} + \mathcal{W}_{B_z}}{\rho} \tag{8.5-4}$$

The mass and molar fluxes are related by

$$N_{i_z} = \frac{\mathcal{W}_{i_z}}{\mathcal{M}_i} \tag{8.5-5}$$

where $\mathcal{M}_i$ is the molecular weight of species i. The use of Eq. (8.5-5) in Eq. (8.5-4) gives

$$v_z = \frac{\mathcal{M}_A N_{A_z} + \mathcal{M}_B N_{B_z}}{\rho} = \frac{N_{A_z}(\mathcal{M}_A - \mathcal{M}_B)}{\rho} \tag{8.5-6}$$

which is nonzero unless $\mathcal{M}_A = \mathcal{M}_B$.

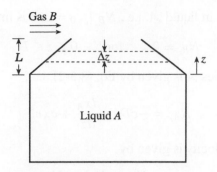

Figure 8.35. Evaporation from a tapered tank.

8.5.1 Diffusion Through a Stagnant Gas

8.5.1.1 *Evaporation from a tapered tank*   Consider a pure liquid $\mathcal{A}$ in an open cylindrical tank with a slightly tapered top as shown in Figure 8.35. The apparatus is arranged in such a manner that the liquid-gas interface remains fixed in space as the evaporation takes place. As engineers, we are interested in the rate of evaporation of $\mathcal{A}$ from the liquid surface into a gas mixture of $\mathcal{A}$ and $\mathcal{B}$. For this purpose, it is necessary to determine the concentration distribution of $\mathcal{A}$ in the gas phase. The problem will be analyzed with the following assumptions:

1. Steady-state conditions prevail.
2. Species $\mathcal{A}$ and $\mathcal{B}$ form an ideal gas mixture.
3. Species $\mathcal{B}$ has a negligible solubility in liquid $\mathcal{A}$.
4. The entire system is maintained at a constant temperature and pressure, i.e., the total molar concentration in the gas phase, $c = P/\mathcal{R}T$, is constant.
5. There is no chemical reaction between species $\mathcal{A}$ and $\mathcal{B}$.

If the taper angle is small, mass transport can be considered one-dimensional in the z-direction, and the conservation statement for species $\mathcal{A}$, Eq. (8.4-1), can be written over a differential volume element of thickness Δz as

$$(AN_{A_z})|_z - (AN_{A_z})|_{z+\Delta z} = 0 \tag{8.5-7}$$

Dividing Eq. (8.5-7) by Δz and letting $\Delta z \to 0$ give

$$\lim_{\Delta z \to 0} \frac{(AN_{A_z})|_z - (AN_{A_z})|_{z+\Delta z}}{\Delta z} = 0 \tag{8.5-8}$$

or,

$$\frac{d(AN_{A_z})}{dz} = 0 \tag{8.5-9}$$

Equation (8.5-9) indicates that

$$A N_{A_z} = \dot{n}_A = \text{constant} \tag{8.5-10}$$

In a similar way, the rate equation for the conservation of species $\mathcal{B}$ leads to

$$A N_{B_z} = \text{constant} \tag{8.5-11}$$

Since species B is insoluble in liquid A, i.e., $N_{B_z}|_{z=0} = 0$, it is implied that

$$N_{B_z} = 0 \qquad \text{for} \qquad 0 \leqslant z \leqslant L \qquad (8.5\text{-}12)$$

The total molar flux of species A is given by Eq. (8.5-2), i.e.,

$$N_{A_z} = -c\mathcal{D}_{AB}\frac{dx_A}{dz} + c_A v_z^* \qquad (8.5\text{-}13)$$

where the molar average velocity is given by

$$v_z^* = \frac{N_{A_z} + N_{B_z}}{c} = \frac{N_{A_z}}{c} \qquad (8.5\text{-}14)$$

which indicates nonzero convective flux. Although there is no bulk motion in the region $0 \leqslant z \leqslant L$, diffusion creates its own convection[9]. The use of Eq. (8.5-14) in Eq. (8.5-13) results in

$$N_{A_z} = -\frac{c\mathcal{D}_{AB}}{1 - x_A}\frac{dx_A}{dz} \qquad (8.5\text{-}15)$$

Substitution of (8.5-15) into Eq. (8.5-10) and rearrangement give

$$\dot{n}_A \int_0^L \frac{dz}{A(z)} = -c\mathcal{D}_{AB} \int_{x_{A_o}}^{x_{A_L}} \frac{dx_A}{1 - x_A} \qquad (8.5\text{-}16)$$

Thus, the rate of evaporation of liquid A is given by

$$\boxed{\dot{n}_A = \frac{c\mathcal{D}_{AB}}{\displaystyle\int_0^L \frac{dz}{A(z)}} \ln\left(\frac{1 - x_{A_L}}{1 - x_{A_o}}\right)} \qquad (8.5\text{-}17)$$

The value of x_A at $z = 0$, x_{A_o}, is the mole fraction of species A in the gas mixture that is in equilibrium with the pure liquid A at the existing temperature and pressure. The use of Dalton's and Raoult's laws at the gas-liquid interface indicates that

$$x_{A_o} = \frac{P_A^{sat}}{P} \qquad (8.5\text{-}18)$$

where P is the total pressure.

When x is small, then $\ln(1 - x) \simeq -x$. Therefore, for small values of x_{A_o} and x_{A_L}, Eq. (8.5-17) reduces to

$$\dot{n}_A = \frac{c\mathcal{D}_{AB}(x_{A_o} - x_{A_L})}{\displaystyle\int_0^L \frac{dz}{A(z)}} \qquad (8.5\text{-}19)$$

Note that Eq. (8.5-19) corresponds to the case when there is no convection, i.e., $v_z^* \simeq 0$.

---

[9]In the literature, this phenomenon is also called *diffusion-induced convection*. This is a characteristic of mass transfer. In the case of heat transfer, for example, conduction does not generate its own convection.

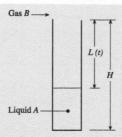

Figure 8.36. The Stefan diffusion tube.

Example 8.20 One way of measuring the diffusion coefficients of vapors is to place a small amount of liquid in a vertical capillary, generally known as the Stefan diffusion tube, and to blow a gas stream of known composition across the top as shown in Figure 8.36. Show how one can estimate the diffusion coefficient by observing the decrease in the liquid-gas interface as a function of time.

Solution

Assumptions

1. Pseudo-steady-state behavior.
2. The system is isothermal.
3. The total pressure remains constant.
4. The mole fraction of species $\mathcal{A}$ at the top of the tube is zero.
5. No turbulence is observed at the top of the tube.

Analysis

System: Liquid in the tube

The inventory rate equation for mass of $\mathcal{A}$ gives

$$- \text{Rate of moles of } \mathcal{A} \text{ out} = \text{Rate of accumulation of moles of } \mathcal{A} \tag{1}$$

or,

$$-\dot{n}_A = \frac{d}{dt}\left[(H - L)A\left(\rho_A^L / \mathcal{M}_A\right)\right] \tag{2}$$

where ρ_A^L is the density of species $\mathcal{A}$ in the liquid phase and A is the cross-sectional area of the tube. The rate of evaporation from the liquid surface, $\dot{n}_A$, can be determined from Eq. (8.5-17). For $A = $ constant and $x_{A_L} = 0$, Eq. (8.5-17) reduces to

$$\dot{n}_A = -\frac{A c \mathcal{D}_{AB}}{L} \ln(1 - x_{A_o}) \tag{3}$$

It should be kept in mind that Eq. (8.5-17) was developed for a steady-state case. For the unsteady problem at hand, the pseudo-steady-state assumption implies that Eq. (3) holds at any given instant, i.e.,

$$\dot{n}_A(t) = -\frac{A c \mathcal{D}_{AB}}{L(t)} \ln(1 - x_{A_o}) \tag{4}$$

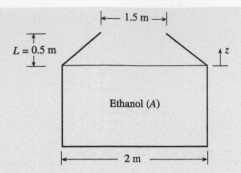

Figure 8.37. Evaporation from a tapered tank.

Substitution of Eq. (4) into Eq. (2) gives

$$-c\,D_{AB}\ln(1-x_{A_o})\int_0^t dt = \frac{\rho_A^L}{\mathcal{M}_A}\int_{L_o}^L L\,dL \tag{5}$$

or,

$$L^2 = -\left[\frac{2\mathcal{M}_A\,cD_{AB}\ln(1-x_{A_o})}{\rho_A^L}\right]t + L_o^2 \tag{6}$$

Therefore, the diffusion coefficient is determined from the slope of the L^2 versus t plot. Alternatively, rearrangement of Eq. (6) yields

$$\frac{t}{L-L_o} = -\left[\frac{\rho_A^L}{2\mathcal{M}_AcD_{AB}\ln(1-x_{A_o})}\right](L-L_o) + \frac{\rho_A^L L_o}{\mathcal{M}_AcD_{AB}\ln(1-x_{A_o})} \tag{7}$$

In this case, the diffusion coefficient is determined from the slope of the $t/(L-L_o)$ versus $(L-L_o)$ plot. What is the advantage of using Eq. (7) over Eq. (6)?

Example 8.21 To decrease the evaporation loss from open storage tanks, it is recommended to use a tapered top as shown in Figure 8.37. Calculate the rate of ethanol loss from the storage tank under steady conditions at 25 °C.

Solution

Physical properties

Diffusion coefficient of ethanol ($\mathcal{A}$) in air ($\mathcal{B}$) at 25 °C (298 K) is

$$(\mathcal{D}_{AB})_{298} = (\mathcal{D}_{AB})_{313}\left(\frac{298}{313}\right)^{3/2} = (1.45 \times 10^{-5})\left(\frac{298}{313}\right)^{3/2} = 1.35 \times 10^{-5}\ \text{m}^2/\text{s}$$

$P_A^{sat} = 58.6$ mmHg

Analysis

In order to determine the molar flow rate of species $\mathcal{A}$ from Eq. (8.5-17), it is first necessary to express the variation in the cross-sectional area in the direction of z. The variation in the diameter as a function of z is

$$D(z) = D_o - \left(\frac{D_o - D_L}{L}\right)z \tag{1}$$

where D_o and D_L are the tank diameters at $z = 0$ and $z = L$, respectively. Therefore, the variation in the cross-sectional area is

$$A(z) = \frac{\pi D^2(z)}{4} = \frac{\pi}{4}\left[D_o - \left(\frac{D_o - D_L}{L}\right)z\right]^2 \tag{2}$$

Substitution of Eq. (2) into Eq. (8.5-17) and integration give the molar rate of evaporation as

$$\dot{n}_A = -\frac{\pi c \mathcal{D}_{AB}(D_o - D_L)\ln(1 - x_{A_o})}{4L\left(\dfrac{1}{D_L} - \dfrac{1}{D_o}\right)} \tag{3}$$

The numerical values are

$$D_o = 2 \text{ m} \qquad D_L = 1.5 \text{ m} \qquad L = 0.5 \text{ m}$$

$$x_{A_o} = \frac{P_A^{sat}}{P} = \frac{58.6}{760} = 0.077$$

$$c = \frac{P}{\mathcal{R}T} = \frac{1}{(0.08205)(25 + 273)} = 41 \times 10^{-3} \text{ kmol/m}^3 = 41 \text{ mol/m}^3$$

Substitution of these values into Eq. (3) gives

$$\dot{n}_A = -\frac{\pi(41)(1.35 \times 10^{-5})(2 - 1.5)\ln(1 - 0.077)}{(4)(0.5)\left(\dfrac{1}{1.5} - \dfrac{1}{2}\right)} \simeq 2.1 \times 10^{-4} \text{ mol/s}$$

Comment: When $D_L \to D_o$, application of L'Hopital's rule gives

$$\lim_{D_L \to D_o} \frac{D_o - D_L}{\dfrac{1}{D_L} - \dfrac{1}{D_o}} = \lim_{D_L \to D_o} \frac{-1}{-\dfrac{1}{D_L^2}} = D_o^2$$

and Eq. (3) reduces to

$$\dot{n}_A = -\frac{(\pi D_o^2/4)c\mathcal{D}_{AB}}{L}\ln(1 - x_{A_o})$$

which is Eq. (4) in Example 8.20.

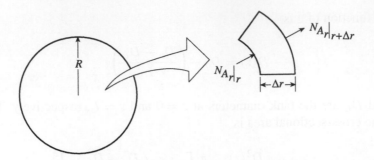

Figure 8.38. Mass transfer from a spherical drop.

8.5.1.2 *Evaporation of a spherical drop*   A liquid ($\mathcal{A}$) droplet of radius R is suspended in a stagnant gas $\mathcal{B}$ as shown in Figure 8.38. We want to determine the rate of evaporation under steady conditions.

Over a differential volume element of thickness Δr, as shown in Figure 8.38, the conservation statement for species $\mathcal{A}$, Eq. (8.4-1), is written as

$$(AN_{A_r})|_r - (AN_{A_r})|_{r+\Delta r} = 0 \tag{8.5-20}$$

Dividing Eq. (8.5-20) by Δr and taking the limit as $\Delta r \to 0$ give

$$\lim_{\Delta r \to 0} \frac{(AN_{A_r})|_r - (AN_{A_r})|_{r+\Delta r}}{\Delta r} = 0 \tag{8.5-21}$$

or,

$$\frac{d(AN_{A_r})}{dr} = 0 \tag{8.5-22}$$

Since flux times area gives the molar transfer rate of species $\mathcal{A}$, $\dot{n}_A$, it is possible to conclude that

$$A N_{A_r} = \text{constant} = \dot{n}_A \tag{8.5-23}$$

Note that the area A in Eq. (8.5-23) is perpendicular to the direction of mass flux and is given by

$$A = 4\pi r^2 \tag{8.5-24}$$

Since the temperature and the total pressure remain constant, the total molar concentration, c, in the gas phase is constant. From Table C.9 in Appendix C, the total molar flux of species $\mathcal{A}$ in the r-direction is given by

$$N_{A_r} = -\mathcal{D}_{AB} \frac{dc_A}{dr} + c_A v_r^* \tag{8.5-25}$$

Since species $\mathcal{B}$ is stagnant, the molar average velocity is expressed as

$$v_r^* = \frac{N_{A_r} + N_{B_r}}{c} = \frac{N_{A_r}}{c} \tag{8.5-26}$$

which indicates nonzero convective flux. Using Eq. (8.5-26) in Eq. (8.5-25) results in

$$N_{A_r} = -\frac{cD_{AB}}{c - c_A} \frac{dc_A}{dr} \qquad (8.5\text{-}27)$$

Substitution of Eqs. (8.5-27) and (8.5-24) into Eq. (8.5-23) and rearrangement give

$$-4\pi cD_{AB} \int_{c_A^*}^{0} \frac{dc_A}{c - c_A} = \dot{n}_A \int_{R}^{\infty} \frac{dr}{r^2} \qquad (8.5\text{-}28)$$

where c_A^* is the saturation concentration of species A in B at $r = R$ in the gas phase. Carrying out the integrations in Eq. (8.5-28) yields

$$\boxed{\dot{n}_A = 4\pi cD_{AB}R \ln\left(\frac{c}{c - c_A^*}\right)} \qquad (8.5\text{-}29)$$

Example 8.22 A benzene droplet with a diameter of 8 mm is suspended by a wire in a laboratory. The temperature and pressure are maintained constant at 25 °C and 1 atm, respectively. Estimate the diffusion coefficient of benzene in air if the variation in the droplet diameter as a function of time is recorded as follows:

t (min)	5	10	15	20	25
D (mm)	7.3	6.5	5.5	4.4	2.9

Solution

Physical properties

For benzene (A): $\begin{cases} \rho_A = 879 \text{ kg/m}^3 \\ M_A = 78 \\ P_A^{sat} = 94.5 \text{ mmHg} \end{cases}$

Assumptions

1. Pseudo-steady-state behavior.
2. Air is insoluble in the droplet.

Analysis

System: Benzene droplet

The inventory rate equation for mass of A gives

$$- \text{Rate of moles of } A \text{ out} = \text{Rate of accumulation of moles of } A \qquad (1)$$

or,

$$-\dot{n}_A = \frac{d}{dt}\left[\frac{4}{3}\pi R^3\left(\frac{\rho_A^L}{M_A}\right)\right] = \frac{4\pi\rho_A^L}{M_A} R^2 \frac{dR}{dt} \qquad (2)$$

where ρ_A^L is the density of species A in the liquid phase.

The rate of evaporation from the droplet surface, $\dot{n}_A$, can be determined from Eq. (8.5-29). However, remember that Eq. (8.5-29) was developed for a steady-state case. For the unsteady problem at hand, the pseudo-steady-state assumption implies that Eq. (8.5-29) holds at any given instant, i.e.,

$$\dot{n}_A(t) = 4\pi c \mathcal{D}_{AB} R(t) \ln\left(\frac{c}{c - c_A^*}\right) \tag{3}$$

Substitution of Eq. (3) into Eq. (2) and rearrangement give

$$-\frac{\rho_A^L}{\mathcal{M}_A} \int_{R_o}^{R} R \, dR = c \mathcal{D}_{AB} \ln\left(\frac{c}{c - c_A^*}\right) \int_0^t dt \tag{4}$$

where R_o is the initial radius of the liquid droplet. Carrying out the integrations in Eq. (4) yields

$$R^2 = R_o^2 - \left[\frac{2c\mathcal{D}_{AB}\mathcal{M}_A}{\rho_A^L} \ln\left(\frac{c}{c - c_A^*}\right)\right] t \tag{5}$$

Since

$$c = \frac{P}{\mathcal{R}T} \quad \text{and} \quad c_A^* = \frac{P_A^{sat}}{\mathcal{R}T} \tag{6}$$

Eq. (5) takes the form

$$R^2 = R_o^2 - \left[\frac{2c\mathcal{D}_{AB}\mathcal{M}_A}{\rho_A^L} \ln\left(\frac{P}{P - P_A^{sat}}\right)\right] t \tag{7}$$

The plot of R^2 versus t is shown below.

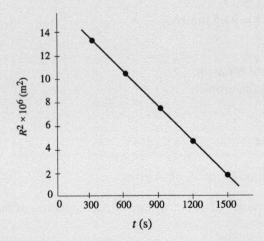

The slope of the straight line is $-9.387 \times 10^{-9} \text{ m}^2/\text{s}$. Hence,

$$\frac{2c\mathcal{D}_{AB}\mathcal{M}_A}{\rho_A^L} \ln\left(\frac{P}{P - P_A^{sat}}\right) = 9.387 \times 10^{-9} \tag{8}$$

The total molar concentration is

$$c = \frac{P}{\mathcal{R}T} = \frac{1}{(0.08205)(25 + 273)} = 0.041 \text{ kmol/m}^3 \tag{9}$$

Substitution of the values into Eq. (8) gives the diffusion coefficient as

$$\mathcal{D}_{AB} = 9.387 \times 10^{-9} \left[\frac{879}{2(0.041)(78) \ln\left(\dfrac{760}{760 - 94.5}\right)} \right] = 9.72 \times 10^{-6} \text{ m}^2/\text{s}$$

8.5.2 Diffusion Through a Stagnant Liquid

Consider a one-dimensional diffusion of liquid $\mathcal{A}$ through a stagnant film of liquid $\mathcal{B}$ with a thickness L as shown in Figure 8.39. The mole fractions of $\mathcal{A}$ at $z = 0$ and $z = L$ are known. As engineers, we are interested in the number of moles of species $\mathcal{A}$ transferring through the film of $\mathcal{B}$ under steady conditions.

Over a differential volume of thickness Δz, the conservation statement for species $\mathcal{A}$, Eq. (8.4-1), is written as

$$N_{A_z}|_z A - N_{A_z}|_{z+\Delta z} A = 0 \tag{8.5-30}$$

Dividing Eq. (8.5-30) by $A \Delta z$ and letting $\Delta z \to 0$ give

$$\lim_{\Delta z \to 0} \frac{N_{A_z}|_z - N_{A_z}|_{z+\Delta z}}{\Delta z} = 0 \tag{8.5-31}$$

or,

$$\frac{dN_{A_z}}{dz} = 0 \quad \Rightarrow \quad N_{A_z} = \text{constant} \tag{8.5-32}$$

To proceed further, it is necessary to express the total molar flux of species $\mathcal{A}$, i.e., N_{A_z}, either by Eq. (8.5-2) or by Eq. (8.5-3).

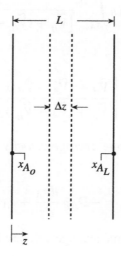

Figure 8.39. Diffusion of liquid $\mathcal{A}$ through a stagnant liquid film $\mathcal{B}$.

8.5.2.1 *Analysis based on the molar average velocity*   From Eq. (8.5-2), the total molar flux of species $\mathcal{A}$ is given as

$$N_{A_z} = -c\mathcal{D}_{AB}\frac{dx_A}{dz} + c_A v_z^*$$ (8.5-33)

It is important to note in this problem that the total molar concentration, c, is not constant but dependent on the mole fractions of species $\mathcal{A}$ and $\mathcal{B}$. Since species $\mathcal{B}$ is stagnant, the expression for the molar average velocity becomes

$$v_z^* = \frac{N_{A_z} + N_{B_z}}{c} = \frac{N_{A_z}}{c}$$ (8.5-34)

Substitution of Eq. (8.5-34) into Eq. (8.5-33) gives the molar flux of species $\mathcal{A}$ as

$$N_{A_z} = -\frac{c\mathcal{D}_{AB}}{1-x_A}\frac{dx_A}{dz}$$ (8.5-35)

Since the total molar concentration, c, is not constant, it is necessary to express c in terms of mole fractions. Assuming ideal solution behavior, i.e., the partial molar volume is equal to the molar volume of the pure substance, the total molar concentration is expressed in the form

$$c = \frac{1}{\widetilde{V}_{mix}} = \frac{1}{x_A\widetilde{V}_A + x_B\widetilde{V}_B}$$ (8.5-36)

Substitution of $x_B = 1 - x_A$ yields

$$c = \frac{1}{\widetilde{V}_B + (\widetilde{V}_A - \widetilde{V}_B)x_A}$$ (8.5-37)

Combining Eqs. (8.5-35) and (8.5-37) and rearrangement give

$$N_{A_z}\int_0^L dz = -\mathcal{D}_{AB}\int_{x_{A_o}}^{x_{A_L}}\frac{dx_A}{[\widetilde{V}_B + (\widetilde{V}_A - \widetilde{V}_B)x_A](1-x_A)}$$ (8.5-38)

Integration of Eq. (8.5-38) results in

$$N_{A_z} = \frac{\mathcal{D}_{AB}}{L\widetilde{V}_A}\left\{\ln\left(\frac{1-x_{A_L}}{1-x_{A_o}}\right) - \ln\left[\frac{\widetilde{V}_B + (\widetilde{V}_A - \widetilde{V}_B)x_{A_L}}{\widetilde{V}_B + (\widetilde{V}_A - \widetilde{V}_B)x_{A_o}}\right]\right\} = \frac{\mathcal{D}_{AB}}{L\widetilde{V}_A}\ln\left(\frac{c_{B_L}}{c_{B_o}}\right)$$ (8.5-39)

8.5.2.2 *Analysis based on the volume average velocity*   The use of Eq. (8.5-3) gives the total molar flux of species $\mathcal{A}$ as

$$N_{A_z} = -\mathcal{D}_{AB}\frac{dc_A}{dz} + c_A v_z^\blacksquare$$ (8.5-40)

From Eq. (C) in Table 8.12, the volume average velocity is expressed as

$$v_z^\blacksquare = \overline{V}_A N_{A_z} + \overline{V}_B N_{B_z} = \overline{V}_A N_{A_z} = \widetilde{V}_A N_{A_z}$$ (8.5-41)

Using Eq. (8.5-41) in Eq. (8.5-40) yields

$$N_{A_z} = -\frac{\mathcal{D}_{AB}}{1 - \widetilde{V}_A c_A} \frac{dc_A}{dz} \tag{8.5-42}$$

Rearrangement of Eq. (8.5-42) results in

$$N_{A_z} \int_0^L dz = -\mathcal{D}_{AB} \int_{c_{A_o}}^{c_{A_L}} \frac{dc_A}{1 - \widetilde{V}_A c_A} \tag{8.5-43}$$

Integration of Eq. (8.5-43) leads to

$$N_{A_z} = \frac{\mathcal{D}_{AB}}{L\widetilde{V}_A} \ln\left(\frac{1 - \widetilde{V}_A c_{A_L}}{1 - \widetilde{V}_A c_{A_o}}\right) \tag{8.5-44}$$

The use of the identity from Eq. (8.5-36), i.e.,

$$1 - \widetilde{V}_A c_A = \widetilde{V}_B c_B \tag{8.5-45}$$

simplifies Eq. (8.5-44) to

$$N_{A_z} = \frac{\mathcal{D}_{AB}}{L\widetilde{V}_A} \ln\left(\frac{c_{B_L}}{c_{B_o}}\right) \tag{8.5-46}$$

which is identical to Eq. (8.5-39).

Example 8.23 Cyclohexane ($\mathcal{A}$) is diffusing through a 1.5 mm thick stagnant benzene ($\mathcal{B}$) film at 25 °C. If $x_{A_o} = 0.15$ and $x_{A_L} = 0.05$, determine the molar flux of cyclohexane under steady conditions. Take $\mathcal{D}_{AB} = 2.09 \times 10^{-5}$ cm^2/s.

Solution

Physical properties

For cyclohexane ($\mathcal{A}$): $\begin{cases} \rho_A = 0.779 \text{ g/cm}^3 \\ \mathcal{M}_A = 84 \end{cases}$

For benzene ($\mathcal{B}$): $\begin{cases} \rho_B = 0.879 \text{ g/cm}^3 \\ \mathcal{M}_B = 78 \end{cases}$

Analysis

The molar volumes of species $\mathcal{A}$ and $\mathcal{B}$ are

$$\widetilde{V}_A = \frac{\mathcal{M}_A}{\rho_A} = \frac{84}{0.779} = 107.8 \text{ cm}^3/\text{mol}$$

$$\widetilde{V}_B = \frac{\mathcal{M}_B}{\rho_B} = \frac{78}{0.879} = 88.7 \text{ cm}^3/\text{mol}$$

The values of the total molar concentration at $z = 0$ and $z = L$ are calculated from Eq. (8.5-37) as

$$c_o = \frac{1}{\widetilde{V}_B + (\widetilde{V}_A - \widetilde{V}_B)x_{A_o}} = \frac{1}{88.7 + (107.8 - 88.7)(0.15)} = 10.9 \times 10^{-3} \text{ mol/cm}^3$$

$$c_L = \frac{1}{\widetilde{V}_B + (\widetilde{V}_A - \widetilde{V}_B)x_{A_L}} = \frac{1}{88.7 + (107.8 - 88.7)(0.05)} = 11.2 \times 10^{-3} \text{ mol/cm}^3$$

Therefore, the use of Eq. (8.5-39) gives the molar flux of cyclohexane through the benzene layer as

$$N_{A_z} = \frac{\mathcal{D}_{AB}}{L\widetilde{V}_A} \ln\left(\frac{c_{B_L}}{c_{B_o}}\right)$$

$$= \frac{2.09 \times 10^{-5}}{(0.15)(107.8)} \ln\left[\frac{(11.2 \times 10^{-3})(1 - 0.05)}{(10.9 \times 10^{-3})(1 - 0.15)}\right] = 1.8 \times 10^{-7} \text{ mol/cm}^2 \cdot \text{s}$$

8.5.3 Diffusion With a Heterogeneous Chemical Reaction

An ideal gas A diffuses at steady-state in the positive z-direction through a flat gas film of thickness δ as shown in Figure 8.40. At $z = \delta$ there is a solid catalytic surface at which A undergoes a first-order heterogeneous dimerization reaction

$$2A \rightarrow B$$

As engineers, we are interested in the determination of the molar flux of species A in the gas film under steady conditions. The gas composition at $z = 0$, i.e., x_{A_o}, is known.

The conservation statement for species A, Eq. (8.4-1), can be written over a differential volume element of thickness Δz as

$$N_{A_z}|_z A - N_{A_z}|_{z+\Delta z} A = 0 \tag{8.5-47}$$

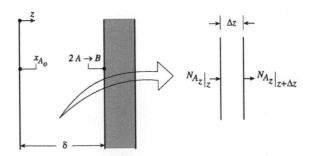

Figure 8.40. Heterogeneous reaction on a catalyst surface.

Dividing Eq. (8.5-47) by $A \, \Delta z$ and letting $\Delta z \to 0$ give

$$\lim_{\Delta z \to 0} \frac{N_{A_z}|_z - N_{A_z}|_{z+\Delta z}}{\Delta z} = 0 \qquad (8.5\text{-}48)$$

or,

$$\frac{dN_{A_z}}{dz} = 0 \quad \Rightarrow \quad N_{A_z} = \text{constant} \qquad (8.5\text{-}49)$$

The total molar flux can be calculated from Eq. (8.5-2) as

$$N_{A_z} = -c\mathcal{D}_{AB} \frac{dx_A}{dz} + c_A v_z^* \qquad (8.5\text{-}50)$$

in which the molar average velocity is given by

$$v_z^* = \frac{N_{A_z} + N_{B_z}}{c} \qquad (8.5\text{-}51)$$

The stoichiometry of the chemical reaction implies that for every 2 moles of $\mathcal{A}$ diffusing in the positive z-direction, 1 mole of $\mathcal{B}$ diffuses back in the negative z-direction. Therefore, the relationship between the fluxes can be expressed as

$$\frac{1}{2} N_{A_z} = -N_{B_z} \qquad (8.5\text{-}52)$$

The use of Eq. (8.5-52) in Eq. (8.5-51) yields

$$v_z^* = \frac{0.5 N_{A_z}}{c} \qquad (8.5\text{-}53)$$

Substitution of Eq. (8.5-53) into Eq. (8.5-50) gives

$$N_{A_z} = -\frac{c\mathcal{D}_{AB}}{1 - 0.5x_A} \frac{dx_A}{dz} \qquad (8.5\text{-}54)$$

Since N_{A_z} is constant, Eq. (8.5-54) can be rearranged as

$$N_{A_z} \int_0^\delta dz = -c\mathcal{D}_{AB} \int_{x_{A_o}}^{x_{A_\delta}} \frac{dx_A}{1 - 0.5x_A} \qquad (8.5\text{-}55)$$

or,

$$N_{A_z} = \frac{2c\mathcal{D}_{AB}}{\delta} \ln\left(\frac{1 - 0.5x_{A_\delta}}{1 - 0.5x_{A_o}}\right) \qquad (8.5\text{-}56)$$

Note that, although x_{A_o} is a known quantity, the mole fraction of species $\mathcal{A}$ in the gas phase at the catalytic surface, x_{A_δ}, is unknown and must be determined from the boundary condition. For heterogeneous reactions, the rate of reaction is empirically specified as

$$\text{at} \quad z = \delta \qquad N_{A_z} = k^s c_A = k^s c x_A \qquad (8.5\text{-}57)$$

where k^s is the surface reaction rate constant. Therefore, x_{A_δ} is expressed from Eq. (8.5-57) as

$$x_{A_\delta} = \frac{N_{A_z}}{ck^s} \tag{8.5-58}$$

Substitution of Eq. (8.5-58) into Eq. (8.5-56) results in

$$\boxed{N_{A_z} = \frac{2c\mathcal{D}_{AB}}{\delta} \ln\left[\frac{1 - 0.5(N_{A_z}/ck^s)}{1 - 0.5x_{A_o}}\right]} \tag{8.5-59}$$

which is a transcendental equation in N_{A_z}. It is interesting to investigate two limiting cases of Eq. (8.5-59).

Case (*i*) k^s is large

Since $\ln(1 - x) \simeq -x$ for small values of x, then

$$\ln\left[1 - 0.5(N_{A_z}/ck^s)\right] \simeq -0.5(N_{A_z}/ck^s) \tag{8.5-60}$$

so that Eq. (8.5-59) reduces to

$$N_{A_z} = \frac{2c\mathcal{D}_{AB}}{\delta} \left(\frac{\Lambda^2}{\Lambda^2 + 1}\right) \ln\left(\frac{1}{1 - 0.5x_{A_o}}\right) \tag{8.5-61}$$

in which Λ represents the ratio of the rate of heterogeneous reaction to the rate of diffusion, i.e., Thiele modulus, and it is given by

$$\Lambda = \sqrt{\frac{k^s \delta}{\mathcal{D}_{AB}}} \tag{8.5-62}$$

Case (*ii*) $k^s = \infty$

This condition implies an instantaneous reaction and Eq. (8.5-59) takes the form

$$N_{A_z} = \frac{2c\mathcal{D}_{AB}}{\delta} \ln\left(\frac{1}{1 - 0.5x_{A_o}}\right) \tag{8.5-63}$$

When $k^s = \infty$, once species A reaches the catalytic surface, it is immediately converted to species B so that $x_{A_\delta} = 0$. Note that Eq. (8.5-63) can also be obtained either from Eq. (8.5-56) by letting $x_{A_\delta} = 0$ or from Eq. (8.5-61) by letting $\Lambda = \infty$.

8.5.3.1 *Comment* The molar average velocity is given by Eq. (8.5-53) and, since both N_{A_z} and c are constants, v_z^* remains constant for $0 \leqslant z \leqslant \delta$. On the other hand, from Eq. (8.5-6) the mass average velocity is

$$v_z = \frac{\mathcal{M}_A N_{A_z} + \mathcal{M}_B N_{B_z}}{\rho} \tag{8.5-64}$$

Expressing N_{B_z} in terms of N_{A_z} by using Eq. (8.5-52) reduces Eq. (8.5-64) to

$$v_z = \frac{N_{A_z}(\mathcal{M}_A - 0.5\mathcal{M}_B)}{\rho} \qquad (8.5\text{-}65)$$

As a result of the dimerization reaction $\mathcal{M}_A = 0.5\mathcal{M}_B$ and we get

$$v_z = 0 \qquad (8.5\text{-}66)$$

In this specific example, therefore, the mass average velocity can be determined on the basis of a solution to a diffusion problem rather than conservation of momentum.

NOTATION

A	area, m^2
a_v	catalyst surface area per unit volume, $1/m$
$\widehat{C}_P$	heat capacity at constant pressure, $kJ/kg{\cdot}K$
c	total concentration, $kmol/m^3$
c_i	concentration of species i, $kmol/m^3$
D	diameter, m
$\mathcal{D}_{AB}$	diffusion coefficient for system $\mathcal{A}\text{-}\mathcal{B}$, m^2/s
e	total energy flux, W/m^2
F_D	drag force, N
H	enthalpy, J
$\mathcal{H}$	partition coefficient
h	heat transfer coefficient, $W/m^2{\cdot}K$
J^*	molecular molar flux, $kmol/m^2{\cdot}s$
k	thermal conductivity, $W/m{\cdot}K$
k^s	surface reaction rate constant, m/s
L	length, m
$\dot{m}$	mass flow rate, kg/s
$\mathcal{M}$	molecular weight, $kg/kmol$
N	total molar flux, $kmol/m^2{\cdot}s$
$\dot{n}$	total molar flow rate, $kmol/s$
$\dot{n}_i$	molar flow rate of species i, $kmol/s$
P	pressure, Pa
$\dot{Q}$	heat transfer rate, W
$\mathcal{Q}$	volumetric flow rate, m^3/s
q	heat flux, W/m^2
R	radius, m; resistance, K/W
$\mathcal{R}$	gas constant, $J/mol{\cdot}K$
T	temperature, $°C$ or K
t	time, s

U	overall heat transfer coefficient, $W/m^2 \cdot K$
V	velocity of the plate in Couette flow, m/s; volume, m^3
v	mass average velocity, m/s
v^*	molar average velocity, m/s
$v^\blacksquare$	volume average velocity, m/s
W	width, m
$\mathcal{W}$	total mass flux, $kg/m^2 \cdot s$
x_i	mole fraction of species i
Δ	difference
η	fin efficiency; effectiveness factor
λ	latent heat of vaporization, J
μ	viscosity, $kg/m \cdot s$
ν	kinematic viscosity, m^2/s
π	total momentum flux, N/m^2
ρ	density, kg/m^3
τ_{ij}	shear stress (flux of j-momentum in the i-direction), N/m^2
ω	mass fraction

Overlines

$\sim$	per mole
$\frown$	per unit mass
$-$	partial molar

Bracket

$\langle a \rangle$	average value of a

Superscript

sat	saturation

Subscripts

A, B	species in binary systems
ch	characteristic
GM	geometric mean
i	species in multicomponent systems
in	inlet
LM	log-mean
mix	mixture
out	outlet
w	wall or surface
∞	free stream

Dimensionless Numbers

Bi_H Biot number for heat transfer
Bi_M Biot number for mass transfer
Nu   Nusselt number
Pr   Prandtl number
Re   Reynolds number
Sc   Schmidt number
Sh   Sherwood number

REFERENCES

Astarita, G., 1997, Dimensional analysis, scaling, and orders of magnitude, Chem. Eng. Sci. 52 (24), 4681.

Bird, R.B., W.E. Stewart and E.N. Lightfoot, 2002, Transport Phenomena, 2nd Ed., Wiley, New York.

Bejan, A., 1984, Convection Heat Transfer, Wiley-Interscience, New York.

Carslaw, H.S. and J.C. Jaeger, 1959, Conduction of Heat in Solids, 2nd Ed., Oxford University Press, London.

Slattery, J.C., 1972, Momentum, Energy, and Mass Transfer in Continua, McGraw-Hill, New York.

Whitaker, S., 1976, Elementary Heat Transfer Analysis, Pergamon Press, New York.

Whitaker, S., 1991, Role of the species momentum equation in the analysis of the Stefan diffusion tube, Ind. Eng. Chem. Research 30, 978.

SUGGESTED REFERENCES FOR FURTHER STUDY

Brodkey, R.S. and H.C. Hershey, 1988, Transport Phenomena: A Unified Approach, McGraw-Hill, New York.

Cussler, E.L., 1997, Diffusion - Mass Transfer in Fluid Systems, 2nd Ed., Cambridge University Press, Cambridge.

Fahien, R.W., 1983, Fundamentals of Transport Phenomena, McGraw-Hill, New York.

Incropera, F.P. and D.P. DeWitt, 2002, Fundamentals of Heat and Mass Transfer, 5th Ed., Wiley, New York.

Middleman, S., 1998, An Introduction to Mass and Heat Transfer – Principles of Analysis and Design, Wiley, New York.

PROBLEMS

8.1   When the ratio of the radius of the inner pipe to that of the outer pipe is close to unity, a concentric annulus may be considered a thin plate slit and its curvature can be neglected. Use this approximation and show that Eqs. (8.1-12) and (8.1-15) can be modified as

$$\frac{v_z}{V} = 1 - \frac{1}{1-\kappa}\left(\frac{r}{R} - 1\right)$$

$$Q = \frac{\pi R^2 V (1 - \kappa^2)}{2}$$

to determine the velocity distribution and volumetric flow rate for Couette flow in a concentric annulus with inner and outer radii of κR and R, respectively.

8.2   The composite wall shown below consists of materials A and B with thermal conductivities $k_A = 10$ W/m·K and $k_B = 0.8$ W/m·K. If the surface area of the wall is 5 m^2, determine the interface temperature between A and B.

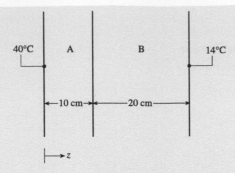

(**Answer:** 39 °C)

8.3 A composite wall consists of a brick of thickness 5 cm with thermal conductivity 1 W/m·K and an insulation of thickness 3 cm with thermal conductivity 0.06 W/m·K. The brick surface is subjected to a uniform heat flux of 400 W/m^2. The surface of the insulation layer dissipates heat by convection to ambient air at 25 °C with an average heat transfer coefficient of 20 W/m^2·K. Determine the surface temperatures under steady conditions.

(**Answer:** 45 °C and 265 °C)

8.4 A printed circuit board (PCB) is a thin plate on which chips and other electronic components are placed. The thin plate is a layered composite consisting of copper foil and a glass-reinforced polymer (FR-4). A cross-sectional view of such a laminated structure is shown in the figure below.

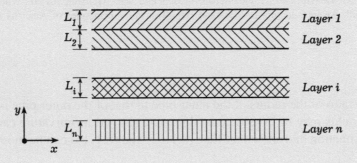

In engineering calculations, it is convenient to treat such a layered structure as a homogeneous material with two different effective thermal conductivities: one describing heat flow within the plane, i.e., in the x-direction, and the other describing heat flow through the thickness of the plate, i.e., in the y-direction.

a) Show that

$$(k_x)_{eff} = \frac{\sum\limits_{i=1}^{n} k_i L_i}{\sum\limits_{i=1}^{n} L_i} \quad \text{and} \quad (k_y)_{eff} = \frac{\sum\limits_{i=1}^{n} L_i}{\sum\limits_{i=1}^{n} \dfrac{L_i}{k_i}}$$

b) Assume that the total PCB thickness is 1.5 mm and that the layers consist only of copper and FR-4, with thermal conductivities 390 and 0.25 W/m·K, respectively. Calculate $(k_x)_{eff}$ and $(k_y)_{eff}$ if the thickness of the copper plate is 30 μm. Repeat the calculations for when the thickness of the copper plate is 100 μm. What is your conclusion?

(**Answer:** $(k_x)_{eff} = 8.05$ W/m·K, $(k_y)_{eff} = 0.26$ W/m·K when $L_{Cu} = 30$ μm; $(k_x)_{eff} = 26.23$ W/m·K, $(k_y)_{eff} = 0.27$ W/m·K when $L_{Cu} = 100$ μm)

8.5 Calculate the steady-state temperature distribution in a long cylindrical rod of thermal conductivity k and radius R. Cooling fluid at a temperature of T_∞ flows over the surface of the cylinder with an average heat transfer coefficient $\langle h \rangle$.

(**Answer:** $T = T_\infty$)

8.6 A spherical tank containing liquid nitrogen at 1 atm pressure is insulated with a material having a thermal conductivity of 1.73×10^{-3} W/m·K. The inside diameter of the tank is 60 cm, and the insulation thickness is 2.5 cm. Estimate the kilograms of nitrogen vaporized per day if the outside surface of the insulation is at 21 °C. The normal boiling point of nitrogen is -196 °C and its latent heat of vaporization is 200 kJ/kg.

(**Answer:** 7.95 kg/day)

8.7 For a rectangular fin in Section 8.2.4 the parameters are given as: $T_\infty = 175$ °C, $T_w = 260$ °C, $k = 105$ W/m·K, $L = 4$ cm, $W = 30$ cm, $B = 5$ mm.

a) Calculate the average heat transfer coefficient and the rate of heat loss through the fin surface for $\Lambda = 0.3, 0.6, 0.8, 1.0, 3.0, 6.0,$ and 8.0.
b) One of your friends claims that as the fin efficiency increases the process becomes more reversible. Do you agree?

8.8 If the length of the rectangular fin described in Section 8.2.4 is infinitely long, then the temperature at the tip of the fin approaches the temperature of the surrounding fluid.

a) Under these circumstances, show that the dimensionless temperature distribution and the rate of heat loss are given by

$$\theta = \exp(-\Lambda\xi) \tag{1}$$

$$\dot{Q} = W(T_w - T_\infty)\sqrt{2kB\langle h \rangle} \tag{2}$$

b) Note that Eq. (8.2-94) reduces to Eq. (2) for large values of Λ. Thus, conclude that the "infinitely long" fin assumption is valid when $\Lambda \geqslant 3$.

8.9 Copper fins of rectangular profile are attached to a plane wall maintained at 180 °C. It is estimated that the heat is transferred to ambient air at 35 °C with an average heat transfer coefficient of 60 W/m²·K. Calculate the steady rate of heat loss if the fins have the dimensions of $B = 1$ mm and $L = 8$ mm and are placed with a fin spacing of 200 fins/m. Heat losses from the edges and the tip of the fin may be considered negligible.

(**Answer:** 34.6 kW/m²)

8.10 Repeat the analysis given in Section 8.2.4 by considering heat losses from the edges as well as from the tip with an average heat transfer coefficient $\langle h \rangle$.

a) Show that the temperature distribution is given by

$$\theta = \frac{\Omega \sinh[\Lambda(1-\xi)] + \Lambda \cosh[\Lambda(1-\xi)]}{\Omega \sinh \Lambda + \Lambda \cosh \Lambda} \tag{1}$$

where the dimensionless quantities are defined as

$$\theta = \frac{\langle T \rangle - T_\infty}{T_w - T_\infty} \qquad \xi = \frac{z}{L} \qquad \Lambda = \sqrt{\left(\frac{1}{B} + \frac{1}{W}\right)\frac{2\langle h \rangle L^2}{k}} \qquad \Omega = \frac{\langle h \rangle L}{k} \tag{2}$$

b) Show that the rate of heat loss from the fin is given by

$$\dot{Q} = \frac{kBW(T_w - T_\infty)\Lambda}{L}\left(\frac{\Omega \cosh \Lambda + \Lambda \sinh \Lambda}{\Omega \sinh \Lambda + \Lambda \cosh \Lambda}\right) \tag{3}$$

c) An aluminum fin with thickness $B = 0.5$ cm, width $W = 3$ cm, and length $L = 20$ cm is attached to a plane wall maintained at $170\,°C$. The fin dissipates heat to ambient air at $25\,°C$ with an average heat transfer coefficient of 40 W/m^2·K. Plot the temperature distribution as a function of position. Also calculate the rate of heat loss from the fin to ambient air.

d) Plot the fin temperature as a function of position and calculate the rate of heat loss if the fin in part (c) is covered with 3 mm thick plastic ($k = 0.07$ W/m·K).

Hint: In this case, Eqs. (1) and (3) are still valid. However, the average heat transfer coefficient $\langle h \rangle$ in the definitions of Λ and Ω must be replaced by the overall heat transfer coefficient U (Why?) defined by

$$U = \left(\frac{L_{plastic}}{k_{plastic}} + \frac{1}{\langle h \rangle}\right)^{-1}$$

e) Calculate the temperature of the plastic surface exposed to air at $\xi = 0.3$.

(**Answer:** c) 45.75 W   d) 22.9 W   e) 68.3 °C)

8.11 A copper fin of rectangular profile is attached to a plane wall maintained at $250\,°C$ and has the dimensions of $B = 3$ mm and $W = 5$ cm. The fin dissipates heat from all of its surfaces to ambient air at $25\,°C$ with an average heat transfer coefficient of 70 W/m^2·K. Estimate the length of the fin if the temperature at the tip of the fin should not exceed $40\,°C$ to avoid burns.

(**Answer:** 30 cm)

8.12 A fin with thickness $B = 6$ mm, width $W = 3$ cm, and length $L = 10$ cm is attached to a plane wall maintained at $200\,°C$. The fin dissipates heat to ambient air at $25\,°C$ with an average heat transfer coefficient of 50 W/m^2·K. What should be the thermal conductivity of the fin material if the temperature at a point 4 cm from the wall should not exceed 120 °C? Assume no heat losses from the sides or the tip of the fin.

(**Answer:** 54.9 W/m·K)

8.13   A solid cylindrical rod of radius R and length L is placed between two walls as shown in the figure below. The surface temperatures of the walls at $z = 0$ and $z = L$ are kept at T_o and T_L, respectively. The rod dissipates heat by convection to ambient air at T_∞ with an average heat transfer coefficient $\langle h \rangle$.

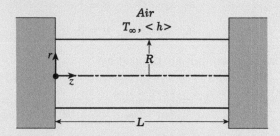

a) Consider a cylindrical differential element of thickness Δr and length Δz within the rod and show that the conservation statement for energy leads to

$$\frac{1}{r}\frac{\partial}{\partial r}\left(r\frac{\partial T}{\partial r}\right) + \frac{\partial^2 T}{\partial z^2} = 0 \tag{1}$$

with the following boundary conditions

$$\text{at}\quad r = 0 \qquad \frac{\partial T}{\partial r} = 0 \tag{2}$$

$$\text{at}\quad r = R \qquad -k\frac{\partial T}{\partial r} = \langle h \rangle (T - T_\infty) \tag{3}$$

$$\text{at}\quad z = 0 \qquad T = T_o \tag{4}$$

$$\text{at}\quad z = L \qquad T = T_L \tag{5}$$

b) The area-averaged temperature is defined by

$$\langle T \rangle = \frac{\displaystyle\int_0^{2\pi}\int_0^R T\,r\,dr\,d\theta}{\displaystyle\int_0^{2\pi}\int_0^R r\,dr\,d\theta} = \frac{2}{R^2}\int_0^R T\,r\,dr \tag{6}$$

Show that the multiplication of Eqs. (1), (4) and (5) by $r\,dr$ and integration from $r = 0$ to $r = R$ give

$$k\frac{d^2\langle T \rangle}{dz^2} - \frac{2\langle h \rangle}{R}\left(T|_{r=R} - T_\infty\right) = 0 \tag{7}$$

$$\text{at} \quad z = 0 \qquad \langle T \rangle = T_o \tag{8}$$

$$\text{at} \quad z = L \qquad \langle T \rangle = T_L \tag{9}$$

c) When $\text{Bi}_\text{H} = \langle h \rangle R / k \ll 1$, $T|_{r=R} \simeq \langle T \rangle$. Under these circumstances, show that Eqs. (7)–(9) become

$$\frac{d^2\theta}{d\xi^2} - \Lambda^2 \theta = 0 \tag{10}$$

$$\text{at} \quad \xi = 0 \qquad \theta = 1 \tag{11}$$

$$\text{at} \quad \xi = 1 \qquad \theta = \theta_L \tag{12}$$

where the dimensionless quantities are defined by

$$\theta = \frac{\langle T \rangle - T_\infty}{T_o - T_\infty} \qquad \theta_L = \frac{T_L - T_\infty}{T_o - T_\infty} \qquad \xi = \frac{z}{L} \qquad \Lambda = \sqrt{\frac{2\langle h \rangle L^2}{kR}} \tag{13}$$

What is the physical significance of Λ?

d) Solve Eq. (10) and show that the temperature distribution within the rod is given by

$$\theta = \frac{\theta_L \sinh(\Lambda \xi) + \sinh\left[\Lambda(1 - \xi)\right]}{\sinh \Lambda} \tag{14}$$

e) Show that the rate of heat loss from the rod to the surrounding fluid is given by

$$\dot{Q} = \frac{2\pi RL \langle h \rangle (T_o - T_\infty)(\cosh \Lambda - 1)(1 + \theta_L)}{\Lambda \sinh \Lambda} \tag{15}$$

f) Calculate the rate of heat loss when $T_o = 120\,°\text{C}$, $T_L = 40\,°\text{C}$, $T_\infty = 25\,°\text{C}$, $\langle h \rangle = 125\ \text{W/m}^2\text{·K}$, $k = 270\ \text{W/m·K}$, $R = 1\ \text{mm}$, and $L = 50\ \text{mm}$.

(**Answer:** f) 1.82 W)

8.14 Consider Problem 8.13 with the following boundary conditions

$$\text{at} \quad z = 0 \qquad -k\frac{d\langle T \rangle}{dz} = q_o \tag{1}$$

$$\text{at} \quad z = L \qquad \langle T \rangle = T_L \tag{2}$$

a) Show that the temperature distribution is given by

$$\theta = \frac{\cosh(\Lambda \xi) + \dfrac{N}{\Lambda} \sinh\left[\Lambda(1 - \xi)\right]}{\cosh \Lambda} \tag{3}$$

where the dimensionless quantities are defined by

$$\theta = \frac{\langle T \rangle - T_\infty}{T_L - T_\infty} \qquad \xi = \frac{z}{L} \qquad \Lambda = \sqrt{\frac{2\langle h \rangle L^2}{kR}} \qquad N = \frac{q_o L}{k(T_L - T_\infty)} \tag{4}$$

b) Show that the rate of heat loss from the rod to the surrounding fluid is given by either

$$\dot{Q} = \frac{2\pi R L \langle h \rangle (T_L - T_\infty)}{\Lambda} \left[\tanh \Lambda + \frac{N}{\Lambda} \left(1 - \frac{1}{\cosh \Lambda} \right) \right] \tag{5}$$

or,

$$\dot{Q} = \pi R^2 \left[q_o + \frac{k(T_L - T_\infty)}{L} \left(\Lambda \tanh \Lambda - \frac{N}{\cosh \Lambda} \right) \right] \tag{6}$$

c) Calculate the rate of heat loss when $q_o = 8 \times 10^4$ W/m^2, $T_L = 40\,°$C, $T_\infty = 25\,°$C, $\langle h \rangle = 125$ W/m^2·K, $k = 270$ W/m·K, $R = 1$ mm, and $L = 50$ mm.

(**Answer:** c) 0.5 W)

8.15 Repeat the analysis given in Section 8.4.4 for a zero-order reaction in the following way:

a) Show that the concentration distribution is given by

$$\theta = 1 + \Lambda^2 \left(\frac{\xi^2}{2} - \xi \right) \tag{1}$$

where the dimensionless quantities are defined by

$$\theta = \frac{\langle c_A \rangle}{c_{A_o}} \qquad \xi = \frac{z}{L} \qquad \Lambda = \sqrt{\frac{2k^s L^2}{R \mathcal{D}_{AB} c_{A_o}}} \tag{2}$$

b) Plot θ versus ξ for $\Lambda = 1$, $\sqrt{2}$, and $\sqrt{3}$. Show why the solution given by Eq. (1) is valid only for $\Lambda \leqslant \sqrt{2}$.

c) For $\Lambda > \sqrt{2}$, only a fraction ϕ ($0 < \phi < 1$) of the surface is available for the chemical reaction. Under these circumstances, show that the concentration distribution is given by

$$\theta = 1 + \Lambda^2 \left(\frac{\xi^2}{2} - \phi \xi \right) \qquad 0 \leqslant \xi \leqslant \phi \tag{3}$$

8.16 Show that the mass average velocity for the Stefan diffusion tube experiment, Example 8.20, is given by

$$v_z = \frac{\mathcal{M}_A \mathcal{D}_{AB}}{\mathcal{M} L} \ln \left(\frac{1}{1 - x_{A_o}} \right)$$

where $\mathcal{M}$ is the molecular weight of the mixture. Note that this result leads to the following interesting conclusions:

i) The mass average velocity is determined on the basis of a solution to a diffusion problem rather than conservation of momentum.

ii) The no-slip boundary condition at the wall of the tube is violated.

For a more thorough analysis of the Stefan diffusion tube problem, see Whitaker (1991).

8.17 Consider diffusion with a heterogeneous chemical reaction as described in Section 8.5.3.

a) Rewrite Eq. (8.5-59) in terms of the dimensionless flux, N_A^*, defined by

$$N_A^* = \frac{N_{A_z}}{ck^s}$$

and calculate its value for $x_{A_o} = 0.7$ and $\Lambda^2 = 6$.

b) Show that the concentration distribution is given by

$$x_A = 2\left[1 - (1 - 0.5x_{A_o})\exp\left(\frac{N_A^*\Lambda^2}{2}\xi\right)\right]$$

where ξ is the dimensionless distance, i.e., $\xi = z/\delta$. Plot x_A versus ξ when $x_{A_o} = 0.7$ and $\Lambda^2 = 6$.

(**Answer:** a) $N_A^* = 0.123$)

8.18 Consider a spherical catalyst particle of radius R over which a first-order heterogeneous reaction

$$A \rightarrow B$$

takes place. The concentration of species A at a distance far from the catalyst particle is c_{A_∞}.

a) Show that the concentration distribution is

$$\frac{c_A}{c_{A_\infty}} = 1 - \left(\frac{\Lambda^2}{1 + \Lambda^2}\right)\frac{R}{r}$$

where Λ is defined by

$$\Lambda = \sqrt{\frac{k^s R}{\mathcal{D}_{AB}}}$$

b) Show that the molar rate of consumption of species A, $\dot{n}_A$, is given by

$$\dot{n}_A = 4\pi\mathcal{D}_{AB}\left(\frac{\Lambda^2}{1 + \Lambda^2}\right)c_{A_\infty}R$$

8.19 Consider a spherical carbon particle of initial radius R_o surrounded by an atmosphere of oxygen. A very rapid heterogeneous reaction

$$2C + O_2 \rightarrow 2CO$$

takes place on the surface of the carbon particle. Show that the time it takes for the carbon particle to disappear completely is

$$t = \frac{1}{48 \ln 2} \frac{R_o^2 \, \rho_C}{c \mathcal{D}_{O_2\text{-}CO}}$$

where ρ_C is the density of carbon.

9

STEADY MICROSCOPIC BALANCES WITH GENERATION

This chapter is the continuation of Chapter 8, with the addition of the generation term to the inventory rate equation. The breakdown of the chapter is the same as that of Chapter 8. Once the governing equations for the velocity, temperature, or concentration are developed, the physical significance of the terms appearing in these equations is explained and the solutions are given in detail. The obtaining of macroscopic level design equations by integrating the microscopic level equations over the volume of the system is also presented.

9.1 MOMENTUM TRANSPORT

For steady transfer of momentum, the inventory rate equation takes the form

$$\begin{pmatrix} \text{Rate of} \\ \text{momentum in} \end{pmatrix} - \begin{pmatrix} \text{Rate of} \\ \text{momentum out} \end{pmatrix} + \begin{pmatrix} \text{Rate of} \\ \text{momentum generation} \end{pmatrix} = 0 \qquad (9.1\text{-}1)$$

In Section 5.1, it was shown that momentum is generated as a result of forces acting on a system, i.e., gravitational and pressure forces. Therefore, Eq. (9.1-1) may also be expressed as

$$\begin{pmatrix} \text{Rate of} \\ \text{momentum in} \end{pmatrix} - \begin{pmatrix} \text{Rate of} \\ \text{momentum out} \end{pmatrix} + \begin{pmatrix} \text{Forces acting} \\ \text{on a system} \end{pmatrix} = 0 \qquad (9.1\text{-}2)$$

As in Chapter 8, our analysis will again be restricted to cases in which the following assumptions hold:

1. Incompressible Newtonian fluid,
2. One-dimensional, fully developed laminar flow,
3. Constant physical properties.

9.1.1 Flow Between Parallel Plates

Consider the flow of a Newtonian fluid between two parallel plates under steady conditions as shown in Figure 9.1. The pressure gradient is imposed in the z-direction while both plates are held stationary.

Velocity components are simplified according to Figure 8.2. Since $v_z = v_z(x)$ and $v_x = v_y = 0$, Table C.1 in Appendix C indicates that the only nonzero shear-stress component is

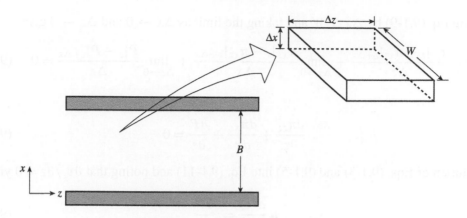

Figure 9.1. Flow between two parallel plates.

τ_{xz}. Hence, the components of the total momentum flux are given by

$$\pi_{xz} = \tau_{xz} + (\rho v_z)\, v_x = \tau_{xz} = -\mu \frac{dv_z}{dx} \tag{9.1-3}$$

$$\pi_{yz} = \tau_{yz} + (\rho v_z)\, v_y = 0 \tag{9.1-4}$$

$$\pi_{zz} = \tau_{zz} + (\rho v_z)\, v_z = \rho v_z^2 \tag{9.1-5}$$

The pressure, on the other hand, may depend on both x and z. Therefore, it is necessary to write the x- and z-components of the equation of motion.

x-component of the equation of motion

For a rectangular differential volume element of thickness Δx, length Δz, and width W, as shown in Figure 9.1, Eq. (9.1-2) is expressed as

$$\left(P|_x - P|_{x+\Delta x}\right) W \Delta z + \rho g W \Delta x \Delta z = 0 \tag{9.1-6}$$

Dividing Eq. (9.1-6) by $W \Delta x \Delta z$ and taking the limit as $\Delta x \to 0$ give

$$\lim_{\Delta x \to 0} \frac{P|_x - P|_{x+\Delta x}}{\Delta x} + \rho g = 0 \tag{9.1-7}$$

or,

$$\frac{\partial P}{\partial x} = \rho g \tag{9.1-8}$$

Note that Eq. (9.1-8) indicates the hydrostatic pressure distribution in the x-direction.

z-component of the equation of motion

Over the differential volume element of thickness Δx, length Δz, and width W, Eq. (9.1-2) takes the form

$$\left(\pi_{zz}|_z W \Delta x + \pi_{xz}|_x W \Delta z\right) - \left(\pi_{zz}|_{z+\Delta z} W \Delta x + \pi_{xz}|_{x+\Delta x} W \Delta z\right)$$

$$+ \left(P|_z - P|_{z+\Delta z}\right) W \Delta x = 0 \tag{9.1-9}$$

Dividing Eq. (9.1-9) by $\Delta x \Delta z W$ and taking the limit as $\Delta x \to 0$ and $\Delta z \to 0$ give

$$\lim_{\Delta z \to 0} \frac{\pi_{zz}|_z - \pi_{zz}|_{z+\Delta z}}{\Delta z} + \lim_{\Delta x \to 0} \frac{\pi_{xz}|_x - \pi_{xz}|_{x+\Delta x}}{\Delta x} + \lim_{\Delta z \to 0} \frac{P|_z - P|_{z+\Delta z}}{\Delta z} = 0 \quad (9.1\text{-}10)$$

or,

$$\frac{\partial \pi_{zz}}{\partial z} + \frac{d\pi_{xz}}{dx} + \frac{\partial P}{\partial z} = 0 \quad (9.1\text{-}11)$$

Substitution of Eqs. (9.1-3) and (9.1-5) into Eq. (9.1-11) and noting that $\partial v_z/\partial z = 0$ yield

$$\mu \underbrace{\frac{d^2 v_z}{dx^2}}_{f(x)} = \underbrace{\frac{\partial P}{\partial z}}_{f(x,z)} \quad (9.1\text{-}12)$$

Since the dependence of P on x is not known, integration of Eq. (9.1-12) with respect to x is not possible at the moment. To circumvent this problem, the effects of the static pressure and the gravitational force are combined in a single term called the *modified pressure*, $\mathcal{P}$. According to Eq. (5.1-16), the modified pressure for this problem is defined as

$$\mathcal{P} = P - \rho g x \quad (9.1\text{-}13)$$

so that

$$\frac{\partial \mathcal{P}}{\partial x} = \frac{\partial P}{\partial x} - \rho g \quad (9.1\text{-}14)$$

and

$$\frac{\partial \mathcal{P}}{\partial z} = \frac{\partial P}{\partial z} \quad (9.1\text{-}15)$$

Combination of Eqs. (9.1-8) and (9.1-14) yields

$$\frac{\partial \mathcal{P}}{\partial x} = 0 \quad (9.1\text{-}16)$$

which implies that $\mathcal{P} = \mathcal{P}(z)$ only. Therefore, the use of Eq. (9.1-15) in Eq. (9.1-12) gives

$$\mu \underbrace{\frac{d^2 v_z}{dx^2}}_{f(x)} = \underbrace{\frac{d\mathcal{P}}{dz}}_{f(z)} \quad (9.1\text{-}17)$$

Note that, while the right-hand side of Eq. (9.1-17) is a function of z only, the left-hand side is dependent only on x. This is possible if and only if both sides of Eq. (9.1-17) are equal to a constant, say λ. Hence,

$$\frac{d\mathcal{P}}{dz} = \lambda \quad \Rightarrow \quad \lambda = -\frac{\mathcal{P}_o - \mathcal{P}_L}{L} \quad (9.1\text{-}18)$$

where $\mathcal{P}_o$ and $\mathcal{P}_L$ are the values of $\mathcal{P}$ at $z = 0$ and $z = L$, respectively. Substitution of Eq. (9.1-18) into Eq. (9.1-17) gives the governing equation for velocity in the form

$$-\mu\frac{d^2 v_z}{dx^2} = \frac{\mathcal{P}_o - \mathcal{P}_L}{L} \qquad (9.1\text{-}19)$$

Integration of Eq. (9.1-19) twice results in

$$v_z = -\frac{\mathcal{P}_o - \mathcal{P}_L}{2\mu L}x^2 + C_1 x + C_2 \qquad (9.1\text{-}20)$$

where C_1 and C_2 are integration constants.

The use of the boundary conditions

$$\text{at} \quad x = 0 \qquad v_z = 0 \qquad (9.1\text{-}21)$$

$$\text{at} \quad x = B \qquad v_z = 0 \qquad (9.1\text{-}22)$$

gives the velocity distribution as

$$v_z = \frac{(\mathcal{P}_o - \mathcal{P}_L)B^2}{2\mu L}\left[\frac{x}{B} - \left(\frac{x}{B}\right)^2\right] \qquad (9.1\text{-}23)$$

The use of the velocity distribution, Eq. (9.1-23), in Eq. (9.1-3) gives the shear stress distribution as

$$\tau_{xz} = \frac{(\mathcal{P}_o - \mathcal{P}_L)B}{2L}\left[2\left(\frac{x}{B}\right) - 1\right] \qquad (9.1\text{-}24)$$

The volumetric flow rate can be determined by integrating the velocity distribution over the cross-sectional area, i.e.,

$$Q = \int_0^W \int_0^B v_z\,dx\,dy \qquad (9.1\text{-}25)$$

Substitution of Eq. (9.1-23) into Eq. (9.1-25) gives the volumetric flow rate in the form

$$Q = \frac{(\mathcal{P}_o - \mathcal{P}_L)WB^3}{12\mu L} \qquad (9.1\text{-}26)$$

Dividing the volumetric flow rate by the flow area gives the average velocity as

$$\langle v_z \rangle = \frac{Q}{WB} = \frac{(\mathcal{P}_o - \mathcal{P}_L)B^2}{12\mu L} \qquad (9.1\text{-}27)$$

9.1.1.1 *Macroscopic balance* Integration of the governing differential equation, Eq. (9.1-19), over the volume of the system gives the macroscopic momentum balance as

$$-\int_0^L \int_0^W \int_0^B \mu \frac{d^2 v_z}{dx^2} \, dx\, dy\, dz = \int_0^L \int_0^W \int_0^B \frac{\mathcal{P}_o - \mathcal{P}_L}{L} \, dx\, dy\, dz \qquad (9.1\text{-}28)$$

or

$$\underbrace{\left(\tau_{xz}|_{x=B} - \tau_{xz}|_{x=0} \right) LW}_{\text{Drag force}} = \underbrace{(\mathcal{P}_o - \mathcal{P}_L)WB}_{\substack{\text{Pressure and gravitational} \\ \text{forces}}} \qquad (9.1\text{-}29)$$

Note that Eq. (9.1-29) is nothing more than Newton's second law of motion. The interaction of the system, i.e., the fluid between the parallel plates, with the surroundings is the drag force, F_D, on the plates and is given by

$$\boxed{F_D = (\mathcal{P}_o - \mathcal{P}_L)WB} \qquad (9.1\text{-}30)$$

On the other hand, the friction factor is the dimensionless interaction of the system with the surroundings and is defined by Eq. (3.1-7), i.e.,

$$F_D = A_{ch} K_{ch} \langle f \rangle \qquad (9.1\text{-}31)$$

or,

$$(\mathcal{P}_o - \mathcal{P}_L)WB = (2WL)\left(\frac{1}{2} \rho \langle v_z \rangle^2 \right)\langle f \rangle \qquad (9.1\text{-}32)$$

Simplification of Eq. (9.1-32) gives

$$\langle f \rangle = \frac{(\mathcal{P}_o - \mathcal{P}_L)B}{\rho L \langle v_z \rangle^2} \qquad (9.1\text{-}33)$$

Elimination of $(\mathcal{P}_o - \mathcal{P}_L)$ between Eqs. (9.1-27) and (9.1-33) leads to

$$\langle f \rangle = 12\left(\frac{\mu}{B \langle v_z \rangle \rho} \right) \qquad (9.1\text{-}34)$$

For flow in noncircular ducts, the Reynolds number based on the hydraulic equivalent diameter was defined in Chapter 4 by Eq. (4.5-37). Since $D_h = 2B$, the Reynolds number is

$$\text{Re}_h = \frac{2B \langle v_z \rangle \rho}{\mu} \qquad (9.1\text{-}35)$$

Therefore, Eq. (9.1-34) takes the final form as

$$\boxed{\langle f \rangle = \frac{24}{\text{Re}_h}} \qquad (9.1\text{-}36)$$

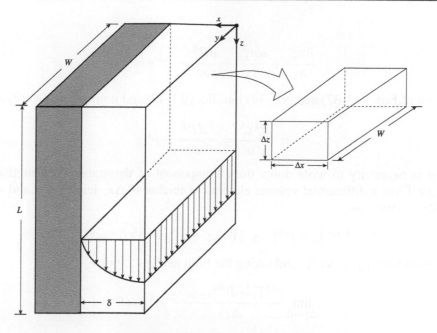

Figure 9.2. Falling film on a vertical plate.

9.1.2   Falling Film on a Vertical Plate

Consider a film of liquid falling down a vertical plate under the action of gravity as shown in Figure 9.2. Since the liquid is in contact with air, it is necessary to consider both phases. Let superscripts L and A represent the liquid and the air, respectively.

For the liquid phase, the velocity components are simplified according to Figure 8.2. Since $v_z = v_z(x)$ and $v_x = v_y = 0$, Table C.1 in Appendix C indicates that the only nonzero shear-stress component is τ_{xz}. Hence, the components of the total momentum flux are given by

$$\pi_{xz}^L = \tau_{xz}^L + \left(\rho^L v_z^L\right)v_x^L = \tau_{xz}^L = -\mu^L \frac{dv_z^L}{dx} \tag{9.1-37}$$

$$\pi_{yz}^L = \tau_{yz}^L + \left(\rho^L v_z^L\right)v_y^L = 0 \tag{9.1-38}$$

$$\pi_{zz}^L = \tau_{zz}^L + \left(\rho^L v_z^L\right)v_z^L = \rho^L \left(v_z^L\right)^2 \tag{9.1-39}$$

The pressure, on the other hand, depends only on z. Therefore, only the z-component of the equation of motion should be considered. For a rectangular differential volume element of thickness Δx, length Δz, and width W, as shown in Figure 9.2, Eq. (9.1-2) is expressed as

$$\left(\pi_{zz}^L|_z W\Delta x + \pi_{xz}^L|_x W\Delta z\right) - \left(\pi_{zz}^L|_{z+\Delta z} W\Delta x + \pi_{xz}^L|_{x+\Delta x} W\Delta z\right)$$
$$+ \left(P^L|_z - P^L|_{z+\Delta z}\right)W\Delta x + \rho^L g W\Delta x\Delta z = 0 \tag{9.1-40}$$

Dividing each term by $W\Delta x\Delta z$ and taking the limit as $\Delta x \to 0$ and $\Delta z \to 0$ give

$$\lim_{\Delta z\to 0} \frac{\pi_{zz}^L|_z - \pi_{zz}^L|_{z+\Delta z}}{\Delta z} + \lim_{\Delta x\to 0} \frac{\pi_{xz}^L|_x - \pi_{xz}^L|_{x+\Delta x}}{\Delta x} + \lim_{\Delta z\to 0} \frac{P^L|_z - P^L|_{z+\Delta z}}{\Delta z} + \rho^L g = 0$$
$$\tag{9.1-41}$$

or,

$$\frac{\partial \pi_{zz}^L}{\partial z} + \frac{d \pi_{xz}^L}{dx} + \frac{\partial P^L}{\partial z} - \rho^L g = 0 \tag{9.1-42}$$

Substitution of Eqs. (9.1-37) and (9.1-39) into Eq. (9.1-42) and noting that $\partial v_z^L / \partial z = 0$ yield

$$-\mu^L \frac{d^2 v_z^L}{dx^2} = -\frac{d P^L}{dz} + \rho^L g \tag{9.1-43}$$

Now, it is necessary to write down the z-component of the equation of motion for the stagnant air. Over a differential volume element of thickness Δx, length Δz, and width W, Eq. (9.1-2) is written as

$$\left(P^A|_z - P^A|_{z+\Delta z}\right) W \Delta x + \rho^A g W \Delta x \Delta z = 0 \tag{9.1-44}$$

Dividing each term by $W \Delta x \Delta z$ and taking the limit as $\Delta z \to 0$ give

$$\lim_{\Delta z \to 0} \frac{P^A|_z - P^A|_{z+\Delta z}}{\Delta z} + \rho^A g = 0 \tag{9.1-45}$$

or,

$$\frac{d P^A}{dz} = \rho^A g \tag{9.1-46}$$

At the liquid-air interface, the jump momentum balance[1] indicates that the normal and tangential components of the total stress tensor are equal to each other, i.e.,

$$\text{at} \quad x = 0 \qquad P^L = P^A \qquad \text{for all } z \tag{9.1-47}$$

$$\text{at} \quad x = 0 \qquad \tau_{xz}^L = \tau_{xz}^A \qquad \text{for all } z \tag{9.1-48}$$

Since both P^L and P^A depend only on z, then

$$\frac{d P^L}{dz} = \frac{d P^A}{dz} \tag{9.1-49}$$

From Eqs. (9.1-46) and (9.1-49), one can conclude that

$$\frac{d P^L}{dz} = \rho^A g \tag{9.1-50}$$

Substitution of Eq. (9.1-50) into Eq. (9.1-43) gives

$$-\mu^L \frac{d^2 v_z^L}{dx^2} = (\rho^L - \rho^A) g \tag{9.1-51}$$

Since $\rho^L \gg \rho^A$, then $\rho^L - \rho^A \approx \rho^L$ and Eq. (9.1-51) takes the form

$$-\mu^L \frac{d^2 v_z^L}{dx^2} = \rho^L g \tag{9.1-52}$$

---

[1] For a thorough discussion on jump balances, see Slattery (1999).

This analysis shows the reason why the pressure term does not appear in the equation of motion when a fluid flows under the action of gravity. This point is usually overlooked in the literature by simply stating that "free surface ⟹ no pressure gradient."

For simplicity, the superscripts in Eq. (9.1-52) will be dropped for the rest of the analysis with the understanding that the properties are those of the liquid. Therefore, the governing equation takes the form

$$-\mu \frac{d^2 v_z}{dx^2} = \rho g \qquad (9.1\text{-}53)$$

Integration of Eq. (9.1-53) twice leads to

$$v_z = -\frac{\rho g}{2\mu} x^2 + C_1 x + C_2 \qquad (9.1\text{-}54)$$

The boundary conditions are

$$\text{at} \quad x = 0 \qquad \frac{dv_z}{dx} = 0 \qquad (9.1\text{-}55)$$

$$\text{at} \quad x = \delta \qquad v_z = 0 \qquad (9.1\text{-}56)$$

Note that Eq. (9.1-55) is a consequence of the equality of shear stresses at the liquid-air interface. Application of the boundary conditions results in

$$v_z = \frac{\rho g \delta^2}{2\mu}\left[1 - \left(\frac{x}{\delta}\right)^2\right] \qquad (9.1\text{-}57)$$

The maximum velocity takes place at the liquid-air interface, i.e., at $x = 0$, as

$$v_{\max} = \frac{\rho g \delta^2}{2\mu} \qquad (9.1\text{-}58)$$

The use of the velocity distribution, Eq. (9.1-57), in Eq. (9.1-37) gives the shear stress distribution as

$$\tau_{xz} = \rho g x \qquad (9.1\text{-}59)$$

Integration of the velocity profile across the flow area gives the volumetric flow rate, i.e.,

$$Q = \int_0^W \int_0^\delta v_z \, dx \, dy \qquad (9.1\text{-}60)$$

Substitution of Eq. (9.1-57) into Eq. (9.1-60) yields

$$Q = \frac{\rho g \delta^3 W}{3\mu} \qquad (9.1\text{-}61)$$

Dividing the volumetric flow rate by the flow area gives the average velocity as

$$\langle v_z \rangle = \frac{Q}{W\delta} = \frac{\rho g \delta^2}{3\mu} \qquad (9.1\text{-}62)$$

9.1.2.1 *Macroscopic balance* Integration of the governing equation, Eq. (9.1-53), over the volume of the system gives the macroscopic equation as

$$-\int_0^L \int_0^W \int_0^\delta \mu \frac{d^2 v_z}{dx^2} \, dx \, dy \, dz = \int_0^L \int_0^W \int_0^\delta \rho g \, dx \, dy \, dz \qquad (9.1\text{-}63)$$

or,

$$\underbrace{\tau_{xz}|_{x=\delta} W L}_{\text{Drag force}} = \underbrace{\rho g \, \delta W L}_{\substack{\text{Mass of the} \\ \text{liquid}}} \qquad (9.1\text{-}64)$$

9.1.3 Flow in a Circular Tube

Consider the flow of a Newtonian fluid in a vertical circular pipe under steady conditions as shown in Figure 9.3. The pressure gradient is imposed in the z-direction.

Simplification of the velocity components according to Figure 8.4 shows that $v_z = v_z(r)$ and $v_r = v_\theta = 0$. Therefore, from Table C.2 in Appendix C, the only nonzero shear stress component is τ_{rz}, and the components of the total momentum flux are given by

$$\pi_{rz} = \tau_{rz} + (\rho v_z) v_r = \tau_{rz} = -\mu \frac{d v_z}{dr} \qquad (9.1\text{-}65)$$

$$\pi_{\theta z} = \tau_{\theta z} + (\rho v_z) v_\theta = 0 \qquad (9.1\text{-}66)$$

$$\pi_{zz} = \tau_{zz} + (\rho v_z) v_z = \rho v_z^2 \qquad (9.1\text{-}67)$$

Since the pressure in the pipe depends on z, it is necessary to consider only the z-component of the equation of motion. For a cylindrical differential volume element of thickness Δr and

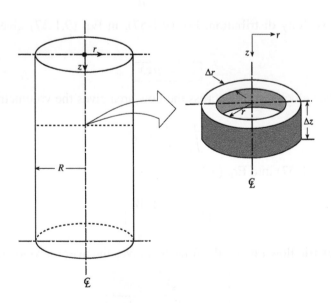

Figure 9.3. Flow in a circular pipe.

length Δz, as shown in Figure 9.3, Eq. (9.1-2) is expressed as

$$\left(\pi_{zz}|_z 2\pi r \Delta r + \pi_{rz}|_r 2\pi r \Delta z\right) - \left[\pi_{zz}|_{z+\Delta z} 2\pi r \Delta r + \pi_{rz}|_{r+\Delta r} 2\pi (r+\Delta r)\Delta z\right]$$

$$+ \left(P|_z - P|_{z+\Delta z}\right) 2\pi r \Delta r + \rho g 2\pi r \Delta r \Delta z = 0 \tag{9.1-68}$$

Dividing Eq. (9.1-68) by $2\pi \Delta r \Delta z$ and taking the limit as $\Delta r \to 0$ and $\Delta z \to 0$ give

$$\lim_{\Delta z \to 0} \frac{\pi_{zz}|_z - \pi_{zz}|_{z+\Delta z}}{\Delta z} + \frac{1}{r} \lim_{\Delta r \to 0} \frac{(r\pi_{rz})|_r - (r\pi_{rz})|_{r+\Delta r}}{\Delta r} + \lim_{\Delta z \to 0} \frac{P|_z - P|_{z+\Delta z}}{\Delta z} + \rho g = 0 \tag{9.1-69}$$

or,

$$\frac{\partial \pi_{zz}}{\partial z} + \frac{1}{r}\frac{d(r\pi_{rz})}{dr} = -\frac{dP}{dz} + \rho g \tag{9.1-70}$$

Substitution of Eqs. (9.1-65) and (9.1-67) into Eq. (9.1-70) and noting that $\partial v_z/\partial z = 0$ give

$$-\frac{\mu}{r}\frac{d}{dr}\left[r\left(\frac{dv_z}{dr}\right)\right] = -\frac{dP}{dz} + \rho g \tag{9.1-71}$$

The modified pressure is defined by

$$\mathcal{P} = P - \rho g z \tag{9.1-72}$$

so that

$$\frac{d\mathcal{P}}{dz} = \frac{dP}{dz} - \rho g \tag{9.1-73}$$

Substitution of Eq. (9.1-73) into Eq. (9.1-71) yields

$$\underbrace{\frac{\mu}{r}\frac{d}{dr}\left[r\left(\frac{dv_z}{dr}\right)\right]}_{f(r)} = \underbrace{\frac{d\mathcal{P}}{dz}}_{f(z)} \tag{9.1-74}$$

Note that, while the right-hand side of Eq. (9.1-74) is a function of z only, the left-hand side is dependent only on r. This is possible if and only if both sides of Eq. (9.1-74) are equal to a constant, say λ. Hence,

$$\frac{d\mathcal{P}}{dz} = \lambda \quad \Rightarrow \quad \lambda = -\frac{\mathcal{P}_o - \mathcal{P}_L}{L} \tag{9.1-75}$$

where $\mathcal{P}_o$ and $\mathcal{P}_L$ are the values of $\mathcal{P}$ at $z = 0$ and $z = L$, respectively. Substitution of Eq. (9.1-75) into Eq. (9.1-74) gives the governing equation for velocity as

$$\boxed{-\frac{\mu}{r}\frac{d}{dr}\left[r\left(\frac{dv_z}{dr}\right)\right] = \frac{\mathcal{P}_o - \mathcal{P}_L}{L}} \tag{9.1-76}$$

Integration of Eq. (9.1-76) twice leads to

$$v_z = -\frac{(\mathcal{P}_o - \mathcal{P}_L)}{4\mu L}r^2 + C_1 \ln r + C_2 \tag{9.1-77}$$

where C_1 and C_2 are integration constants.

The center of the tube, i.e., $r = 0$, is included in the flow domain. However, the presence of the term $\ln r$ makes $v_z \to -\infty$ as $r \to 0$. Therefore, a physically possible solution exists only if $C_1 = 0$. This condition is usually expressed as "v_z is finite at $r = 0$." Alternatively, the use of the symmetry condition, i.e., $dv_z/dr = 0$ at $r = 0$, also leads to $C_1 = 0$. The constant C_2 can be evaluated by using the no-slip boundary condition on the surface of the tube, i.e.,

$$\text{at} \quad r = R \qquad v_z = 0 \tag{9.1-78}$$

so that the velocity distribution becomes

$$\boxed{v_z = \frac{(\mathcal{P}_o - \mathcal{P}_L)R^2}{4\mu L}\left[1 - \left(\frac{r}{R}\right)^2\right]} \tag{9.1-79}$$

The maximum velocity takes place at the center of the tube, i.e.,

$$v_{\max} = \frac{(\mathcal{P}_o - \mathcal{P}_L)R^2}{4\mu L} \tag{9.1-80}$$

The use of Eq. (9.1-79) in Eq. (9.1-65) gives the shear stress distribution as

$$\boxed{\tau_{rz} = \frac{(\mathcal{P}_o - \mathcal{P}_L)r}{2L}} \tag{9.1-81}$$

The volumetric flow rate can be determined by integrating the velocity distribution over the cross-sectional area, i.e.,

$$Q = \int_0^{2\pi} \int_0^R v_z r\, dr\, d\theta \tag{9.1-82}$$

Substitution of Eq. (9.1-79) into Eq. (9.1-82) and integration give

$$\boxed{Q = \frac{\pi(\mathcal{P}_o - \mathcal{P}_L)R^4}{8\mu L}} \tag{9.1-83}$$

which is known as the *Hagen-Poiseuille law*. Dividing the volumetric flow rate by the flow area gives the average velocity as

$$\boxed{\langle v_z \rangle = \frac{Q}{\pi R^2} = \frac{(\mathcal{P}_o - \mathcal{P}_L)R^2}{8\mu L}} \tag{9.1-84}$$

9.1.3.1 *Macroscopic balance*   Integration of the governing differential equation, Eq. (9.1-76), over the volume of the system gives

$$-\int_0^L \int_0^{2\pi} \int_0^R \frac{\mu}{r}\frac{d}{dr}\left[r\left(\frac{dv_z}{dr}\right)\right] r\, dr\, d\theta\, dz = \int_0^L \int_0^{2\pi} \int_0^R \frac{(\mathcal{P}_o - \mathcal{P}_L)}{L} r\, dr\, d\theta\, dz \tag{9.1-85}$$

or,

$$\underbrace{\tau_{rz}|_{r=R} 2\pi RL}_{\text{Drag force}} = \underbrace{\pi R^2 (\mathcal{P}_o - \mathcal{P}_L)}_{\substack{\text{Pressure and gravitational} \\ \text{forces}}} \qquad (9.1\text{-}86)$$

The interaction of the system, i.e., the fluid in the tube, with the surroundings manifests itself as the drag force, F_D, on the wall and is given by

$$\boxed{F_D = \pi R^2 (\mathcal{P}_o - \mathcal{P}_L)} \qquad (9.1\text{-}87)$$

On the other hand, the dimensionless interaction of the system with the surroundings, i.e., the friction factor, is given by Eq. (3.1-7), i.e.,

$$F_D = A_{ch} K_{ch} \langle f \rangle \qquad (9.1\text{-}88)$$

or,

$$\pi R^2 (\mathcal{P}_o - \mathcal{P}_L) = (2\pi RL) \left(\frac{1}{2} \rho \langle v_z \rangle^2 \right) \langle f \rangle \qquad (9.1\text{-}89)$$

Expressing the average velocity in terms of the volumetric flow rate by using Eq. (9.1-84) reduces Eq. (9.1-89) to

$$\langle f \rangle = \frac{\pi^2 D^5 (\mathcal{P}_o - \mathcal{P}_L)}{32 \rho L Q^2} \qquad (9.1\text{-}90)$$

which is nothing more than Eq. (4.5-6).

Elimination of $(\mathcal{P}_o - \mathcal{P}_L)$ between Eqs. (9.1-84) and (9.1-89) leads to

$$\boxed{\langle f \rangle = 16 \left(\frac{\mu}{D \langle v_z \rangle \rho} \right) = \frac{16}{\text{Re}}} \qquad (9.1\text{-}91)$$

9.1.4   Axial Flow in an Annulus

Consider the flow of a Newtonian fluid in a vertical concentric annulus under steady conditions as shown in Figure 9.4. A constant pressure gradient is imposed in the positive z-direction while the inner rod is stationary.

The development of the velocity distribution follows the same lines for flow in a circular tube with the result

$$\boxed{-\frac{\mu}{r} \frac{d}{dr} \left[r \left(\frac{dv_z}{dr} \right) \right] = \frac{\mathcal{P}_o - \mathcal{P}_L}{L}} \qquad (9.1\text{-}92)$$

Integration of Eq. (9.1-92) twice leads to

$$v_z = -\frac{(\mathcal{P}_o - \mathcal{P}_L)}{4\mu L} r^2 + C_1 \ln r + C_2 \qquad (9.1\text{-}93)$$

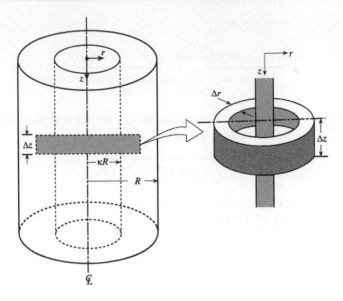

Figure 9.4. Flow in a concentric annulus.

In this case, however, $r = 0$ is not within the flow field. The use of the boundary conditions

$$\text{at} \quad r = R \qquad v_z = 0 \tag{9.1-94}$$

$$\text{at} \quad r = \kappa R \qquad v_z = 0 \tag{9.1-95}$$

gives the velocity distribution as

$$v_z = \frac{(\mathcal{P}_o - \mathcal{P}_L)R^2}{4\mu L}\left[1 - \left(\frac{r}{R}\right)^2 - \left(\frac{1 - \kappa^2}{\ln \kappa}\right)\ln\left(\frac{r}{R}\right)\right] \tag{9.1-96}$$

The use of Eq. (9.1-96) in Eq. (9.1-65) gives the shear stress distribution as

$$\tau_{rz} = \frac{(\mathcal{P}_o - \mathcal{P}_L)R}{2L}\left[\frac{r}{R} + \frac{1 - \kappa^2}{2\ln \kappa}\left(\frac{R}{r}\right)\right] \tag{9.1-97}$$

The volumetric flow rate can be determined by integrating the velocity distribution over the annular cross-sectional area, i.e.,

$$\mathcal{Q} = \int_0^{2\pi}\int_{\kappa R}^R v_z r\, dr\, d\theta \tag{9.1-98}$$

Substitution of Eq. (9.1-96) into Eq. (9.1-98) and integration give

$$\mathcal{Q} = \frac{\pi(\mathcal{P}_o - \mathcal{P}_L)R^4}{8\mu L}\left[1 - \kappa^4 + \frac{(1 - \kappa^2)^2}{\ln \kappa}\right] \tag{9.1-99}$$

Dividing the volumetric flow rate by the flow area gives the average velocity as

$$\langle v_z \rangle = \frac{Q}{\pi R^2 (1 - \kappa^2)} = \frac{(\mathcal{P}_o - \mathcal{P}_L)R^2}{8\mu L}\left(1 + \kappa^2 + \frac{1 - \kappa^2}{\ln \kappa}\right) \qquad (9.1\text{-}100)$$

9.1.4.1 *Macroscopic balance* Integration of the governing differential equation, Eq. (9.1-92), over the volume of the system gives

$$-\int_0^L \int_0^{2\pi} \int_{\kappa R}^R \frac{\mu}{r}\frac{d}{dr}\left[r\left(\frac{dv_z}{dr}\right)\right] r\, dr\, d\theta\, dz = \int_0^L \int_0^{2\pi} \int_{\kappa R}^R \frac{(\mathcal{P}_o - \mathcal{P}_L)}{L} r\, dr\, d\theta\, dz \qquad (9.1\text{-}101)$$

or,

$$\underbrace{\tau_{rz}|_{r=R}2\pi RL - \tau_{rz}|_{r=\kappa R}2\pi \kappa RL}_{\text{Drag force}} = \underbrace{\pi R^2(1 - \kappa^2)(\mathcal{P}_o - \mathcal{P}_L)}_{\substack{\text{Pressure and gravitational} \\ \text{forces}}} \qquad (9.1\text{-}102)$$

Note that Eq. (9.1-102) is nothing more than Newton's second law of motion. The interaction of the system, i.e., the fluid in the concentric annulus, with the surroundings is the drag force, F_D, on the walls and is given by

$$F_D = \pi R^2 (1 - \kappa^2)(\mathcal{P}_o - \mathcal{P}_L) \qquad (9.1\text{-}103)$$

On the other hand, the friction factor is defined by Eq. (3.1-7) as

$$F_D = A_{ch} K_{ch} \langle f \rangle \qquad (9.1\text{-}104)$$

or,

$$\pi R^2 (1 - \kappa^2)(\mathcal{P}_o - \mathcal{P}_L) = \left[2\pi R(1 + \kappa)L\right]\left(\frac{1}{2}\rho \langle v_z \rangle^2\right)\langle f \rangle \qquad (9.1\text{-}105)$$

Elimination of $(\mathcal{P}_o - \mathcal{P}_L)$ between Eqs. (9.1-100) and (9.1-105) gives

$$\langle f \rangle = \frac{8\mu}{R\langle v_z \rangle \rho}\frac{(1 - \kappa)}{\left(1 + \kappa^2 + \dfrac{1 - \kappa^2}{\ln \kappa}\right)} \qquad (9.1\text{-}106)$$

Since $D_h = 2R(1 - \kappa)$, the Reynolds number based on the hydraulic equivalent diameter is

$$\text{Re}_h = \frac{2R(1 - \kappa)\langle v_z \rangle \rho}{\mu} \qquad (9.1\text{-}107)$$

so that Eq. (9.1-106) becomes

$$\langle f \rangle = \frac{16}{\text{Re}_h}\left[\frac{(1 - \kappa)^2}{1 + \kappa^2 + \dfrac{1 - \kappa^2}{\ln \kappa}}\right] \qquad (9.1\text{-}108)$$

9.1.4.2 *Investigation of the limiting cases*

■ **Case** (i) $\kappa \to 1$

When the ratio of the radius of the inner pipe to that of the outer pipe is close to unity, i.e., $\kappa \to 1$, a concentric annulus may be considered a thin-plane slit and its curvature can be neglected. Approximation of a concentric annulus as a parallel plate requires the width, W, and the length, L, of the plate to be defined as

$$W = \pi R(1 + \kappa) \qquad (9.1\text{-}109)$$

$$B = R(1 - \kappa) \qquad (9.1\text{-}110)$$

Therefore, the product WB^3 is equal to

$$WB^3 = \pi R^4 (1 - \kappa^2)(1 - \kappa)^2 \implies \pi R^4 = \frac{WB^3}{(1 - \kappa^2)(1 - \kappa)^2} \qquad (9.1\text{-}111)$$

so that Eq. (9.1-99) becomes

$$\mathcal{Q} = \frac{(\mathcal{P}_o - \mathcal{P}_L)WB^3}{8\mu L} \lim_{\kappa \to 1} \left[\frac{1 + \kappa^2}{(1 - \kappa)^2} + \frac{1 + \kappa}{(1 - \kappa)\ln\kappa} \right] \qquad (9.1\text{-}112)$$

Substitution of $\psi = 1 - \kappa$ into Eq. (9.1-112) gives

$$\mathcal{Q} = \frac{(\mathcal{P}_o - \mathcal{P}_L)WB^3}{8\mu L} \lim_{\psi \to 0} \left[\frac{\psi^2 - 2\psi + 2}{\psi^2} + \frac{2 - \psi}{\psi \ln(1 - \psi)} \right] \qquad (9.1\text{-}113)$$

The Taylor series expansion of the term $\ln(1 - \psi)$ is

$$\ln(1 - \psi) = -\psi - \frac{1}{2}\psi^2 - \frac{1}{3}\psi^3 - \cdots \qquad (9.1\text{-}114)$$

Using Eq. (9.1-114) in Eq. (9.1-113) and carrying out the divisions yield

$$\mathcal{Q} = \frac{(\mathcal{P}_o - \mathcal{P}_L)WB^3}{8\mu L} \lim_{\psi \to 0} \left[1 - \frac{2}{\psi} + \frac{2}{\psi^2} + \left(-\frac{2}{\psi^2} + \frac{2}{\psi} - \frac{1}{3} - \frac{\psi}{2} + \cdots \right) \right] \qquad (9.1\text{-}115)$$

or,

$$\mathcal{Q} = \frac{(\mathcal{P}_o - \mathcal{P}_L)WB^3}{8\mu L} \lim_{\psi \to 0} \left(\frac{2}{3} - \frac{\psi}{2} + \cdots \right) = \frac{(\mathcal{P}_o - \mathcal{P}_L)WB^3}{12\mu L} \qquad (9.1\text{-}116)$$

which is equivalent to Eq. (9.1-26).

■ **Case** (ii) $\kappa \to 0$

When the ratio of the radius of the inner pipe to that of the outer pipe is close to zero, i.e., $\kappa \to 0$, a concentric annulus may be considered a circular pipe of radius R. In this case, Eq. (9.1-99) becomes

$$\mathcal{Q} = \frac{\pi (\mathcal{P}_o - \mathcal{P}_L) R^4}{8\mu L} \lim_{\kappa \to 0} \left[1 - \kappa^4 + \frac{(1 - \kappa^2)^2}{\ln\kappa} \right] \qquad (9.1\text{-}117)$$

Since $\ln 0 = -\infty$, Eq. (9.1-117) reduces to

$$Q = \frac{\pi (\mathcal{P}_o - \mathcal{P}_L) R^4}{8 \mu L} \tag{9.1-118}$$

which is identical to Eq. (9.1-83).

9.1.5   Physical Significance of the Reynolds Number

The physical significance attributed to the Reynolds number for both laminar and turbulent flows is that it is the ratio of the inertial forces to the viscous forces. However, examination of the governing equations for fully developed laminar flow: (*i*) between parallel plates, Eq. (9.1-19), (*ii*) in a circular pipe, Eq. (9.1-76), and (*iii*) in a concentric annulus, Eq. (9.1-92), indicates that the only forces present are the pressure and the viscous forces. Inertial forces do not exist in these problems. Since both pressure and viscous forces are kept in the governing equation for velocity, they must, more or less, have the same order of magnitude. Therefore, the ratio of pressure to viscous forces, which is a dimensionless number, has an order of magnitude of unity.

On the other hand, the use of the $\frac{1}{2}(\rho \langle v_z \rangle^2)$ term instead of pressure is not appropriate since this term comes from the Bernoulli equation, which is developed for no-friction (or reversible) flows.

Therefore, in the case of a fully developed laminar flow, attributing a physical significance of "inertial force/viscous force" to the Reynolds number is not correct. A more appropriate approach may be given in terms of the time scales discussed in Section 3.4.1. For the flow of a liquid through a circular pipe of length L with an average velocity of $\langle v_z \rangle$, the convective time scale for momentum transport is the mean residence time, i.e.,

$$(t_{ch})_{conv} = \frac{L}{\langle v_z \rangle} \tag{9.1-119}$$

On the other hand, the viscous time scale is given by

$$(t_{ch})_{mol} = \frac{L^2}{\nu} \tag{9.1-120}$$

Therefore, the Reynolds number is given by

$$\text{Re} = \frac{\text{Viscous time scale}}{\text{Convective time scale for momentum transport}} = \frac{L \langle v_z \rangle}{\nu} \tag{9.1-121}$$

For a more thorough discussion on the subject, see Bejan (1984).

9.2   ENERGY TRANSPORT WITHOUT CONVECTION

For steady transport of energy, the inventory rate equation takes the form

$$\begin{pmatrix} \text{Rate of} \\ \text{energy in} \end{pmatrix} - \begin{pmatrix} \text{Rate of} \\ \text{energy out} \end{pmatrix} + \begin{pmatrix} \text{Rate of} \\ \text{energy generation} \end{pmatrix} = 0 \tag{9.2-1}$$

As stated in Section 5.2, generation of energy may occur as a result of chemical and nuclear reactions, absorption radiation, presence of magnetic fields, and viscous dissipation. It is of industrial importance to know the temperature distribution resulting from the internal generation of energy because exceeding of the maximum allowable temperature may lead to deterioration of the material of construction.

9.2.1 Conduction in Rectangular Coordinates

Consider one-dimensional transfer of energy in the z-direction through a plane wall of thickness L and surface area A as shown in Figure 9.5. Let $\Re$ be the position-dependent rate of energy generation per unit volume within the wall.

Since $T = T(z)$, Table C.4 in Appendix C indicates that the only nonzero energy flux component is e_z, and it is given by

$$e_z = q_z = -k \frac{dT}{dz} \tag{9.2-2}$$

For a rectangular volume element of thickness Δz as shown in Figure 9.5, Eq. (9.2-1) is expressed as

$$q_z|_z A - q_z|_{z+\Delta z} A + \Re A \Delta z = 0 \tag{9.2-3}$$

Dividing each term by $A \Delta z$ and taking the limit as $\Delta z \to 0$ give

$$\lim_{\Delta z \to 0} \frac{q_z|_z - q_z|_{z+\Delta z}}{\Delta z} + \Re = 0 \tag{9.2-4}$$

or,

$$\frac{dq_z}{dz} = \Re \tag{9.2-5}$$

Substitution of Eq. (9.2-2) into Eq. (9.2-5) gives the governing equation for temperature as

$$\boxed{-\frac{d}{dz}\left(k \frac{dT}{dz}\right) = \Re} \tag{9.2-6}$$

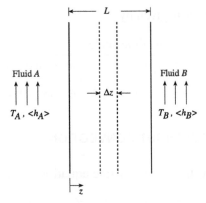

Figure 9.5. Conduction through a plane wall with generation.

Integration of Eq. (9.2-6) gives

$$k\frac{dT}{dz} = -\int_0^z \Re(u)\,du + C_1 \tag{9.2-7}$$

where u is a dummy variable of integration and C_1 is an integration constant. Integration of Eq. (9.2-7) once more leads to

$$\boxed{\int_0^T k(T)\,dT = -\int_0^z \left[\int_0^z \Re(u)\,du\right]dz + C_1 z + C_2} \tag{9.2-8}$$

Evaluation of the constants C_1 and C_2 requires the boundary conditions to be specified. The solution of Eq. (9.2-8) will be presented for two types of boundary conditions, namely, Type I and Type II. In the case of the Type I boundary condition, the temperatures at both surfaces are specified. On the other hand, the Type II boundary condition implies that while the temperature is specified at one of the surfaces the other surface is subjected to a constant wall heat flux.

Type I boundary condition

The solution of Eq. (9.2-8) subject to the boundary conditions

$$\text{at} \quad z = 0 \qquad T = T_o \tag{9.2-9a}$$

$$\text{at} \quad z = L \qquad T = T_L \tag{9.2-9b}$$

is given by

$$\int_{T_o}^T k(T)\,dT = -\int_0^z \left[\int_0^z \Re(u)\,du\right]dz + \left\{\int_{T_o}^{T_L} k(T)\,dT + \int_0^L \left[\int_0^z \Re(u)\,du\right]dz\right\}\frac{z}{L} \tag{9.2-10}$$

Note that, when $\Re = 0$, Eq. (9.2-10) reduces to Eq. (G) in Table 8.1. Equation (9.2-10) may be further simplified depending on whether the thermal conductivity and/or energy generation per unit volume are constant.

■ **Case** (i) $k = $ **constant**

In this case, Eq. (9.2-10) reduces to

$$k(T - T_o) = -\int_0^z \left[\int_0^z \Re(u)\,du\right]dz + \left\{k(T_L - T_o) + \int_0^L \left[\int_0^z \Re(u)\,du\right]dz\right\}\frac{z}{L} \tag{9.2-11}$$

When $\Re = 0$, Eq. (9.2-11) reduces to Eq. (H) in Table 8.1.

■ **Case** (ii) $k = $ **constant;** $\Re = $ **constant**

In this case, Eq. (9.2-10) simplifies to

$$T = T_o + \frac{\Re L^2}{2k}\left[\frac{z}{L} - \left(\frac{z}{L}\right)^2\right] - (T_o - T_L)\frac{z}{L} \tag{9.2-12}$$

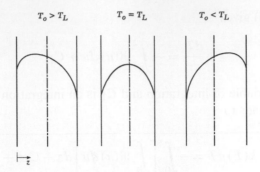

Figure 9.6. Representative temperature distributions in a rectangular wall with constant generation.

The location of the maximum temperature can be obtained from $dT/dz = 0$ as

$$\left(\frac{z}{L}\right)_{T=T_{\max}} = \frac{1}{2} - \frac{k}{\Re L^2}(T_o - T_L) \tag{9.2-13}$$

Substitution of Eq. (9.2-13) into Eq. (9.2-12) gives the value of the maximum temperature as

$$T_{\max} = \frac{T_o + T_L}{2} + \frac{\Re L^2}{8k} + \frac{k(T_o - T_L)^2}{2\Re L^2} \tag{9.2-14}$$

The representative temperature profiles depending on the values of T_o and T_L are shown in Figure 9.6.

Type II boundary condition

The solution of Eq. (9.2-8) subject to the boundary conditions

$$\text{at} \quad z = 0 \qquad -k\frac{dT}{dz} = q_o \tag{9.2-15a}$$

$$\text{at} \quad z = L \qquad T = T_L \tag{9.2-15b}$$

is given by

$$\int_{T_L}^{T} k(T)\,dT = \int_{z}^{L}\left[\int_{0}^{z}\Re(u)\,du\right]dz + q_o L\left(1 - \frac{z}{L}\right) \tag{9.2-16}$$

When $\Re = 0$, Eq. (9.2-16) reduces to Eq. (G) in Table 8.2. Further simplifications of Eq. (9.2-16) depending on whether k and/or $\Re$ are constant are given below.

■ Case (*i*) $k =$ constant

In this case, Eq. (9.2-16) reduces to

$$k(T - T_L) = \int_{z}^{L}\left[\int_{0}^{z}\Re(u)\,du\right]dz + q_o L\left(1 - \frac{z}{L}\right) \tag{9.2-17}$$

When $\Re = 0$, Eq. (9.2-17) reduces to Eq. (H) in Table 8.2.

■ **Case** (*ii*) $k = $ **constant;** $\Re = $ **constant**

In this case, Eq. (9.2-16) reduces to

$$T = T_L + \frac{\Re L^2}{2k}\left[1 - \left(\frac{z}{L}\right)^2\right] + \frac{q_o L}{k}\left(1 - \frac{z}{L}\right) \qquad (9.2\text{-}18)$$

9.2.1.1 *Macroscopic equation* The integration of the governing equation, Eq. (9.2-6), over the volume of the system gives

$$-\int_0^L \int_0^W \int_0^H \frac{d}{dz}\left(k\frac{dT}{dz}\right)dx\,dy\,dz = \int_0^L \int_0^W \int_0^H \Re\,dx\,dy\,dz \qquad (9.2\text{-}19)$$

Integration of Eq. (9.2-19) yields

$$\underbrace{WH\left[\left(-k\frac{dT}{dz}\right)_{z=L} + \left(k\frac{dT}{dz}\right)_{z=0}\right]}_{\text{Net rate of energy out}} = \underbrace{WH\int_0^L \Re\,dz}_{\substack{\text{Rate of energy} \\ \text{generation}}} \qquad (9.2\text{-}20)$$

which is simply the macroscopic energy balance under steady conditions by considering the plane wall as a system. Note that energy must leave the system from at least one of the surfaces to maintain steady conditions. The "net rate of energy out" in Eq. (9.2-20) implies that the rate of energy leaving the system is in excess of the rate of energy entering it.

It is also possible to make use of Newton's law of cooling to express the rate of heat loss from the system. If heat is lost from both surfaces to the surroundings, Eq. (9.2-20) can be written as

$$\langle h_A \rangle (T_o - T_A) + \langle h_B \rangle (T_L - T_B) = \int_0^L \Re\,dz \qquad (9.2\text{-}21)$$

where T_o and T_L are the surface temperatures at $z = 0$ and $z = L$, respectively.

Example 9.1 Energy generation rate as a result of an exothermic reaction is 1×10^4 W/m³ in a 50 cm thick wall of thermal conductivity 20 W/m·K. The left face of the wall is insulated while the right side is held at 45 °C by a coolant. Calculate the maximum temperature in the wall under steady conditions.

Solution

Let z be the distance measured from the left face. The use of Eq. (9.2-18) with $q_o = 0$ gives the temperature distribution as

$$T = T_L + \frac{\Re L^2}{2k}\left[1 - \left(\frac{z}{L}\right)^2\right] = 45 + \frac{(1 \times 10^4)(0.5)^2}{2(20)}\left[1 - \left(\frac{z}{0.5}\right)^2\right] \qquad (1)$$

Simplification of Eq. (1) leads to

$$T = 107.5 - 250z^2 \qquad (2)$$

Since $dT/dz = 0$ at $z = 0$, the maximum temperature occurs at the insulated surface and its value is 107.5 °C.

Example 9.2 Consider a composite solid of materials A and B, shown in the figure below. An electrical resistance heater embedded in solid B generates heat at a constant volumetric rate of $\Re$ (W/m^3). The composite solid is cooled from both sides to avoid excessive heating.

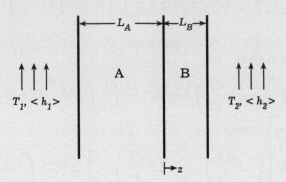

a) Obtain expressions for the steady temperature distributions in solids A and B.
b) Calculate the rate of heat loss from the surfaces located at $z = -L_A$ and $z = L_B$.
c) For the following numerical values

$$T_1 = -5\,^\circ C \quad T_2 = 25\,^\circ C \quad \langle h_1 \rangle = 500 \text{ W/m}^2 \cdot \text{K} \quad \langle h_2 \rangle = 10 \text{ W/m}^2 \cdot \text{K}$$

$$k_A = 180 \text{ W/m} \cdot \text{K} \quad k_B = 1.2 \text{ W/m} \cdot \text{K} \quad L_A = 36 \text{ cm} \quad L_B = 3 \text{ cm}$$

calculate the value of $\Re$ to keep the surface temperature of the wall at $z = -L_A$ constant at 15 °C.
d) Obtain the temperature distribution in solid A when the thickness of solid B is very small, and draw the electrical analog. A practical application of this case is the use of a surface heater, i.e., a very thin plastic film containing electrical resistance, to clear condensation and ice from the rear window of your car or condensation from the mirror in your bathroom.

Solution

a) Since area is constant, the governing equation for temperature in solid A can be easily obtained from Eq. (8.2-5) as

$$\frac{dq_z^A}{dz} = 0 \quad \Rightarrow \quad \frac{d^2 T_A}{dz^2} = 0 \tag{1}$$

The solution of Eq. (1) gives

$$T_A = C_1 z + C_2 \tag{2}$$

The governing equation for temperature in solid B is obtained from Eqs. (9.2-5) and (9.2-6) as

$$-\frac{dq_z^B}{dz} + \Re = 0 \quad \Rightarrow \quad \frac{d^2 T_B}{dz^2} = -\frac{\Re}{k_B} \tag{3}$$

The solution of Eq. (3) yields

$$T_B = -\frac{\Re}{2k_B} z^2 + C_3 z + C_4 \tag{4}$$

Evaluation of the constants C_1, C_2, C_3, and C_4 requires four boundary conditions. They are expressed as

$$\text{at} \quad z = -L_A \qquad k_A \frac{dT_A}{dz} = \langle h_1 \rangle (T_A - T_1) \tag{5}$$

$$\text{at} \quad z = L_B \qquad -k_B \frac{dT_B}{dz} = \langle h_2 \rangle (T_B - T_2) \tag{6}$$

$$\text{at} \quad z = 0 \qquad T_A = T_B \tag{7}$$

$$\text{at} \quad z = 0 \qquad k_A \frac{dT_A}{dz} = k_B \frac{dT_B}{dz} \tag{8}$$

Application of the boundary conditions leads to the following temperature distributions within solids A and B

$$T_A = T_1 + \left[\frac{\Re L_B \left(\dfrac{1}{\langle h_2 \rangle} + \dfrac{L_B}{2k_B} \right) + T_2 - T_1}{k_A \left(\dfrac{1}{\langle h_1 \rangle} + \dfrac{L_A}{k_A} + \dfrac{L_B}{k_B} + \dfrac{1}{\langle h_2 \rangle} \right)} \right] \left(z + \frac{k_A}{\langle h_1 \rangle} + L_A \right) \tag{9}$$

$$T_B = T_1 - \frac{\Re}{2k_B} z^2 + \left[\frac{\Re L_B \left(\dfrac{1}{\langle h_2 \rangle} + \dfrac{L_B}{2k_B} \right) + T_2 - T_1}{k_A \left(\dfrac{1}{\langle h_1 \rangle} + \dfrac{L_A}{k_A} + \dfrac{L_B}{k_B} + \dfrac{1}{\langle h_2 \rangle} \right)} \right] \left(\frac{k_A}{k_B} z + \frac{k_A}{\langle h_1 \rangle} + L_A \right) \tag{10}$$

b) The rate of heat transfer per unit area through the surface at $z = -L_A$ is given by

$$\frac{\dot{Q}|_{z=-L_A}}{A} = k_A \frac{dT_A}{dz} \bigg|_{z=-L_A} = \frac{\Re L_B \left(\dfrac{1}{\langle h_2 \rangle} + \dfrac{L_B}{2k_B} \right) + T_2 - T_1}{\dfrac{1}{\langle h_1 \rangle} + \dfrac{L_A}{k_A} + \dfrac{L_B}{k_B} + \dfrac{1}{\langle h_2 \rangle}} \tag{11}$$

On the other hand, the rate of heat transfer per unit area through the surface at $z = L_B$ is given by

$$\frac{\dot{Q}|_{z=L_B}}{A} = -k_B \frac{dT_B}{dz} \bigg|_{z=L_B} = \Re L_B - \frac{\Re L_B \left(\dfrac{1}{\langle h_2 \rangle} + \dfrac{L_B}{2k_B} \right) + T_2 - T_1}{\dfrac{1}{\langle h_1 \rangle} + \dfrac{L_A}{k_A} + \dfrac{L_B}{k_B} + \dfrac{1}{\langle h_2 \rangle}} \tag{12}$$

Note that the addition of Eqs. (11) and (12) results in

$$\underbrace{\dot{Q}|_{z=-L_A} + \dot{Q}|_{z=L_B}}_{\text{Rate of energy out}} = \underbrace{\Re A L_B}_{\substack{\text{Rate of energy} \\ \text{generation}}} \tag{13}$$

which is nothing more than the steady-state macroscopic energy balance by considering a composite solid as a system.

c) Evaluation of Eq. (9) at $z = -L_A$ leads to

$$T_A|_{z=-L_A} = T_1 + \left[\frac{\Re L_B \left(\dfrac{1}{\langle h_2 \rangle} + \dfrac{L_B}{2k_B} \right) + T_2 - T_1}{\langle h_1 \rangle \left(\dfrac{1}{\langle h_1 \rangle} + \dfrac{L_A}{k_A} + \dfrac{L_B}{k_B} + \dfrac{1}{\langle h_2 \rangle} \right)} \right] \tag{14}$$

Solving Eq. (14) for $\Re$ leads to

$$\Re = \frac{(T_A|_{z=-L_A} - T_1)\langle h_1 \rangle \left(\dfrac{1}{\langle h_1 \rangle} + \dfrac{L_A}{k_A} + \dfrac{L_B}{k_B} + \dfrac{1}{\langle h_2 \rangle} \right) + T_1 - T_2}{L_B \left(\dfrac{1}{\langle h_2 \rangle} + \dfrac{L_B}{2k_B} \right)}$$

$$= \frac{(15+5)(500) \left(\dfrac{1}{500} + \dfrac{0.36}{180} + \dfrac{0.03}{1.2} + \dfrac{1}{10} \right) - 5 - 25}{0.03 \left[\dfrac{1}{10} + \dfrac{0.03}{2(1.2)} \right]} = 3.73 \times 10^5 \text{ W/m}^3$$

d) When the thickness of solid B is very small, then it is possible to assume that the temperature in solid B is constant and equal to the temperature in solid A at $z = 0$. Moreover, the heat generation is expressed in terms of the heat generation rate per unit area, i.e., $\overline{\Re} = \Re L_B$. Thus, Eq. (9) becomes

$$T_A = T_1 + \left[\frac{\overline{\Re} + \langle h_2 \rangle (T_2 - T_1)}{k_A \dfrac{\langle h_2 \rangle}{\langle h_1 \rangle} + L_A \langle h_2 \rangle + k_A} \right] \left(z + \frac{k_A}{\langle h_1 \rangle} + L_A \right) \tag{15}$$

The electrical circuit analog of this case is shown in the figure below:

Comment: When Eq. (3) is integrated in the z-direction, the result is

$$\int_0^{L_B} k_B \frac{d^2 T_B}{dz^2} \, dz + \int_0^{L_B} \Re \, dz = 0 \tag{16}$$

or,

$$\underbrace{k_B \frac{dT_B}{dz}\bigg|_{z=L_B}}_{-\langle h_2\rangle(T_A|_{z=0} - T_2)} \underbrace{- k_B \frac{dT_B}{dz}\bigg|_{z=0}}_{k_A \frac{dT_A}{dz}\bigg|_{z=0}} + \underbrace{\Re L_B}_{\overline{\Re}} = 0 \tag{17}$$

Thus, the solution of Eq. (1) with the following boundary conditions

$$\text{at} \quad z = -L_A \qquad k_A \frac{dT_A}{dz} = \langle h_1\rangle(T_A - T_1) \tag{18}$$

$$\text{at} \quad z = 0 \qquad -k_A \frac{dT_A}{dz} + \overline{\Re} = \langle h_2\rangle(T_A - T_2) \tag{19}$$

also results in Eq. (15).

9.2.2 Conduction in Cylindrical Coordinates

9.2.2.1 *Hollow cylinder*   Consider one-dimensional transfer of energy in the r-direction through a hollow cylinder of inner and outer radii of R_1 and R_2, respectively, as shown in Figure 9.7. Let $\Re$ be the rate of energy generation per unit volume within the cylinder.

Since $T = T(r)$, Table C.5 in Appendix C indicates that the only nonzero energy flux component is e_r, and it is given by

$$e_r = q_r = -k \frac{dT}{dr} \tag{9.2-22}$$

For a cylindrical differential volume element of thickness Δr as shown in Figure 9.7, the inventory rate equation for energy, Eq. (9.2-1), is expressed as

$$2\pi L (r q_r)|_r - 2\pi L (r q_r)|_{r+\Delta r} + 2\pi r \Delta r L \, \Re = 0 \tag{9.2-23}$$

Dividing each term by $2\pi L \Delta r$ and taking the limit as $\Delta r \to 0$ give

$$\lim_{\Delta r \to 0} \frac{(r q_r)|_r - (r q_r)|_{r+\Delta r}}{\Delta r} + r \, \Re = 0 \tag{9.2-24}$$

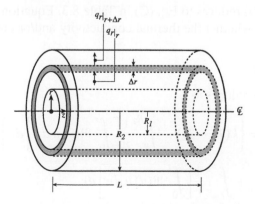

Figure 9.7. One-dimensional conduction through a hollow cylinder with internal generation.

or,

$$\frac{1}{r}\frac{d}{dr}(rq_r) = \Re \tag{9.2-25}$$

Substitution of Eq. (9.2-22) into Eq. (9.2-25) gives the governing equation for temperature as

$$\boxed{-\frac{1}{r}\frac{d}{dr}\left(rk\frac{dT}{dr}\right) = \Re} \tag{9.2-26}$$

Integration of Eq. (9.2-26) gives

$$k\frac{dT}{dr} = -\frac{1}{r}\int_0^r \Re(u)\,u\,du + \frac{C_1}{r} \tag{9.2-27}$$

where u is a dummy variable of integration and C_1 is an integration constant. Integration of Eq. (9.2-27) once more leads to

$$\boxed{\int_0^T k(T)\,dT = -\int_0^r \frac{1}{r}\left[\int_0^r \Re(u)\,u\,du\right]dr + C_1\ln r + C_2} \tag{9.2-28}$$

Evaluation of the constants C_1 and C_2 requires the boundary conditions to be specified.

Type I boundary condition

The solution of Eq. (9.2-28) subject to the boundary conditions

$$\text{at} \quad r = R_1 \qquad T = T_1 \tag{9.2-29a}$$

$$\text{at} \quad r = R_2 \qquad T = T_2 \tag{9.2-29b}$$

is given by

$$\int_{T_2}^T k(T)\,dT = \left\{\int_{T_2}^{T_1} k(T)\,dT - \int_{R_1}^{R_2}\frac{1}{r}\left[\int_0^r \Re(u)\,u\,du\right]dr\right\}\frac{\ln(r/R_2)}{\ln(R_1/R_2)}$$

$$+ \int_r^{R_2}\frac{1}{r}\left[\int_0^r \Re(u)u\,du\right]dr \tag{9.2-30}$$

When $\Re = 0$, Eq. (9.2-30) reduces to Eq. (C) in Table 8.3. Equation (9.2-30) may be further simplified depending on whether the thermal conductivity and/or energy generation per unit volume are constant.

■ **Case** (*i*) $k = $ **constant**

In this case, Eq. (9.2-30) reduces to

$$k(T - T_2) = \left\{k(T_1 - T_2) - \int_{R_1}^{R_2}\frac{1}{r}\left[\int_0^r \Re(u)\,u\,du\right]dr\right\}\frac{\ln(r/R_2)}{\ln(R_1/R_2)}$$

$$+ \int_r^{R_2}\frac{1}{r}\left[\int_0^r \Re(u)\,u\,du\right]dr \tag{9.2-31}$$

When $\Re = 0$, Eq. (9.2-31) simplifies to Eq. (D) in Table 8.3.

■ **Case** (ii) $k =$ **constant;** $\Re =$ **constant**

In this case, Eq. (9.2-30) reduces to

$$T = T_2 + \frac{\Re R_2^2}{4k}\left[1 - \left(\frac{r}{R_2}\right)^2\right] + \left\{T_1 - T_2 - \frac{\Re R_2^2}{4k}\left[1 - \left(\frac{R_1}{R_2}\right)^2\right]\right\}\frac{\ln(r/R_2)}{\ln(R_1/R_2)} \quad (9.2\text{-}32)$$

The location of maximum temperature can be obtained from $dT/dr = 0$ as

$$\left(\frac{r}{R_2}\right)_{T=T_{\max}} = \left\{\frac{\dfrac{2k(T_1 - T_2)}{\Re R_2^2} - \dfrac{1}{2}\left[1 - \left(\dfrac{R_1}{R_2}\right)^2\right]}{\ln\left(\dfrac{R_1}{R_2}\right)}\right\}^{1/2} \quad (9.2\text{-}33)$$

Type II boundary condition

The solution of Eq. (9.2-28) subject to the boundary conditions

$$\text{at} \quad r = R_1 \quad -k\frac{dT}{dz} = q_1 \quad (9.2\text{-}34a)$$

$$\text{at} \quad r = R_2 \quad T = T_2 \quad (9.2\text{-}34b)$$

is given by

$$\int_{T_2}^{T} k(T)\, dT = \int_{r}^{R_2}\frac{1}{r}\left[\int_0^r \Re(u)\, u\, du\right]dr + \left[\int_0^{R_1}\Re(u)\, u\, du - q_1 R_1\right]\ln\left(\frac{r}{R_2}\right) \quad (9.2\text{-}35)$$

When $\Re = 0$, Eq. (9.2-35) reduces to Eq. (C) in Table 8.4.

■ **Case** (i) $k =$ **constant**

In this case, Eq. (9.2-35) reduces to

$$k(T - T_2) = \int_{r}^{R_2}\frac{1}{r}\left[\int_0^r \Re(u)u\, du\right]dr + \left[\int_0^{R_1}\Re(u)u\, du - q_1 R_1\right]\ln\left(\frac{r}{R_2}\right) \quad (9.2\text{-}36)$$

When $\Re = 0$, Eq. (9.2-36) simplifies to Eq. (D) in Table 8.4.

■ **Case** (ii) $k =$ **constant;** $\Re =$ **constant**

In this case, Eq. (9.2-35) simplifies to

$$T = T_2 + \frac{\Re R_2^2}{4k}\left[1 - \left(\frac{r}{R_2}\right)^2\right] + \left(\frac{\Re R_1^2}{2k} - \frac{q_1 R_1}{k}\right)\ln\left(\frac{r}{R_2}\right) \quad (9.2\text{-}37)$$

Macroscopic equation

The integration of the governing equation, Eq. (9.2-26), over the volume of the system gives

$$-\int_0^L \int_0^{2\pi} \int_{R_1}^{R_2}\frac{1}{r}\frac{d}{dr}\left(rk\frac{dT}{dr}\right)r\, dr\, d\theta\, dz = \int_0^L \int_0^{2\pi} \int_{R_1}^{R_2}\Re\, r\, dr\, d\theta\, dz \quad (9.2\text{-}38)$$

Integration of Eq. (9.2-38) yields

$$
\underbrace{\left(-k\frac{dT}{dr}\right)_{r=R_2} 2\pi R_2 L + \left(k\frac{dT}{dr}\right)_{r=R_1} 2\pi R_1 L}_{\text{Net rate of energy out}} = \underbrace{2\pi L \int_{R_1}^{R_2} \Re\, r\, dr}_{\substack{\text{Rate of energy}\\ \text{generation}}} \qquad (9.2\text{-}39)
$$

which is the macroscopic energy balance under steady conditions by considering the hollow cylinder as a system.

It is also possible to make use of Newton's law of cooling to express the rate of heat loss from the system. If heat is lost from both surfaces to the surroundings, Eq. (9.2-39) can be written as

$$
R_1 \langle h_A \rangle (T_1 - T_A) + R_2 \langle h_B \rangle (T_2 - T_B) = \int_{R_1}^{R_2} \Re\, r\, dr \qquad (9.2\text{-}40)
$$

where T_1 and T_2 are the surface temperatures at $r = R_1$ and $r = R_2$, respectively.

Example 9.3 A catalytic reaction is being carried out in a packed bed in the annular space between two concentric cylinders with inner radius $R_1 = 1.5$ cm and outer radius $R_2 = 1.8$ cm. The entire surface of the inner cylinder is insulated. The rate of generation of energy per unit volume as a result of a chemical reaction is 5×10^6 W/m^3 and it is uniform throughout the annular reactor. The effective thermal conductivity of the bed is 0.5 W/m·K. If the inner surface temperature is measured as 280 °C, calculate the temperature of the outer surface.

Solution

The temperature distribution is given by Eq. (9.2-37). Since $q_1 = 0$, it reduces to

$$
T = T_2 + \frac{\Re R_2^2}{4k}\left[1 - \left(\frac{r}{R_2}\right)^2\right] + \frac{\Re R_1^2}{2k}\ln\left(\frac{r}{R_2}\right) \qquad (1)
$$

The temperature, T_1, at $r = R_1$ is given by

$$
T_1 = T_2 + \frac{\Re R_2^2}{4k}\left[1 - \left(\frac{R_1}{R_2}\right)^2\right] + \frac{\Re R_1^2}{2k}\ln\left(\frac{R_1}{R_2}\right) \qquad (2)
$$

Substitution of the numerical values into Eq. (2) gives

$$
280 = T_2 + \frac{(5 \times 10^6)(1.8 \times 10^{-2})^2}{4(0.5)}\left[1 - \left(\frac{1.5}{1.8}\right)^2\right] + \frac{(5 \times 10^6)(1.5 \times 10^{-2})^2}{2(0.5)}\ln\left(\frac{1.5}{1.8}\right) \qquad (3)
$$

or,

$$
T_2 = 237.6\,°C \qquad (4)
$$

9.2.2.2 *Solid cylinder* Consider a solid cylinder of radius R with a constant surface temperature of T_R. The solution obtained for a hollow cylinder, Eq. (9.2-28), is also valid for this case. However, since the temperature must have a finite value at the center, i.e., $r = 0$, then C_1 must be zero and the temperature distribution becomes

$$\int_0^T k(T)\, dT = -\int_0^r \frac{1}{r}\left[\int_0^r \Re(u)u\, du\right] dr + C_2 \tag{9.2-41}$$

The use of the boundary condition

$$\text{at} \quad r = R \qquad T = T_R \tag{9.2-42}$$

gives the solution in the form

$$\int_{T_R}^T k(T)\, dT = \int_r^R \frac{1}{r}\left[\int_0^r \Re(u)\, u\, du\right] dr \tag{9.2-43}$$

■ **Case** (*i*) $k = $ **constant**

Simplification of Eq. (9.2-43) gives

$$k(T - T_R) = \int_r^R \frac{1}{r}\left[\int_0^r \Re(u)\, u\, du\right] dr \tag{9.2-44}$$

■ **Case** (*ii*) $k = $ **constant**; $\Re = $ **constant**

In this case, Eq. (9.2-43) simplifies to

$$T = T_R + \frac{\Re R^2}{4k}\left[1 - \left(\frac{r}{R}\right)^2\right] \tag{9.2-45}$$

which implies that the variation in temperature with respect to the radial position is parabolic with the maximum temperature at the center of the cylinder.

Macroscopic equation

The integration of the governing equation, Eq. (9.2-26), over the volume of the system gives

$$-\int_0^L \int_0^{2\pi} \int_0^R \frac{1}{r}\frac{d}{dr}\left(rk\frac{dT}{dr}\right) r\, dr\, d\theta\, dz = \int_0^L \int_0^{2\pi} \int_0^R \Re r\, dr\, d\theta\, dz \tag{9.2-46}$$

Integration of Eq. (9.2-46) yields

$$\underbrace{\left(-k\frac{dT}{dr}\right)_{r=R} 2\pi RL}_{\text{Rate of energy out}} = \underbrace{2\pi L \int_0^R \Re r\, dr}_{\text{Rate of energy generation}} \tag{9.2-47}$$

which is the macroscopic energy balance under steady conditions by considering the solid cylinder as a system. It is also possible to make use of Newton's law of cooling to express the rate of heat loss from the system to the surroundings at T_∞ with an average heat transfer coefficient $\langle h \rangle$. In this case, Eq. (9.2-47) reduces to

$$R\langle h \rangle(T_R - T_\infty) = \int_0^R \Re r\, dr \tag{9.2-48}$$

Example 9.4 Rate of heat generation per unit volume, $\Re_e$, during the transmission of an electric current through wires is given by

$$\Re_e = \frac{1}{k_e}\left(\frac{I}{\pi R^2}\right)^2$$

where I is the current, k_e is the electrical conductivity, and R is the radius of the wire.

a) Obtain an expression for the difference between the maximum and the surface temperatures of the wire.

b) Develop a correlation that will permit the selection of the electric current and the wire diameter if the difference between the maximum and the surface temperatures is specified. If the wire must carry a larger current, should the wire have a larger or smaller diameter?

Solution

Assumption

1. The thermal and electrical conductivities of the wire are constant.

Analysis

a) The temperature distribution is given by Eq. (9.2-45) as

$$T = T_R + \frac{\Re_e R^2}{4k}\left[1 - \left(\frac{r}{R}\right)^2\right] \tag{1}$$

where T_R is the surface temperature. The maximum temperature occurs at $r = 0$, i.e.,

$$T_{\max} - T_R = \frac{\Re_e R^2}{4k} \tag{2}$$

b) Expressing $\Re_e$ in terms of I and k_e gives

$$T_{\max} - T_R = \left(\frac{1}{4\pi k k_e}\right)\frac{I^2}{R^2} \tag{3}$$

Therefore, if I increases, R must be increased in order to keep $T_{\max} - T_R$ constant.

Example 9.5 Energy is generated in a cylindrical nuclear fuel element of radius R_F at a rate of

$$\Re = \Re_o(1 + \beta r^2)$$

It is clad in a material of radius R_C and the outside surface temperature is kept constant at T_o by a coolant. Determine the steady temperature distribution in the fuel element.

Solution

The temperature distribution within the fuel element can be determined from Eq. (9.2-44), i.e.,

$$k_F(T^F - T_i) = \Re_o \int_r^{R_F} \frac{1}{r}\left[\int_0^r (1 + \beta u^2) u\, du\right] dr \tag{1}$$

or,

$$T^F = T_i + \frac{\Re_o R_F^2}{4k_F}\left\{1 - \left(\frac{r}{R_F}\right)^2 + \frac{\beta R_F^2}{4}\left[1 - \left(\frac{r}{R_F}\right)^4\right]\right\} \tag{2}$$

in which the interface temperature T_i at $r = R_F$ is not known. To express T_i in terms of known quantities, consider the temperature distribution in the cladding. Since there is no internal generation within the cladding, the use of Eq. (D) in Table 8.3 gives

$$\frac{T_o - T^C}{T_o - T_i} = \frac{\ln(r/R_C)}{\ln(R_F/R_C)} \tag{3}$$

The energy flux at $r = R_F$ is continuous, i.e.,

$$k_F \frac{dT^F}{dr} = k_C \frac{dT^C}{dr} \tag{4}$$

Substitution of Eqs. (2) and (3) into Eq. (4) gives

$$T_i = T_o + \frac{\Re_o R_F^2 \ln(R_C/R_F)}{2k_C}\left(1 + \frac{\beta R_F^2}{2}\right) \tag{5}$$

Therefore, the temperature distribution given by Eq. (2) becomes

$$T^F - T_o = \frac{\Re_o R_F^2}{4k_F}\left\{1 - \left(\frac{r}{R_F}\right)^2 + \frac{\beta R_F^2}{4}\left[1 - \left(\frac{r}{R_F}\right)^4\right]\right\}$$
$$+ \frac{\Re_o R_F^2 \ln(R_C/R_F)}{2k_C}\left(1 + \frac{\beta R_F^2}{2}\right) \tag{6}$$

9.2.3 Conduction in Spherical Coordinates

9.2.3.1 *Hollow sphere*   Consider one-dimensional transfer of energy in the r-direction through a hollow sphere of inner and outer radii of R_1 and R_2, respectively, as shown in Figure 9.8. Let $\Re$ be the rate of generation per unit volume within the sphere.

Since $T = T(r)$, Table C.6 in Appendix C indicates that the only nonzero energy flux component is e_r, and it is given by

$$e_r = q_r = -k\frac{dT}{dr} \tag{9.2-49}$$

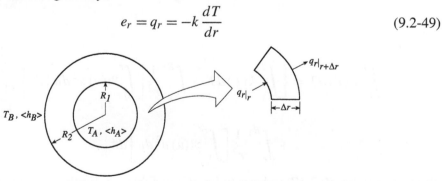

Figure 9.8. One-dimensional conduction through a hollow sphere with internal generation.

For a spherical differential volume of thickness Δr as shown in Figure 9.8, the inventory rate equation for energy, Eq. (9.2-1), is expressed as

$$4\pi (r^2 q_r)\big|_r - 4\pi (r^2 q_r)\big|_{r+\Delta r} + 4\pi r^2 \Delta r\, \Re = 0 \tag{9.2-50}$$

Dividing each term by $4\pi\, \Delta r$ and taking the limit as $\Delta r \to 0$ give

$$\lim_{\Delta r \to 0} \frac{(r^2 q_r)|_r - (r^2 q_r)|_{r+\Delta r}}{\Delta r} + r^2 \Re = 0 \tag{9.2-51}$$

or,

$$\frac{1}{r^2} \frac{d}{dr}(r^2 q_r) = \Re \tag{9.2-52}$$

Substitution of Eq. (9.2-49) into Eq. (9.2-52) gives the governing equation for temperature as

$$\boxed{-\frac{1}{r^2} \frac{d}{dr}\left(r^2 k \frac{dT}{dr}\right) = \Re} \tag{9.2-53}$$

Integration of Eq. (9.2-53) gives

$$k\frac{dT}{dr} = -\frac{1}{r^2} \int_0^r \Re(u)\, u^2 du + \frac{C_1}{r^2} \tag{9.2-54}$$

where u is the dummy variable of integration. Integration of Eq. (9.2-54) once more leads to

$$\boxed{\int_0^T k(T)\, dT = -\int_0^r \frac{1}{r^2}\left[\int_0^r \Re(u)\, u^2\, du\right] dr - \frac{C_1}{r} + C_2} \tag{9.2-55}$$

Evaluation of the constants C_1 and C_2 requires the boundary conditions to be specified.

Type I boundary condition

The solution of Eq. (9.2-55) subject to the boundary conditions

$$\text{at} \quad r = R_1 \qquad T = T_1 \tag{9.2-56a}$$

$$\text{at} \quad r = R_2 \qquad T = T_2 \tag{9.2-56b}$$

is given by

$$\int_{T_2}^T k(T)\, dT = \left\{\int_{T_2}^{T_1} k(T)\, dT - \int_{R_1}^{R_2} \frac{1}{r^2}\left[\int_0^r \Re(u)\, u^2\, du\right] dr\right\} \frac{\dfrac{1}{R_2} - \dfrac{1}{r}}{\dfrac{1}{R_2} - \dfrac{1}{R_1}}$$

$$+ \int_r^{R_2} \frac{1}{r^2}\left[\int_0^r \Re(u)\, u^2\, du\right] dr \tag{9.2-57}$$

When $\Re = 0$, Eq. (9.2-57) reduces to Eq. (C) in Table 8.5. Further simplification of Eq. (9.2-57) depends on the functional forms of k and $\Re$.

■ **Case** (*i*) $k = $ **constant**

In this case, Eq. (9.2-57) reduces to

$$k(T - T_2) = \left\{ k(T_1 - T_2) - \int_{R_1}^{R_2} \frac{1}{r^2} \left[\int_0^r \Re(u)\, u^2\, du \right] dr \right\} \frac{\frac{1}{R_2} - \frac{1}{r}}{\frac{1}{R_2} - \frac{1}{R_1}}$$

$$+ \int_r^{R_2} \frac{1}{r^2} \left[\int_0^r \Re(u)\, u^2\, du \right] dr \tag{9.2-58}$$

When $\Re = 0$, Eq. (9.2-58) reduces to Eq. (D) in Table 8.5.

■ **Case** (*ii*) $k = $ **constant**; $\Re = $ **constant**

In this case, Eq. (9.2-57) simplifies to

$$T = T_2 + \left\{ T_1 - T_2 - \frac{\Re R_2^2}{6k} \left[1 - \left(\frac{R_1}{R_2} \right)^2 \right] \right\} \frac{\frac{1}{R_2} - \frac{1}{r}}{\frac{1}{R_2} - \frac{1}{R_1}} + \frac{\Re R_2^2}{6k} \left[1 - \left(\frac{r}{R_2} \right)^2 \right] \tag{9.2-59}$$

Type II boundary condition

The solution of Eq. (9.2-55) subject to the boundary conditions

$$\text{at} \quad r = R_1 \qquad -k \frac{dT}{dz} = q_1 \tag{9.2-60a}$$

$$\text{at} \quad r = R_2 \qquad T = T_2 \tag{9.2-60b}$$

is given by

$$\int_{T_2}^{T} k(T)\, dT = \int_r^{R_2} \frac{1}{r^2} \left[\int_0^r \Re(u)\, u^2\, du \right] dr + \left[q_1 R_1^2 - \int_0^{R_1} \Re(u)\, u^2\, du \right] \left(\frac{1}{r} - \frac{1}{R_2} \right) \tag{9.2-61}$$

When $\Re = 0$, Eq. (9.2-61) reduces to Eq. (C) in Table 8.6. Further simplification of Eq. (9.2-61) depends on the functional forms of k and $\Re$.

■ **Case** (*i*) $k = $ **constant**

In this case, Eq. (9.2-61) reduces to

$$k(T - T_2) = \int_r^{R_2} \frac{1}{r^2} \left[\int_0^r \Re(u) u^2\, du \right] dr + \left[q_1 R_1^2 - \int_0^{R_1} \Re(u) u^2\, du \right] \left(\frac{1}{r} - \frac{1}{R_2} \right) \tag{9.2-62}$$

When $\Re = 0$, Eq. (9.2-62) reduces to Eq. (D) in Table 8.6.

■ **Case** (*ii*) $k = $ **constant**; $\Re = $ **constant**

In this case, Eq. (9.2-61) simplifies to

$$T = T_2 + \frac{\Re R_2^2}{6k}\left[1 - \left(\frac{r}{R_2}\right)^2\right] + \left(\frac{q_1 R_1^2}{k} - \frac{\Re R_1^3}{3k}\right)\left(\frac{1}{r} - \frac{1}{R_2}\right) \tag{9.2-63}$$

Macroscopic equation

The integration of the governing equation, Eq. (9.2-53), over the volume of the system gives

$$-\int_0^{2\pi}\int_0^{\pi}\int_{R_1}^{R_2}\frac{1}{r^2}\frac{d}{dr}\left(r^2 k\frac{dT}{dr}\right)r^2\sin\theta\,dr\,d\theta\,d\phi = \int_0^{2\pi}\int_0^{\pi}\int_{R_1}^{R_2}\Re r^2\sin\theta\,dr\,d\theta\,d\phi \tag{9.2-64}$$

Integration of Eq. (9.2-64) yields

$$\underbrace{\left(-k\frac{dT}{dr}\right)_{r=R_2}4\pi R_2^2 + \left(k\frac{dT}{dr}\right)_{r=R_1}4\pi R_1^2}_{\text{Net rate of energy out}} = \underbrace{4\pi\int_{R_1}^{R_2}\Re r^2\,dr}_{\substack{\text{Rate of energy}\\\text{generation}}} \tag{9.2-65}$$

which is the macroscopic energy balance under steady conditions by considering the hollow sphere as a system.

It is also possible to make use of Newton's law of cooling to express the rate of heat loss from the system. If heat is lost from both surfaces, Eq. (9.2-65) can be written as

$$R_1^2\langle h_A\rangle(T_1 - T_A) + R_2^2\langle h_B\rangle(T_2 - T_B) = \int_{R_1}^{R_2}\Re r^2\,dr \tag{9.2-66}$$

where T_1 and T_2 are the surface temperatures at $r = R_1$ and $r = R_2$, respectively.

9.2.3.2 *Solid sphere* Consider a solid sphere of radius R with a constant surface temperature of T_R. The solution obtained for a hollow sphere, Eq. (9.2-55), is also valid for this case. However, since the temperature must have a finite value at the center, i.e., $r = 0$, then C_1 must be zero and the temperature distribution becomes

$$\int_0^T k(T)\,dT = -\int_0^r\frac{1}{r^2}\left[\int_0^r\Re(u)\,u^2\,du\right]dr + C_2 \tag{9.2-67}$$

The use of the boundary condition

$$\text{at}\quad r = R \quad T = T_R \tag{9.2-68}$$

gives the solution in the form

$$\int_{T_R}^T k(T)\,dT = \int_r^R\frac{1}{r^2}\left[\int_0^r\Re(u)\,u^2\,du\right]dr \tag{9.2-69}$$

■ **Case** (i) $k =$ **constant**

Simplification of Eq. (9.2-69) gives

$$k(T - T_R) = \int_r^R \frac{1}{r^2}\left[\int_0^r \Re(u)\, u^2\, du\right] dr \qquad (9.2\text{-}70)$$

■ **Case** (ii) $k =$ **constant**; $\Re =$ **constant**

In this case, Eq. (9.2-69) simplifies to

$$T = T_R + \frac{\Re R^2}{6k}\left[1 - \left(\frac{r}{R}\right)^2\right] \qquad (9.2\text{-}71)$$

which implies that the variation in temperature with respect to the radial position is parabolic with the maximum temperature at the center of the sphere.

Macroscopic equation

The integration of the governing equation, Eq. (9.2-53), over the volume of the system gives

$$-\int_0^{2\pi}\int_0^\pi\int_0^R \frac{1}{r^2}\frac{d}{dr}\left(r^2 k \frac{dT}{dr}\right) r^2 \sin\theta\, dr\, d\theta\, d\phi = \int_0^{2\pi}\int_0^\pi\int_0^R \Re r^2 \sin\theta\, dr\, d\theta\, d\phi \qquad (9.2\text{-}72)$$

Integration of Eq. (9.2-72) yields

$$\underbrace{\left(-k\frac{dT}{dr}\right)_{r=R} 4\pi R^2}_{\text{Rate of energy out}} = \underbrace{4\pi \int_0^R \Re r^2\, dr}_{\substack{\text{Rate of energy}\\\text{generation}}} \qquad (9.2\text{-}73)$$

which is the macroscopic energy balance under steady conditions by considering the solid sphere as a system. It is also possible to make use of Newton's law of cooling to express the rate of heat loss from the system to the surroundings at T_∞ with an average heat transfer coefficient $\langle h\rangle$. In this case, Eq. (9.2-73) reduces to

$$R^2 \langle h\rangle (T_R - T_\infty) = \int_0^R \Re r^2\, dr \qquad (9.2\text{-}74)$$

Example 9.6 Consider Example 3.2 in which energy generation as a result of fission within a spherical reactor of radius R is given as

$$\Re = \Re_o\left[1 - \left(\frac{r}{R}\right)^2\right]$$

Cooling fluid at a temperature of T_∞ flows over a reactor with an average heat transfer coefficient of $\langle h\rangle$. Determine the temperature distribution and the rate of heat loss from the reactor surface.

Solution

The temperature distribution within the reactor can be calculated from Eq. (9.2-70). Note that

$$\int_0^r \Re(u)\, u^2\, du = \Re_o \int_0^r \left[1 - \left(\frac{u}{R} \right)^2 \right] u^2\, du = \Re_o \left(\frac{r^3}{3} - \frac{r^5}{5R^2} \right) \tag{1}$$

Substitution of Eq. (1) into Eq. (9.2-70) gives

$$k(T - T_R) = \Re_o \int_r^R \frac{1}{r^2} \left(\frac{r^3}{3} - \frac{r^5}{5R^2} \right) dr \tag{2}$$

Evaluation of the integration gives the temperature distribution as

$$T = T_R + \frac{7}{60} \frac{\Re_o R^2}{k} - \frac{\Re_o R^2}{2k} \left[\frac{1}{3} \left(\frac{r}{R} \right)^2 - \frac{1}{10} \left(\frac{r}{R} \right)^4 \right] \tag{3}$$

This result, however, contains an unknown quantity, T_R. Therefore, it is necessary to express T_R in terms of the known quantities, i.e., T_∞ and $\langle h \rangle$.

One way of calculating the surface temperature, T_R, is to use the macroscopic energy balance given by Eq. (9.2-74), i.e.,

$$R^2 \langle h \rangle (T_R - T_\infty) = \Re_o \int_0^R \left[1 - \left(\frac{r}{R} \right)^2 \right] r^2\, dr \tag{4}$$

Equation (4) gives the surface temperature as

$$T_R = T_\infty + \frac{2}{15} \frac{\Re_o R}{\langle h \rangle} \tag{5}$$

Another way of calculating the surface temperature is to equate Newton's law of cooling and Fourier's law of heat conduction at the surface of the sphere, i.e.,

$$\langle h \rangle (T_R - T_\infty) = -k \left. \frac{dT}{dr} \right|_{r=R} \tag{6}$$

From Eq. (3)

$$\left. \frac{dT}{dr} \right|_{r=R} = -\frac{2 \Re_o R^2}{15k} \tag{7}$$

Substituting Eq. (7) into Eq. (6) and solving for T_R result in Eq. (5).

Therefore, the temperature distribution within the reactor in terms of the known quantities is given by

$$T = T_\infty + \frac{2}{15} \frac{\Re_o R}{\langle h \rangle} + \frac{7}{60} \frac{\Re_o R^2}{k} - \frac{\Re_o R^2}{2k} \left[\frac{1}{3} \left(\frac{r}{R} \right)^2 - \frac{1}{10} \left(\frac{r}{R} \right)^4 \right] \tag{8}$$

The rate of heat loss can be calculated from Eq. (9.2-73) as

$$\dot{Q}_{loss} = 4\pi \Re_o \int_0^R \left[1 - \left(\frac{r}{R} \right)^2 \right] r^2\, dr = \frac{8\pi}{15} \Re_o R^3 \tag{9}$$

Note that the calculation of the rate of heat loss does not require the temperature distribution to be known.

9.3 ENERGY TRANSPORT WITH CONVECTION

9.3.1 Laminar Flow Forced Convection in a Pipe

Consider the laminar flow of an incompressible Newtonian fluid in a circular pipe under the action of a pressure gradient as shown in Figure 9.9. The velocity distribution is given by Eqs. (9.1-79) and (9.1-84) as

$$v_z = 2\langle v_z \rangle \left[1 - \left(\frac{r}{R} \right)^2 \right] \tag{9.3-1}$$

Suppose that the fluid, which is at a uniform temperature of T_o for $z < 0$, is started to be heated for $z > 0$ and we want to develop the governing equation for temperature.

In general, $T = T(r, z)$ and, from Table C.5 in Appendix C, the nonzero energy flux components are

$$e_r = -k \frac{\partial T}{\partial r} \tag{9.3-2}$$

$$e_z = -k \frac{\partial T}{\partial z} + (\rho \widehat{C}_P T) v_z \tag{9.3-3}$$

Since there is no generation of energy, Eq. (9.2-1) simplifies to

$$\text{(Rate of energy in)} - \text{(Rate of energy out)} = 0 \tag{9.3-4}$$

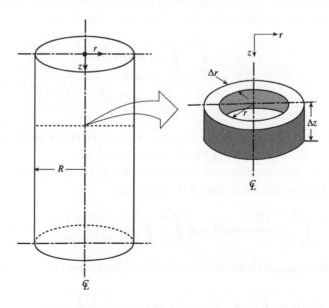

Figure 9.9. Forced convection heat transfer in a pipe.

For a cylindrical differential volume element of thickness Δr and length Δz, as shown in Figure 9.9, Eq. (9.3-4) is expressed as

$$\left(e_r|_r 2\pi r \Delta z + e_z|_z 2\pi r \Delta r \right) - \left[e_r|_{r+\Delta r} 2\pi (r+\Delta r)\Delta z + e_z|_{z+\Delta z} 2\pi r \Delta r \right] = 0 \qquad (9.3\text{-}5)$$

Dividing Eq. (9.3-5) by $2\pi \Delta r \Delta z$ and taking the limit as $\Delta r \to 0$ and $\Delta z \to 0$ give

$$\lim_{\Delta r \to 0} \frac{(re_r)|_r - (re_r)|_{r+\Delta r}}{\Delta r} + \lim_{\Delta z \to 0} r \frac{e_z|_z - e_z|_{z+\Delta z}}{\Delta z} = 0 \qquad (9.3\text{-}6)$$

or,

$$\frac{1}{r}\frac{\partial (re_r)}{\partial r} + \frac{\partial e_z}{\partial z} = 0 \qquad (9.3\text{-}7)$$

Substitution of Eqs. (9.3-2) and (9.3-3) into Eq. (9.3-7) yields

$$\underbrace{\rho \widehat{C}_P v_z \frac{\partial T}{\partial z}}_{\substack{\text{Convection in}\\ z\text{-direction}}} = \underbrace{\frac{k}{r}\frac{\partial}{\partial r}\left(r \frac{\partial T}{\partial r}\right)}_{\substack{\text{Conduction in}\\ r\text{-direction}}} + \underbrace{k\frac{\partial^2 T}{\partial z^2}}_{\substack{\text{Conduction in}\\ z\text{-direction}}} \qquad (9.3\text{-}8)$$

In the z-direction, energy is transported by both convection and conduction. As stated by Eq. (2.4-8), conduction can be considered negligible with respect to convection when $Pe_H \gg 1$. Under these circumstances, Eq. (9.3-8) reduces to

$$\boxed{\rho \widehat{C}_P v_z \frac{\partial T}{\partial z} = \frac{k}{r}\frac{\partial}{\partial r}\left(r \frac{\partial T}{\partial r}\right)} \qquad (9.3\text{-}9)$$

As engineers, we are interested in the variation in the bulk fluid temperature, T_b, rather than the local temperature, T. For forced convection heat transfer in a circular pipe of radius R, the bulk fluid temperature defined by Eq. (4.1-1) takes the form

$$T_b = \frac{\displaystyle\int_0^{2\pi}\int_0^R v_z T r\, dr\, d\theta}{\displaystyle\int_0^{2\pi}\int_0^R v_z r\, dr\, d\theta} \qquad (9.3\text{-}10)$$

Note that, while the fluid temperature, T, depends on both the radial and the axial coordinates, the bulk temperature, T_b, depends only on the axial direction.

To determine the governing equation for the bulk temperature, it is necessary to integrate Eq. (9.3-9) over the cross-sectional area of the pipe, i.e.,

$$\rho \widehat{C}_P \int_0^{2\pi}\int_0^R v_z \frac{\partial T}{\partial z} r\, dr\, d\theta = k \int_0^{2\pi}\int_0^R \frac{1}{r}\frac{\partial}{\partial r}\left(r \frac{\partial T}{\partial r}\right) r\, dr\, d\theta \qquad (9.3\text{-}11)$$

Since $v_z \neq v_z(z)$, the integral on the left-hand side of Eq. (9.3-11) can be rearranged as

$$\int_0^{2\pi}\int_0^R v_z \frac{\partial T}{\partial z} r\, dr\, d\theta = \int_0^{2\pi}\int_0^R \frac{\partial (v_z T)}{\partial z} r\, dr\, d\theta = \frac{d}{dz}\left(\int_0^{2\pi}\int_0^R v_z T r\, dr\, d\theta \right) \qquad (9.3\text{-}12)$$

Substitution of Eq. (9.3-10) into Eq. (9.3-12) yields

$$\int_0^{2\pi} \int_0^R v_z \frac{\partial T}{\partial z} r \, dr \, d\theta = \frac{d}{dz} \left(T_b \underbrace{\int_0^{2\pi} \int_0^R v_z r \, dr \, d\theta}_{\langle v_z \rangle \pi R^2} \right) = \frac{\dot{m}}{\rho} \frac{dT_b}{dz} \tag{9.3-13}$$

where $\dot{m}$ is the mass flow rate given by

$$\dot{m} = \rho \langle v_z \rangle \pi R^2 \tag{9.3-14}$$

On the other hand, since $\partial T / \partial r = 0$ as a result of the symmetry condition at the center of the tube, the integral on the right-hand side of Eq. (9.3-11) takes the form

$$\int_0^{2\pi} \int_0^R \frac{1}{r} \frac{\partial}{\partial r} \left(r \frac{\partial T}{\partial r} \right) r \, dr \, d\theta = 2\pi R \left. \frac{\partial T}{\partial r} \right|_{r=R} \tag{9.3-15}$$

Substitution of Eqs. (9.3-13) and (9.3-15) into Eq. (9.3-11) gives the governing equation for the bulk temperature in the form

$$\boxed{\dot{m} \widehat{C}_P \frac{dT_b}{dz} = \pi D k \left. \frac{\partial T}{\partial r} \right|_{r=R}} \tag{9.3-16}$$

The solution of Eq. (9.3-16) requires the boundary conditions associated with the problem to be known. The two most commonly used boundary conditions are the *constant wall temperature* and *constant wall heat flux*.

Constant wall temperature

Constant wall temperature occurs in evaporators and condensers in which phase change takes place on one side of the surface. The heat flux at the wall can be represented either by Fourier's law of heat conduction or by Newton's law of cooling, i.e.,

$$q_r|_{r=R} = k \left. \frac{\partial T}{\partial r} \right|_{r=R} = h(T_w - T_b) \tag{9.3-17}$$

It is implicitly implied in writing Eq. (9.3-17) that the temperature increases in the radial direction. Substitution of Eq. (9.3-17) into Eq. (9.3-16) and rearrangement yield

$$\dot{m} \widehat{C}_P \int_{T_{b_{in}}}^{T_b} \frac{dT_b}{T_w - T_b} = \pi D \int_0^z h \, dz \tag{9.3-18}$$

Since the wall temperature, T_w, is constant, integration of Eq. (9.3-18) yields

$$\dot{m} \widehat{C}_P \ln\left(\frac{T_w - T_{b_{in}}}{T_w - T_b} \right) = \pi D \langle h \rangle_z z \tag{9.3-19}$$

in which $\langle h \rangle_z$ is the average heat transfer coefficient from the entrance to the point z defined by

$$\langle h \rangle_z = \frac{1}{z} \int_0^z h \, dz \tag{9.3-20}$$

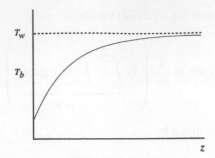

Figure 9.10. Variation in the bulk temperature with the axial direction for a constant wall temperature.

If Eq. (9.3-19) is solved for T_b, the result is

$$T_b = T_w - (T_w - T_{b_{in}}) \exp\left[-\left(\frac{\pi D \langle h \rangle_z}{\dot{m} \widehat{C}_P}\right) z\right] \tag{9.3-21}$$

which indicates that the bulk fluid temperature varies exponentially with the axial direction as shown in Figure 9.10.

Evaluation of Eq. (9.3-19) over the total length, L, of the pipe gives

$$\dot{m} \widehat{C}_P \ln\left(\frac{T_w - T_{b_{in}}}{T_w - T_{b_{out}}}\right) = \pi D \langle h \rangle L \tag{9.3-22}$$

where

$$\langle h \rangle = \frac{1}{L} \int_0^L h \, dz \tag{9.3-23}$$

If Eq. (9.3-22) is solved for $T_{b_{out}}$, the result is

$$T_{b_{out}} = T_w - (T_w - T_{b_{in}}) \exp\left[-\left(\frac{\pi D \langle h \rangle}{\dot{m} \widehat{C}_P}\right) L\right] \tag{9.3-24}$$

Equation (9.3-24) can be expressed in terms of dimensionless numbers with the help of Eq. (3.4-5), i.e.,

$$\mathrm{St_H} = \frac{\mathrm{Nu}}{\mathrm{Re\,Pr}} = \frac{\langle h \rangle}{\rho \langle v_z \rangle \widehat{C}_P} = \frac{\langle h \rangle}{[\dot{m}/(\pi D^2/4)]\widehat{C}_P} \tag{9.3-25}$$

The use of Eq. (9.3-25) in Eq. (9.3-24) gives

$$T_{b_{out}} = T_w - (T_w - T_{b_{in}}) \exp\left[-\frac{4\,\mathrm{Nu}(L/D)}{\mathrm{Re\,Pr}}\right] \tag{9.3-26}$$

As engineers, we are interested in the rate of heat transferred to the fluid, i.e.,

$$\dot{Q} = \dot{m} \widehat{C}_P (T_{b_{out}} - T_{b_{in}}) = \dot{m} \widehat{C}_P \left[(T_w - T_{b_{in}}) - (T_w - T_{b_{out}})\right] \tag{9.3-27}$$

Substitution of Eq. (9.3-22) into Eq. (9.3-27) results in

$$\dot{Q} = (\pi D L)\langle h \rangle \left[\frac{(T_w - T_{b_{in}})(T_w - T_{b_{out}})}{\ln\left(\dfrac{T_w - T_{b_{in}}}{T_w - T_{b_{out}}}\right)} \right] \qquad (9.3\text{-}28)$$

Note that Eq. (9.3-28) can be expressed in the form

$$\dot{Q} = A_H \langle h \rangle (\Delta T)_{ch} = (\pi D L)\langle h \rangle \Delta T_{LM} \qquad (9.3\text{-}29)$$

which is identical to Eqs. (3.2-7) and (4.5-29).

Constant wall heat flux

The constant wall heat flux type boundary condition is encountered when electrical resistance is wrapped around the pipe. Since the heat flux at the wall is constant, then

$$q_r|_{r=R} = k \left.\frac{\partial T}{\partial r}\right|_{r=R} = q_w = \text{constant} \qquad (9.3\text{-}30)$$

Substitution of Eq. (9.3-30) into Eq. (9.3-16) gives

$$\frac{dT_b}{dz} = \frac{\pi D q_w}{\dot{m}\widehat{C}_P} = \text{constant} \qquad (9.3\text{-}31)$$

Integration of Eq. (9.3-31) gives the variation in the bulk temperature in the axial direction as

$$T_b = T_{b_{in}} + \left(\frac{\pi D q_w}{\dot{m}\widehat{C}_P}\right) z \qquad (9.3\text{-}32)$$

Therefore, the bulk fluid temperature varies linearly in the axial direction as shown in Figure 9.11.

Evaluation of Eq. (9.3-32) over the total length gives the bulk temperature at the exit of the pipe as

$$T_{b_{out}} = T_{b_{in}} + \left(\frac{\pi D q_w}{\dot{m}\widehat{C}_P}\right) L = T_{b_{in}} + \frac{4 q_w L}{k \,\text{Re}\,\text{Pr}} \qquad (9.3\text{-}33)$$

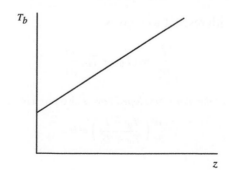

Figure 9.11. Variation in the bulk temperature with the axial direction for a constant wall heat flux.

The rate of heat transferred to the fluid is given by

$$\dot{Q} = \dot{m}\widehat{C}_P(T_{b_{out}} - T_{b_{in}}) \tag{9.3-34}$$

Substitution of Eq. (9.3-33) into Eq. (9.3-34) yields

$$\dot{Q} = (\pi DL)q_w \tag{9.3-35}$$

9.3.1.1 *Thermally developed flow*   As stated in Section 8.1, when the fluid velocity is no longer dependent on the axial direction z, the flow is said to be hydrodynamically fully developed. In the case of heat transfer, if the ratio

$$\frac{T - T_b}{T_w - T_b} \tag{9.3-36}$$

does not vary along the axial direction, then the temperature profile is said to be thermally fully developed.

It is important to note that, although the fluid temperature, T, bulk fluid temperature, T_b, and wall temperature, T_w, may change along the axial direction, the ratio given in Eq. (9.3-36) is independent of the axial coordinate[2], i.e.,

$$\frac{\partial}{\partial z}\left(\frac{T - T_b}{T_w - T_b}\right) = 0 \tag{9.3-37}$$

Equation (9.3-37) indicates that

$$\frac{\partial T}{\partial z} = \left(\frac{T_w - T}{T_w - T_b}\right)\frac{dT_b}{dz} + \left(\frac{T - T_b}{T_w - T_b}\right)\frac{dT_w}{dz} \tag{9.3-38}$$

Example 9.7 For a thermally developed flow of a fluid with constant physical properties, show that the local heat transfer coefficient is a constant.

Solution

For a thermally developed flow, the ratio given in Eq. (9.3-36) depends only on the radial coordinate r, i.e.,

$$\frac{T - T_b}{T_w - T_b} = f(r) \tag{1}$$

Differentiation of Eq. (1) with respect to r gives

$$\frac{\partial T}{\partial r} = (T_w - T_b)\frac{df}{dr} \tag{2}$$

---

[2]In the literature, the condition for *thermally developed flow* is also given in the form

$$\frac{\partial}{\partial z}\left(\frac{T_w - T}{T_w - T_b}\right) = 0$$

Note that

$$\frac{T_w - T}{T_w - T_b} = 1 - \frac{T - T_b}{T_w - T_b}.$$

which is valid at all points within the flow field. Evaluation of Eq. (2) at the surface of the pipe yields

$$\left.\frac{\partial T}{\partial r}\right|_{r=R} = (T_w - T_b)\left.\frac{df}{dr}\right|_{r=R} \tag{3}$$

On the other hand, the heat flux at the wall is expressed as

$$q_r|_{r=R} = k \left.\frac{\partial T}{\partial r}\right|_{r=R} = h(T_w - T_b) \tag{4}$$

Substitution of Eq. (3) into Eq. (4) gives

$$h = k \left.\frac{df}{dr}\right|_{r=R} = \text{constant} \tag{5}$$

Example 9.8 For a thermally developed flow, show that the temperature gradient in the axial direction, $\partial T/\partial z$, remains constant for a constant wall heat flux.

Solution

The heat flux at the wall is given by

$$q_r|_{r=R} = h(T_w - T_b) = \text{constant} \tag{1}$$

Since h is constant for a thermally developed flow, Eq. (1) implies that

$$T_w - T_b = \text{constant} \tag{2}$$

or,

$$\frac{dT_w}{dz} = \frac{dT_b}{dz} \tag{3}$$

Therefore, Eq. (9.3-38) simplifies to

$$\frac{\partial T}{\partial z} = \frac{dT_b}{dz} = \frac{dT_w}{dz} \tag{4}$$

Since dT_b/dz is constant according to Eq. (9.3-31), $\partial T/\partial z$ also remains constant, i.e.,

$$\frac{\partial T}{\partial z} = \frac{dT_b}{dz} = \frac{dT_w}{dz} = \frac{\pi D q_w}{\dot{m}\widehat{C}_P} = \text{constant} \tag{5}$$

9.3.1.2 *Nusselt number for a thermally developed flow*   Substitution of Eq. (9.3-1) into Eq. (9.3-9) gives

$$2\rho\widehat{C}_P\langle v_z\rangle\left[1 - \left(\frac{r}{R}\right)^2\right]\frac{\partial T}{\partial z} = \frac{k}{r}\frac{\partial}{\partial r}\left(r\frac{\partial T}{\partial r}\right) \tag{9.3-39}$$

It should always be kept in mind that the purpose of solving the above equation for temperature distribution is to obtain a correlation to use in the design of heat transfer equipment, such as heat exchangers and evaporators. As shown in Chapter 4, heat transfer correlations are expressed in terms of the Nusselt number. Therefore, Eq. (9.3-39) will be solved for a thermally developed flow for two different types of boundary conditions, i.e., constant wall heat flux and constant wall temperature, to determine the Nusselt number.

Constant wall heat flux

In the case of a constant wall heat flux, as shown in Example 9.8, the temperature gradient in the axial direction is constant and expressed in the form

$$\frac{\partial T}{\partial z} = \frac{\pi D q_w}{\dot{m}\widehat{C}_P} = \frac{\pi D q_w}{[\rho\langle v_z\rangle(\pi R^2)]\widehat{C}_P} = \text{constant} \tag{9.3-40}$$

Since we are interested in the determination of the Nusselt number, it is appropriate to express $\partial T/\partial z$ in terms of the Nusselt number. Note that the Nusselt number is given by

$$\text{Nu} = \frac{hD}{k} = \frac{[q_w/(T_w - T_b)]D}{k} \tag{9.3-41}$$

Therefore, Eq. (9.3-40) reduces to

$$\frac{\partial T}{\partial z} = \frac{\text{Nu}(T_w - T_b)\,k}{\rho\widehat{C}_P R^2\langle v_z\rangle} \tag{9.3-42}$$

Substitution of Eq. (9.3-42) into Eq. (9.3-39) yields

$$\frac{2}{R^2}\left[1 - \left(\frac{r}{R}\right)^2\right]\text{Nu}(T_w - T_b) = \frac{1}{r}\frac{\partial}{\partial r}\left(r\frac{\partial T}{\partial r}\right) \tag{9.3-43}$$

In terms of the dimensionless variables

$$\theta = \frac{T - T_b}{T_w - T_b} \tag{9.3-44}$$

$$\xi = \frac{r}{R} \tag{9.3-45}$$

Eq. (9.3-43) takes the form

$$2\,\text{Nu}(1 - \xi^2) = \frac{1}{\xi}\frac{d}{d\xi}\left(\xi\frac{d\theta}{d\xi}\right) \tag{9.3-46}$$

It is important to note that θ depends only on ξ (or r).

The boundary conditions associated with Eq. (9.3-46) are

$$\text{at} \quad \xi = 0 \qquad \frac{d\theta}{d\xi} = 0 \tag{9.3-47}$$

$$\text{at} \quad \xi = 1 \qquad \theta = 1 \tag{9.3-48}$$

Integration of Eq. (9.3-46) with respect to ξ gives

$$\xi\frac{d\theta}{d\xi} = \left(\xi^2 - \frac{\xi^4}{2}\right)\text{Nu} + C_1 \tag{9.3-49}$$

where C_1 is an integration constant. Application of Eq. (9.3-47) indicates that $C_1 = 0$. Integration of Eq. (9.3-49) once more with respect to ξ and the use of the boundary condition given by Eq. (9.3-48) give

$$\theta = 1 - \frac{\text{Nu}}{8}(3 - 4\xi^2 + \xi^4) \tag{9.3-50}$$

On the other hand, the bulk temperature in dimensionless form can be expressed as

$$\theta_b = \frac{T_b - T_b}{T_w - T_b} = 0 = \frac{\displaystyle\int_0^1 (1 - \xi^2)\,\theta\,\xi\,d\xi}{\displaystyle\int_0^1 (1 - \xi^2)\,\xi\,d\xi} \tag{9.3-51}$$

Substitution of Eq. (9.3-50) into Eq. (9.3-51) and integration give the Nusselt number as

$$\boxed{\text{Nu} = \frac{48}{11}} \tag{9.3-52}$$

Constant wall temperature

When the wall temperature is constant, Eq. (9.3-38) indicates that

$$\frac{\partial T}{\partial z} = \left(\frac{T_w - T}{T_w - T_b}\right)\frac{dT_b}{dz} \tag{9.3-53}$$

The variation in T_b as a function of the axial position can be obtained from Eq. (9.3-21) as

$$\frac{dT_b}{dz} = \frac{\pi D \langle h \rangle_z}{\dot{m}\,\widehat{C}_P} \underbrace{(T_w - T_{b_{in}})\exp\left[-\left(\frac{\pi D \langle h \rangle_z}{\dot{m}\,\widehat{C}_P}\right)z\right]}_{(T_w - T_b)} \tag{9.3-54}$$

Since the heat transfer coefficient is constant for a thermally developed flow, Eq. (9.3-54) becomes

$$\frac{dT_b}{dz} = \frac{\pi D\,h(T_w - T_b)}{\dot{m}\widehat{C}_P} = \frac{4h(T_w - T_b)}{D\langle v_z \rangle \rho \widehat{C}_P} \tag{9.3-55}$$

The use of Eq. (9.3-55) in Eq. (9.3-53) yields

$$\frac{\partial T}{\partial z} = \frac{4h(T_w - T)}{D\langle v_z \rangle \rho \widehat{C}_P} \tag{9.3-56}$$

Substitution of Eq. (9.3-56) into Eq. (9.3-39) gives

$$\frac{8}{D^2}\left(\frac{hD}{k}\right)\left[1 - \left(\frac{r}{R}\right)^2\right](T_w - T) = \frac{1}{r}\frac{\partial}{\partial r}\left(r\frac{\partial T}{\partial r}\right) \tag{9.3-57}$$

In terms of the dimensionless variables defined by Eqs. (9.3-44) and (9.3-45), Eq. (9.3-57) becomes

$$2\operatorname{Nu}(1 - \xi^2)(1 - \theta) = \frac{1}{\xi}\frac{d}{d\xi}\left(\xi\frac{d\theta}{d\xi}\right) \tag{9.3-58}$$

The boundary conditions associated with Eq. (9.3-58) are

$$\text{at} \quad \xi = 0 \qquad \frac{d\theta}{d\xi} = 0 \tag{9.3-59}$$

$$\text{at} \quad \xi = 1 \qquad \theta = 1 \tag{9.3-60}$$

Note that the use of the substitution

$$u = 1 - \theta \tag{9.3-61}$$

reduces Eqs. (9.3-58)–(9.3-60) to

$$-2\operatorname{Nu}(1 - \xi^2)u = \frac{1}{\xi}\frac{d}{d\xi}\left(\xi\frac{du}{d\xi}\right) \tag{9.3-62}$$

$$\text{at} \quad \xi = 0 \qquad \frac{du}{d\xi} = 0 \tag{9.3-63}$$

$$\text{at} \quad \xi = 1 \qquad u = 0 \tag{9.3-64}$$

Equation (9.3-62) can be solved for Nu by the method of Stodola and Vianello as explained in Section B.3.4.1 in Appendix B.

A reasonable first guess for u that satisfies the boundary conditions is

$$u_1 = 1 - \xi^2 \tag{9.3-65}$$

Substitution of Eq. (9.3-65) into the left-hand side of Eq. (9.3-62) gives

$$\frac{d}{d\xi}\left(\xi\frac{du}{d\xi}\right) = -2\operatorname{Nu}(\xi - 2\xi^3 + \xi^5) \tag{9.3-66}$$

The solution of Eq. (9.3-66) is

$$u = \operatorname{Nu}\underbrace{\left(\frac{11 - 18\xi^2 + 9\xi^4 - 2\xi^6}{36}\right)}_{f_1(\xi)} \tag{9.3-67}$$

Therefore, the first approximation to the Nusselt number is

$$\operatorname{Nu}^{(1)} = \frac{\displaystyle\int_0^1 \xi(1 - \xi^2)^2 f_1(\xi)\,d\xi}{\displaystyle\int_0^1 \xi(1 - \xi^2) f_1^2(\xi)\,d\xi} \tag{9.3-68}$$

Substitution of $f_1(\xi)$ from Eq. (9.3-67) into Eq. (9.3-68) and evaluation of the integrals give

$$\text{Nu} = 3.663 \qquad (9.3\text{-}69)$$

On the other hand, the value of the Nusselt number, as calculated by Graetz (1883, 1885) and later independently by Nusselt (1910), is 3.66. Therefore, for a thermally developed laminar flow in a circular pipe with constant wall temperature, Nu = 3.66 for all practical purposes.

Example 9.9 Water flows through a circular pipe of 5 cm internal diameter with an average velocity of 0.01 m/s. Determine the length of the pipe to increase the water temperature from 20 °C to 60 °C for the following conditions:

a) Steam condenses on the outer surface of the pipe so as to keep the surface temperature at 100 °C.
b) Electrical wires are wrapped around the outer surface of the pipe to provide a constant wall heat flux of 1500 W/m².

Solution

Physical properties

The mean bulk temperature is $(20 + 60)/2 = 40\,°\text{C}$ (313 K).

For water at 313 K: $\begin{cases} \rho = 992 \text{ kg/m}^3 \\ \mu = 654 \times 10^{-6} \text{ kg/m·s} \\ k = 632 \times 10^{-3} \text{ W/m·K} \\ \text{Pr} = 4.32 \end{cases}$

Assumptions

1. Steady-state conditions prevail.
2. Flow is hydrodynamically and thermally fully developed.

Analysis

The Reynolds number is

$$\text{Re} = \frac{D\langle v_z \rangle \rho}{\mu} = \frac{(0.05)(0.01)(992)}{654 \times 10^{-6}} = 758 \quad \Rightarrow \quad \text{Laminar flow}$$

a) Since the wall temperature is constant, from Eq. (9.3-26)

$$L = \frac{D\,\text{Re}\,\text{Pr}}{4\,\text{Nu}} \ln\left(\frac{T_w - T_{b_{in}}}{T_w - T_{b_{out}}}\right) = \frac{(0.05)(758)(4.32)}{4(3.66)} \ln\left(\frac{100-20}{100-60}\right) = 7.8 \text{ m}$$

b) For a constant heat flux at the wall, the use of Eq. (9.3-33) gives

$$L = \frac{(T_{b_{out}} - T_{b_{in}})\,k\,\text{Re}\,\text{Pr}}{4 q_w} = \frac{(60-20)(632 \times 10^{-3})(758)(4.32)}{4(1500)} = 13.8 \text{ m}$$

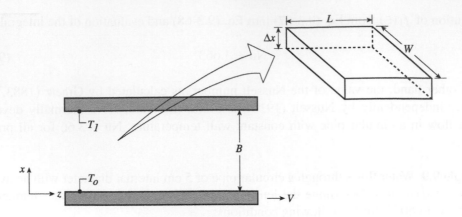

Figure 9.12. Couette flow with heat transfer.

9.3.2 Viscous Heating in a Couette Flow

Viscous heating becomes an important problem during flow of liquids in lubrication, viscometry, and extrusion. Let us consider Couette flow of a Newtonian fluid between two large parallel plates as shown in Figure 9.12. The surfaces at $x = 0$ and $x = B$ are maintained at T_o and T_1, respectively, with $T_o > T_1$.

Rate of energy generation per unit volume as a result of viscous dissipation is given by[3]

$$\Re = \mu \left(\frac{dv_z}{dx} \right)^2 \qquad (9.3\text{-}70)$$

The velocity distribution for this problem is given by Eq. (8.1-12) as

$$\frac{v_z}{V} = 1 - \frac{x}{B} \qquad (9.3\text{-}71)$$

The use of Eq. (9.3-71) in Eq. (9.3-70) gives the rate of energy generation per unit volume as

$$\Re = \frac{\mu V^2}{B^2} \qquad (9.3\text{-}72)$$

The boundary conditions for the temperature, i.e.,

$$\text{at} \quad x = 0 \qquad T = T_o \qquad (9.3\text{-}73)$$

$$\text{at} \quad x = B \qquad T = T_1 \qquad (9.3\text{-}74)$$

suggest that $T = T(x)$. Therefore, Table C.4 in Appendix C indicates that the only nonzero energy flux component is e_x, and it is given by

$$e_x = q_x = -k \frac{dT}{dx} \qquad (9.3\text{-}75)$$

---

[3]The origin of this term comes from $-(\tau : \nabla v)$, which represents the irreversible degradation of mechanical energy into thermal energy in the equation of energy. For a more detailed discussion on the subject, see Bird *et al.* (2002).

For a rectangular volume element of thickness Δx, as shown in Figure 9.12, Eq. (9.2-1) is expressed as

$$q_x|_x WL - q_x|_{x+\Delta x} WL + \left(\frac{\mu V^2}{B^2}\right) WL\Delta x = 0 \qquad (9.3\text{-}76)$$

Dividing each term by $WL\Delta x$ and taking the limit as $\Delta x \to 0$ give

$$\lim_{\Delta x \to 0} \frac{q_x|_x - q_x|_{x+\Delta x}}{\Delta x} + \frac{\mu V^2}{B^2} = 0 \qquad (9.3\text{-}77)$$

or,

$$-\frac{dq_x}{dx} + \frac{\mu V^2}{B^2} = 0 \qquad (9.3\text{-}78)$$

Substitution of Eq. (9.3-75) into Eq. (9.3-78) gives the governing equation for temperature as

$$\boxed{k\frac{d^2T}{dx^2} + \frac{\mu V^2}{B^2} = 0} \qquad (9.3\text{-}79)$$

in which both viscosity and thermal conductivity are assumed to be independent of temperature. The physical significance and the order of magnitude of the terms in Eq. (9.3-79) are given in Table 9.1. Therefore, the ratio of the viscous dissipation to conduction, which is known as the *Brinkman number*, is given by

$$\text{Br} = \frac{\text{Viscous dissipation}}{\text{Conduction}} = \frac{\mu V^2/B^2}{k(T_o - T_1)/B^2} = \frac{\mu V^2}{k(T_o - T_1)} \qquad (9.3\text{-}80)$$

Before solving Eq. (9.3-79), it is convenient to express the governing equation and the boundary conditions in dimensionless form. Introduction of the dimensionless quantities

$$\theta = \frac{T - T_1}{T_o - T_1} \qquad \xi = \frac{x}{B} \qquad (9.3\text{-}81)$$

reduces Eqs. (9.3-79), (9.3–73), and (9.3-74) to

$$\frac{d^2\theta}{d\xi^2} = -\,\text{Br} \qquad (9.3\text{-}82)$$

Table 9.1. The physical significance and the order of magnitude of the terms in Eq. (9.3-79)

Term	Physical Significance	Order of Magnitude
$k\dfrac{d^2T}{dx^2}$	Conduction	$\dfrac{k(T_o - T_1)}{B^2}$
$\dfrac{\mu V^2}{B^2}$	Viscous dissipation	$\dfrac{\mu V^2}{B^2}$

$$\text{at}\quad \xi = 0 \qquad \theta = 1 \tag{9.3-83}$$

$$\text{at}\quad \xi = 1 \qquad \theta = 0 \tag{9.3-84}$$

Integration of Eq. (9.3-82) twice gives

$$\theta = -\frac{\text{Br}}{2}\xi^2 + C_1\xi + C_2 \tag{9.3-85}$$

Application of the boundary conditions, Eqs. (9.3-83) and (9.3-84), gives the solution as

$$\boxed{\theta = -\frac{\text{Br}}{2}\xi^2 + \left(\frac{\text{Br}}{2} - 1\right)\xi + 1} \tag{9.3-86}$$

Note that, when Br $= 0$, i.e., no viscous dissipation, Eq. (9.3-86) reduces to Eq. (8.3-10). The variation in θ as a function of ξ with Br as a parameter is shown in Figure 9.13.

In engineering calculations, it is more appropriate to express the solution in terms of the Nusselt number. Calculation of the Nusselt number, on the other hand, requires the evaluation of the bulk temperature defined by

$$T_b = \frac{\displaystyle\int_0^W \int_0^B v_z T\,dx\,dy}{\displaystyle\int_0^W \int_0^B v_z\,dx\,dy} = \frac{\displaystyle\int_0^B v_z T\,dx}{\displaystyle\int_0^B v_z\,dx} \tag{9.3-87}$$

In dimensionless form, Eq. (9.3-87) becomes

$$\theta_b = \frac{T_b - T_1}{T_o - T_1} = \frac{\displaystyle\int_0^1 \phi\theta\,d\xi}{\displaystyle\int_0^1 \phi\,d\xi} \tag{9.3-88}$$

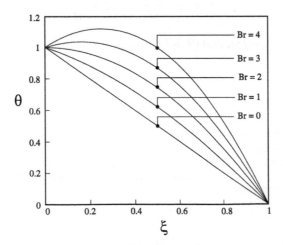

Figure 9.13. Variation in θ as a function of ξ with Br as a parameter.

where

$$\phi = \frac{v_z}{V} \tag{9.3-89}$$

Substitution of Eqs. (9.3-71) and (9.3-86) into Eq. (9.3-88) gives

$$\boxed{\theta_b = \frac{Br+8}{12}} \tag{9.3-90}$$

Calculation of the Nusselt number for the bottom plate

The heat flux at the bottom plate is expressed as

$$-k\frac{dT}{dx}\bigg|_{x=0} = \langle h\rangle_o (T_o - T_b) \tag{9.3-91}$$

Therefore, the Nusselt number becomes

$$Nu_o = \frac{\langle h\rangle_o (2B)}{k} = 2B\frac{-(dT/dx)_{x=0}}{T_o - T_b} \tag{9.3-92}$$

The term $2B$ in the definition of the Nusselt number represents the hydraulic equivalent diameter for parallel plates. In dimensionless form, Eq. (9.3-92) becomes

$$Nu_o = \frac{2(d\theta/d\xi)_{\xi=0}}{\theta_b - 1} \tag{9.3-93}$$

The use of Eq. (9.3-86) in Eq. (9.3-93) gives

$$\boxed{Nu_o = 12\left(\frac{Br-2}{Br-4}\right)} \tag{9.3-94}$$

Note that Nu_o takes the following values depending on the value of Br:

$$Nu_o = \begin{cases} 0 & Br = 2 \\ < 0 & 2 < Br < 4 \\ \infty & Br = 4 \end{cases} \tag{9.3-95}$$

When $Br = 2$, the temperature gradient at the lower plate is zero, i.e., it is an adiabatic surface. When $2 < Br < 4$, as can be seen from Figure 9.13, temperature reaches a maximum within the flow field. For example, for $Br = 3$, θ reaches the maximum value of 1.042 at $\xi = 0.167$ and heat transfer takes place from the fluid to the lower plate. When $Br = 4$, $\theta_b = 1$ from Eq. (9.3-90) and, as a result of very high viscous dissipation, T_b becomes uniform at the value of T_o. Since the driving force, i.e., $T_o - T_b$, is zero, Nu_o is undefined under these circumstances.

Calculation of the Nusselt number for the upper plate

The heat flux at the upper plate is

$$k\frac{dT}{dx}\bigg|_{x=B} = \langle h \rangle_1 (T_1 - T_b) \tag{9.3-96}$$

Therefore, the Nusselt number becomes

$$\text{Nu}_1 = \frac{\langle h \rangle_1 (2B)}{k} = 2B\frac{(dT/dx)_{x=B}}{T_1 - T_b} = -\frac{2(d\theta/d\xi)_{\xi=1}}{\theta_b} \tag{9.3-97}$$

Substitution of Eq. (9.3-86) into Eq. (9.3-97) gives

$$\boxed{\text{Nu}_1 = 12\left(\frac{\text{Br}+2}{\text{Br}+8}\right)} \tag{9.3-98}$$

9.4 MASS TRANSPORT WITHOUT CONVECTION

Under steady conditions, the conservation statement for species $\mathcal{A}$ is expressed by

$$\begin{pmatrix} \text{Rate of} \\ \text{species } \mathcal{A} \text{ in} \end{pmatrix} - \begin{pmatrix} \text{Rate of} \\ \text{species } \mathcal{A} \text{ out} \end{pmatrix} + \begin{pmatrix} \text{Rate of} \\ \text{species } \mathcal{A} \text{ generation} \end{pmatrix} = 0 \tag{9.4-1}$$

In this section, we restrict our analysis to cases in which convection is negligible and mass transfer takes place mainly by diffusion.

9.4.1 Diffusion in a Liquid with a Homogeneous Reaction

Gas $\mathcal{A}$ dissolves in liquid $\mathcal{B}$ and diffuses into the liquid phase as shown in Figure 9.14. As it diffuses, species $\mathcal{A}$ undergoes an irreversible chemical reaction with species $\mathcal{B}$ to form $\mathcal{AB}$, i.e.,

$$A + B \rightarrow AB$$

The rate of reaction is expressed by

$$r = kc_A$$

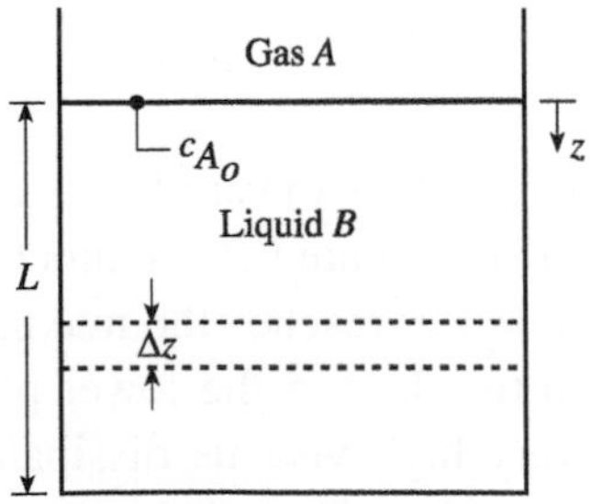

Figure 9.14. Diffusion and reaction in a liquid.

We are interested in the determination of the concentration distribution within the liquid phase and the rate of depletion of species $\mathcal{A}$.

The problem will be analyzed with the following assumptions:

1. Steady-state conditions prevail.
2. The convective flux is negligible with respect to the molecular flux.
3. The total concentration is constant, i.e.,

$$c = c_A + c_B + c_{AB} \simeq c_B$$

4. The concentration of $\mathcal{AB}$ does not interfere with the diffusion of $\mathcal{A}$ through $\mathcal{B}$, i.e., $\mathcal{A}$ molecules, for the most part, hit $\mathcal{B}$ molecules and hardly ever hit $\mathcal{AB}$ molecules. This is known as *pseudo-binary behavior.*

Since $c_A = c_A(z)$, Table C.8 in Appendix C indicates that the only nonzero molar flux component is N_{A_z}, and it is given by

$$N_{A_z} = J_{A_z}^* = -\mathcal{D}_{AB} \frac{dc_A}{dz} \tag{9.4-2}$$

For a differential volume element of thickness Δz, as shown in Figure 9.14, Eq. (9.4-1) is expressed as

$$N_{A_z}|_z A - N_{A_z}|_{z+\Delta z} A + \Re_A A \Delta z = 0 \tag{9.4-3}$$

Dividing Eq. (9.4-3) by $A \Delta z$ and taking the limit as $\Delta z \to 0$ give

$$\lim_{\Delta z \to 0} \frac{N_{A_z}|_z - N_{A_z}|_{z+\Delta z}}{\Delta z} + \Re_A = 0 \tag{9.4-4}$$

or,

$$-\frac{dN_{A_z}}{dz} + \Re_A = 0 \tag{9.4-5}$$

The use of Eq. (5.3-26) gives the rate of depletion of species $\mathcal{A}$ per unit volume as

$$\Re_A = -kc_A \tag{9.4-6}$$

Substitution of Eqs. (9.4-2) and (9.4-6) into Eq. (9.4-5) yields

$$\boxed{\mathcal{D}_{AB} \frac{d^2 c_A}{dz^2} - kc_A = 0} \tag{9.4-7}$$

The boundary conditions associated with the problem are

$$\text{at} \quad z = 0 \qquad c_A = c_{A_o} \tag{9.4-8}$$

$$\text{at} \quad z = L \qquad \frac{dc_A}{dz} = 0 \tag{9.4-9}$$

The value of c_{A_o} in Eq. (9.4-8) can be determined from Henry's law. The boundary condition given by Eq. (9.4-9) indicates that since species $\mathcal{A}$ cannot diffuse through the bottom of the

Table 9.2. The physical significance and the order of magnitude of the terms in Eq. (9.4-7)

Term	Physical Significance	Order of Magnitude
$\mathcal{D}_{AB}\dfrac{d^2 c_A}{dz^2}$	Rate of diffusion	$\mathcal{D}_{AB}\dfrac{c_{A_o}}{L^2}$
$k c_A$	Rate of reaction	$k c_{A_o}$

container, i.e., impermeable wall, then the molar flux and the concentration gradient of species $\mathcal{A}$ are zero.

The physical significance and the order of magnitude of the terms in Eq. (9.4-7) are given in Table 9.2. Therefore, the ratio of the rate of reaction to the rate of diffusion is given by

$$\frac{\text{Rate of reaction}}{\text{Rate of diffusion}} = \frac{k c_{A_o}}{\mathcal{D}_{AB} c_{A_o}/L^2} = \frac{kL^2}{\mathcal{D}_{AB}} \tag{9.4-10}$$

and the Thiele modulus[4], Λ, is defined by

$$\Lambda = \sqrt{\frac{kL^2}{\mathcal{D}_{AB}}} \tag{9.4-11}$$

Introduction of the dimensionless quantities

$$\theta = \frac{c_A}{c_{A_o}} \qquad \xi = \frac{z}{L} \tag{9.4-12}$$

reduces Eqs. (9.4-7)–(9.4-9) to the form

$$\frac{d^2\theta}{d\xi^2} = \Lambda^2 \theta \tag{9.4-13}$$

$$\text{at} \quad \xi = 0 \qquad \theta = 1 \tag{9.4-14}$$

$$\text{at} \quad \xi = 1 \qquad \frac{d\theta}{d\xi} = 0 \tag{9.4-15}$$

Note that Eqs. (9.4-13)–(9.4-15) are similar to Eqs. (8.2-82)–(8.2-84). Therefore, the solution is given by Eq. (8.2-88), i.e.,

$$\boxed{\theta = \frac{\cosh[\Lambda(1-\xi)]}{\cosh \Lambda}} \tag{9.4-16}$$

It is interesting to observe how the Thiele modulus affects the concentration distribution. Figure 9.15 shows variation in θ as a function of ξ with Λ being a parameter. Since the Thiele

---

[4] Since the reaction rate constant, k, has the unit of s^{-1}, the characteristic time, or time scale, for the reaction is given by

$$(t_{ch})rxn = \frac{1}{k}$$

Thus, the Thiele modulus can also be interpreted as the ratio of diffusive time scale to reaction time scale.

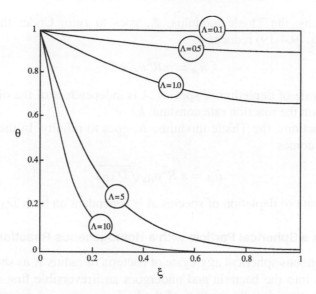

Figure 9.15. Variation in θ as a function of ξ with Λ being a parameter.

modulus indicates the rate of reaction with respect to the rate of diffusion, $\Lambda = 0$ implies no chemical reaction and hence $\theta = 1$ ($c_A = c_{A_o}$) for all ξ. Therefore, for very small values of Λ, θ is almost unity throughout the liquid. On the other hand, for large values of Λ, i.e., rate of reaction $\gg$ rate of diffusion, as soon as species $\mathcal{A}$ enters the liquid phase, it undergoes a homogeneous reaction with species $\mathcal{B}$. As a result, species $\mathcal{A}$ is depleted before it reaches the bottom of the container. Note that the slope of the tangent to the curve drawn at $\xi = 1$ has a zero slope, i.e., parallel to the ξ-axis.

9.4.1.1 *Macroscopic equation* Integration of the governing equation, Eq. (9.4-7), over the volume of the system gives

$$\int_0^L \int_0^{2\pi} \int_0^R \mathcal{D}_{AB} \frac{d^2 c_A}{dz^2} r \, dr \, d\theta \, dz = \int_0^L \int_0^{2\pi} \int_0^R k c_A r \, dr \, d\theta \, dz \qquad (9.4\text{-}17)$$

Carrying out the integrations yields

$$\underbrace{\pi R^2 \left(-\mathcal{D}_{AB} \frac{dc_A}{dz} \bigg|_{z=0} \right)}_{\substack{\text{Rate of moles of species } \mathcal{A} \\ \text{entering the liquid}}} = \underbrace{\pi R^2 k \int_0^L c_A \, dz}_{\substack{\text{Rate of depletion of species } \mathcal{A} \\ \text{by homogeneous chem. rxn.}}} \qquad (9.4\text{-}18)$$

which is the macroscopic inventory rate equation for species $\mathcal{A}$ by considering the liquid in the tank as a system. Substitution of Eq. (9.4-16) into Eq. (9.4-18) gives the molar rate of depletion of species $\mathcal{A}$, $\dot{n}_A$, as

$$\boxed{\dot{n}_A = \frac{\pi R^2 \mathcal{D}_{AB} c_{A_o} \Lambda \tanh \Lambda}{L}} \qquad (9.4\text{-}19)$$

For slow reactions, the Thiele modulus, Λ, goes to zero. Under these circumstances, $\tanh \Lambda \to \Lambda$ and Eq. (9.4-19) reduces to

$$\dot{n}_A = \pi R^2 c_{A_o} k L \tag{9.4-20}$$

indicating that the rate of depletion of species $\mathcal{A}$ is independent of the diffusion coefficient, $\mathcal{D}_{AB}$, and depends on the reaction rate constant, k.

For very fast reactions, the Thiele modulus, Λ, goes to infinity. In this case, $\tanh \Lambda \to 1$ and Eq. (9.4-19) becomes

$$\dot{n}_A = \pi R^2 c_{A_o} \sqrt{\mathcal{D}_{AB} k} \tag{9.4-21}$$

indicating that the rate of depletion of species $\mathcal{A}$ is dependent on both $\mathcal{D}_{AB}$ and k.

9.4.2 Diffusion in a Spherical Particle with a Homogeneous Reaction

Consider a homogeneous spherical aggregate of bacteria of radius R as shown in Figure 9.16. Species $\mathcal{A}$ diffuses into the bacteria and undergoes an irreversible first-order reaction. The concentration of species $\mathcal{A}$ at the surface of the bacteria, c_{A_R}, is known. We want to determine the rate of consumption of species $\mathcal{A}$. The problem will be analyzed with the following assumptions:

1. Steady-state conditions prevail.
2. Convective flux is negligible with respect to the molecular flux.
3. The total concentration is constant.

Since $c_A = c_A(r)$, Table C.9 in Appendix C indicates that the only nonzero molar flux component is N_{A_r}, and it is given by

$$N_{A_r} = J_{A_r}^* = -\mathcal{D}_{AB} \frac{dc_A}{dr} \tag{9.4-22}$$

For a spherical differential volume element of thickness Δr, as shown in Figure 9.16, Eq. (9.4-1) is expressed in the form

$$N_{A_r}|_r 4\pi r^2 - N_{A_r}|_{r+\Delta r} 4\pi (r + \Delta r)^2 + 4\pi r^2 \Delta r \, \Re_A = 0 \tag{9.4-23}$$

Dividing Eq. (9.4-23) by $4\pi \Delta r$ and taking the limit as $\Delta r \to 0$ give

$$\lim_{\Delta r \to 0} \frac{(r^2 N_{A_r})|_r - (r^2 N_{A_r})|_{r+\Delta r}}{\Delta r} + r^2 \Re_A = 0 \tag{9.4-24}$$

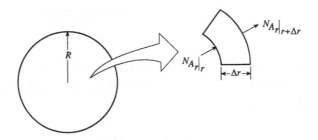

Figure 9.16. Diffusion and homogeneous reaction inside a spherical particle.

or,

$$-\frac{d(r^2 N_{A_r})}{dr} + r^2 \Re_A = 0 \qquad (9.4\text{-}25)$$

The use of Eq. (5.3-26) gives the rate of depletion of species $\mathcal{A}$ per unit volume as

$$\Re_A = -kc_A \qquad (9.4\text{-}26)$$

Substitution of Eqs. (9.4-22) and (9.4-26) into Eq. (9.4-25) gives

$$\boxed{\frac{\mathcal{D}_{AB}}{r^2} \frac{d}{dr}\left(r^2 \frac{dc_A}{dr} \right) - kc_A = 0} \qquad (9.4\text{-}27)$$

in which the diffusion coefficient is considered constant. The boundary conditions associated with Eq. (9.4-27) are

$$\text{at} \quad r = 0 \qquad \frac{dc_A}{dr} = 0 \qquad (9.4\text{-}28)$$

$$\text{at} \quad r = R \qquad c_A = c_{A_R} \qquad (9.4\text{-}29)$$

The physical significance and the order of magnitude of the terms in Eq. (9.4-27) are given in Table 9.3. Therefore, the ratio of the rate of reaction to the rate of diffusion is given by

$$\frac{\text{Rate of reaction}}{\text{Rate of diffusion}} = \frac{kc_{A_R}}{\mathcal{D}_{AB}c_{A_R}/R^2} = \frac{k\,R^2}{\mathcal{D}_{AB}} \qquad (9.4\text{-}30)$$

Introduction of the dimensionless quantities

$$\theta = \frac{c_A}{c_{A_R}} \qquad \xi = \frac{r}{R} \qquad \Lambda = \sqrt{\frac{k\,R^2}{\mathcal{D}_{AB}}} \qquad (9.4\text{-}31)$$

reduces Eqs. (9.4-27)–(9.4-29) to

$$\frac{1}{\xi^2} \frac{d}{d\xi}\left(\xi^2 \frac{d\theta}{d\xi} \right) - \Lambda^2 \theta = 0 \qquad (9.4\text{-}32)$$

$$\text{at} \quad \xi = 0 \qquad \frac{d\theta}{d\xi} = 0 \qquad (9.4\text{-}33)$$

$$\text{at} \quad \xi = 1 \qquad \theta = 1 \qquad (9.4\text{-}34)$$

Table 9.3. The physical significance and the order of magnitude of the terms in Eq. (9.4-27)

Term	Physical Significance	Order of Magnitude
$\dfrac{\mathcal{D}_{AB}}{r^2} \dfrac{d}{dr}\left(r^2 \dfrac{dc_A}{dr} \right)$	Rate of diffusion	$\mathcal{D}_{AB} \dfrac{c_{A_R}}{R^2}$
kc_A	Rate of reaction	kc_{A_R}

Problems in spherical coordinates are converted to rectangular coordinates by the use of the following transformation

$$\theta = \frac{u(\xi)}{\xi} \tag{9.4-35}$$

From Eq. (9.4-35), note that

$$\frac{d\theta}{d\xi} = \frac{1}{\xi}\frac{du}{d\xi} - \frac{u}{\xi^2} \tag{9.4-36}$$

$$\xi^2\frac{d\theta}{d\xi} = \xi\frac{du}{d\xi} - u \tag{9.4-37}$$

$$\frac{d}{d\xi}\left(\xi^2\frac{d\theta}{d\xi}\right) = \frac{du}{d\xi} + \xi\frac{d^2u}{d\xi^2} - \frac{du}{d\xi} = \xi\frac{d^2u}{d\xi^2} \tag{9.4-38}$$

Substitution of Eqs. (9.4-35) and (9.4-38) into Eq. (9.4-32) yields

$$\frac{d^2u}{d\xi^2} - \Lambda^2 u = 0 \tag{9.4-39}$$

On the other hand, the boundary conditions, Eqs. (9.4-33) and (9.4-34), become

$$\text{at} \quad \xi = 0 \quad u = 0 \tag{9.4-40}$$

$$\text{at} \quad \xi = 1 \quad u = 1 \tag{9.4-41}$$

The solution of Eq. (9.4-39) is

$$u = K_1 \sinh(\Lambda\xi) + K_2\cosh(\Lambda\xi) \tag{9.4-42}$$

where K_1 and K_2 are constants. Application of the boundary conditions, Eqs. (9.4-40) and (9.4-41), gives the solution as

$$u = \frac{\sinh(\Lambda\xi)}{\sinh\Lambda} \tag{9.4-43}$$

or,

$$\boxed{\frac{c_A}{c_{A_R}} = \frac{R}{r}\frac{\sinh[\Lambda(r/R)]}{\sinh\Lambda}} \tag{9.4-44}$$

9.4.2.1 *Macroscopic equation* Integration of the governing differential equation, Eq. (9.4-27), over the spherical aggregate of bacteria gives

$$\int_0^{2\pi}\int_0^{\pi}\int_0^{R}\frac{\mathcal{D}_{AB}}{r^2}\frac{d}{dr}\left(r^2\frac{dc_A}{dr}\right)r^2\sin\theta\,dr\,d\theta\,d\phi = \int_0^{2\pi}\int_0^{\pi}\int_0^{R}kc_A r^2\sin\theta\,dr\,d\theta\,d\phi \tag{9.4-45}$$

Carrying out the integrations yields

$$\underbrace{4\pi R^2 \mathcal{D}_{AB} \frac{dc_A}{dr}\bigg|_{r=R}}_{\substack{\text{Rate of moles of species } \mathcal{A} \\ \text{entering the bacteria}}} = \underbrace{4\pi k \int_0^R c_A r^2 \, dr}_{\substack{\text{Rate of consumption of species } \mathcal{A} \\ \text{by homogeneous chem. rxn.}}} \qquad (9.4\text{-}46)$$

Substitution of Eq. (9.4-44) into Eq. (9.4-46) gives the molar rate of consumption of species $\mathcal{A}$, $\dot{n}_A$, as

$$\boxed{\dot{n}_A = -4\pi R \, \mathcal{D}_{AB} \, c_{A_R} \left(1 - \Lambda \coth \Lambda\right)} \qquad (9.4\text{-}47)$$

The minus sign in Eq. (9.4-47) indicates that the flux is in the negative r-direction, i.e., towards the center of the sphere.

9.5 MASS TRANSPORT WITH CONVECTION

9.5.1 Laminar Forced Convection in a Pipe

Consider the laminar flow of an incompressible Newtonian liquid ($\mathcal{B}$) in a circular pipe under the action of a pressure gradient as shown in Figure 9.17. The velocity distribution is given by Eqs. (9.1-79) and (9.1-84) as

$$v_z = 2\langle v_z \rangle \left[1 - \left(\frac{r}{R}\right)^2\right] \qquad (9.5\text{-}1)$$

Suppose that the liquid has a uniform species $\mathcal{A}$ concentration of c_{A_o} for $z < 0$. For $z > 0$, species $\mathcal{A}$ concentration starts to change as a function of r and z as a result of mass transfer from the walls of the pipe. We want to develop the governing equation for species $\mathcal{A}$ concentration. Liquid viscosity is assumed to be unaffected by mass transfer.

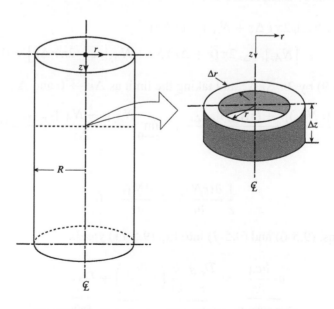

Figure 9.17. Forced convection mass transfer in a pipe.

From Table C.8 in Appendix C, the nonzero mass flux components for species $\mathcal{A}$ are

$$W_{A_r} = -\rho \mathcal{D}_{AB} \frac{\partial \omega_A}{\partial r} \tag{9.5-2}$$

$$W_{A_z} = -\rho \mathcal{D}_{AB} \frac{\partial \omega_A}{\partial z} + \rho_A v_z \tag{9.5-3}$$

For a dilute liquid solution, the total density is almost constant and Eqs. (9.5-2) and (9.5-3) become

$$W_{A_r} = -\mathcal{D}_{AB} \frac{\partial \rho_A}{\partial r} \tag{9.5-4}$$

$$W_{A_z} = -\mathcal{D}_{AB} \frac{\partial \rho_A}{\partial z} + \rho_A v_z \tag{9.5-5}$$

Dividing Eqs. (9.5-4) and (9.5-5) by the molecular weight of species $\mathcal{A}$, $\mathcal{M}_A$, gives

$$N_{A_r} = -\mathcal{D}_{AB} \frac{\partial c_A}{\partial r} \tag{9.5-6}$$

$$N_{A_z} = -\mathcal{D}_{AB} \frac{\partial c_A}{\partial z} + c_A v_z \tag{9.5-7}$$

Since there is no generation of species $\mathcal{A}$, Eq. (9.4-1) simplifies to

$$\begin{pmatrix} \text{Rate of} \\ \text{species } \mathcal{A} \text{ in} \end{pmatrix} - \begin{pmatrix} \text{Rate of} \\ \text{species } \mathcal{A} \text{ out} \end{pmatrix} = 0 \tag{9.5-8}$$

For a cylindrical differential volume element of thickness Δr and length Δz, as shown in Figure 9.17, Eq. (9.5-8) is expressed as

$$\left(N_{A_r}|_r 2\pi r \Delta z + N_{A_z}|_z 2\pi r \Delta r \right)$$
$$- \left[N_{A_r}|_{r+\Delta r} 2\pi (r + \Delta r) \Delta z + N_{A_z}|_{z+\Delta z} 2\pi r \Delta r \right] = 0 \tag{9.5-9}$$

Dividing Eq. (9.5-9) by $2\pi \Delta r \Delta z$ and taking the limit as $\Delta r \to 0$ and $\Delta z \to 0$ give

$$\frac{1}{r} \lim_{\Delta r \to 0} \frac{(r N_{A_r})|_r - (r N_{A_r})|_{r+\Delta r}}{\Delta r} + \lim_{\Delta z \to 0} \frac{N_{A_z}|_z - N_{A_z}|_{z+\Delta z}}{\Delta z} = 0 \tag{9.5-10}$$

or,

$$\frac{1}{r} \frac{\partial (r N_{A_r})}{\partial r} + \frac{\partial N_{A_z}}{\partial z} = 0 \tag{9.5-11}$$

Substitution of Eqs. (9.5-6) and (9.5-7) into Eq. (9.5-11) yields

$$\underbrace{v_z \frac{\partial c_A}{\partial z}}_{\substack{\text{Convection in} \\ z\text{-direction}}} = \underbrace{\frac{\mathcal{D}_{AB}}{r} \frac{\partial}{\partial r} \left(r \frac{\partial c_A}{\partial r} \right)}_{\text{Diffusion in } r\text{-direction}} + \underbrace{\mathcal{D}_{AB} \frac{\partial^2 c_A}{\partial z^2}}_{\substack{\text{Diffusion in} \\ z\text{-direction}}} \tag{9.5-12}$$

In the z-direction, the mass of species $\mathcal{A}$ is transported by both convection and diffusion. As stated by Eq. (2.4-8), diffusion can be considered negligible with respect to convection when $\text{Pe}_M \gg 1$. Under these circumstances, Eq. (9.5-12) reduces to

$$v_z \frac{\partial c_A}{\partial z} = \frac{\mathcal{D}_{AB}}{r} \frac{\partial}{\partial r}\left(r \frac{\partial c_A}{\partial r}\right) \tag{9.5-13}$$

As engineers, we are interested in the variation in the bulk concentration of species $\mathcal{A}$, c_{A_b}, rather than the local concentration, c_A. For forced convection mass transfer in a circular pipe of radius R, the bulk concentration defined by Eq. (4.1-1) takes the form

$$c_{A_b} = \frac{\displaystyle\int_0^{2\pi}\int_0^R v_z c_A r\, dr\, d\theta}{\displaystyle\int_0^{2\pi}\int_0^R v_z r\, dr\, d\theta} \tag{9.5-14}$$

In general, the concentration of species $\mathcal{A}$, c_A, may depend on both the radial and axial coordinates. However, the bulk concentration of species $\mathcal{A}$, c_{A_b}, depends only on the axial direction.

To determine the governing equation for the bulk concentration of species $\mathcal{A}$, it is necessary to integrate Eq. (9.5-13) over the cross-sectional area of the tube, i.e.,

$$\int_0^{2\pi}\int_0^R v_z \frac{\partial c_A}{\partial z} r\, dr\, d\theta = \mathcal{D}_{AB}\int_0^{2\pi}\int_0^R \frac{1}{r}\frac{\partial}{\partial r}\left(r\frac{\partial c_A}{\partial r}\right) r\, dr\, d\theta \tag{9.5-15}$$

Since $v_z \neq v_z(z)$, the integral on the left-hand side of Eq. (9.5-15) can be rearranged as

$$\int_0^{2\pi}\int_0^R v_z \frac{\partial c_A}{\partial z} r\, dr\, d\theta = \int_0^{2\pi}\int_0^R \frac{\partial(v_z c_A)}{\partial z} r\, dr\, d\theta = \frac{d}{dz}\left(\int_0^{2\pi}\int_0^R v_z c_A r\, dr\, d\theta\right) \tag{9.5-16}$$

Substitution of Eq. (9.5-14) into Eq. (9.5-16) yields

$$\int_0^{2\pi}\int_0^R v_z \frac{\partial c_A}{\partial z} r\, dr\, d\theta = \frac{d}{dz}\left(c_{A_b}\underbrace{\int_0^{2\pi}\int_0^R v_z r\, dr\, d\theta}_{\mathcal{Q}}\right) = \mathcal{Q}\frac{dc_{A_b}}{dz} \tag{9.5-17}$$

where $\mathcal{Q}$ is the volumetric flow rate.

On the other hand, since $\partial c_A/\partial r = 0$ as a result of the symmetry condition at the center of the tube, the integral on the right-hand side of Eq. (9.5-15) takes the form

$$\int_0^{2\pi}\int_0^R \frac{1}{r}\frac{\partial}{\partial r}\left(r\frac{\partial c_A}{\partial r}\right) r\, dr\, d\theta = \pi D \frac{\partial c_A}{\partial r}\bigg|_{r=R} \tag{9.5-18}$$

Substitution of Eqs. (9.5-17) and (9.5-18) into Eq. (9.5-15) gives the governing equation for the bulk concentration in the form

$$\mathcal{Q}\frac{dc_{A_b}}{dz} = \pi D\, \mathcal{D}_{AB}\frac{\partial c_A}{\partial r}\bigg|_{r=R} \tag{9.5-19}$$

The solution of Eq. (9.5-19) requires the boundary conditions associated with the problem to be known.

Constant wall concentration

If the inner surface of the pipe is coated with species $\mathcal{A}$, the molar flux of species $\mathcal{A}$ on the surface can be represented by

$$\mathcal{D}_{AB}\frac{\partial c_A}{\partial r}\bigg|_{r=R} = k_c(c_{A_w} - c_{A_b}) \tag{9.5-20}$$

It is implicitly implied in writing Eq. (9.5-20) that the concentration increases in the radial direction. Substitution of Eq. (9.5-20) into Eq. (9.5-19) and rearrangement yield

$$\mathcal{Q}\int_{c_{A_{b_{in}}}}^{c_{A_b}} \frac{dc_{A_b}}{c_{A_w} - c_{A_b}} = \pi D \int_0^z k_c\, dz \tag{9.5-21}$$

Since the wall concentration, c_{A_w}, is constant, integration of Eq. (9.5-21) yields

$$\mathcal{Q}\ln\left(\frac{c_{A_w} - c_{A_{b_{in}}}}{c_{A_w} - c_{A_b}}\right) = \pi D \langle k_c\rangle_z z \tag{9.5-22}$$

in which $\langle k_c\rangle_z$ is the average mass transfer coefficient from the entrance to the point z defined by

$$\langle k_c\rangle_z = \frac{1}{z}\int_0^z k_c\, dz \tag{9.5-23}$$

If Eq. (9.5-22) is solved for c_{A_b}, the result is

$$c_{A_b} = c_{A_w} - (c_{A_w} - c_{A_{b_{in}}})\exp\left[-\left(\frac{\pi D \langle k_c\rangle_z}{\mathcal{Q}}\right)z\right] \tag{9.5-24}$$

which indicates that the bulk concentration of species $\mathcal{A}$ varies exponentially with the axial direction as shown in Figure 9.18.

Evaluation of Eq. (9.5-22) over the total length, L, of the pipe gives

$$\mathcal{Q}\ln\left(\frac{c_{A_w} - c_{A_{b_{in}}}}{c_{A_w} - c_{A_{b_{out}}}}\right) = \pi D \langle k_c\rangle L \tag{9.5-25}$$

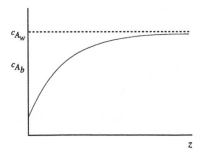

Figure 9.18. Variation in the bulk concentration of species $\mathcal{A}$ with the axial direction for a constant wall concentration.

where

$$\langle k_c \rangle = \frac{1}{L} \int_0^L k_c \, dz \tag{9.5-26}$$

If Eq. (9.5-25) is solved for $c_{A_{bout}}$, the result is

$$c_{A_{bout}} = c_{A_w} - (c_{A_w} - c_{A_{bin}}) \exp\left[-\left(\frac{\pi D \langle k_c \rangle}{Q} \right) L \right] \tag{9.5-27}$$

Equation (9.5-27) can be expressed in terms of dimensionless numbers with the help of Eq. (3.4-6). The result is

$$\text{St}_M = \frac{\text{Sh}}{\text{Re}\,\text{Sc}} = \frac{\langle k_c \rangle}{\langle v_z \rangle} = \frac{\langle k_c \rangle}{Q/(\pi D^2/4)} \tag{9.5-28}$$

The use of Eq. (9.5-28) in Eq. (9.5-27) gives

$$c_{A_{bout}} = c_{A_w} - (c_{A_w} - c_{A_{bin}}) \exp\left[-\frac{4\,\text{Sh}(L/D)}{\text{Re}\,\text{Sc}} \right] \tag{9.5-29}$$

As engineers, we are interested in the rate of moles of species A transferred to the fluid, i.e.,

$$\dot{n}_A = Q\,(c_{A_{bout}} - c_{A_{bin}}) = Q\big[(c_{A_w} - c_{A_{bin}}) - (c_{A_w} - c_{A_{bout}})\big] \tag{9.5-30}$$

Substitution of Eq. (9.5-25) into Eq. (9.5-30) results in

$$\dot{n}_A = (\pi D L)\langle k_c \rangle \left[\frac{(c_{A_w} - c_{A_{bin}}) - (c_{A_w} - c_{A_{bout}})}{\ln\left(\dfrac{c_{A_w} - c_{A_{bin}}}{c_{A_w} - c_{A_{bout}}} \right)} \right] \tag{9.5-31}$$

Note that Eq. (9.5-31) can be expressed in the form

$$\dot{n}_A = A_M \langle k_c \rangle (\Delta c_A)_{ch} = (\pi D L)\langle k_c \rangle (\Delta c_A)_{LM} \tag{9.5-32}$$

which is identical to Eqs. (3.3-7) and (4.5-34).

Constant wall mass flux

Consider a circular pipe with a porous wall. If species A is forced through the porous wall at a specified rate per unit area, then the molar flux of species A on the pipe surface remains constant, i.e.,

$$N_{A_r}|_{r=R} = D_{AB} \left. \frac{\partial c_A}{\partial r} \right|_{r=R} = N_{A_w} = \text{constant} \tag{9.5-33}$$

Substitution of Eq. (9.5-33) into Eq. (9.5-19) gives

$$\frac{dc_{A_b}}{dz} = \frac{\pi D N_{A_w}}{Q} = \text{constant} \tag{9.5-34}$$

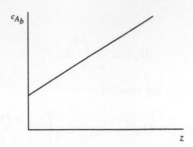

Figure 9.19. Variation in the bulk concentration of species $\mathcal{A}$ with the axial direction for a constant wall heat flux.

Integration of Eq. (9.5-34) gives the variation in the bulk concentration of species $\mathcal{A}$ in the axial direction as

$$c_{A_b} = c_{A_{b_{in}}} + \left(\frac{\pi D N_{A_w}}{Q}\right)z \tag{9.5-35}$$

Therefore, the bulk concentration of species $\mathcal{A}$ varies linearly in the axial direction as shown in Figure 9.19.

Evaluation of Eq. (9.5-35) over the total length gives the bulk concentration of species $\mathcal{A}$ at the exit of the pipe as

$$c_{A_{b_{out}}} = c_{A_{b_{in}}} + \left(\frac{\pi D N_{A_w}}{Q}\right)L = c_{A_{b_{in}}} + \frac{4 N_{A_w} L}{\mathcal{D}_{AB}\,\mathrm{Re}\,\mathrm{Sc}} \tag{9.5-36}$$

The rate of moles of species $\mathcal{A}$ transferred is given by

$$\dot{n}_A = Q\left(c_{A_{b_{out}}} - c_{A_{b_{in}}}\right) \tag{9.5-37}$$

Substitution of Eq. (9.5-36) into Eq. (9.5-37) yields

$$\boxed{\dot{n}_A = (\pi D L) N_{A_w}} \tag{9.5-38}$$

9.5.1.1 *Fully developed concentration profile*　　If the ratio

$$\frac{c_A - c_{A_b}}{c_{A_w} - c_{A_b}} \tag{9.5-39}$$

does not vary along the axial direction, then the concentration profile is said to be fully developed.

It is important to note that, although the local concentration, c_A, the bulk concentration, c_{A_b}, and the wall concentration, c_{A_w}, may change along the axial direction, the ratio given in

Eq. (9.5-39) is independent of the axial coordinate[5], i.e.,

$$\frac{\partial}{\partial z}\left(\frac{c_A - c_{A_b}}{c_{A_w} - c_{A_b}}\right) = 0 \tag{9.5-40}$$

Equation (9.5-40) indicates that

$$\frac{\partial c_A}{\partial z} = \left(\frac{c_{A_w} - c_A}{c_{A_w} - c_{A_b}}\right)\frac{dc_{A_b}}{dz} + \left(\frac{c_A - c_{A_b}}{c_{A_w} - c_{A_b}}\right)\frac{dc_{A_w}}{dz} \tag{9.5-41}$$

Example 9.10 Consider the flow of a fluid with constant physical properties. Show that the local mass transfer coefficient is a constant when the concentration profile is fully developed.

Solution

For a fully developed concentration profile, the ratio given in Eq. (9.5-39) depends only on the radial coordinate r, i.e.,

$$\frac{c_A - c_{A_b}}{c_{A_w} - c_{A_b}} = f(r) \tag{1}$$

Differentiation of Eq. (1) with respect to r gives

$$\frac{\partial c_A}{\partial r} = (c_{A_w} - c_{A_b})\frac{df}{dr} \tag{2}$$

which is valid at all points within the flow field. Evaluation of Eq. (2) at the surface of the pipe yields

$$\left.\frac{\partial c_A}{\partial r}\right|_{r=R} = (c_{A_w} - c_{A_b})\left.\frac{df}{dr}\right|_{r=R} \tag{3}$$

On the other hand, the molar flux of species $\mathcal{A}$ at the pipe surface is expressed as

$$N_{A_w} = \mathcal{D}_{AB}\left.\frac{\partial c_A}{\partial r}\right|_{r=R} = k_c(c_{A_w} - c_{A_b}) \tag{4}$$

Substitution of Eq. (3) into Eq. (4) gives

$$k_c = \mathcal{D}_{AB}\left(\frac{df}{dr}\right)_{r=R} = \text{constant} \tag{5}$$

Example 9.11 When the concentration profile is fully developed, show that the concentration gradient in the axial direction, $\partial c_A/\partial z$, remains constant for a constant wall mass flux.

---

[5]In the literature, the condition for *the fully developed concentration profile* is also given in the form

$$\frac{\partial}{\partial z}\left(\frac{c_{A_w} - c_A}{c_{A_w} - c_{A_b}}\right) = 0$$

Note that

$$\frac{c_{A_w} - c_A}{c_{A_w} - c_{A_b}} = 1 - \frac{c_A - c_{A_b}}{c_{A_w} - c_{A_b}}.$$

Solution

The molar flux of species $\mathcal{A}$ at the surface of the pipe is given by

$$N_{A_r}|_{r=R} = k_c(c_{A_w} - c_{A_b}) = \text{constant} \tag{1}$$

Since k_c is constant for a fully developed concentration profile, Eq. (1) implies that

$$c_{A_w} - c_{A_b} = \text{constant} \tag{2}$$

or,

$$\frac{dc_{A_w}}{dz} = \frac{dc_{A_b}}{dz} \tag{3}$$

Therefore, Eq. (9.5-41) simplifies to

$$\frac{\partial c_A}{\partial z} = \frac{dc_{A_b}}{dz} = \frac{dc_{A_w}}{dz} \tag{4}$$

Since dc_{A_b}/dz is constant according to Eq. (9.5-34), $\partial c_A/\partial z$ also remains constant, i.e.,

$$\frac{\partial c_A}{\partial z} = \frac{dc_{A_b}}{dz} = \frac{dc_{A_w}}{dz} = \frac{\pi D N_{A_w}}{Q} = \text{constant} \tag{5}$$

9.5.1.2 *Sherwood number for a fully developed concentration profile* Substitution of Eq. (9.5-1) into Eq. (9.5-13) gives

$$2\langle v_z\rangle\left[1 - \left(\frac{r}{R}\right)^2\right]\frac{\partial c_A}{\partial z} = \frac{\mathcal{D}_{AB}}{r}\frac{\partial}{\partial r}\left(r\frac{\partial c_A}{\partial r}\right) \tag{9.5-42}$$

It should always be kept in mind that the purpose of solving the above equation for concentration distribution is to obtain a correlation to calculate the number of moles of species $\mathcal{A}$ transferred between the phases. As shown in Chapter 4, mass transfer correlations are expressed in terms of the Sherwood number. Therefore, Eq. (9.5-42) will be solved for a fully developed concentration profile for two different types of boundary conditions, i.e., constant wall mass flux and constant wall concentration, to determine the Sherwood number.

Constant wall mass flux

As shown in Example 9.11, in the case of a constant wall mass flux, the concentration gradient in the axial direction is constant and expressed in the form

$$\frac{\partial c_A}{\partial z} = \frac{\pi D N_{A_w}}{Q} = \text{constant} \tag{9.5-43}$$

Since we are interested in the determination of the Sherwood number, it is appropriate to express $\partial c_A/\partial z$ in terms of the Sherwood number. Note that the Sherwood number is given by

$$\text{Sh} = \frac{k_c D}{\mathcal{D}_{AB}} = \frac{[N_{A_w}/(c_{A_w} - c_{A_b})]D}{\mathcal{D}_{AB}} \tag{9.5-44}$$

Therefore, Eq. (9.5-43) reduces to

$$\frac{\partial c_A}{\partial z} = \frac{\text{Sh}(c_{A_w} - c_{A_b})\,\mathcal{D}_{AB}}{R^2\langle v_z \rangle} \tag{9.5-45}$$

Substitution of Eq. (9.5-45) into Eq. (9.5-42) yields

$$\frac{2}{R^2}\left[1 - \left(\frac{r}{R}\right)^2\right]\text{Sh}(c_{A_w} - c_{A_b}) = \frac{1}{r}\frac{\partial}{\partial r}\left(r\,\frac{\partial c_A}{\partial r}\right) \tag{9.5-46}$$

In terms of the dimensionless variables

$$\theta = \frac{c_A - c_{A_b}}{c_{A_w} - c_{A_b}} \tag{9.5-47}$$

$$\xi = \frac{r}{R} \tag{9.5-48}$$

Eq. (9.5-46) takes the form

$$2\,\text{Sh}(1 - \xi^2) = \frac{1}{\xi}\frac{d}{d\xi}\left(\xi\,\frac{d\theta}{d\xi}\right) \tag{9.5-49}$$

It is important to note that θ depends only on ξ (or r).

The boundary conditions associated with Eq. (9.5-49) are

$$\text{at} \quad \xi = 0 \qquad \frac{d\theta}{d\xi} = 0 \tag{9.5-50}$$

$$\text{at} \quad \xi = 1 \qquad \theta = 1 \tag{9.5-51}$$

Note that Eqs. (9.5-49)–(9.5-51) are identical to Eqs. (9.3-46)–(9.3-48) with the only exception that Nu is replaced by Sh. Therefore, the solution is given by Eq. (9.3-50), i.e.,

$$\theta = 1 - \frac{\text{Sh}}{8}(3 - 4\xi^2 + \xi^4) \tag{9.5-52}$$

On the other hand, the bulk concentration in dimensionless form can be expressed as

$$\theta_b = \frac{c_{A_b} - c_{A_b}}{c_{A_w} - c_{A_b}} = 0 = \frac{\displaystyle\int_0^1 (1 - \xi^2)\,\theta\,\xi\,d\xi}{\displaystyle\int_0^1 (1 - \xi^2)\,\xi\,d\xi} \tag{9.5-53}$$

Substitution of Eq. (9.5-52) into Eq. (9.5-53) gives the Sherwood number as

$$\boxed{\text{Sh} = \frac{48}{11}} \tag{9.5-54}$$

Constant wall concentration

When the wall concentration is constant, Eq. (9.5-41) indicates that

$$\frac{\partial c_A}{\partial z} = \left(\frac{c_{A_w} - c_A}{c_{A_w} - c_{A_b}}\right)\frac{dc_{A_b}}{dz} \tag{9.5-55}$$

The variation in c_{A_b} as a function of the axial position can be obtained from Eq. (9.5-24) as

$$\frac{dc_{A_b}}{dz} = \frac{\pi D \langle k_c\rangle_z}{Q} \underbrace{(c_{A_w} - c_{A_{b_{in}}}) \exp\left[-\left(\frac{\pi D \langle k_c\rangle_z}{Q}\right)z\right]}_{(c_{A_w} - c_{A_b})} \tag{9.5-56}$$

Since the mass transfer coefficient is constant for a fully developed concentration profile, Eq. (9.5-56) becomes

$$\frac{dc_{A_b}}{dz} = \frac{\pi D k_c (c_{A_w} - c_{A_b})}{Q} = \frac{4k_c (c_{A_w} - c_{A_b})}{D \langle v_z\rangle} \tag{9.5-57}$$

The use of Eq. (9.5-57) in Eq. (9.5-55) yields

$$\frac{\partial c_A}{\partial z} = \frac{4k_c (c_{A_w} - c_A)}{D \langle v_z\rangle} \tag{9.5-58}$$

Substitution of Eq. (9.5-58) into Eq. (9.5-42) gives

$$\frac{8}{D^2}\left(\frac{k_c D}{\mathcal{D}_{AB}}\right)\left[1 - \left(\frac{r}{R}\right)^2\right](c_{A_w} - c_A) = \frac{1}{r}\frac{\partial}{\partial r}\left(r\frac{\partial c_A}{\partial r}\right) \tag{9.5-59}$$

In terms of the dimensionless variables defined by Eqs. (9.5-47) and (9.5-48), Eq. (9.5-59) becomes

$$2\,\mathrm{Sh}(1 - \xi^2)(1 - \theta) = \frac{1}{\xi}\frac{d}{d\xi}\left(\xi\frac{d\theta}{d\xi}\right) \tag{9.5-60}$$

The boundary conditions associated with Eq. (9.5-60) are

$$\text{at}\quad \xi = 0 \qquad \frac{d\theta}{d\xi} = 0 \tag{9.5-61}$$

$$\text{at}\quad \xi = 1 \qquad \theta = 1 \tag{9.5-62}$$

The use of the substitution

$$u = 1 - \theta \tag{9.5-63}$$

reduces Eqs. (9.5-60)–(9.5-62) to

$$-2\,\mathrm{Sh}(1 - \xi^2)\,u = \frac{1}{\xi}\frac{d}{d\xi}\left(\xi\frac{du}{d\xi}\right) \tag{9.5-64}$$

$$\text{at} \quad \xi = 0 \qquad \frac{du}{d\xi} = 0 \tag{9.5-65}$$

$$\text{at} \quad \xi = 1 \qquad u = 0 \tag{9.5-66}$$

Equation (9.5-64) can be solved for Sh by the method of Stodola and Vianello as explained in Section B.3.4.1 in Appendix B.

A reasonable first guess for u that satisfies the boundary conditions is

$$u_1 = 1 - \xi^2 \tag{9.5-67}$$

Substitution of Eq. (9.5-67) into the left-hand side of Eq. (9.5-64) gives

$$\frac{d}{d\xi}\left(\xi\frac{du}{d\xi}\right) = -2\,\mathrm{Sh}(\xi - 2\xi^3 + \xi^5) \tag{9.5-68}$$

The solution of Eq. (9.5-68) is

$$u = \mathrm{Sh}\underbrace{\left(\frac{11 - 18\xi^2 + 9\xi^4 - 2\xi^6}{36}\right)}_{f_1(\xi)} \tag{9.5-69}$$

Therefore, the first approximation to the Sherwood number is

$$\mathrm{Sh}^{(1)} = \frac{\displaystyle\int_0^1 \xi(1-\xi^2)^2 f_1(\xi)\,d\xi}{\displaystyle\int_0^1 \xi(1-\xi^2) f_1^2(\xi)\,d\xi} \tag{9.5-70}$$

Substitution of $f_1(\xi)$ from Eq. (9.5-69) into Eq. (9.5-70) and evaluation of the integrals give

$$\mathrm{Sh} = 3.663 \tag{9.5-71}$$

On the other hand, the value of the Sherwood number, as calculated by Graetz (1883, 1885) and Nusselt (1910), is 3.66. Therefore, for a fully developed concentration profile in a circular pipe with a constant wall concentration, $\mathrm{Sh} = 3.66$ for all practical purposes.

9.5.1.3 *Sherwood number for a fully developed velocity profile*   For water flowing in a circular pipe of diameter D at a Reynolds number of 100 and at a temperature of $20\,°C$, Skelland (1974) calculated the length of the tube, L, required for the velocity, temperature, and concentration distributions to reach a fully developed profile as

$$L = \begin{cases} 5D & \text{fully developed velocity profile} \\ 35D & \text{fully developed temperature profile} \\ 6000D & \text{fully developed concentration profile} \end{cases} \tag{9.5-72}$$

Therefore, a fully developed concentration profile is generally not attained for fluids with high Schmidt numbers, and the use of Eqs. (9.5-54) and (9.5-71) may lead to erroneous results.

When the velocity profile is fully developed, it is recommended to use the following semi-empirical correlations suggested by Hausen (1943):

$$\boxed{Sh = 3.66 + \frac{0.668[(D/L)\,Re\,Sc]}{1 + 0.04[(D/L)\,Re\,Sc]^{2/3}}} \quad c_{A_w} = \text{constant} \qquad (9.5\text{-}73)$$

$$\boxed{Sh = 4.36 + \frac{0.023[(D/L)\,Re\,Sc]}{1 + 0.0012[(D/L)\,Re\,Sc]}} \quad N_{A_w} = \text{constant} \qquad (9.5\text{-}74)$$

In the calculation of the mass transfer rates by the use of Eqs. (9.5-73) and (9.5-74), the appropriate driving force is the log-mean concentration difference.

Example 9.12 Pure water at 25 °C flows through a smooth metal pipe of 6 cm internal diameter with an average velocity of 1.5×10^{-3} m/s. Once the fully developed velocity profile is established, the metal pipe is replaced by a pipe, cast from benzoic acid, of the same inside diameter. If the length of the pipe made of a benzoic acid is 2 m, calculate the concentration of benzoic acid in water at the exit of the pipe.

Solution

Physical properties

From Example 4.8:

For water ($\mathcal{B}$) at 25 °C (298 K): $\begin{cases} \rho = 1000 \text{ kg/m}^3 \\ \mu = 892 \times 10^{-6} \text{ kg/m·s} \\ \mathcal{D}_{AB} = 1.21 \times 10^{-9} \text{ m}^2/\text{s} \end{cases}$

$Sc = 737$

Saturation solubility of benzoic acid ($\mathcal{A}$) in water $= 3.412$ kg/m^3.

Analysis

The Reynolds number is

$$Re = \frac{D\langle v_z \rangle \rho}{\mu} = \frac{(6 \times 10^{-2})(1.5 \times 10^{-3})(1000)}{892 \times 10^{-6}} = 101 \quad \Rightarrow \quad \text{Laminar flow} \qquad (1)$$

Note that the term $(D/L)\,Re\,Sc$ becomes

$$\left(\frac{D}{L}\right) Re\,Sc = \left(\frac{6 \times 10^{-2}}{2}\right)(101)(737) = 2233 \qquad (2)$$

Since the concentration at the surface of the pipe is constant, the use of Eq. (9.5-73) gives

$$Sh = 3.66 + \frac{0.668[(D/L)\,Re\,Sc]}{1 + 0.04[(D/L)\,Re\,Sc]^{2/3}} = 3.66 + \frac{0.0668(2233)}{1 + 0.04(2233)^{2/3}} = 22.7 \qquad (3)$$

Considering the water in the pipe as a system, a macroscopic mass balance on benzoic acid gives

$$\underbrace{(\pi D^2/4)\langle v_z\rangle}_{Q}\big[(c_{A_b})_{out} - (c_{A_b})_{in}\big] = \underbrace{(\pi D L)}_{A_M}\langle k_c\rangle \underbrace{\frac{[c_{A_w} - (c_{A_b})_{out}] - [c_{A_w} - (c_{A_b})_{in}]}{\ln\left[\dfrac{c_{A_w} - (c_{A_b})_{out}}{c_{A_w} - (c_{A_b})_{in}}\right]}}_{(\Delta c_A)_{LM}} \tag{4}$$

Since $(c_{A_b})_{in} = 0$, Eq. (4) simplifies to

$$(c_{A_b})_{out} = c_{A_w}\left[1 - \exp\left(-\frac{4L}{D}\frac{\langle k_c\rangle}{\langle v_z\rangle}\right)\right] = c_{A_w}\left[1 - \exp\left(-\frac{4L}{D}\frac{\text{Sh}}{\text{Re}\,\text{Sc}}\right)\right] \tag{5}$$

Substitution of the numerical values into Eq. (5) gives

$$(c_{A_b})_{out} = 3.412\left\{1 - \exp\left[-\frac{4(2)(22.7)}{(6\times10^{-2})(101)(737)}\right]\right\} = 0.136 \text{ kg/m}^3 \tag{6}$$

Comment: One could also use Eq. (4.5-31) to calculate the Sherwood number, i.e.,

$$\text{Sh} = 1.86[\text{Re}\,\text{Sc}(D/L)]^{1/3} = 1.86(2233)^{1/3} = 24.3$$

which is not very different from 22.7.

9.5.2   Diffusion into a Falling Liquid Film

Consider gas absorption in a wetted-wall column as shown in Figure 9.20. An incompressible Newtonian liquid ($\mathcal{B}$) flows in laminar flow over a flat plate of width W and length L as a thin film of thickness δ under the action of gravity. Gas $\mathcal{A}$ flows in a countercurrent direction to the liquid and we want to determine the amount of $\mathcal{A}$ absorbed by the liquid. The fully developed velocity distribution is given by Eqs. (9.1-57) and (9.1-58) as

$$v_z = v_{\max}\left[1 - \left(\frac{x}{\delta}\right)^2\right] \tag{9.5-75}$$

where

$$v_{\max} = \frac{3}{2}\langle v_z\rangle = \frac{\rho g \delta^2}{2\mu} \tag{9.5-76}$$

Liquid viscosity is assumed to be unaffected by mass transfer.

In general, the concentration of species $\mathcal{A}$ in the liquid phase changes as a function of x and z. Therefore, from Table C.7 in Appendix C, the nonzero mass flux components are

$$\mathcal{W}_{A_x} = -\rho\mathcal{D}_{AB}\frac{\partial \omega_A}{\partial x} \tag{9.5-77}$$

$$\mathcal{W}_{A_z} = -\rho\mathcal{D}_{AB}\frac{\partial \omega_A}{\partial z} + \rho_A v_z \tag{9.5-78}$$

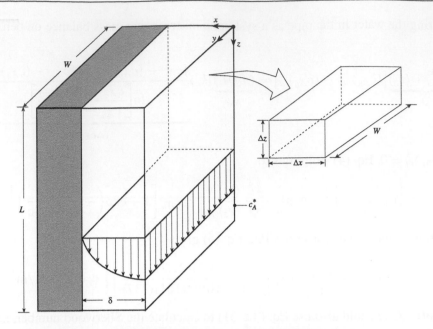

Figure 9.20. Diffusion into a falling liquid film.

For a dilute liquid solution, the total density is almost constant and Eqs. (9.5-77) and (9.5-78) become

$$\mathcal{W}_{A_x} = -\mathcal{D}_{AB}\frac{\partial \rho_A}{\partial x} \tag{9.5-79}$$

$$\mathcal{W}_{A_z} = -\mathcal{D}_{AB}\frac{\partial \rho_A}{\partial z} + \rho_A v_z \tag{9.5-80}$$

Dividing Eqs. (9.5-79) and (9.5-80) by the molecular weight of species $\mathcal{A}$, $\mathcal{M}_A$, gives

$$N_{A_x} = -\mathcal{D}_{AB}\frac{\partial c_A}{\partial x} \tag{9.5-81}$$

$$N_{A_z} = -\mathcal{D}_{AB}\frac{\partial c_A}{\partial z} + c_A v_z \tag{9.5-82}$$

Since there is no generation of species $\mathcal{A}$, Eq. (9.4-1) simplifies to

$$\left(\begin{array}{c}\text{Rate of}\\ \text{species }\mathcal{A}\text{ in}\end{array}\right) - \left(\begin{array}{c}\text{Rate of}\\ \text{species }\mathcal{A}\text{ out}\end{array}\right) = 0 \tag{9.5-83}$$

For a rectangular differential volume element of thickness Δx, length Δz, and width W, as shown in Figure 9.20, Eq. (9.5-83) is expressed as

$$\left(N_{A_x}|_x W\Delta z + N_{A_z}|_z W\Delta x\right) - \left(N_{A_x}|_{x+\Delta x} W\Delta z + N_{A_z}|_{z+\Delta z} W\Delta x\right) = 0 \tag{9.5-84}$$

Dividing Eq. (9.5-84) by $W\Delta x \Delta z$ and taking the limit as $\Delta x \to 0$ and $\Delta z \to 0$ give

$$\lim_{\Delta x\to 0}\frac{N_{A_x}|_x - N_{A_x}|_{x+\Delta x}}{\Delta x} + \lim_{\Delta z\to 0}\frac{N_{A_z}|_z - N_{A_z}|_{z+\Delta z}}{\Delta z} = 0 \tag{9.5-85}$$

or,

$$\frac{\partial N_{A_x}}{\partial x} + \frac{\partial N_{A_z}}{\partial z} = 0 \tag{9.5-86}$$

Substitution of Eqs. (9.5-81) and (9.5-82) into Eq. (9.5-86) yields

$$\underbrace{v_z \frac{\partial c_A}{\partial z}}_{\substack{\text{Convection in} \\ z\text{-direction}}} = \underbrace{\mathcal{D}_{AB} \frac{\partial^2 c_A}{\partial x^2}}_{\substack{\text{Diffusion in} \\ x\text{-direction}}} + \underbrace{\mathcal{D}_{AB} \frac{\partial^2 c_A}{\partial z^2}}_{\substack{\text{Diffusion in} \\ z\text{-direction}}} \tag{9.5-87}$$

In the z-direction, the mass of species $\mathcal{A}$ is transported by both convection and diffusion. As stated by Eq. (2.4-8), diffusion can be considered negligible with respect to convection when $\text{Pe}_M \gg 1$. Under these circumstances, Eq. (9.5-87) reduces to

$$\boxed{v_{\max}\left[1 - \left(\frac{x}{\delta}\right)^2\right]\frac{\partial c_A}{\partial z} = \mathcal{D}_{AB}\frac{\partial^2 c_A}{\partial x^2}} \tag{9.5-88}$$

The boundary conditions associated with Eq. (9.5-88) are

$$\text{at} \quad z = 0 \qquad c_A = c_{A_o} \tag{9.5-89}$$

$$\text{at} \quad x = 0 \qquad c_A = c_A^* \tag{9.5-90}$$

$$\text{at} \quad x = \delta \qquad \frac{\partial c_A}{\partial x} = 0 \tag{9.5-91}$$

It is assumed that the liquid has a uniform concentration of c_{A_o} for $z < 0$. At the liquid-gas interface, the value of c_A^* is determined from the solubility data, i.e., Henry's law. Equation (9.5-91) indicates that species $\mathcal{A}$ cannot diffuse through the wall.

The problem will be analyzed for two cases, namely, for long and short contact times.

9.5.2.1 *Long contact times* The solution of Eq. (9.5-88) subject to the boundary conditions given by Eqs. (9.5-89)–(9.5-91) was first obtained by Johnstone and Pigford (1942). Their series solution expresses the bulk concentration of species $\mathcal{A}$ at $z = L$ as

$$\frac{c_A^* - (c_{A_b})_L}{c_A^* - c_{A_o}} = 0.7857e^{-5.1213\eta} + 0.1001e^{-39.318\eta} + 0.03599e^{-105.64\eta} + \cdots \tag{9.5-92}$$

where

$$\eta = \frac{\mathcal{D}_{AB}L}{\delta^2 v_{\max}} = \frac{2\mathcal{D}_{AB}L}{3\delta^2 \langle v_z \rangle} \tag{9.5-93}$$

As engineers, we are interested in expressing the results in the form of a mass transfer correlation. For this purpose, it is first necessary to obtain an expression for the mass transfer coefficient.

For a rectangular differential volume element of thickness Δx, length Δz, and width W, as shown in Figure 9.20, the conservation statement given by Eq. (9.5-83) is also expressed as

$$\left[Qc_{A_b}\big|_z + k_c(c_A^* - c_{A_b})W\Delta z\right] - Qc_{A_b}\big|_{z+\Delta z} = 0 \tag{9.5-94}$$

Dividing Eq. (9.5-94) by Δz and taking the limit as $\Delta z \to 0$ give

$$\mathcal{Q} \lim_{\Delta z \to 0} \frac{c_{A_b}|_z - c_{A_b}|_{z+\Delta z}}{\Delta z} + k_c(c_A^* - c_{A_b})W = 0 \qquad (9.5\text{-}95)$$

or,

$$\mathcal{Q}\frac{dc_{A_b}}{dz} = k_c(c_A^* - c_{A_b})W \qquad (9.5\text{-}96)$$

Equation (9.5-96) is a separable equation and rearrangement gives

$$\mathcal{Q}\int_{c_{A_o}}^{(c_{A_b})_L} \frac{dc_{A_b}}{c_A^* - c_{A_b}} = W \int_0^L k_c\, dz \qquad (9.5\text{-}97)$$

Carrying out the integrations yields

$$\boxed{\langle k_c \rangle = \frac{\mathcal{Q}}{WL} \ln\left[\frac{c_A^* - c_{A_o}}{c_A^* - (c_{A_b})_L} \right]} \qquad (9.5\text{-}98)$$

where the average mass transfer coefficient, $\langle k_c \rangle$, is defined by

$$\langle k_c \rangle = \frac{1}{L}\int_0^L k_c\, dz \qquad (9.5\text{-}99)$$

The rate of moles of species $\mathcal{A}$ transferred to the liquid is

$$\dot{n}_A = \mathcal{Q}[(c_{A_b})_L - c_{A_o}] = \mathcal{Q}\left\{ (c_A^* - c_{A_o}) - [c_A^* - (c_{A_b})_L] \right\} \qquad (9.5\text{-}100)$$

Elimination of $\mathcal{Q}$ between Eqs. (9.5-98) and (9.5-100) leads to

$$\dot{n}_A = (WL)\langle k_c \rangle \underbrace{\frac{(c_A^* - c_{A_o}) - [c_A^* - (c_{A_b})_L]}{\ln\left[\dfrac{c_A^* - c_{A_o}}{c_A^* - (c_{A_b})_L} \right]}}_{(\Delta c_A)_{LM}} \qquad (9.5\text{-}101)$$

When $\eta > 0.1$, all the terms in Eq. (9.5-92), excluding the first, become almost zero, i.e.,

$$\frac{c_A^* - (c_{A_b})_L}{c_A^* - c_{A_o}} = 0.7857 e^{-5.1213\eta} \qquad (9.5\text{-}102)$$

The use of Eq. (9.5-102) in Eq. (9.5-98) gives

$$\langle k_c \rangle = \frac{\mathcal{Q}}{WL}(5.1213\eta + 0.241) \qquad (9.5\text{-}103)$$

Since we restrict our analysis to long contact times, i.e., η is large, then Eq. (9.5-103) simplifies to

$$\langle k_c \rangle = \frac{\mathcal{Q}}{WL}(5.1213\eta) \qquad (9.5\text{-}104)$$

Substitution of Eq. (9.5-93) into Eq. (9.5-104) and the use of $\mathcal{Q} = \langle v_z \rangle W \delta$ give

$$\langle k_c \rangle = 3.41 \frac{\mathcal{D}_{AB}}{\delta} \qquad (9.5\text{-}105)$$

Therefore, the average value of the Sherwood number becomes

$$\boxed{\text{Sh} = \frac{\langle k_c \rangle \delta}{\mathcal{D}_{AB}} = 3.41} \qquad (9.5\text{-}106)$$

It is also possible to arrive at this result using a different approach (see Problem 9.28). Equation (9.5-106) is usually recommended when

$$\text{Re} = \frac{4\delta \langle v_z \rangle \rho}{\mu} = \frac{4\dot{m}}{\mu W} < 100 \qquad (9.5\text{-}107)$$

Note that the term 4δ in the definition of the Reynolds number represents the hydraulic equivalent diameter.

9.5.2.2 *Short contact times* If the solubility of species $\mathcal{A}$ in liquid $\mathcal{B}$ is low, for short contact times, species $\mathcal{A}$ penetrates only a short distance into the falling liquid film. Under these circumstances, species $\mathcal{A}$, for the most part, has the impression that the film is moving throughout with a velocity equal to $v_{\max}$. Furthermore, species $\mathcal{A}$ does not feel the presence of the solid wall at $x = \delta$. Hence, if the film were of infinite thickness moving with the velocity $v_{\max}$, species $\mathcal{A}$ would not know the difference.

In light of the above discussion, Eqs. (9.5-88)–(9.5-91) take the following form

$$v_{\max} \frac{\partial c_A}{\partial z} = \mathcal{D}_{AB} \frac{\partial^2 c_A}{\partial x^2} \qquad (9.5\text{-}108)$$

$$\text{at} \quad z = 0 \qquad c_A = c_{A_o} \qquad (9.5\text{-}109)$$

$$\text{at} \quad x = 0 \qquad c_A = c_A^* \qquad (9.5\text{-}110)$$

$$\text{at} \quad x = \infty \qquad c_A = c_{A_o} \qquad (9.5\text{-}111)$$

Introduction of the dimensionless concentration ϕ as

$$\phi = \frac{c_A - c_{A_o}}{c_A^* - c_{A_o}} \qquad (9.5\text{-}112)$$

reduces Eqs. (9.5-108)–(9.5-111) to

$$v_{\max} \frac{\partial \phi}{\partial z} = \mathcal{D}_{AB} \frac{\partial^2 \phi}{\partial x^2} \qquad (9.5\text{-}113)$$

$$\text{at} \quad z = 0 \qquad \phi = 0 \qquad (9.5\text{-}114)$$

$$\text{at} \quad x = 0 \qquad \phi = 1 \qquad (9.5\text{-}115)$$

$$\text{at} \quad x = \infty \qquad \phi = 0 \qquad (9.5\text{-}116)$$

Since Eqs. (9.5-114) and (9.5-116) are the same and there is no length scale, this parabolic partial differential equation can be solved by the similarity solution as explained in Section B.3.6.2 in Appendix B. The solution is sought in the form

$$\phi = f(\Psi) \tag{9.5-117}$$

where

$$\Psi = \frac{x}{\sqrt{4\mathcal{D}_{AB}z/v_{max}}} \tag{9.5-118}$$

The chain rule of differentiation gives

$$\frac{\partial \phi}{\partial z} = \frac{df}{d\Psi}\frac{\partial \Psi}{\partial z} = -\frac{1}{2}\frac{\Psi}{z}\frac{df}{d\Psi} \tag{9.5-119}$$

$$\frac{\partial^2 \phi}{\partial x^2} = \frac{d^2 f}{d\Psi^2}\left(\frac{\partial \Psi}{\partial x}\right)^2 + \frac{df}{d\Psi}\frac{\partial^2 \Psi}{\partial x^2} = \frac{v_{max}}{4\mathcal{D}_{AB}z}\frac{d^2 f}{d\Psi^2} \tag{9.5-120}$$

Substitution of Eqs. (9.5-119) and (9.5-120) into Eq. (9.5-113) yields

$$\frac{d^2 f}{d\Psi^2} + 2\Psi \frac{df}{d\Psi} = 0 \tag{9.5-121}$$

The boundary conditions associated with Eq. (9.5-121) are

$$\text{at} \quad \Psi = 0 \qquad \phi = 1 \tag{9.5-122}$$

$$\text{at} \quad \Psi = \infty \qquad \phi = 0 \tag{9.5-123}$$

The integrating factor for Eq. (9.5-121) is $\exp(\Psi^2)$. Multiplication of Eq. (9.5-121) by the integrating factor gives

$$\frac{d}{d\Psi}\left(e^{\Psi^2}\frac{df}{d\Psi}\right) = 0 \quad \Rightarrow \quad \frac{df}{d\Psi} = K_1 e^{-\Psi^2} \tag{9.5-124}$$

Integration of Eq. (9.5-124) leads to

$$f = K_1 \int_0^{\Psi} e^{-u^2}\, du + K_2 \tag{9.5-125}$$

where u is a dummy variable of integration. Application of the boundary condition defined by Eq. (9.5-122) gives $K_2 = 1$. On the other hand, the use of the boundary condition defined by Eq. (9.5-123) gives

$$K_1 = -\frac{1}{\displaystyle\int_0^{\infty} e^{-u^2}\, du} = -\frac{2}{\sqrt{\pi}} \tag{9.5-126}$$

Therefore, the solution becomes

$$f = 1 - \frac{2}{\sqrt{\pi}} \int_0^{\Psi} e^{-u^2}\, du \tag{9.5-127}$$

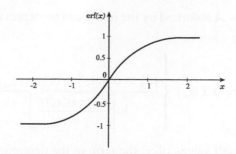

Figure 9.21. The error function.

or,

$$\boxed{\frac{c_A - c_{A_o}}{c_A^* - c_{A_o}} = 1 - \mathrm{erf}\left(\frac{x}{\sqrt{4\mathcal{D}_{AB}z/v_{\max}}}\right)} \qquad (9.5\text{-}128)$$

where $\mathrm{erf}(x)$ is the error function defined by

$$\mathrm{erf}(x) = \frac{2}{\sqrt{\pi}} \int_0^x e^{-u^2} du \qquad (9.5\text{-}129)$$

The plot of the error function is shown in Figure 9.21.

Macroscopic equation

Integration of the governing equation, Eq. (9.5-108), over the volume of the system gives

$$\int_0^L \int_0^W \int_0^\delta v_{\max} \frac{\partial c_A}{\partial z} \, dx \, dy \, dz = \int_0^L \int_0^W \int_0^\delta \mathcal{D}_{AB} \frac{\partial^2 c_A}{\partial x^2} \, dx \, dy \, dz \qquad (9.5\text{-}130)$$

Evaluation of the integrations yields

$$\underbrace{v_{\max} W \int_0^\delta \left(c_A|_{z=L} - c_{A_o}\right) dx}_{\substack{\text{Net molar rate of species } \mathcal{A} \\ \text{entering the liquid}}} = \underbrace{W \int_0^L \left(-\mathcal{D}_{AB} \frac{\partial c_A}{\partial x}\right)_{x=0} dz}_{\substack{\text{Molar rate of species } \mathcal{A} \text{ entering} \\ \text{the liquid through the interface}}} \qquad (9.5\text{-}131)$$

which is the macroscopic inventory rate equation for the mass of species $\mathcal{A}$ by considering the falling liquid film as a system. The use of Eq. (9.5-128) in Eq. (9.5-131) gives the rate of moles of species $\mathcal{A}$ absorbed in the liquid as

$$\boxed{\dot{n}_A = WL(c_A^* - c_{A_o})\sqrt{\frac{4\mathcal{D}_{AB}v_{\max}}{\pi L}}} \qquad (9.5\text{-}132)$$

The rate of moles of species $\mathcal{A}$ absorbed by the liquid can be expressed in terms of the average mass transfer coefficient as

$$\dot{n}_A = WL\langle k_c\rangle \left\{ \frac{[c_A^* - (c_{A_b})_L] - (c_A^* - c_{A_o})}{\ln\left[\dfrac{c_A^* - (c_{A_b})_L}{c_A^* - c_{A_o}}\right]} \right\} \tag{9.5-133}$$

Since $\ln(1+x) \simeq x$ for small values of x, the term in the denominator of Eq. (9.5-133) can be approximated as

$$\ln\left[\frac{c_A^* - (c_{A_b})_L}{c_A^* - c_{A_o}}\right] = \ln\left[1 + \frac{c_{A_o} - (c_{A_b})_L}{c_A^* - c_{A_o}}\right] \simeq \frac{c_{A_o} - (c_{A_b})_L}{c_A^* - c_{A_o}} \tag{9.5-134}$$

The use of Eq. (9.5-134) in Eq. (9.5-133) gives

$$\dot{n}_A = WL\langle k_c\rangle(c_A^* - c_{A_o}) \tag{9.5-135}$$

The average mass transfer coefficient can be calculated from Eqs. (9.5-132) and (9.5-135) as

$$\langle k_c\rangle = \sqrt{\frac{4D_{AB}v_{\max}}{\pi L}} \tag{9.5-136}$$

Therefore, the Sherwood number is

$$Sh = \frac{\langle k_c\rangle\delta}{D_{AB}} = \sqrt{\frac{4\delta^2 v_{\max}}{\pi D_{AB} L}} = 0.691\left(\frac{\delta}{L}\right)^{1/2} Re^{1/2} Sc^{1/2} \tag{9.5-137}$$

Equation (9.5-137) is recommended when

$$1200 > Re = \frac{4\delta\langle v_z\rangle\rho}{\mu} = \frac{4\dot{m}}{\mu W} > 100$$

It should be kept in mind that the calculated mass of species $\mathcal{A}$ absorbed by the liquid based on Eq. (9.5-132) usually underestimates the actual amount. This is due to the increase in the mass transfer area as a result of ripple formation even at very small values of Re.

In the literature, Eq. (9.5-136) is also expressed in the form

$$\langle k_c\rangle = \sqrt{\frac{4D_{AB}}{\pi t_{\exp}}} \tag{9.5-138}$$

where the *exposure time*, or *gas-liquid contact time*, is defined by

$$t_{\exp} = \frac{L}{v_{\max}} \tag{9.5-139}$$

Equation (9.5-138) is also applicable to gas absorption to laminar liquid jets and mass transfer from ascending bubbles, if the penetration distance of the solute is small.

Example 9.13 A laminar liquid jet issuing at a volumetric flow rate of Q is used for absorption of gas A. If the jet has a diameter D and a length L, derive an expression for the rate of absorption of species A.

Solution

The time of exposure can be defined by

$$t_{\text{exp}} = \frac{L}{\langle v \rangle} = \frac{L}{4Q/\pi D^2} \tag{1}$$

Therefore, Eq. (9.5-138) becomes

$$\langle k_c \rangle = \frac{4}{\pi D} \sqrt{\frac{Q\mathcal{D}_{AB}}{L}} \tag{2}$$

The rate of moles of species A absorbed by the jet is

$$\dot{n}_A = (\pi DL)\langle k_c \rangle (c_A^* - c_{A_o}) \tag{3}$$

where c_{A_o} is the initial concentration of species A in the jet and c_A^* is the equilibrium solubility of species A in the liquid. Substitution of Eq. (2) into Eq. (3) gives

$$\dot{n}_A = 4(c_A^* - c_{A_o})\sqrt{Q\mathcal{D}_{AB}L} \tag{4}$$

9.5.3 Analysis of a Plug Flow Reactor

A plug flow reactor consists of a cylindrical pipe in which concentration, temperature, and reaction rate are assumed to vary only along the axial direction. Analysis of these reactors is usually done with the following assumptions:

- Steady-state conditions prevail.
- Reactor is isothermal.
- There is no mixing in the axial direction.

The conservation statement for species i over a differential volume element of thickness Δz, as shown in Figure 9.22, is expressed as

$$(Q c_i)|_z - (Q c_i)|_{z+\Delta z} + \alpha_i r A \Delta z = 0 \tag{9.5-140}$$

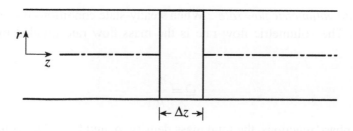

Figure 9.22. Plug flow reactor.

where α_i is the stoichiometric coefficient of species i, and r is the chemical reaction rate expression. Dividing Eq. (9.5-140) by Δz and taking the limit as $\Delta z \to 0$ give

$$\lim_{\Delta z \to 0} \frac{(Qc_i)|_z - (Qc_i)|_{z+\Delta z}}{\Delta z} + \alpha_i r A = 0 \tag{9.5-141}$$

or,

$$\frac{d(Qc_i)}{dz} = \alpha_i r A \tag{9.5-142}$$

It is customary to write Eq. (9.5-142) in terms of $dV = A\,dz$ rather than dz, so that Eq. (9.5-142) becomes

$$\frac{d(Qc_i)}{dV} = \alpha_i r \tag{9.5-143}$$

Equation (9.5-143) can also be expressed in the form

$$\frac{d\dot{n}_i}{dV} = \alpha_i r \tag{9.5-144}$$

where $\dot{n}_i$ is the molar flow rate of species i.

The variation in the number of moles of species i as a function of the molar extent of the reaction is given by Eq. (5.3-10). It is also possible to express this equation as

$$\dot{n}_i = \dot{n}_{i_o} + \alpha_i \dot{\varepsilon} \tag{9.5-145}$$

Let us assume that the rate of reaction has the form

$$r = k c_i^n = k \left(\frac{\dot{n}_i}{Q}\right)^n \tag{9.5-146}$$

Substitution of Eq. (9.5-146) into Eq. (9.5-144) gives

$$\boxed{\frac{d\dot{n}_i}{dV} = \alpha_i k \left(\frac{\dot{n}_i}{Q}\right)^n} \tag{9.5-147}$$

Integration of Eq. (9.5-147) depends on whether the volumetric flow rate is constant or not.

9.5.3.1 *Constant volumetric flow rate* When steady-state conditions prevail, the mass flow rate is constant. The volumetric flow rate is the mass flow rate divided by the total mass density, i.e.,

$$Q = \frac{\dot{m}}{\rho} \tag{9.5-148}$$

For most liquid phase reactions, the total mass density, ρ, and hence the volumetric flow rate are constant.

Table 9.4. Requirements for the constant volumetric flow rate for a plug flow reactor operating under steady and isothermal conditions

Liquid Phase Reactions	Gas Phase Reactions
Constant total mass density	■ No change in the total number of moles during the reaction ($\overline{\alpha} = 0$)
	■ Negligible pressure drop across the reactor

For gas phase reactions, on the other hand, the total mass density is given by the ideal gas equation of state as

$$\rho = \frac{P\mathcal{M}}{\mathcal{R}T} \tag{9.5-149}$$

where $\mathcal{M}$ is the molecular weight of the reacting mixture. Substitution of Eq. (9.5-149) into Eq. (9.5-148) gives

$$\mathcal{Q} = \frac{\dot{n}\mathcal{R}T}{P} \tag{9.5-150}$$

Therefore, $\mathcal{Q}$ remains constant when $\dot{n}$ and P do not change along the reactor. The conditions for the constancy of $\mathcal{Q}$ are summarized in Table 9.4.

When $\mathcal{Q}$ is constant, Eq. (9.5-147) can be rearranged as

$$\int_0^V dV = \frac{\mathcal{Q}^n}{\alpha_i k} \int_{\dot{n}_{i_o}}^{\dot{n}_i} \frac{d\dot{n}_i}{\dot{n}_i^n} \tag{9.5-151}$$

Depending on the values of n the results are

$$V = \begin{cases} \dfrac{\mathcal{Q}}{\alpha_i k}(c_i - c_{i_o}) & n = 0 \\[2ex] \dfrac{\mathcal{Q}}{\alpha_i k}\ln\left(\dfrac{c_i}{c_{i_o}}\right) & n = 1 \\[2ex] \dfrac{\mathcal{Q}}{\alpha_i k}\dfrac{1}{1-n}\left(\dfrac{1}{c_i^{1-n}} - \dfrac{1}{c_{i_o}^{1-n}}\right) & n \geqslant 2 \end{cases} \tag{9.5-152}$$

9.5.3.2 *Variable volumetric flow rate* When the volumetric flow rate is not constant, integration of Eq. (9.5-147) is possible only after expressing both $\dot{n}_i$ and $\mathcal{Q}$ in terms of $\dot{\varepsilon}$. The following example explains the procedure in detail.

Example 9.14 The irreversible gas phase reaction

$$A \rightarrow B + C$$

is carried out in a constant pressure batch reactor at 400 °C and 5 atm pressure. The reaction is first-order and the time required to achieve 60% conversion was found to be 50 min.

Suppose that this reaction is to be carried out in a plug flow reactor that operates isothermally at 400 °C and at a pressure of 10 atm. The volumetric flow rate of the feed entering the reactor is 0.05 m^3/h and it consists of pure A. Calculate the volume of the reactor required to achieve 80% conversion.

Solution

First, it is necessary to determine the rate constant by using the data given for the batch reactor. The conservation statement for the number of moles of species $\mathcal{A}$, Eq. (7.2-5), reduces to

$$\alpha_A r \, V = \frac{dn_A}{dt} \tag{1}$$

or,

$$-k c_A V = \frac{dn_A}{dt} \tag{2}$$

Substitution of the identity $n_A = c_A V$ into Eq. (2) and rearrangement give

$$-k \int_0^t dt = \int_{n_{A_o}}^{n_A} \frac{dn_A}{dt} \tag{3}$$

Integration gives the rate constant, k, as

$$k = -\frac{1}{t} \ln\left(\frac{n_A}{n_{A_o}}\right) \tag{4}$$

The fractional conversion, X, is

$$X = \frac{n_{A_o} - n_A}{n_{A_o}} = 1 - \frac{n_A}{n_{A_o}} \tag{5}$$

Therefore, Eq. (4) can be expressed in terms of the fractional conversion as

$$k = -\frac{\ln(1 - X)}{t} \tag{6}$$

Substitution of the numerical values into Eq. (6) gives

$$k = -\frac{\ln(1 - 0.6)}{(50/60)} = 1.1 \text{ h}^{-1} \tag{7}$$

For a plug flow reactor, Eq. (9.5-147) takes the form

$$\frac{d\dot{n}_A}{dV} = -k \frac{\dot{n}_A}{\mathcal{Q}} \tag{8}$$

Since the volumetric flow rate is not constant, i.e., $\bar{\alpha} = 1$, it is necessary to express $\mathcal{Q}$ in terms of $\dot{\varepsilon}$. The use of Eq. (9.5-145) gives

$$\dot{n}_A = \dot{n}_{A_o} - \dot{\varepsilon} \tag{9}$$

$$\dot{n}_B = \dot{\varepsilon} \tag{10}$$

$$\dot{n}_C = \dot{\varepsilon} \tag{11}$$

Therefore, the total molar flow rate, $\dot{n}$, is

$$\dot{n} = \dot{n}_{A_o} + \dot{\varepsilon} \tag{12}$$

Substitution of Eq. (12) into Eq. (9.5-150) gives the volumetric flow rate as

$$\mathcal{Q} = \frac{RT}{P}(\dot{n}_{A_o} + \dot{\varepsilon}) = \frac{RT\,\dot{n}_{A_o}}{P}\left(1 + \frac{\dot{\varepsilon}}{\dot{n}_{A_o}}\right) = \mathcal{Q}_o\left(1 + \frac{\dot{\varepsilon}}{\dot{n}_{A_o}}\right) \tag{13}$$

where $\mathcal{Q}_o$ is the volumetric flow rate at the inlet of the reactor.
  Substitution of Eqs. (9) and (13) into Eq. (8) gives

$$\frac{d\dot{\varepsilon}}{dV} = \frac{k\,\dot{n}_{A_o}}{\mathcal{Q}_o}\frac{[1 - (\dot{\varepsilon}/\dot{n}_{A_o})]}{(1 + \dot{\varepsilon}/\dot{n}_{A_o})} \tag{14}$$

The fractional conversion expression for a plug flow reactor is similar to Eq. (5), and so

$$X = \frac{\dot{n}_{A_o} - \dot{n}_A}{\dot{n}_{A_o}} = 1 - \frac{\dot{n}_A}{\dot{n}_{A_o}} \tag{15}$$

Substitution of Eq. (9) into Eq. (15) yields

$$X = \frac{\dot{\varepsilon}}{\dot{n}_{A_o}} \tag{16}$$

The use of Eq. (16) in Eq. (14) and rearrangement give

$$V = \frac{\mathcal{Q}_o}{k}\int_0^{0.8}\left(\frac{1 + X}{1 - X}\right)dX = 2.42\left(\frac{\mathcal{Q}_o}{k}\right) \tag{17}$$

Substitution of the numerical values into Eq. (17) gives

$$V = \frac{(2.42)(0.05)}{1.1} = 0.11\ \text{m}^3 \tag{18}$$

NOTATION

A	area, m^2
A_H	heat transfer area, m^2
A_M	mass transfer area, m^2
$\widehat{C}_P$	heat capacity at constant pressure, kJ/kg·K
c	total concentration, kmol/m^3
c_i	concentration of species i, kmol/m^3
D	pipe diameter, m
$\mathcal{D}_{AB}$	diffusion coefficient for system $\mathcal{A}\text{-}\mathcal{B}$, m^2/s
e	total energy flux, W/m^2
F_D	drag force, N
f	friction factor
g	acceleration of gravity, m/s^2

$\mathcal{H}$	partition coefficient
h	heat transfer coefficient, $W/m^2 \cdot K$
J^*	molecular molar flux, $kmol/m^2 \cdot s$
K	kinetic energy per unit volume, J/m^3
k	reaction rate constant; thermal conductivity, $W/m \cdot K$
k_c	mass transfer coefficient, m/s
L	length, m
$\dot{m}$	mass flow rate, kg/s
$\mathcal{M}$	molecular weight, kg/kmol
N	total molar flux, $kmol/m^2 \cdot s$
$\dot{n}$	molar flow rate, kmol/s
P	pressure, Pa
$\mathcal{P}$	modified pressure, Pa
$\dot{Q}$	heat transfer rate, W
$\mathcal{Q}$	volumetric flow rate, m^3/s
q	heat flux, W/m^2
r	rate of a chemical reaction, $kmol/m^3 \cdot s$
$\mathfrak{R}$	Rate of generation per unit volume
$\mathcal{R}$	gas constant, $J/mol \cdot K$
T	temperature, °C or K
t	time, s
V	velocity of the plate in Couette flow, m/s; volume, m^3
v	velocity, m/s
W	width, m
$\mathcal{W}$	total mass flux, $kg/m^2 \cdot s$
X	fractional conversion
x	rectangular coordinate, m
z	rectangular coordinate, m
α_i	stoichiometric coefficient of species i
Δ	difference
ΔH_{rxn}	heat of reaction, J
$\dot{\varepsilon}$	time rate of change of molar extent, kmol/s
λ	latent heat of vaporization, J
v	kinematic viscosity, m^2/s
μ	viscosity, $kg/m \cdot s$
π	total momentum flux, N/m^2
ρ	density, kg/m^3
τ_{ij}	shear stress (flux of j-momentum in the i-direction), N/m^2
ω	mass fraction

Overlines

$\sim$	per mole
$\frown$	per unit mass
$-$	partial molar

Bracket

$\langle a \rangle$ average value of a

Superscripts

A air
L liquid
o standard state
sat saturation

Subscripts

A, B species in binary systems
b bulk
ch characteristic
exp   exposure
i species in multicomponent systems
in inlet
int interphase
LM log-mean
max   maximum
out outlet
ref reference
sys system
w wall or surface
∞ free-stream

Dimensionless Numbers

Br   Brinkman number
Nu   Nusselt number
Pr   Prandtl number
Re   Reynolds number
Re_h Reynolds number based on the hydraulic equivalent diameter
Sc   Schmidt number
Sh   Sherwood number
St_H Stanton number for heat transfer
St_M Stanton number for mass transfer

REFERENCES

Astarita, G., 1967, Mass Transfer With Chemical Reaction, Elsevier, Amsterdam.
Bejan, A., 1984, Convection Heat Transfer, Wiley, New York.
Bird, R.B., W.E. Stewart and E.N. Lightfoot, 2002, Transport Phenomena, 2nd Ed., Wiley, New York.
Breidenbach, R.W., M.J. Saxton, L.D. Hansen and R.S. Criddle, 1997, Heat generation and dissipation in plants: Can the alternative oxidative phosphorylation pathway serve a thermoregulatory role in plant tissues other than specialized organs?, Plant Physiol. 117, 1177.

Gavis, J. and R.L. Laurence, 1968, Viscous heating in plane and circular flow between moving surfaces, Ind. Eng. Chem. Fundamentals 7, 232.

Graetz, L., 1883, On the heat capacity of fluids – Part 1, Ann. Phys. Chem. 18, 79.

Graetz, L., 1885, On the heat capacity of fluids – Part 2, Ann. Phys. Chem. 25, 337.

Hausen, H., 1943, Verfahrenstechnik Beih. Z. Ver. deut. Ing. 4, 91.

Johnstone, H.F. and R.L. Pigford, 1942, Distillation in a wetted-wall column, Trans. AIChE 38, 25.

Nusselt, W., 1910, The dependence of the heat transfer coefficient on tube length, Z. Ver. Dt. Ing., 54, 1154.

Skelland, A.H.P., 1974, Diffusional Mass Transfer, Wiley, New York.

Slattery, J.C., 1999, Advanced Transport Phenomena, Cambridge University Press, Cambridge.

Whitaker, S., 1968, Introduction to Fluid Mechanics, Prentice-Hall, Englewood Cliffs, New Jersey.

SUGGESTED REFERENCES FOR FURTHER STUDY

Brodkey, R.S. and H.C. Hershey, 1988, Transport Phenomena: A Unified Approach, McGraw-Hill, New York.

Cussler, E.L., 1997, Diffusion – Mass Transfer in Fluid Systems, 2nd Ed., Cambridge University Press, Cambridge.

Fahien, R.W., 1983, Fundamentals of Transport Phenomena, McGraw-Hill, New York.

Geankoplis, J., 1983, Transport Processes – Momentum, Heat, and Mass, Allyn and Bacon, Boston.

Hines, A.L. and R.N. Maddox, 1985, Mass Transfer – Fundamentals and Applications, Prentice-Hall, Englewood Cliffs, New Jersey.

Incropera, F.P. and D.P. DeWitt, 2002, Fundamentals of Heat and Mass Transfer, 5th Ed., Wiley, New York.

Middleman, S., 1998, An Introduction to Mass and Heat Transfer – Principles of Analysis and Design, Wiley, New York.

Seader, J.D. and E.J. Henley, 1998, Separation Process Principles, Wiley, New York.

Shah, R.K. and A.L. London, 1978, Laminar Flow Forced Convection in Ducts, Advances in Heat Transfer, Academic Press, New York.

PROBLEMS

9.1 The hydrostatic pressure distribution in fluids can be calculated from the equation

$$\frac{dP}{dz} = \rho g_z$$

where

$$g_z = \begin{cases} g & \text{if positive } z \text{ is in the direction of gravity} \\ -g & \text{if positive } z \text{ is in the direction opposite to gravity} \end{cases}$$

a) If the systolic pressure at the aorta is 120 mmHg, what is the pressure in the neck 25 cm higher and at a position in the legs 90 cm lower? The density of blood is 1.05 g/cm^3.

b) The lowest point on the earth's surface is located in the western Pacific Ocean, in the Marianas Trench. It is about 11 km below sea level. Estimate the pressure at the bottom of the ocean. Take the density of seawater as 1025 kg/m^3.

c) The highest point on the earth's surface is the top of Mount Everest, located in the Himalayas on the border of Nepal and China. It is approximately 8900 m above sea level. If the average rate of decrease in air temperature with altitude is 6.5 °C/km, estimate the air pressure at the top of Mount Everest. Assume that the temperature at sea level is 15 °C. Why is it difficult to breathe at high altitudes?

(**Answer:** a) P_{neck} = 100.7 mmHg, P_{leg} = 189.5 mmHg b) 1090 atm c) 0.31 atm)

9.2 When the ratio of the radius of the inner pipe to that of the outer pipe is close to unity, a concentric annulus may be considered a thin plate slit and its curvature can be neglected. Use this approximation and show that the modifications of Eqs. (9.1-23) and (9.1-26) for the axial flow in a concentric annulus with inner and outer radii of κR and R, respectively, lead to

$$v_z^{approx} = \frac{(\mathcal{P}_o - \mathcal{P}_L)R^2}{4\mu L}\left\{2(1-\kappa)\left[\frac{r}{R} - \kappa - \frac{1}{1-\kappa}\left(\frac{r}{R} - \kappa\right)^2\right]\right\}$$

$$\mathcal{Q}_{approx} = \frac{\pi(\mathcal{P}_o - \mathcal{P}_L)R^4}{8\mu L}\left[\frac{2(1-\kappa^2)(1-\kappa)^2}{3}\right]$$

Also calculate v_z^{approx}/v_z^{exact} as a function of ξ and $\mathcal{Q}_{approx}/\mathcal{Q}_{exact}$ for κ values of 0.6, 0.7, 0.8, and 0.9.

9.3 Oil spills on water can be removed by lowering a moving belt of width W into the water. The belt moves upward and skims the oil into a reservoir aboard the ship as shown in the figure below.

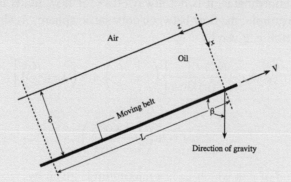

a) Show that the velocity profile and the volumetric flow rate are given by

$$v_z = \frac{\rho g \delta^2 \cos\beta}{2\mu}\left[1 - \left(\frac{x}{\delta}\right)^2\right] - V$$

$$\mathcal{Q} = \frac{W\rho g \delta^3 \cos\beta}{3\mu} - WV\delta$$

b) Determine the belt speed that will give a zero volumetric flow rate and specify the design criteria for positive and negative flow rates.

9.4 For laminar flow of a Newtonian fluid in a circular pipe, the velocity profile is parabolic and Eqs. (9.1-80) and (9.1-84) indicate that

$$\frac{\langle v_z \rangle}{v_{max}} = 0.5$$

In the case of a turbulent flow, experimentally determined velocity profiles can be represented in the form

$$v_z = v_{max}\left(1 - \frac{r}{R}\right)^{1/n}$$

where n depends on the value of the Reynolds number. Show that the ratio $\langle v_z \rangle / v_{max}$ is given as (Whitaker, 1968)

Re	n	$\langle v_z \rangle / v_{max}$
4×10^3	6	0.79
1×10^5	7	0.82
3×10^6	10	0.87

This is the reason why the velocity profile for a turbulent flow is generally considered "flat" in engineering analysis.

9.5 One of the misconceptions of students studying fluid mechanics is that "the shear stress is always zero at the point of maximum velocity." While this is a valid statement for flow fields in rectangular coordinates, it is not always true for flow fields in curvilinear coordinates. Consider, for example, the flow between concentric spheres as shown in Example 1.3. The velocity distribution is given by

$$v_\theta = \frac{R|\Delta P|}{2\mu\, E(\varepsilon) \sin\theta}\left[\left(1 - \frac{r}{R}\right) + \kappa\left(1 - \frac{R}{r}\right)\right] \tag{1}$$

where

$$E(\varepsilon) = \ln\left(\frac{1 + \cos\varepsilon}{1 - \cos\varepsilon}\right) \tag{2}$$

a) Show that the value of r at which v_θ is a maximum is given by

$$r = \sqrt{\kappa}\, R \tag{3}$$

b) Show that the value of r at which $\tau_{r\theta}$ is zero is given by

$$r = \frac{2\kappa R}{1 + \kappa} \tag{4}$$

and conclude that, in this particular case, the maximum value of the velocity occurs at a different value of r than that for zero momentum flux.

9.6 The steady temperature distribution within a solid body is given by

$$T = 50 + e^x(10 - y^2)\sin 3z$$

If the thermal conductivity of the solid is 380 W/m·K, express the rate of energy generation per unit volume as a function of position.

(**Answer:** $380 e^x (82 - 8y^2)\sin 3z$)

9.7 Heat is generated in a slab at a constant volumetric rate of $\Re$ (W/m^3). For cooling purposes, both sides of the slab are exposed to fluids A and B having different temperatures and velocities as shown in the figure below.

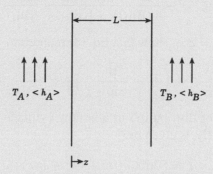

a) Consider a differential volume of thickness Δz within the slab and show that the governing equation for temperature under steady conditions is given by

$$k \frac{d^2 T}{dz^2} + \Re = 0 \tag{1}$$

which is subject to the following boundary conditions

$$\text{at} \quad z = 0 \quad -k \frac{dT}{dz} = \langle h_A \rangle (T_A - T) \tag{2}$$

$$\text{at} \quad z = L \quad -k \frac{dT}{dz} = \langle h_B \rangle (T - T_B) \tag{3}$$

b) In terms of the following dimensionless quantities

$$\theta = \frac{T - T_B}{T_A - T_B} \quad \xi = \frac{z}{L} \quad (\text{Bi}_\text{H})_A = \frac{\langle h_A \rangle L}{k} \quad (\text{Bi}_\text{H})_B = \frac{\langle h_B \rangle L}{k} \quad \Lambda = \frac{\Re L^2}{k(T_A - T_B)} \tag{4}$$

show that Eqs. (1)–(3) take the form

$$\frac{d^2 \theta}{d\xi^2} + \Lambda = 0 \tag{5}$$

$$\text{at} \quad \xi = 0 \quad -\frac{d\theta}{d\xi} = (\text{Bi}_\text{H})_A (1 - \theta) \tag{6}$$

$$\text{at} \quad \xi = 1 \quad -\frac{d\theta}{d\xi} = (\text{Bi}_\text{H})_B \theta \tag{7}$$

c) Show that the solution is given by

$$\theta = -\frac{\Lambda}{2} \xi^2 + \left\{ \frac{\Lambda[1 + 0.5(\text{Bi}_\text{H})_B] - (\text{Bi}_\text{H})_B}{1 + (\text{Bi}_\text{H})_B + [(\text{Bi}_\text{H})_B/(\text{Bi}_\text{H})_A]} \right\} \xi$$

$$+ \frac{1 + (\text{Bi}_\text{H})_B + [\Lambda/(\text{Bi}_\text{H})_A][1 + 0.5(\text{Bi}_\text{H})_B]}{1 + (\text{Bi}_\text{H})_B + [(\text{Bi}_\text{H})_B/(\text{Bi}_\text{H})_A]} \tag{8}$$

d) Show that the dimensionless location of the maximum temperature within the slab, ξ^*, is given by

$$\xi^* = \frac{1 + (Bi_H)_B[0.5 - (1/\Lambda)]}{1 + (Bi_H)_B + [(Bi_H)_B/(Bi_H)_A]} \qquad (9)$$

e) When $(Bi_H)_A = (Bi_H)_B = Bi_H$, show that Eq. (8) reduces to

$$\theta = -\frac{\Lambda}{2}\xi^2 + \left(\frac{\Lambda}{2} - \frac{Bi_H}{Bi_H + 2}\right)\xi + \frac{\Lambda}{2\,Bi_H} + \frac{1 + Bi_H}{2 + Bi_H} \qquad (10)$$

9.8 The steady temperature distribution in a hollow cylinder of inner and outer radii of 50 cm and 80 cm, respectively, is given by

$$T = 5000(4.073 - 6r^2 + \ln r)$$

where T is in degrees Celsius and r is in meters. If the thermal conductivity is 5 W/m·K, find the rate of energy generation per unit volume.

(**Answer:** 6×10^5 W/m^3)

9.9 Energy generation within a hollow cylinder of inside and outside radii of 60 cm and 80 cm, respectively, is 10^6 W/m^3. If both surfaces are maintained at 55 °C and the thermal conductivity is 15 W/m·K, calculate the maximum temperature under steady conditions.

(**Answer:** 389.6 °C)

9.10 Consider a long cylindrical rod of radius R and thermal conductivity k in which there is a uniform heat generation rate per unit volume $\Re$. Cooling fluid at a temperature of T_∞ flows over the surface of the cylinder with an average heat transfer coefficient $\langle h \rangle$.

a) Show that the steady-state temperature distribution within the rod is given by

$$T = \frac{\Re R^2}{4k}\left[1 - \left(\frac{r}{R}\right)^2\right] + \frac{\Re R}{2\langle h \rangle} + T_\infty$$

b) Show that the surface temperature of the rod, T_s, is given by

$$T_s = \frac{\Re R}{2\langle h \rangle} + T_\infty$$

c) Show that the rate of heat loss from the rod to the cooling fluid is given by

$$\dot{Q} = \pi R^2 L \Re$$

d) Is it possible to solve parts (b) and (c) of the problem without calculating the temperature distribution?

9.11 Consider a cylindrical rod with two concentric regions, A and B, as shown in the figure below.

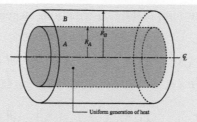

Uniform generation of heat

Heat is generated uniformly in region A at a rate $\Re$ (W/m^3). Cooling fluid at a temperature of T_∞ flows over the surface of the cylinder with an average heat transfer coefficient $\langle h \rangle$.

a) Show that the steady-state temperature distributions in regions A and B are expressed as

$$T_A = \frac{\Re R_A^2}{4k_A}\left[1 - \left(\frac{r}{R_A}\right)^2\right] + \frac{\Re R_A^2}{2}\left[\frac{1}{\langle h \rangle R_B} + \frac{1}{k_B}\ln\left(\frac{R_B}{R_A}\right)\right] + T_\infty$$

$$T_B = -\frac{\Re R_A^2}{2k_B}\ln\left(\frac{r}{R_B}\right) + \frac{\Re R_A^2}{2\langle h \rangle R_B} + T_\infty$$

where k_A and k_B are the thermal conductivities of regions A and B, respectively.

b) Show that the outer surface temperature, T_s, which is in contact with the cooling fluid, and the interface temperature, T_{AB}, are given by

$$T_s = \frac{\Re R_A^2}{2\langle h \rangle R_B} + T_\infty$$

$$T_{AB} = \frac{\Re R_A^2}{2k_B}\ln\left(\frac{R_B}{R_A}\right) + \frac{\Re R_A^2}{2\langle h \rangle R_B} + T_\infty$$

c) Show that the rate of heat loss to the cooling fluid is given by

$$\dot{Q} = \pi R_A^2 L \Re$$

d) Is it possible to solve parts (b) and (c) of the problem without calculating the temperature distributions?

9.12 Consider a cylindrical rod with two concentric regions, A and B, as shown in the figure below.

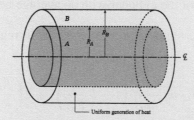

Uniform generation of heat

Heat is generated uniformly in region B at a rate $\Re$ (W/m^3). Cooling fluid at a temperature of T_∞ flows over the surface of the cylinder with an average heat transfer coefficient $\langle h \rangle$.

a) Show that the steady-state temperature distributions in regions A and B are expressed as

$$T_A = \frac{\Re R_B^2}{4k_B}\left[1-\left(\frac{R_A}{R_B}\right)^2 + 2\left(\frac{R_A}{R_B}\right)^2 \ln\left(\frac{R_A}{R_B}\right)\right] + \frac{\Re R_B}{2\langle h\rangle}\left[1-\left(\frac{R_A}{R_B}\right)^2\right] + T_\infty$$

$$T_B = \frac{\Re R_B^2}{4k_B}\left[1-\left(\frac{r}{R_B}\right)^2 + 2\left(\frac{R_A}{R_B}\right)^2 \ln\left(\frac{r}{R_B}\right)\right] + \frac{\Re R_B}{2\langle h\rangle}\left[1-\left(\frac{R_A}{R_B}\right)^2\right] + T_\infty$$

where k_A and k_B are the thermal conductivities of regions A and B, respectively.

b) Show that the outer surface temperature, T_s, which is in contact with the cooling fluid, is given by

$$T_s = \frac{\Re R_B}{2\langle h\rangle}\left[1-\left(\frac{R_A}{R_B}\right)^2\right] + T_\infty$$

c) Show that the rate of heat loss to the cooling fluid is given by

$$\dot{Q} = \pi(R_B^2 - R_A^2)L\Re$$

d) Is it possible to solve parts (b) and (c) of the problem without calculating the temperature distributions?

9.13 Consider a cylindrical rod with four concentric regions, A, B, C, and D, as shown in the figure below.

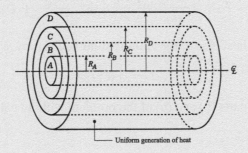

Heat is generated uniformly in region D at a rate $\Re$ (W/m^3). Cooling fluid at a temperature of T_∞ flows over the surface of the cylinder with an average heat transfer coefficient $\langle h\rangle$.

a) Show that the steady-state temperature distributions in regions A, B, C, and D are expressed as

$$T_A = T_B = T_C = \frac{\Re R_D^2}{4k_D}\left[1-\left(\frac{R_C}{R_D}\right)^2 + 2\left(\frac{R_C}{R_D}\right)^2 \ln\left(\frac{R_C}{R_D}\right)\right]$$

$$+ \frac{\Re R_D}{2\langle h\rangle}\left[1-\left(\frac{R_C}{R_D}\right)^2\right] + T_\infty$$

$$T_D = \frac{\Re R_D^2}{4k_D}\left[1 - \left(\frac{r}{R_D}\right)^2 + 2\left(\frac{R_C}{R_D}\right)^2 \ln\left(\frac{r}{R_D}\right)\right] + \frac{\Re R_D}{2\langle h\rangle}\left[1 - \left(\frac{R_C}{R_D}\right)^2\right] + T_\infty$$

where k_A, k_B, k_C, and k_D are the thermal conductivities of regions A, B, C, and D, respectively.

b) Show that the outer surface temperature, T_s, which is in contact with the cooling fluid, is given by

$$T_s = \frac{\Re R_D}{2\langle h\rangle}\left[1 - \left(\frac{R_C}{R_D}\right)^2\right] + T_\infty$$

c) Show that the rate of heat loss to the cooling fluid is given by

$$\dot{Q} = \pi\left(R_D^2 - R_C^2\right)L\Re$$

d) Is it possible to solve parts (b) and (c) of the problem without calculating the temperature distributions?

9.14   The rate of generation per unit volume is sometimes expressed as a function of temperature rather than of position. Consider the transmission of an electric current through a wire of radius R. If the surface temperature is constant at T_R and the rate of generation per unit volume is given as

$$\Re = \Re_o(1 + aT) \tag{1}$$

a) Show that the governing equation for temperature is given by

$$\frac{d}{dr}\left(r\frac{dT}{dr}\right) + \frac{\Re_o}{k}(1 + aT)r = 0 \tag{2}$$

b) Use the transformation

$$u = 1 + aT \tag{3}$$

and reduce Eq. (2) to the form

$$\frac{d}{dr}\left(r\frac{du}{dr}\right) + \phi ru = 0 \tag{4}$$

where

$$\phi = \frac{\Re_o a}{k} \tag{5}$$

c) Solve Eq. (4) to get

$$\frac{T + (1/a)}{T_R + (1/a)} = \frac{J_o(\sqrt{\phi}\,r)}{J_o(\sqrt{\phi}R)} \tag{6}$$

d) What happens to Eq. (6) when $\sqrt{\phi}R = 2.4048$?

9.15 To estimate the increase in tissue temperature of plants resulting from metabolic heat, Breidenbach *et al.* (1997) considered a long cylindrical stem of radius R insulated by a boundary layer of stagnant air of thickness δ. Outside the boundary layer, the air is assumed to be well mixed and uniform at temperature T_∞. The stem tissue produces heat at a constant volumetric rate of $\Re$ (W/m^3).

a) Show that the steady temperature distribution within the stem is given by

$$T - T_\infty = \frac{\Re R^2}{4k}\left[1 - \left(\frac{r}{R}\right)^2\right] + \frac{\Re R^2}{2k_a}\ln\left(1 + \frac{\delta}{R}\right)$$

where k and k_a are the thermal conductivities of the stem and air, respectively.

b) Calculate the temperature in excess of the ambient air at the center and the surface of the spadix for *Philodendron selloum* for the following numerical values:

$$R = 1 \text{ cm} \qquad \delta = 1 \text{ mm} \qquad \Re = 50 \times 10^3 \text{ W/m}^3$$

$$k = 0.6 \text{ W/m·K} \qquad k_a = 0.0257 \text{ W/m·K}$$

(**Answer:** $\Delta T = 11.4$ K $\Delta T = 9.3$ K)

9.16 For laminar flow forced convection in a circular pipe with a constant wall temperature, the governing equation for temperature, Eq. (9.3-9), is integrated over the cross-sectional area of the tube in Section 9.3.1 to obtain Eq. (9.3-18), i.e.,

$$\dot{m}\widehat{C}_P\frac{dT_b}{dz} = \pi D h(T_w - T_b) \tag{1}$$

a) Now let us assume that the flow is turbulent. Over a differential volume element of thickness Δz, as shown in the figure below, write down the inventory rate equation for energy and show that the result is identical to Eq. (1).

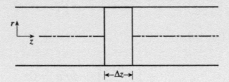

Integrate Eq. (1) to get

$$\dot{m}\widehat{C}_P\ln\left(\frac{T_w - T_{b_{in}}}{T_w - T_{b_{out}}}\right) = \pi D \langle h\rangle L \tag{2}$$

b) Water enters the inner pipe ($D = 23$ mm) of a double-pipe heat exchanger at 15 °C with a mass flow rate of 0.3 kg/s. Steam condenses in the annular region so as to keep the wall temperature almost constant at 112 °C. Determine the length of the heat exchanger if the outlet water temperature is 35 °C.

(**Answer:** b) 1.13 m)

9.17   Consider the heating of fluid A by fluid B in a countercurrent double-pipe heat exchanger as shown in the figure below.

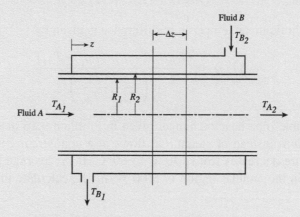

a) Show from the macroscopic energy balance that the rate of heat transferred is given by

$$\dot{Q} = (\dot{m}\widehat{C}_P)_A(T_{A_2} - T_{A_1}) = (\dot{m}\,\widehat{C}_P)_B(T_{B_2} - T_{B_1}) \tag{1}$$

where T_A and T_B are the bulk temperatures of the fluids A and B, respectively. State your assumptions.

b) Over the differential volume element of thickness Δz, write down the inventory rate equation for energy for the fluids A and B separately and show that

$$(\dot{m}\widehat{C}_P)_A\frac{dT_A}{dz} = \pi D_1 U_A(T_B - T_A) \tag{2}$$

$$(\dot{m}\widehat{C}_P)_B\frac{dT_B}{dz} = \pi D_1 U_A(T_B - T_A) \tag{3}$$

where U_A is the overall heat transfer coefficient based on the inside radius of the inner pipe given by Eq. (8.2-42), i.e.,

$$U_A = \left[\frac{1}{\langle h_A\rangle} + \frac{R_1\ln(R_2/R_1)}{k_w} + \frac{R_1}{\langle h_B\rangle R_2}\right]^{-1} \tag{4}$$

in which k_w represents the thermal conductivity of the inner pipe.

c) Subtract Eq. (2) from Eq. (3) to obtain

$$\frac{d(T_B - T_A)}{dz} = \left[\frac{1}{(\dot{m}\widehat{C}_P)_B} - \frac{1}{(\dot{m}\widehat{C}_P)_A}\right]\pi D_1 U_A(T_B - T_A) \tag{5}$$

d) Combine Eqs. (1) and (5) to get

$$\frac{d(T_B - T_A)}{dz} = \left[\frac{(T_{B_2} - T_{A_2}) - (T_{B_1} - T_{A_1})}{\dot{Q}}\right]\pi D_1 U_A(T_B - T_A) \tag{6}$$

e) Integrate Eq. (6) and show that the rate of heat transferred is given as

$$\dot{Q} = (\pi D_1 L) U_A \Delta T_{LM} \tag{7}$$

where the logarithmic mean temperature difference is given by

$$\Delta T_{LM} = \frac{(T_{B_2} - T_{A_2}) - (T_{B_1} - T_{A_1})}{\ln\left(\dfrac{T_{B_2} - T_{A_2}}{T_{B_1} - T_{A_1}}\right)} \tag{8}$$

f) Consider the double-pipe heat exchanger given in Problem 9.16 in which oil is used as the heating medium instead of steam. Oil flows in a countercurrent direction to water and its temperature decreases from 130 °C to 80 °C. If the average heat transfer coefficient for the oil in the annular region is 1100 W/m^2·K, calculate the length of the heat exchanger.

(**Answer:** f) 5.2 m)

9.18 You are a design engineer in a petroleum refinery. Oil is cooled from 60 °C to 40 °C in the inner pipe of a double-pipe heat exchanger. Cooling water flows countercurrently to the oil, entering at 15 °C and leaving at 35 °C. The oil tube has an inside diameter of 22 mm and an outside diameter of 25 mm with the inside and outside heat transfer coefficients of 600 and 1400 W/m^2·K, respectively. It is required to increase the oil flow rate by 40%. Estimate the exit temperatures of both oil and water at the increased flow rate.

(**Answer:** $T_{oil} = 43$ °C, $T_{water} = 39$ °C)

9.19 Water flowing at a mass flow rate of 0.4 kg/s is to be cooled from 82 °C to 42 °C in a thin-walled stainless-steel pipe of 10 cm diameter. Cooling is accomplished by an external air stream at 20 °C flowing perpendicular to the pipe with a velocity of 25 m/s. It is required to calculate the length of the pipe

a) Show that Eq. (6) in Example 4.15 is applicable to this problem with the following modification:

$$L = \frac{\dot{m}\widehat{C}_P}{\pi D U} \ln\left(\frac{T_{b_{in}} - T_\infty}{T_{b_{out}} - T_\infty}\right) \tag{1}$$

where $\dot{m}$ is the mass flow rate of water, T_∞ is the temperature of air, and U is the overall heat transfer coefficient defined by

$$U = \left(\frac{1}{h_{water}} + \frac{1}{h_{air}}\right)^{-1} \tag{2}$$

b) Estimate h_{water} and h_{air} from Dittus-Boelter and Zhukauskas correlations, respectively, and evaluate U.

c) Substitute the numerical values into Eq. (1) and calculate L.

(**Answer:** b) $U = 65.2$ W/m^2·K   c) 84.7 m)

9.20   Repeat the analysis given in Section 9.3.2 if both surfaces are maintained at T_o. Show that the maximum temperature rise in the fluid is given by

$$T_{\max} - T_o = \frac{\mu V^2}{8k}$$

9.21   Repeat the analysis given in Section 9.3.2 if the upper surface is adiabatic while the lower surface is maintained at T_o. Show that the temperature distribution is given by

$$T - T_o = \frac{\mu V^2}{k}\left[\frac{x}{B} - \frac{1}{2}\left(\frac{x}{B}\right)^2\right]$$

9.22   Repeat the analysis given in Section 9.3.2 for laminar flow of a Newtonian fluid between two fixed parallel plates under the action of a pressure gradient. The temperatures of the surfaces at $x = 0$ and $x = B$ are kept constant at T_o.

a) Obtain the temperature distribution as

$$T - T_o = \left(\frac{\mathcal{P}_o - \mathcal{P}_L}{L}\right)^2 \frac{B^4}{12\mu k}\left[\frac{1}{2}\left(\frac{x}{B}\right) - \frac{3}{2}\left(\frac{x}{B}\right)^2 + 2\left(\frac{x}{B}\right)^3 - \left(\frac{x}{B}\right)^4\right]$$

b) Show that the Nusselt numbers for the upper and lower plates are the same and equal to

$$\mathrm{Nu} = \frac{2B\langle h\rangle}{k} = \frac{35}{2}$$

in which the term $2B$ represents the hydraulic equivalent diameter.

9.23   Consider Couette flow of a Newtonian liquid between two large parallel plates as shown in the figure below. As a result of the viscous dissipation, liquid temperature varies in the x-direction. Although the thermal conductivity and density of the liquid are assumed to be independent of temperature, the variation in the liquid viscosity with temperature is given as

$$\mu = \mu_o \exp\left[-\beta\left(\frac{T - T_o}{T_o}\right)\right] \tag{1}$$

a) Show that the equations of motion and energy reduce to

$$\frac{d}{dx}\left(\mu \frac{dv_z}{dx}\right) = 0 \tag{2}$$

$$k\frac{d^2 T}{dx^2} + \mu\left(\frac{dv_z}{dx}\right)^2 = 0 \tag{3}$$

b) Integrate Eq. (2) and obtain the velocity distribution in the form

$$\frac{v_z}{V} = \frac{\displaystyle\int_0^x \frac{dx}{\mu}}{\displaystyle\int_0^B \frac{dx}{\mu}} \tag{4}$$

c) Substitute Eq. (4) into Eq. (3) to get

$$\frac{d^2\theta}{d\xi^2} + \Lambda\, e^{\theta} = 0 \tag{5}$$

where the dimensionless quantities are defined by

$$\theta = \frac{\beta(T - T_o)}{T_o} \qquad \xi = \frac{x}{B} \qquad Br = \frac{\mu_o V^2}{k\,T_o} \qquad \Lambda = \frac{Br\,\beta}{\left(\displaystyle\int_0^1 e^{\theta}\, d\xi\right)^2} \tag{6}$$

d) Multiply Eq. (5) by $2(d\theta/d\xi)$ and integrate the resulting equation to get

$$\frac{d\theta}{d\xi} = \pm\sqrt{2\Lambda}\sqrt{C - e^{\theta}} \tag{7}$$

where C is an integration constant.

e) Note that θ reaches a maximum value at $\ln C$. Therefore, the plus sign must be used in Eq. (7) when $0 \leqslant \theta \leqslant \ln C$. On the other hand, the negative sign must be used when $\ln C \leqslant \theta \leqslant 1$. Show that the integration of Eq. (7) leads to

$$\sqrt{2\Lambda}\,\xi = \int_0^{\ln C} \frac{d\theta}{\sqrt{C - e^{\theta}}} - \int_{\ln C}^{\theta} \frac{d\theta}{\sqrt{C - e^{\theta}}} \tag{8}$$

Solve Eq. (8) to obtain

$$\theta = \ln\left\{ C\,\mathrm{sech}^2\left[\sqrt{\frac{\Lambda C}{8}}(2\xi - 1)\right]\right\} \tag{9}$$

where C is the solution of

$$C = \cosh^2\left(\sqrt{\frac{\Lambda C}{8}}\right) \tag{10}$$

f) Substitute Eq. (9) into Eq. (4) and show that the velocity distribution is given by

$$\frac{v_z}{V} = \frac{1}{2}\left\{ 1 + \frac{\tanh\left[\sqrt{\dfrac{\Lambda C}{8}}(2\xi - 1)\right]}{\tanh\left(\sqrt{\dfrac{\Lambda C}{8}}\right)}\right\} \tag{11}$$

For more detailed information on this problem, see Gavis and Laurence (1968).

9.24   Two large porous plates are separated by a distance B as shown in the figure below. The temperatures of the lower and the upper plates are T_o and T_1, respectively, with $T_1 > T_o$. Air at a temperature of T_o is blown in the x-direction with a velocity of V.

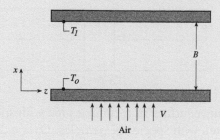

a) Show that the inventory rate equation for energy becomes

$$\rho \widehat{C}_P V \frac{dT}{dx} = k \frac{d^2 T}{dx^2} \tag{1}$$

b) Show that the introduction of the dimensionless variables

$$\theta = \frac{T - T_o}{T_1 - T_o} \qquad \xi = \frac{x}{B} \qquad \lambda = \frac{\rho \widehat{C}_P V B}{k} \tag{2}$$

reduces Eq. (1) to

$$\frac{d^2 \theta}{d\xi^2} - \lambda \frac{d\theta}{d\xi} = 0 \tag{3}$$

c) Solve Eq. (3) and show that the velocity distribution is given as

$$\theta = \frac{1 - e^{\lambda \xi}}{1 - e^{\lambda}} \tag{4}$$

d) Show that the heat flux at the lower plate is given by

$$q_x|_{x=0} = \frac{\lambda k (T_1 - T_o)}{B(1 - e^{\lambda})} \tag{5}$$

9.25   Rework the problem given in Section 9.4.1 for a zeroth-order chemical reaction, i.e., $r = k_o$, and show that the concentration profile is given by

$$c_A - c_{A_o} = \frac{k_o L^2}{2 \mathcal{D}_{AB}} \left[\left(\frac{z}{L} \right)^2 - 2 \left(\frac{z}{L} \right) \right]$$

9.26   For laminar flow forced convection in a circular pipe with a constant wall concentration, the governing equation for the concentration of species $\mathcal{A}$, Eq. (9.5-13), is integrated over the cross-sectional area of the tube in Section 9.5.1 to obtain Eq. (9.5-21), i.e.,

$$Q \frac{dc_{A_b}}{dz} = \pi D k_c (c_{A_w} - c_{A_b}) \tag{1}$$

a) Now assume that the flow is turbulent. Over a differential volume element of thickness Δz, as shown in the figure below, write down the inventory rate equation for the mass of species $\mathcal{A}$ and show that the result is identical to Eq. (1).

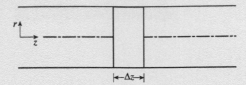

b) Instead of coating the inner surface of a circular pipe with species $\mathcal{A}$, let us assume that the circular pipe is packed with species $\mathcal{A}$ particles. Over a differential volume element of thickness Δz, write down the inventory rate equation for the mass of species $\mathcal{A}$ and show that the result is

$$Q\frac{dc_{A_b}}{dz} = a_V A k_c (c_{A_w} - c_{A_b}) \tag{2}$$

where A is the cross-sectional area of the pipe and a_v is the packing surface area per unit volume. Note that for a circular pipe $a_v = 4/D$ and $A = \pi D^2/4$ so that Eq. (2) reduces to Eq. (1).

9.27 A liquid is being transported in a circular plastic tube of inner and outer radii R_1 and R_2, respectively. The dissolved O_2 (species $\mathcal{A}$) concentration in the liquid is c_{A_o}. Develop an expression relating the increase in O_2 concentration as a function of the tubing length as follows:

a) Over a differential volume element of thickness Δz, write down the inventory rate equation for the mass of species $\mathcal{A}$ and show that the governing equation is

$$Q\frac{dc_{A_b}}{dz} = \frac{2\pi \mathcal{D}_{AB}}{\ln(R_2/R_1)}(c_{A_\infty} - c_{A_b}) \tag{1}$$

where $\mathcal{D}_{AB}$ is the diffusion coefficient of O_2 in a plastic tube and c_{A_∞} is the concentration of O_2 in the air surrounding the tube. In the development of Eq. (1), note that the molar rate of O_2 transfer through the tubing can be represented by Eq. (B) in Table 8.9.

b) Show that the integration of Eq. (1) leads to

$$c_{A_b} = c_{A_\infty} - (c_{A_\infty} - c_{A_o})\exp\left[-\frac{2\pi \mathcal{D}_{AB}z}{Q\ln(R_2/R_1)}\right] \tag{2}$$

9.28 Using the solution given by Johnstone and Pigford (1942), the Sherwood number is calculated as 3.41 for long contact times in Section 9.5.2.1. Obtain the same result by using an alternative approach as follows:

a) In terms of the following dimensionless quantities

$$\phi = \frac{c_A^* - c_A}{c_A^* - c_{A_o}} \qquad \xi = \frac{x}{\delta} \qquad \eta = \frac{\mathcal{D}_{AB}z}{v_{max}\delta^2} \tag{1}$$

show that Eqs. (9.5-88)–(9.5-91) reduce to

$$(1 - \xi^2) \frac{\partial \phi}{\partial \eta} = \frac{\partial^2 \phi}{\partial \xi^2} \tag{2}$$

$$\text{at} \quad \eta = 0 \qquad \phi = 1 \tag{3}$$

$$\text{at} \quad \xi = 0 \qquad \phi = 0 \tag{4}$$

$$\text{at} \quad \xi = 1 \qquad \frac{\partial \phi}{\partial \xi} = 0 \tag{5}$$

b) Use the method of separation of variables by proposing a solution in the form

$$\phi(\eta, \xi) = F(\eta) \, G(\xi) \tag{6}$$

and show that the solution is given by

$$\phi = \sum_{n=1}^{\infty} A_n e^{-\lambda_n^2 \eta} G_n(\xi) \tag{7}$$

where

$$A_n = \frac{\displaystyle\int_0^1 (1 - \xi^2) \, G_n(\xi) \, d\xi}{\displaystyle\int_0^1 (1 - \xi^2) \, G_n^2(\xi) \, d\xi} \tag{8}$$

and $G_n(\xi)$ are the eigenfunctions of the equation

$$\frac{d^2 G_n}{d\xi^2} + (1 - \xi^2) \lambda_n^2 G_n = 0 \tag{9}$$

c) Show that the Sherwood number is given by

$$\text{Sh} = \frac{k_c \delta}{\mathcal{D}_{AB}} = \frac{(\partial \phi / \partial \xi)_{\xi=0}}{\phi_b} \tag{10}$$

in which ϕ_b is the dimensionless bulk temperature defined by

$$\phi_b = \frac{3}{2} \int_0^1 (1 - \xi^2) \, \phi \, d\xi \tag{11}$$

d) Substitute Eq. (7) into Eq. (10) to get

$$\text{Sh} = \frac{2}{3} \frac{\displaystyle\sum_{n=1}^{\infty} A_n e^{-\lambda_n^2 \eta} (dG_n/d\xi)_{\xi=0}}{\displaystyle\sum_{n=1}^{\infty} (A_n/\lambda_n^2) \, e^{-\lambda_n^2 \eta} (dG_n/d\xi)_{\xi=0}} \tag{12}$$

For large values of η, show that Eq. (12) reduces to

$$Sh = \frac{2}{3}\lambda_1^2 \tag{13}$$

e) Use the method of Stodola and Vianello and show that the first approximation gives

$$\lambda_1^2 = 5.122 \tag{14}$$

Hint: Use $G_1 = \xi(\xi - 2)$ as a trial function.

9.29 Use Eq. (9.5-128) and show that $c_A \simeq c_{A_o}$ when

$$\frac{x}{\sqrt{4\mathcal{D}_{AB}z/v_{\max}}} = 2$$

Therefore, conclude that the penetration distance for concentration, δ_c, is given by

$$\delta_c(z) = 4\sqrt{\frac{\mathcal{D}_{AB}z}{v_{\max}}}$$

9.30 H_2S is being absorbed by pure water flowing down a vertical wall with a volumetric flow rate of 6.5×10^{-6} m^3/s at 20°C. The height and the width of the plate are 2 m and 0.8 m, respectively. If the diffusion coefficient of H_2S in water is 1.3×10^{-9} m^2/s and its solubility is 0.1 kmol/m^3, calculate the rate of absorption of H_2S into water.

(**Answer:** 6.5×10^{-7} kmol/s)

9.31 Water at 25°C flows down a wetted wall column of 5 cm diameter and 1.5 m height at a volumetric flow rate of 8.5×10^{-6} m^3/s. Pure CO_2 at a pressure of 1 atm flows in the countercurrent direction. If the solubility of CO_2 is 0.0336 kmol/m^3, determine the rate of absorption of CO_2 into water.

(**Answer:** 1.87×10^{-7} kmol/s)

9.32 Consider an industrial absorber in which gas bubbles ($\mathcal{A}$) rise through a liquid ($\mathcal{B}$) column. Bubble diameters usually range from 0.2 to 0.6 cm, while bubble velocities range from 15 to 35 cm/s (Astarita, 1967). Making use of Eq. (9.5-138) show that the range for the average mass transfer coefficient is

$$0.018 < \langle k_c \rangle < 0.047 \text{ cm/s}$$

Hint: A reasonable estimate for $\mathcal{D}_{AB}$ is 10^{-5} cm^2/s.

9.33 Consider a gas film of thickness δ, composed of species $\mathcal{A}$ and $\mathcal{B}$, adjacent to a flat catalyst particle in which gas $\mathcal{A}$ diffuses at steady-state through the film to the catalyst surface (positive z-direction) where the isothermal first-order heterogeneous reaction $\mathcal{A} \rightarrow \mathcal{B}$ occurs. As $\mathcal{B}$ leaves the surface it decomposes by an isothermal first-order heterogeneous reaction, $\mathcal{B} \rightarrow \mathcal{A}$. The gas composition at $z = 0$, i.e., x_{A_o} and x_{B_o}, is known.

a) Show that the equations representing the conservation of mass for species $\mathcal{A}$ and $\mathcal{B}$ are given by

$$\frac{dN_{A_z}}{dz} = \Re_A \tag{1}$$

$$\frac{dN_{B_z}}{dz} = \Re_B \tag{2}$$

where

$$\Re_A = -\Re_B = kcx_B \tag{3}$$

b) Using the heterogeneous reaction rate expression at the surface of the catalyst, conclude that

$$N_{A_z} = -N_{B_z} \qquad 0 \leqslant z \leqslant \delta \tag{4}$$

c) Since $x_A + x_B = 1$ everywhere in $0 \leqslant z \leqslant \delta$, solution of one of the conservation equations is sufficient to determine the concentration distribution within the film. Show that the governing equation for the mole fraction of species $\mathcal{B}$ is

$$\frac{d^2 x_B}{dz^2} - \left(\frac{k}{\mathcal{D}_{AB}}\right)x_B = 0 \tag{5}$$

subject to the boundary conditions

$$\text{at} \quad z = 0 \qquad x_B = x_{B_o} \tag{6}$$

$$\text{at} \quad z = \delta \qquad x_B = 1 + \frac{N_{B_z}}{ck^s} \tag{7}$$

where k^s is the surface reaction rate constant.

d) Show that the solution of Eq. (5) is given by

$$x_B = x_{B_o} \cosh(\Lambda \xi) + \phi \sinh(\Lambda \xi) \tag{8}$$

where

$$\phi = \frac{1 - x_{B_o}\cosh\Lambda + \dfrac{x_{B_o} - \cosh\Lambda}{(\lambda/\Lambda)\sinh\Lambda + \cosh\Lambda}}{\sinh\Lambda} \qquad \Lambda = \sqrt{\frac{k\delta^2}{\mathcal{D}_{AB}}} \qquad \lambda = \frac{k^s\delta}{\mathcal{D}_{AB}} \tag{9}$$

e) For an instantaneous heterogeneous reaction, show that Eq. (8) reduces to

$$x_B = x_{B_o}\cosh(\Lambda\xi) + \left(\frac{1 - x_{B_o}\cosh\Lambda}{\sinh\Lambda}\right)\sinh(\Lambda\xi) \tag{10}$$

f) If there is no homogeneous reaction, show that Eq. (8) takes the form

$$x_B = x_{B_o} + \left(\frac{\lambda}{\lambda + 1}\right)(1 - x_{B_o})\xi \tag{11}$$

9.34 A long polymeric rod of radius R is subjected to high temperatures for a prolonged period and, as a result, the following depolymerization reaction takes place within the rod:

$$P \rightarrow M + A$$

a) If the reaction is irreversible and zero-order, show that the dimensionless concentration distribution of species A under steady conditions is given by

$$\theta = 1 + \Lambda(1 - \xi^2)$$

The dimensionless quantities are defined by

$$\theta = \frac{c_A}{c_A^*} \qquad \xi = \frac{r}{R} \qquad \Lambda = \frac{k_o R^2}{4 \mathcal{D}_{AB} c_A^*}$$

where k_o is the zero-order reaction rate constant in mol/m^3·s, and c_A^* is the known equilibrium concentration of species A at the surface of the rod.

b) What is the physical significance of the term Λ? What would be the concentration distribution of species A when $\Lambda \ll 1$?

9.35 For diffusion in a spherical particle with a first-order homogeneous reaction, the governing equation in dimensionless form is given by Eq. (9.4-32), i.e.,

$$\frac{d}{d\xi}\left(\xi^2 \frac{d\theta}{d\xi}\right) - \Lambda^2 \xi^2 \theta = 0 \tag{1}$$

with the following boundary conditions

$$\text{at} \quad \xi = 0 \qquad \frac{d\theta}{d\xi} = 0 \tag{2}$$

$$\text{at} \quad \xi = 1 \qquad \theta = 0 \tag{3}$$

a) Comparison of Eq. (1) with Eq. (B.2-16) indicates that Eq. (1) is reducible to Bessel's equation. Show that the solution is given by

$$\theta = \frac{1}{\sqrt{\xi}}\left[K_1' I_{1/2}(\Lambda\xi) + K_2' I_{-1/2}(\Lambda\xi)\right] \tag{4}$$

where K_1' and K_2' are constants.

b) Use the following identities

$$I_{1/2}(x) = \sqrt{\frac{2}{\pi x}} \sinh x \qquad \text{and} \qquad I_{-1/2}(x) = \sqrt{\frac{2}{\pi x}} \cosh x \tag{5}$$

and show that Eq. (4) reduces to Eq. (9.4-44).

10

UNSTEADY-STATE MICROSCOPIC BALANCES WITHOUT GENERATION

The presence of the accumulation term in the inventory rate equation leads to a partial differential equation even if the transport takes place in one direction. The solution of partial differential equations not only depends on the structure of the equation itself, but also on the boundary conditions. Systematic treatment of momentum, energy, and mass transport based on the different types of partial differential equations as well as the initial and boundary conditions is a formidable task and beyond the scope of this text. Therefore, only some representative examples on momentum, energy, and mass transport will be covered in this chapter.

10.1 MOMENTUM TRANSPORT

Consider an incompressible Newtonian fluid contained between two large parallel plates of area A, separated by a distance B as shown in Figure 10.1. The system is initially at rest, but at time $t = 0$ the lower plate is set in motion in the z-direction at a constant velocity V while the upper plate is kept stationary. The development of a laminar velocity profile is required as a function of position and time.

Postulating $v_z = v_z(t, x)$ and $v_x = v_y = 0$, Table C.1 in Appendix C indicates that the only nonzero shear-stress component is τ_{xz}. Therefore, the components of the total momentum flux are expressed as

$$\pi_{xz} = \tau_{xz} + (\rho v_z)v_x = \tau_{xz} = -\mu \frac{\partial v_z}{\partial x} \tag{10.1-1}$$

$$\pi_{yz} = \tau_{yz} + (\rho v_z)v_y = 0 \tag{10.1-2}$$

$$\pi_{zz} = \tau_{zz} + (\rho v_z)v_z = \rho v_z^2 \tag{10.1-3}$$

The conservation statement for momentum is expressed as

$$\begin{pmatrix} \text{Rate of} \\ \text{momentum in} \end{pmatrix} - \begin{pmatrix} \text{Rate of} \\ \text{momentum out} \end{pmatrix} = \begin{pmatrix} \text{Rate of momentum} \\ \text{accumulation} \end{pmatrix} \tag{10.1-4}$$

For a rectangular differential volume element of thickness Δx, length Δz, and width W, as shown in Figure 10.1, Eq. (10.1-4) is expressed as

$$\left(\pi_{zz}|_z W \Delta x + \pi_{xz}|_x W \Delta z\right) - \left(\pi_{zz}|_{z+\Delta z} W \Delta x + \pi_{xz}|_{x+\Delta x} W \Delta z\right) = \frac{\partial}{\partial t}(W \Delta x \Delta z \rho v_z) \tag{10.1-5}$$

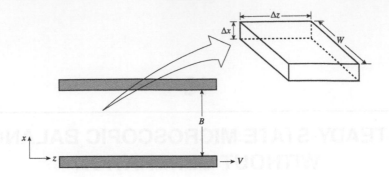

Figure 10.1. Unsteady Couette flow between parallel plates.

Dividing Eq. (10.1-5) by $W \Delta x \, \Delta z$ and taking the limit as $\Delta x \to 0$ and $\Delta z \to 0$ give

$$\rho \frac{\partial v_z}{\partial t} = \lim_{\Delta x \to 0} \frac{\pi_{xz}|_x - \pi_{xz}|_{x+\Delta x}}{\Delta x} + \lim_{\Delta z \to 0} \frac{\pi_{zz}|_z - \pi_{zz}|_{z+\Delta z}}{\Delta z} \qquad (10.1\text{-}6)$$

or,

$$\rho \frac{\partial v_z}{\partial t} = -\frac{\partial \pi_{xz}}{dx} - \frac{\partial \pi_{zz}}{\partial z} \qquad (10.1\text{-}7)$$

Substitution of Eqs. (10.1-1) and (10.1-3) into Eq. (10.1-7) and noting that $\partial v_z / \partial z = 0$ yield

$$\boxed{\rho \frac{\partial v_z}{\partial t} = \mu \frac{\partial^2 v_z}{\partial x^2}} \qquad (10.1\text{-}8)$$

The initial and boundary conditions associated with Eq. (10.1-8) are

$$\text{at} \quad t = 0 \qquad v_z = 0 \qquad (10.1\text{-}9)$$

$$\text{at} \quad x = 0 \qquad v_z = V \qquad (10.1\text{-}10)$$

$$\text{at} \quad x = B \qquad v_z = 0 \qquad (10.1\text{-}11)$$

The physical significance and the order of magnitude of the terms in Eq. (10.1-8) are given in Table 10.1. Note that the ratio of the viscous force to the rate of momentum accumulation is given by

$$\frac{\text{Viscous force}}{\text{Rate of momentum accumulation}} = \frac{\mu V / B^2}{\rho V / t} = \frac{\nu t}{B^2} \qquad (10.1\text{-}12)$$

In the literature, the term $\nu t / B^2$ is usually referred to as the *Fourier number*.

Introduction of the dimensionless quantities

$$\theta = \frac{v_z}{V} \qquad \xi = \frac{x}{B} \qquad \tau = \frac{\nu t}{B^2} \qquad (10.1\text{-}13)$$

reduces Eqs. (10.1-8)–(10.1-11) to

$$\frac{\partial \theta}{\partial \tau} = \frac{\partial^2 \theta}{\partial \xi^2} \qquad (10.1\text{-}14)$$

Table 10.1. The physical significance and the order of magnitude of the terms in Eq. (10.1-8)

Term	Physical Significance	Order of Magnitude
$\mu \dfrac{\partial^2 v_z}{\partial x^2}$	Viscous force	$\dfrac{\mu V}{B^2}$
$\rho \dfrac{\partial v_z}{\partial t}$	Rate of momentum accumulation	$\dfrac{\rho V}{t}$

$$\text{at} \quad \tau = 0 \qquad \theta = 0 \tag{10.1-15}$$

$$\text{at} \quad \xi = 0 \qquad \theta = 1 \tag{10.1-16}$$

$$\text{at} \quad \xi = 1 \qquad \theta = 0 \tag{10.1-17}$$

Since the boundary condition at $\xi = 0$ is not homogeneous, the method of separation of variables cannot be directly applied to obtain the solution. To circumvent this problem, propose a solution in the form

$$\theta(\tau, \xi) = \theta_\infty(\xi) - \theta_t(\tau, \xi) \tag{10.1-18}$$

in which $\theta_\infty(\xi)$ is the steady-state solution, i.e.,

$$\frac{d^2 \theta_\infty}{d\xi^2} = 0 \tag{10.1-19}$$

with the following boundary conditions

$$\text{at} \quad \xi = 0 \qquad \theta_\infty = 1 \tag{10.1-20}$$

$$\text{at} \quad \xi = 1 \qquad \theta_\infty = 0 \tag{10.1-21}$$

The steady-state solution is given by

$$\theta_\infty = 1 - \xi \tag{10.1-22}$$

which is identical to Eq. (8.1-12).

Substitution of Eq. (10.1-22) into Eq. (10.1-18) yields

$$\theta_t = 1 - \xi - \theta \tag{10.1-23}$$

The use of Eq. (10.1-23) in Eq. (10.1-14) gives the governing equation for the transient contribution as

$$\frac{\partial \theta_t}{\partial \tau} = \frac{\partial^2 \theta_t}{\partial \xi^2} \tag{10.1-24}$$

The initial and the boundary conditions associated with Eq. (10.1-24) become

$$\text{at} \quad \tau = 0 \qquad \theta_t = 1 - \xi \tag{10.1-25}$$

$$\text{at} \quad \xi = 0 \qquad \theta_t = 0 \tag{10.1-26}$$

$$\text{at} \quad \xi = 1 \qquad \theta_t = 0 \tag{10.1-27}$$

The boundary conditions at $\xi = 0$ and $\xi = 1$ are homogeneous and the parabolic partial differential equation given by Eq. (10.1-24) can now be solved by the method of separation of variables as described in Section B.3.6.1 in Appendix B.

The separation of variables method assumes that the solution can be represented as a product of two functions of the form

$$\theta_t(\tau, \xi) = F(\tau)G(\xi) \tag{10.1-28}$$

Substitution of Eq. (10.1-28) into Eq. (10.1-24) and rearrangement give

$$\frac{1}{F}\frac{dF}{d\tau} = \frac{1}{G}\frac{d^2G}{d\xi^2} \tag{10.1-29}$$

While the left-hand side of Eq. (10.1-29) is a function of τ only, the right-hand side is dependent only on ξ. This is possible only if both sides of Eq. (10.1-29) are equal to a constant, say $-\lambda^2$, i.e.,

$$\frac{1}{F}\frac{dF}{d\tau} = \frac{1}{G}\frac{d^2G}{d\xi^2} = -\lambda^2 \tag{10.1-30}$$

The choice of a negative constant is due to the fact that the solution will decay to zero as time increases, i.e., $\theta_t \to 0$ as $\tau \to \infty$. The choice of a positive constant would give a solution that becomes infinite as time increases.

Equation (10.1-30) results in two ordinary differential equations. The equation for F is given by

$$\frac{dF}{d\tau} + \lambda^2 F = 0 \tag{10.1-31}$$

The solution of Eq. (10.1-31) is

$$F(\tau) = e^{-\lambda^2 \tau} \tag{10.1-32}$$

On the other hand, the equation for G is

$$\frac{d^2G}{d\xi^2} + \lambda^2 G = 0 \tag{10.1-33}$$

and it is subject to the following boundary conditions

$$\text{at} \quad \xi = 0 \qquad G = 0 \tag{10.1-34}$$

$$\text{at} \quad \xi = 1 \qquad G = 0 \tag{10.1-35}$$

Note that Eq. (10.1-33) is a Sturm-Liouville equation with a weight function of unity[1]. The solution of Eq. (10.1-33) is given by

$$G(\xi) = C_1 \sin(\lambda\xi) + C_2 \cos(\lambda\xi) \tag{10.1-36}$$

---

[1] See Section B.3.4 in Appendix B.

where C_1 and C_2 are constants. While the application of Eq. (10.1-34) gives $C_2 = 0$, the use of Eq. (10.1-35) results in

$$C_1 \sin \lambda = 0 \tag{10.1-37}$$

For a nontrivial solution, the eigenvalues are given by

$$\sin \lambda = 0 \quad \Rightarrow \quad \lambda_n = n\pi \qquad n = 1, 2, 3, \ldots \tag{10.1-38}$$

Therefore, the transient solution is expressed as

$$\theta_t = \sum_{n=1}^{\infty} A_n \exp(-n^2\pi^2\tau) \sin(n\pi\xi) \tag{10.1-39}$$

The unknown coefficients A_n can be determined by using the initial condition, i.e., Eq. (10.1-25). The result is

$$1 - \xi = \sum_{n=0}^{\infty} A_n \sin(n\pi\xi) \tag{10.1-40}$$

Since the eigenfunctions are simply orthogonal[2], multiplication of Eq. (10.1-40) by $\sin(m\pi\xi)\,d\xi$ and integration from $\xi = 0$ to $\xi = 1$ give

$$\int_0^1 (1 - \xi) \sin(m\pi\xi)\,d\xi = \sum_{n=1}^{\infty} A_n \int_0^1 \sin(m\pi\xi) \sin(n\pi\xi)\,d\xi \tag{10.1-41}$$

The integral on the right-hand side of Eq. (10.1-41) is zero when $n \neq m$ and nonzero when $n = m$. Therefore, the summation drops out when $n = m$, and Eq. (10.1-41) reduces to the form

$$\int_0^1 (1 - \xi) \sin(n\pi\xi)\,d\xi = A_n \int_0^1 \sin^2(n\pi\xi)\,d\xi \tag{10.1-42}$$

Evaluation of the integrals gives

$$A_n = \frac{2}{n\pi} \tag{10.1-43}$$

and the transient solution takes the form

$$\theta_t = \frac{2}{\pi} \sum_{n=1}^{\infty} \frac{1}{n} \exp(-n^2\pi^2\tau) \sin(n\pi\xi) \tag{10.1-44}$$

Substitution of the steady-state and the transient solutions, Eqs. (10.1-22) and (10.1-44), into Eq. (10.1-18) gives the dimensionless velocity distribution as

$$\boxed{\theta = 1 - \xi - \frac{2}{\pi} \sum_{n=1}^{\infty} \frac{1}{n} \exp(-n^2\pi^2\tau) \sin(n\pi\xi)} \tag{10.1-45}$$

---

[2]See Section B.3.2 in Appendix B.

The volumetric flow rate can be determined by integrating the velocity distribution over the cross-sectional area of the plate, i.e.,

$$Q = \int_0^W \int_0^B v_z \, dx \, dy = WBV \int_0^1 \theta \, d\xi \tag{10.1-46}$$

Substitution of Eq. (10.1-45) into Eq. (10.1-46) gives

$$Q = \frac{WBV}{2} \left\{ 1 - \frac{8}{\pi^2} \sum_{k=0}^{\infty} \frac{1}{(2k+1)^2} \exp\left[-(2k+1)^2 \pi^2 \tau\right] \right\} \tag{10.1-47}$$

Note that $Q = WBV/2$ under steady conditions, i.e., $\tau = \infty$, which is identical to Eq. (8.1-15).

10.1.1 Solution for Short Times

Once the lower plate is set in motion, only the thin layer adjacent to the lower plate feels the motion of the plate during the initial stages. This thin layer does not feel the presence of the stationary plate at $x = B$ at all. For a fluid particle within this layer, the upper plate is at infinity. Therefore, the governing equation, together with the initial and boundary conditions, is expressed as

$$\rho \frac{\partial v_z}{\partial t} = \mu \frac{\partial^2 v_z}{\partial x^2} \tag{10.1-48}$$

$$\text{at} \quad t = 0 \qquad v_z = 0 \tag{10.1-49}$$

$$\text{at} \quad x = 0 \qquad v_z = V \tag{10.1-50}$$

$$\text{at} \quad x = \infty \qquad v_z = 0 \tag{10.1-51}$$

In the literature, this problem is generally known as *Stokes' first problem*[3]. Note that there is no length scale in this problem and the boundary condition at $x = \infty$ is the same as the initial condition. Therefore, Eq. (10.1-48) can be solved by similarity analysis as described in Section B.3.6.2 in Appendix B. The solution is given by

$$\frac{v_z}{V} = 1 - \text{erf}\left(\frac{x}{\sqrt{4\nu t}}\right) \tag{10.1-52}$$

where $\text{erf}(y)$ is the error function defined by

$$\text{erf}(y) = \frac{2}{\sqrt{\pi}} \int_0^y e^{-u^2} \, du \tag{10.1-53}$$

The drag force exerted on the plate is given by

$$F_D = A \left(-\mu \frac{\partial v_z}{\partial x}\Big|_{x=0}\right) \tag{10.1-54}$$

---

[3] Some authors refer to this problem as the *Rayleigh problem*.

The use of Eq. (10.1-52) in Eq. (10.1-54) leads to

$$F_D = \frac{A\mu V}{\sqrt{\pi \nu t}} \tag{10.1-55}$$

When $x/\sqrt{4\nu t} = 2$, Eq. (10.1-52) becomes

$$\frac{v_z}{V} = 1 - \text{erf}(2) = 1 - 0.995 = 0.005$$

indicating that $v_z \simeq 0$. Therefore, the *viscous penetration depth*, δ, is given by

$$\delta = 4\sqrt{\nu t} \tag{10.1-56}$$

The penetration depth changes with the square root of the momentum diffusivity and is independent of the plate velocity. The momentum diffusivities for water and air at $20\,°C$ are 1×10^{-6} and 15.08×10^{-6} m^2/s, respectively. The viscous penetration depths for water and air after one minute are 3.1 cm and 12 cm, respectively.

10.2 ENERGY TRANSPORT

The conservation statement for energy reduces to

$$\begin{pmatrix} \text{Rate of} \\ \text{energy in} \end{pmatrix} - \begin{pmatrix} \text{Rate of} \\ \text{energy out} \end{pmatrix} = \begin{pmatrix} \text{Rate of energy} \\ \text{accumulation} \end{pmatrix} \tag{10.2-1}$$

As in Section 8.2, our analysis will be restricted to the application of Eq. (10.2-1) to conduction in solids and stationary liquids. The solutions of almost all imaginable conduction problems in different coordinate systems with various initial and boundary conditions are given by Carslaw and Jaeger (1959). For this reason, only some representative problems will be presented in this section.

Using Eq. (7.1-14), the Biot number for heat transfer is expressed in the form

$$\text{Bi}_H = \frac{\text{Temperature difference in solid}}{\text{Temperature difference in fluid}} \tag{10.2-2}$$

Thus, the temperature distribution is considered uniform within the solid phase when $\text{Bi}_H \ll 1$. This obviously raises the question "What should the value of Bi_H be so that the condition $\text{Bi}_H \ll 1$ is satisfied?" In the literature, it is generally assumed that the internal resistance to heat transfer is negligible and the temperature distribution within the solid is almost uniform when $\text{Bi}_H < 0.1$. Under these conditions, the so-called *lumped-parameter analysis* is possible as can be seen in the solution of the problems in Section 7.5. When $0.1 < \text{Bi}_H < 40$, the internal and external resistances to heat transfer have almost the same order of magnitude. The external resistance to heat transfer is considered negligible when $\text{Bi}_H > 40$. Representative temperature profiles within the solid and fluid phases depending on the value of the Biot number are shown in Figure 10.2.

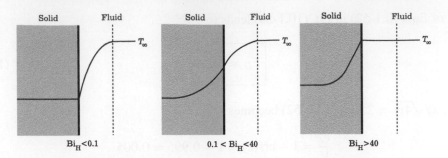

Figure 10.2. Effect of Bi_H on the temperature distribution.

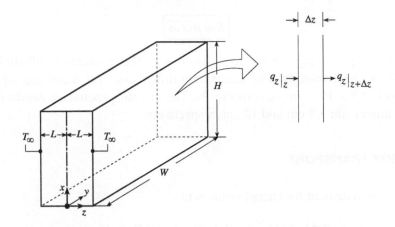

Figure 10.3. Unsteady-state conduction through a rectangular slab.

10.2.1 Heating of a Rectangular Slab

Consider a rectangular slab of thickness $2L$ as shown in Figure 10.3. Initially the slab temperature is uniform at a value of T_o. At $t = 0$, the temperatures of the surfaces at $z = \pm L$ are exposed to a fluid at a constant temperature of T_∞ ($T_\infty > T_o$). Let us assume $Bi_H > 40$ so that the resistance to heat transfer in the fluid phase is negligible and the temperatures of the slab surfaces are almost equal to T_∞.

As engineers, we are interested in the amount of heat transferred into the slab. For this purpose, it is first necessary to determine the temperature profile within the slab as a function of position and time.

If $L/H \ll 1$ and $L/W \ll 1$, then it is possible to assume that the conduction is one-dimensional (see Problem 10.1) and to postulate that $T = T(t, z)$. In that case, Table C.4 in Appendix C indicates that the only nonzero energy flux component is e_z, and it is given by

$$e_z = q_z = -k \frac{\partial T}{\partial z} \tag{10.2-3}$$

For a rectangular differential volume element of thickness Δz, as shown in Figure 10.3, Eq. (10.2-1) is expressed as

$$q_z|_z WH - q_z|_{z+\Delta z} WH = \frac{\partial}{\partial t}\Big[WH\Delta z \rho \widehat{C}_P (T - T_o)\Big] \tag{10.2-4}$$

Following the notation introduced by Bird *et al.* (2002), "in" and "out" directions are taken in the positive z-direction. Dividing Eq. (10.2-4) by $WH\Delta z$ and letting $\Delta z \to 0$ give

$$\rho \widehat{C}_P \frac{\partial T}{\partial t} = \lim_{\Delta z \to 0} \frac{q_z|_z - q_z|_{z+\Delta z}}{\Delta z} \tag{10.2-5}$$

or,

$$\rho \widehat{C}_P \frac{\partial T}{\partial t} = -\frac{\partial q_z}{\partial z} \tag{10.2-6}$$

Substitution of Eq. (10.2-3) into Eq. (10.2-6) gives the governing equation for temperature as

$$\boxed{\rho \widehat{C}_P \frac{\partial T}{\partial t} = k \frac{\partial^2 T}{\partial z^2}} \tag{10.2-7}$$

All physical properties are assumed to be independent of temperature in the development of Eq. (10.2-7). The initial and boundary conditions associated with Eq. (10.2-7) are

$$\text{at} \quad t = 0 \qquad T = T_o \tag{10.2-8}$$

$$\text{at} \quad z = L \qquad T = T_\infty \tag{10.2-9}$$

$$\text{at} \quad z = -L \qquad T = T_\infty \tag{10.2-10}$$

Note that $z = 0$ represents a plane of symmetry across which there is no net flux, i.e., $\partial T/\partial z = 0$. Therefore, it is also possible to express the initial and boundary conditions as

$$\text{at} \quad t = 0 \qquad T = T_o \tag{10.2-11}$$

$$\text{at} \quad z = 0 \qquad \frac{\partial T}{\partial z} = 0 \tag{10.2-12}$$

$$\text{at} \quad z = L \qquad T = T_\infty \tag{10.2-13}$$

The boundary condition at $z = 0$ can also be interpreted as an insulated surface. As a result, the governing equation for temperature, Eq. (10.2-7), together with the initial and boundary conditions given by Eqs. (10.2-11)–(10.2-13), represents the following problem statement: "A slab of thickness L is initially at a uniform temperature of T_o. One side of the slab is perfectly insulated while the other surface is exposed to a fluid at constant temperature of T_∞ with $T_\infty > T_o$ for $t > 0$."

The physical significance and the order of magnitude of the terms in Eq. (10.2-7) are given in Table 10.2. Note that the ratio of the rate of conduction to the rate of energy accumulation is given by

$$\frac{\text{Rate of conduction}}{\text{Rate of energy accumulation}} = \frac{k(T_\infty - T_o)/L^2}{\rho \widehat{C}_P (T_\infty - T_o)/t} = \frac{\alpha t}{L^2} \tag{10.2-14}$$

In the literature, the term $\alpha t/L^2$ is usually referred to as the *Fourier number*.

Introduction of the dimensionless quantities

$$\theta = \frac{T_\infty - T}{T_\infty - T_o} \qquad \xi = \frac{z}{L} \qquad \tau = \frac{\alpha t}{L^2} \tag{10.2-15}$$

Table 10.2. The physical significance and the order of magnitude of the terms in Eq. (10.2-7)

Term	Physical Significance	Order of Magnitude
$k \dfrac{\partial^2 \langle T \rangle}{\partial z^2}$	Rate of conduction in z-direction	$\dfrac{k(T_\infty - T_o)}{L^2}$
$\rho \widehat{C}_P \dfrac{\partial T}{\partial t}$	Rate of energy accumulation	$\dfrac{\rho \widehat{C}_P (T_\infty - T_o)}{t}$

reduces Eqs. (10.2-7)–(10.2-10) to

$$\frac{\partial \theta}{\partial \tau} = \frac{\partial^2 \theta}{\partial \xi^2} \tag{10.2-16}$$

$$\text{at} \quad \tau = 0 \qquad \theta = 1 \tag{10.2-17}$$

$$\text{at} \quad \xi = 1 \qquad \theta = 0 \tag{10.2-18}$$

$$\text{at} \quad \xi = -1 \qquad \theta = 0 \tag{10.2-19}$$

Since the governing equation, as well as the boundary conditions in the ξ-direction, is homogeneous, this parabolic partial differential equation can be solved by the method of separation of variables as explained in Section B.3.6.1 in Appendix B.

The solution can be represented as a product of two functions of the form

$$\theta(\tau, \xi) = F(\tau)G(\xi) \tag{10.2-20}$$

so that Eq. (10.2-16) reduces to

$$\frac{1}{F} \frac{dF}{d\tau} = \frac{1}{G} \frac{d^2 G}{d\xi^2} \tag{10.2-21}$$

While the left-hand side of Eq. (10.2-21) is a function of τ only, the right-hand side is dependent only on ξ. This is possible only if both sides of Eq. (10.2-21) are equal to a constant, say $-\lambda^2$, i.e.,

$$\frac{1}{F} \frac{dF}{d\tau} = \frac{1}{G} \frac{d^2 G}{d\xi^2} = -\lambda^2 \tag{10.2-22}$$

The choice of a negative constant is due to the fact that the dimensionless temperature, θ, will decay to zero, i.e., $T \to T_\infty$, as time increases. The choice of a positive constant would give a solution that becomes infinite as time increases.

Equation (10.2-22) results in two ordinary differential equations. The equation for F is given by

$$\frac{dF}{d\tau} + \lambda^2 F = 0 \tag{10.2-23}$$

The solution of Eq. (10.2-23) is

$$F(\tau) = e^{-\lambda^2 \tau} \tag{10.2-24}$$

On the other hand, the equation for G is

$$\frac{d^2 G}{d\xi^2} + \lambda^2 G = 0 \tag{10.2-25}$$

and it is subject to the boundary conditions

$$\text{at} \quad \xi = 1 \qquad G = 0 \tag{10.2-26}$$

$$\text{at} \quad \xi = -1 \qquad G = 0 \tag{10.2-27}$$

Note that Eq. (10.2-25) is a Sturm-Liouville equation with a weight function of unity. The solution of Eq. (10.2-25) is given by

$$G(\xi) = C_1 \sin(\lambda\xi) + C_2 \cos(\lambda\xi) \tag{10.2-28}$$

where C_1 and C_2 are constants. Since the problem is symmetric around the z-axis, then θ, and hence G, must be even functions[4] of ξ, i.e., $C_1 = 0$. Application of Eq. (10.2-26) gives

$$C_2 \cos \lambda = 0 \tag{10.2-29}$$

For a nontrivial solution, the eigenvalues are given by

$$\cos \lambda = 0 \quad \Rightarrow \quad \lambda_n = \left(n + \tfrac{1}{2}\right)\pi \qquad n = 0, 1, 2, \ldots \tag{10.2-30}$$

Therefore, the general solution, which is the summation of all possible solutions, becomes

$$\theta = \sum_{n=0}^{\infty} A_n \exp\left[-\left(n + \tfrac{1}{2}\right)^2 \pi^2 \tau\right] \cos\left[\left(n + \tfrac{1}{2}\right)\pi\xi\right] \tag{10.2-31}$$

The unknown coefficients A_n can be determined by using the initial condition, i.e., Eq. (10.2-17). The result is

$$1 = \sum_{n=0}^{\infty} A_n \cos\left[\left(n + \tfrac{1}{2}\right)\pi\xi\right] \tag{10.2-32}$$

Since the eigenfunctions are simply orthogonal, multiplication of Eq. (10.2-32) by $\cos\left[\left(m + \tfrac{1}{2}\right)\pi\xi\right] d\xi$ and integration from $\xi = 0$ to $\xi = 1$ give

$$\int_0^1 \cos\left[\left(m + \tfrac{1}{2}\right)\pi\xi\right] d\xi = \sum_{n=0}^{\infty} A_n \int_0^1 \cos\left[\left(n + \tfrac{1}{2}\right)\pi\xi\right] \cos\left[\left(m + \tfrac{1}{2}\right)\pi\xi\right] d\xi \tag{10.2-33}$$

The integral on the right-hand side of Eq. (10.2-33) is zero when $n \neq m$ and nonzero when $n = m$. Therefore, the summation drops out when $n = m$, and Eq. (10.2-33) reduces to the form

$$\int_0^1 \cos\left[\left(n + \tfrac{1}{2}\right)\pi\xi\right] d\xi = A_n \int_0^1 \cos^2\left[\left(n + \tfrac{1}{2}\right)\pi\xi\right] d\xi \tag{10.2-34}$$

---

[4]A function $f(x)$ is said to be an *odd function* if $f(-x) = -f(x)$ and an *even function* if $f(-x) = f(x)$.

Evaluation of the integrals yields

$$A_n = 2 \frac{(-1)^n}{(n + \frac{1}{2})\pi} \tag{10.2-35}$$

and the solution representing the dimensionless temperature distribution is expressed as

$$\theta = \frac{2}{\pi} \sum_{n=0}^{\infty} \frac{(-1)^n}{(n + \frac{1}{2})} \exp\left[-(n + \tfrac{1}{2})^2 \pi^2 \tau\right] \cos\left[(n + \tfrac{1}{2})\pi \xi\right] \tag{10.2-36}$$

Example 10.1 Show that the series solution given by Eq. (10.2-36) can be approximated by the first term of the series when $\tau \geqslant 0.2$.

Solution

The cosine function appearing in Eq. (10.2-36) varies between ± 1. Let X be the function defined by

$$X = \frac{(-1)^n}{(n + \frac{1}{2})} \exp\left[-(n + \tfrac{1}{2})^2 \pi^2 \tau\right]$$

For various values of τ and n, the calculated values of X are given in the following table:

		X		
n	$\tau = 0.1$	$\tau = 0.2$	$\tau = 0.5$	$\tau = 1$
0	1.563	1.221	0.582	0.170
1	−0.072	−7.854 × 10⁻³	−1.004 × 10⁻⁵	−1.513 × 10⁻¹⁰
2	8.377 × 10⁻⁴	1.755 × 10⁻⁶	1.612 × 10⁻¹⁴	0
3	−1.604 × 10⁻⁶	−9.005 × 10⁻¹²	0	0

Note that when $\tau \geqslant 0.2$, X values become negligible for $n \geqslant 1$. Under these circumstances, the dominant term of the series is the first term.

Example 10.2 A copper slab ($\alpha = 117 \times 10^{-6}$ m²/s) 10 cm thick is initially at a temperature of 20 °C. The slab is dipped in boiling water at atmospheric pressure.

a) Estimate the time it takes for the center of the slab to reach 40 °C.
b) Calculate the time to reach steady-state conditions.

Solution

Assumption

1. The external resistance to heat transfer is negligible, i.e., $Bi_H > 40$, so that the surface temperature of the slab is equal to the boiling point temperature of water under atmospheric pressure, i.e., 100 °C.

Analysis

a) The temperature at the center of the slab, T_c, can be found by evaluating Eq. (10.2-36) at $\xi = 0$. The result is given by

$$\frac{T_\infty - T_c}{T_\infty - T_o} = \frac{2}{\pi} \sum_{n=0}^{\infty} \frac{(-1)^n}{\left(n + \frac{1}{2}\right)} \exp\left[-\left(n + \tfrac{1}{2}\right)^2 \pi^2 \tau\right] \tag{1}$$

Substitution of the numerical values into Eq. (1) gives

$$\frac{100 - 40}{100 - 20} = \frac{2}{\pi} \sum_{n=0}^{\infty} \frac{(-1)^n}{\left(n + \frac{1}{2}\right)} \exp\left[-\left(n + \tfrac{1}{2}\right)^2 \pi^2 \tau\right] \tag{2}$$

The solution of Eq. (2) results in

$$\tau = 0.212 \quad \Rightarrow \quad t = \frac{\tau L^2}{\alpha} = \frac{(0.212)(0.05)^2}{117 \times 10^{-6}} = 4.5 \text{ s}$$

b) Under steady conditions, the slab temperature will be at T_∞, i.e., $100\,°C$, throughout. Mathematically speaking, steady-state conditions are reached when $t \to \infty$. The practical question to ask at this point is "what does $t = \infty$ indicate?" If the time to reach steady-state, t_∞, is defined as the time for the center temperature to reach 99% of the surface temperature, i.e., $0.99T_\infty$, then Eq. (1) becomes

$$\frac{0.01 T_\infty}{T_\infty - T_o} = \frac{2}{\pi} \sum_{n=0}^{\infty} \frac{(-1)^n}{\left(n + \frac{1}{2}\right)} \exp\left[-\left(n + \tfrac{1}{2}\right)^2 \pi^2 \tau_\infty\right] \tag{3}$$

Substitution of the numerical values into Eq. (3) leads to

$$\tau_\infty = 1.874 \quad \Rightarrow \quad t_\infty = \frac{\tau_\infty L^2}{\alpha} = \frac{(1.874)(0.05)^2}{117 \times 10^{-6}} = 40 \text{ s}$$

Comment: In Section 3.4.1, the time it takes for a given process to reach steady-state is defined by

$$t = \frac{L_{ch}^2}{\alpha} \tag{4}$$

For the problem at hand, L_{ch} is half the thickness of the slab. Substitution of the numerical values into Eq. (4) gives

$$t = \frac{(0.05)^2}{117 \times 10^{-6}} = 21 \text{ s}$$

As far as the orders of magnitude are concerned, such an estimate is not very off the exact value.

Example 10.3 Repeat part (a) of Example 10.2 for a 10 cm thick stainless steel slab ($\alpha = 3.91 \times 10^{-6} \text{ m}^2/\text{s}$).

Solution

Note that Eq. (2) in Example 10.2 is also valid for this problem. Thus,

$$\tau = 0.212 \quad \Rightarrow \quad t = \frac{\tau L^2}{\alpha} = \frac{(0.212)(0.05)^2}{3.91 \times 10^{-6}} = 136 \text{ s}$$

Comment: Once the problem for a copper slab is solved, is it possible to estimate the time for the center of a stainless steel slab to reach $40\,°C$ without solving Eq. (1) in Example 10.2? To answer this question, note that the orders of magnitude of the accumulation and conduction terms in Eq. (10.2-7) must be, more or less, equal to each other. This leads to the fact that the order of magnitude of the Fourier number is unity. Thus, it is possible to equate the Fourier numbers, i.e.,

$$\left(\frac{\alpha t}{L^2}\right)_{\text{copper}} = \left(\frac{\alpha t}{L^2}\right)_{\text{stainless steel}} \tag{2}$$

Simplification of Eq. (2) leads to

$$t_{\text{stainless steel}} = t_{\text{copper}}\left(\frac{\alpha_{\text{copper}}}{\alpha_{\text{stainless steel}}}\right) = 4.5\left(\frac{117 \times 10^{-6}}{3.91 \times 10^{-6}}\right) = 135 \text{ s}$$

which is almost equal to the exact value.

10.2.1.1 *Macroscopic equation* Integration of the governing equation for temperature, Eq. (10.2-7), over the volume of the system gives

$$\int_{-L}^{L}\int_{0}^{W}\int_{0}^{H} \rho \widehat{C}_P \frac{\partial T}{\partial t}\, dx\, dy\, dz = \int_{-L}^{L}\int_{0}^{W}\int_{0}^{H} k \frac{\partial^2 T}{\partial z^2}\, dx\, dy\, dz \tag{10.2-37}$$

or,

$$\underbrace{\frac{d}{dt}\left[\int_{-L}^{L}\int_{0}^{W}\int_{0}^{H} \rho \widehat{C}_P (T - T_o)\, dx\, dy\, dz\right]}_{\text{Rate of accumulation of energy}} = \underbrace{2WH\left(k \frac{\partial T}{\partial z}\bigg|_{z=L}\right)}_{\substack{\text{Rate of energy entering} \\ \text{from surfaces at } z=\pm L}} \tag{10.2-38}$$

which is the macroscopic energy balance by considering the rectangular slab as a system. The rate of energy entering the slab, $\dot{Q}$, is given by

$$\dot{Q} = 2WH\left(k \frac{\partial T}{\partial z}\bigg|_{z=L}\right) = -\frac{2WHk(T_\infty - T_o)}{L}\frac{\partial \theta}{\partial \xi}\bigg|_{\xi=1} \tag{10.2-39}$$

The use of Eq. (10.2-36) in Eq. (10.2-39) gives

$$\dot{Q} = \frac{4WHk(T_\infty - T_o)}{L}\sum_{n=0}^{\infty}\exp\left[-\left(n + \tfrac{1}{2}\right)^2 \pi^2 \tau\right] \tag{10.2-40}$$

The amount of heat transferred can be calculated from

$$Q = \int_0^t \dot{Q}\, dt = \frac{L^2}{\alpha} \int_0^\tau \dot{Q}\, d\tau \qquad (10.2\text{-}41)$$

Substitution of Eq. (10.2-40) into Eq. (10.2-41) yields

$$\boxed{\frac{Q}{Q_\infty} = 1 - \frac{2}{\pi^2} \sum_{n=0}^{\infty} \frac{1}{\left(n + \frac{1}{2}\right)^2} \exp\!\left[-\left(n + \tfrac{1}{2}\right)^2 \pi^2 \tau\right]} \qquad (10.2\text{-}42)$$

where Q_∞ is the amount of heat transferred to the slab until steady-state conditions are reached, i.e.,

$$Q_\infty = 2LWH\rho \widehat{C}_P (T_\infty - T_o) \qquad (10.2\text{-}43)$$

Example 10.4 Estimate the amount of heat transferred to the copper slab in Example 10.2 until the center temperature reaches 40 °C. Express your answer as a fraction of the total heat transfer that would be transferred until the steady conditions are reached.

Solution

Substitution of the numerical values into Eq. (10.2-42) yields

$$\frac{Q}{Q_\infty} = 1 - \frac{2}{\pi^2} \sum_{n=0}^{\infty} \frac{1}{\left(n + \frac{1}{2}\right)^2} \exp\!\left[-\left(n + \tfrac{1}{2}\right)^2 \pi^2 (0.212)\right] = 0.519$$

Comment: In Example 10.2, the time to reach steady-state is defined as the time for the center temperature to reach 99% of the surface temperature and τ_∞ is calculated as 1.874. When $\tau_\infty = 1.874$, then

$$\frac{Q}{Q_\infty} = 1 - \frac{2}{\pi^2} \sum_{n=0}^{\infty} \frac{1}{\left(n + \frac{1}{2}\right)^2} \exp\!\left[-\left(n + \tfrac{1}{2}\right)^2 \pi^2 (1.874)\right] = 0.992$$

Therefore, the time to reach steady-state can also be defined as the time when $Q = 0.99 Q_\infty$.

10.2.1.2 *Solution for short times* Let s be the distance measured from the surface of the slab, i.e.,

$$s = L - z \qquad (10.2\text{-}44)$$

so that Eq. (10.2-7) reduces to

$$\frac{\partial T}{\partial t} = \alpha \frac{\partial^2 T}{\partial s^2} \qquad (10.2\text{-}45)$$

At small values of time, the heat does not penetrate very far into the slab. Under these circumstances, it is possible to consider the slab a semi-infinite medium in the s-direction. The initial and boundary conditions associated with Eq. (10.2-45) become

$$\text{at} \quad t = 0 \qquad T = T_o \tag{10.2-46}$$

$$\text{at} \quad s = 0 \qquad T = T_\infty \tag{10.2-47}$$

$$\text{at} \quad s = \infty \qquad T = T_o \tag{10.2-48}$$

Introduction of the dimensionless temperature

$$\phi = \frac{T - T_o}{T_\infty - T_o} \tag{10.2-49}$$

reduces Eqs. (10.2-45)–(10.2-48) to

$$\frac{\partial \phi}{\partial t} = \alpha \frac{\partial^2 \phi}{\partial s^2} \tag{10.2-50}$$

$$\text{at} \quad t = 0 \qquad \phi = 0 \tag{10.2-51}$$

$$\text{at} \quad s = 0 \qquad \phi = 1 \tag{10.2-52}$$

$$\text{at} \quad s = \infty \qquad \phi = 0 \tag{10.2-53}$$

Since there is no length scale in the problem and the boundary condition at $s = \infty$ is the same as the initial condition, this parabolic partial differential equation can be solved by the similarity solution as explained in Section B.3.6.2 in Appendix B. The solution is sought in the form

$$\phi = f(\eta) \qquad \text{where} \qquad \eta = \frac{s}{\sqrt{4\alpha t}} \tag{10.2-54}$$

The chain rule of differentiation gives

$$\frac{\partial \phi}{\partial t} = \frac{df}{d\eta} \frac{\partial \eta}{\partial t} = -\frac{1}{2} \frac{\eta}{t} \frac{df}{d\eta} \tag{10.2-55}$$

$$\frac{\partial^2 \phi}{\partial s^2} = \frac{d^2 f}{d\eta^2} \left(\frac{\partial \eta}{\partial s} \right)^2 + \frac{df}{d\eta} \frac{\partial^2 \eta}{\partial s^2} = \frac{1}{4\alpha t} \frac{d^2 f}{d\eta^2} \tag{10.2-56}$$

Substitution of Eqs. (10.2-55) and (10.2-56) into Eq. (10.2-50) gives

$$\frac{d^2 f}{d\eta^2} + 2\eta \frac{df}{d\eta} = 0 \tag{10.2-57}$$

The boundary conditions associated with Eq. (10.2-57) are

$$\text{at} \quad \eta = 0 \qquad f = 1 \tag{10.2-58}$$

$$\text{at} \quad \eta = \infty \qquad f = 0 \tag{10.2-59}$$

The integrating factor for Eq. (10.2-57) is $\exp(\eta^2)$. Multiplication of Eq. (10.2-57) by the integrating factor yields

$$\frac{d}{d\eta} \left(e^{\eta^2} \frac{df}{d\eta} \right) = 0 \tag{10.2-60}$$

which implies that

$$\frac{df}{d\eta} = C_1 e^{-\eta^2} \tag{10.2-61}$$

Integration of Eq. (10.2-61) gives

$$f = C_1 \int_0^\eta e^{-u^2}\, du + C_2 \tag{10.2-62}$$

where u is a dummy variable of integration. Application of Eq. (10.2-58) gives $C_2 = 1$. On the other hand, application of Eq. (10.2-59) leads to

$$C_1 = -\frac{1}{\displaystyle\int_0^\infty e^{-u^2}\, du} = -\frac{2}{\sqrt{\pi}} \tag{10.2-63}$$

Therefore, the solution becomes

$$f = 1 - \frac{2}{\sqrt{\pi}} \int_0^\eta e^{-u^2}\, du = 1 - \mathrm{erf}(\eta) \tag{10.2-64}$$

or,

$$\boxed{\frac{T_\infty - T}{T_\infty - T_o} = \mathrm{erf}\left(\frac{s}{\sqrt{4\alpha t}}\right)} \tag{10.2-65}$$

The rate of heat transfer into the semi-infinite slab of cross-sectional area A is

$$\dot{Q} = A\left(-k\left.\frac{\partial T}{\partial s}\right|_{s=0}\right) \tag{10.2-66}$$

The use of Eq. (10.2-65) in Eq. (10.2-66) gives

$$\boxed{\dot{Q} = \frac{Ak(T_\infty - T_o)}{\sqrt{\pi \alpha t}}} \tag{10.2-67}$$

The amount of heat transferred is determined from

$$Q = \int_0^t \dot{Q}\, dt \tag{10.2-68}$$

Substitution of Eq. (10.2-67) into Eq. (10.2-68) leads to

$$\boxed{Q = \frac{2Ak(T_\infty - T_o)\sqrt{t}}{\sqrt{\pi \alpha}}} \tag{10.2-69}$$

When $s/\sqrt{4\alpha t} = 2$, Eq. (10.2-65) becomes

$$\frac{T_\infty - T}{T_\infty - T_o} = \mathrm{erf}(2) = 0.995$$

indicating that the temperature practically drops to the initial temperature, i.e., $T \simeq T_o$. Therefore, the *thermal penetration depth*, δ_t, is given by

$$\boxed{\delta_t = 4\sqrt{\alpha t}} \tag{10.2-70}$$

The assumption of a semi-infinite medium (or short time solution) is no longer valid when the temperature at $s = L$ becomes equal to or greater than T_o. Therefore, the solution given by Eq. (10.2-65) holds as long as

$$1 \leqslant \mathrm{erf}\left(\frac{L}{\sqrt{4\alpha t}}\right) \tag{10.2-71}$$

Since erf(2) $\simeq 1$, Eq. (10.2-71) simplifies to

$$\boxed{t \leqslant \frac{L^2}{16\alpha}} \quad \text{Criterion for semi-infinite medium assumption} \tag{10.2-72}$$

Example 10.5 One of the surfaces of a thick wall is exposed to gases at 350 °C. If the initial wall temperature is uniform at 20 °C, determine the time required for a point 5 cm below the surface to reach 280 °C. The thermal diffusivity of the wall is 4×10^{-7} m^2/s.

Solution

Assumptions

1. The Biot number is large enough for the external resistance to heat transfer to be neglected so that the surface temperature of the wall is almost equal to the gas temperature.
2. Since the wall thickness is large, it may be considered a semi-infinite medium.

Analysis

Equation (10.2-65) is written as

$$\frac{350 - 280}{350 - 20} = \mathrm{erf}\left(\frac{s}{\sqrt{4\alpha t}}\right) \quad \Rightarrow \quad \frac{s}{\sqrt{4\alpha t}} = 0.19$$

Therefore, the time required is

$$t = \frac{1}{4(4 \times 10^{-7})}\left(\frac{0.05}{0.19}\right)^2 = 43,283 \text{ s} \simeq 12 \text{ h}$$

Comment: In this particular example, the thermal penetration depth after 12 hours is

$$\delta_t = 4\sqrt{(4 \times 10^{-7})(12)(3600)} = 0.53 \text{ m}$$

Note that the temperature distribution is confined to the region $0 \leqslant s < 53$ cm. For $s \geqslant 53$ cm, the temperature is 20 °C.

Example 10.6 A concrete wall ($\alpha = 6.6 \times 10^{-7}$ m^2/s) of thickness 15 cm is initially at a temperature of 20 °C. Both sides of the wall are exposed to hot gases at 180 °C.

a) How long will it take for the center temperature to start to rise?
b) When does the temperature profile reach steady-state?

Solution

Assumption

1. The Biot number is large enough for the external resistance to heat transfer to be neglected so that the surface temperature of the wall is almost equal to the gas temperature.

Analysis

a) The center temperature will start to rise when the thermal penetration depth reaches half the thickness of the slab. Thus, from Eq. (10.2-70)

$$t = \frac{L^2}{16\alpha} = \frac{(7.5 \times 10^{-2})^2}{16(6.6 \times 10^{-7})} = 533 \text{ s} \simeq 9 \text{ min} \tag{1}$$

b) From Section 3.4.1, the time scale to reach steady-state is given by

$$t = \frac{L^2}{\alpha} = \frac{(7.5 \times 10^{-2})^2}{6.6 \times 10^{-7}} = 8523 \text{ s} \simeq 2 \text{ h } 22 \text{ min} \tag{2}$$

In other words, the system reaches steady-state when $\tau = 1$. Although this is not the exact value, it gives an engineering estimate of the time it takes to reach steady-state.

Comment: The concrete wall reaches steady-state when the center temperature becomes 180 °C. Now, let us calculate the center temperature after 8523 s using Eq. (1) in Example 10.2:

$$\frac{180 - T_c}{180 - 20} = \frac{2}{\pi} \sum_{n=0}^{\infty} \frac{(-1)^n}{\left(n + \frac{1}{2}\right)} \exp\left[-\left(n + \tfrac{1}{2}\right)^2 \pi^2 (1)\right] \quad \Rightarrow \quad T_c = 163\,°\text{C}$$

Therefore, the use of Eq. (2) to predict time to steady-state is quite satisfactory.

10.2.2 Heating of a Rectangular Slab: Revisited

In Section 10.2.1, the temperatures of the surfaces at $z = \pm L$ are assumed to be constant at T_∞. This boundary condition is only applicable when the external resistance to heat transfer is negligible, i.e., $\text{Bi}_\text{H} > 40$. When $0.1 < \text{Bi}_\text{H} < 40$, however, the external resistance to heat transfer should be taken into consideration and the surface temperature will be different from the fluid temperature surrounding the slab. Under these circumstances, the previously defined boundary condition at the fluid-solid interface, Eq. (10.2-13), has to be replaced by

$$\text{at} \quad z = L \qquad k \frac{\partial T}{\partial z} = \langle h \rangle (T_\infty - T) \tag{10.2-73}$$

In terms of the dimensionless quantities defined by Eq. (10.2-15), Eq. (10.2-73) becomes

$$\text{at} \quad \xi = 1 \qquad -\frac{\partial \theta}{\partial \xi} = \text{Bi}_\text{H}\, \theta \tag{10.2-74}$$

where the Biot number for heat transfer is defined by

$$\text{Bi}_{\text{H}} = \frac{\langle h \rangle L}{k} \tag{10.2-75}$$

The solution procedure given in Section 10.2.1 is also applicable to this problem. In other words, the use of the method of separation of variables in which the solution is sought in the form

$$\theta(\tau, \xi) = F(\tau)G(\xi) \tag{10.2-76}$$

reduces the governing equation, Eq. (10.2-16), to two ordinary differential equations of the form

$$\frac{dF}{d\tau} + \lambda^2 F = 0 \quad \Rightarrow \quad F(\tau) = e^{-\lambda^2 \tau} \tag{10.2-77}$$

$$\frac{d^2 G}{d\xi^2} + \lambda^2 G = 0 \quad \Rightarrow \quad G(\xi) = C_1 \sin(\lambda \xi) + C_2 \cos(\lambda \xi) \tag{10.2-78}$$

Therefore, the solution becomes

$$\theta = e^{-\lambda^2 \tau} \big[C_1 \sin(\lambda \xi) + C_2 \cos(\lambda \xi) \big] \tag{10.2-79}$$

Since the problem is symmetric around the z-axis, $C_1 = 0$. Application of Eq. (10.2-74) gives

$$\lambda \tan \lambda = \text{Bi}_{\text{H}} \tag{10.2-80}$$

The transcendental equation given by Eq. (10.2-80) determines an infinite number of eigenvalues for a particular value of Bi_{H}. Designating any particular value of an eigenvalue by λ_n, Eq. (10.2-80) takes the form

$$\boxed{\lambda_n \tan \lambda_n = \text{Bi}_{\text{H}}} \qquad n = 1, 2, 3, \ldots \tag{10.2-81}$$

The first five roots of Eq. (10.2-81) are given as a function of Bi_{H} in Table 10.3.

The general solution is the summation of all possible solutions, i.e.,

$$\theta = \sum_{n=1}^{\infty} A_n \exp\left(-\lambda_n^2 \tau\right) \cos(\lambda_n \xi) \tag{10.2-82}$$

Table 10.3. The roots of Eq. (10.2-81)

Bi_{H}	λ_1	λ_2	λ_3	λ_4	λ_5
0	0.000	3.142	6.283	9.425	12.566
0.1	0.311	3.173	6.299	9.435	12.574
0.5	0.653	3.292	6.362	9.477	12.606
1.0	0.860	3.426	6.437	9.529	12.645
2.0	1.077	3.644	6.578	9.630	12.722
10.0	1.429	4.306	7.228	10.200	13.214

The unknown coefficients A_n can be determined by using the initial condition given by Eq. (10.2-17). The result is

$$A_n = \frac{\int_0^1 \cos(\lambda_n \xi)\, d\xi}{\int_0^1 \cos^2(\lambda_n \xi)\, d\xi} = \frac{4 \sin \lambda_n}{2\lambda_n + \sin 2\lambda_n} \tag{10.2-83}$$

Therefore, the dimensionless temperature distribution is expressed as

$$\boxed{\theta = 4 \sum_{n=1}^{\infty} \frac{\sin \lambda_n}{2\lambda_n + \sin 2\lambda_n} \exp\left(-\lambda_n^2 \tau\right) \cos(\lambda_n \xi)} \tag{10.2-84}$$

When $\tau \geqslant 0.2$, the series solution given by Eq. (10.2-84) can be approximated by the first term of the series.

Example 10.7 Estimate the value of the dimensionless temperature at the surface of the slab as a function of the Biot number.

Solution

The dimensionless temperature at the slab surface, θ_s, can be found by evaluating Eq. (10.2-84) at $\xi = 1$. The result is

$$\theta_s = 2 \sum_{n=1}^{\infty} \frac{\sin 2\lambda_n}{2\lambda_n + \sin 2\lambda_n} \exp\left(-\lambda_n^2 \tau\right) \tag{1}$$

The calculated values of θ_s as a function of the dimensionless time, τ, are given in the table below for three different Biot numbers. The series in Eq. (1) converges by considering at most 6 terms when $\tau > 0$. When $\tau = 0$, however, approximately 300 terms are required for the convergence. When the surface temperature, T_s, approaches the ambient temperature, T_∞, the dimensionless surface temperature becomes zero. Note that, for small values of Biot numbers, the surface temperature is different from the ambient temperature and varies as a function of time. However, for large values of Biot numbers, i.e., $\mathrm{Bi_H} = 40$, the surface temperature is almost equal to the ambient temperature for $\tau > 0$.

	θ_s		
τ	$\mathrm{Bi_H} = 1$	$\mathrm{Bi_H} = 10$	$\mathrm{Bi_H} = 40$
0	0.999	0.993	0.973
0.1	0.724	0.171	0.044
0.2	0.643	0.122	0.031
0.3	0.589	0.097	0.024
0.4	0.544	0.079	0.019
0.5	0.505	0.064	0.015
0.6	0.468	0.052	0.012
0.7	0.435	0.043	0.009
0.8	0.404	0.035	0.007
0.9	0.375	0.028	0.006
1.0	0.348	0.023	0.005

Comment: Rearrangement of Eq. (10.2-81) gives

$$\cos \lambda_n = \frac{\lambda_n \sin \lambda_n}{\text{Bi}_\text{H}} \tag{2}$$

When $\text{Bi}_\text{H} \to \infty$, from Eq. (2) we have

$$\cos \lambda_n = 0 \quad \Rightarrow \quad \lambda_n = \left(n + \tfrac{1}{2}\right)\pi \qquad n = 0, 1, 2, \ldots \tag{3}$$

Noting that

$$\sin\left[\left(n + \tfrac{1}{2}\right)\pi\right] = (-1)^n \qquad \text{and} \qquad \sin 2\lambda_n = 0$$

the dimensionless temperature distribution expressed by Eq. (10.2-84) reduces to Eq. (10.2-36). Mathematically speaking, the surface temperature is equal to the exposed ambient temperature when $\text{Bi}_\text{H} \to \infty$.

Example 10.8 A cake baked at 175 °C for half an hour is taken out of the oven and inverted on a rack to cool. The thickness of the cake is 6 cm, the kitchen temperature is 20 °C, and the average heat transfer coefficient is 12 W/m²·K.

a) Estimate the time it takes for the center to reach 40 °C. Take $k = 0.18$ W/m·K and $\alpha = 1.2 \times 10^{-7}$ m²/s for the cake.

b) Calculate the time to reach steady-state conditions.

Solution

a) The temperature at the center of the cake, T_c, can be found by evaluating Eq. (10.2-84) at $\xi = 0$. Considering only the first term of the series, the result is

$$\frac{T_\infty - T_c}{T_\infty - T_o} = \frac{4 \sin \lambda_1}{2\lambda_1 + \sin 2\lambda_1} \exp\left(-\lambda_1^2 \tau\right) \tag{1}$$

The Biot number is

$$\text{Bi}_\text{H} = \frac{\langle h \rangle L}{k} = \frac{(12)(0.03)}{(0.18)} = 2$$

From Table 10.3, $\lambda_1 = 1.077$. Substitution of the numerical values into Eq. (1) gives

$$\frac{20 - 40}{20 - 175} = \frac{4 \sin 61.7}{2(1.077) + \sin 123.4} \exp\left[-(1.077)^2 \tau\right] \tag{2}$$

in which 1.077 rad $= 61.7°$. Solving Eq. (2) for τ yields

$$\tau = 1.907 \quad \Rightarrow \quad t = \frac{\tau L^2}{\alpha} = \frac{(1.907)(0.03)^2}{1.2 \times 10^{-7}} = 14{,}303 \text{ s} \simeq 4 \text{ h}$$

Note that since $\tau = 1.907 > 0.2$ the approximation of the series solution by the first term is justified.

b) If the time to reach steady-state, t_∞, is defined as the time for the center temperature to drop to $1.01T_\infty$, then Eq. (1) becomes

$$\frac{-0.01T_\infty}{T_\infty - T_o} = \frac{4\sin\lambda_1}{2\lambda_1 + \sin 2\lambda_1}\exp\left(-\lambda_1^2\tau_\infty\right) \tag{3}$$

or,

$$\frac{-(0.01)(20)}{20 - 175} = \frac{4\sin 61.7}{2(1.077) + \sin 123.4}\exp\left[-(1.077)^2\tau_\infty\right] \tag{4}$$

The solution of Eq. (4) gives

$$\tau_\infty = 5.877 \quad\Rightarrow\quad t_\infty = \frac{\tau_\infty L^2}{\alpha} = \frac{(5.877)(0.03)^2}{1.2\times 10^{-7}} = 44{,}078\text{ s} \simeq 12.2\text{ h}$$

Comment: The actual cooling time is obviously less than 4 h as a result of the heat loss from the edges as well as the heat transfer to the rack by conduction. Moreover, since the temperature is assumed to change only along the thickness of the cake, i.e., the axial direction, the shape of the cake (square or cylindrical) is irrelevant in the solution of the problem.

10.2.2.1 *Macroscopic equation* Equation (10.2-38) represents the macroscopic energy balance for the rectangular slab, and the rate of energy entering the slab, $\dot{Q}$, is given by Eq. (10.2-39). The use of Eq. (10.2-84) in Eq. (10.2-39) gives

$$\boxed{\dot{Q} = \frac{8WHk(T_\infty - T_o)}{L}\sum_{n=1}^{\infty}\frac{\lambda_n\sin^2\lambda_n}{2\lambda_n + \sin 2\lambda_n}\exp\left(-\lambda_n^2\tau\right)} \tag{10.2-85}$$

The amount of heat transferred can be calculated from

$$Q = \int_0^t \dot{Q}\,dt = \frac{L^2}{\alpha}\int_0^\tau \dot{Q}\,d\tau \tag{10.2-86}$$

Substitution of Eq. (10.2-85) into Eq. (10.2-86) yields

$$\boxed{\frac{Q}{Q_\infty} = 1 - 4\sum_{n=1}^{\infty}\frac{\sin^2\lambda_n}{\lambda_n(2\lambda_n + \sin 2\lambda_n)}\exp\left(-\lambda_n^2\tau\right)} \tag{10.2-87}$$

where Q_∞ is defined by Eq. (10.2-43).

Example 10.9 Estimate the amount of heat transferred from the cake in Example 10.8 until the center temperature drops to 40 °C. Express your answer as a fraction of the total heat transferred until steady conditions are reached.

Solution

Considering only the first term of the series in Eq. (10.2-87), we have

$$\frac{Q}{Q_\infty} = 1 - \frac{4\sin^2\lambda_1}{\lambda_1(2\lambda_1 + \sin 2\lambda_1)}\exp\left(-\lambda_1^2\tau\right) \tag{1}$$

Substitution of the numerical values into Eq. (1) yields

$$\frac{Q}{Q_\infty} = 1 - \frac{4\sin^2 61.7}{(1.077)[2(1.077) + \sin 123.4]}\exp\left[-(1.077)^2(1.907)\right] = 0.895$$

10.2.3 Heating of a Solid Cylinder

A solid cylinder of radius R and length L is initially at a uniform temperature of T_o. At $t = 0$, it is exposed to a fluid at constant temperature T_∞ ($T_\infty > T_o$). The Biot number is not very large and so the external fluid resistance to heat transfer has to be taken into consideration. The average heat transfer coefficient, $\langle h \rangle$, between the surface of the cylinder and the fluid is known. To determine the amount of heat transferred into the solid cylinder, it is first necessary to determine the temperature profile within the cylinder as a function of position and time.

In general, $T = T(t, r, z)$ and Table C.5 in Appendix C indicates that the nonzero energy flux components are given as

$$e_r = q_r = -k\frac{\partial T}{\partial r} \qquad \text{and} \qquad e_z = q_z = -k\frac{\partial T}{\partial z} \tag{10.2-88}$$

For a cylindrical differential volume element of thickness Δr and length Δz, as shown in Figure 10.4, Eq. (10.2-1) is expressed in the form

$$\left(q_r|_r 2\pi r \Delta z + q_z|_z 2\pi r \Delta r\right) - \left[q_r|_{r+\Delta r} 2\pi (r + \Delta r)\Delta z + q_z|_{z+\Delta z} 2\pi r \Delta r\right]$$

$$= \frac{\partial}{\partial t}\left[2\pi r \Delta r \Delta z \rho \widehat{C}_P(T - T_o)\right] \tag{10.2-89}$$

Dividing Eq. (10.2-89) by $2\pi \Delta r \Delta z$ and taking the limit as $\Delta r \to 0$ and $\Delta z \to 0$ give

$$\rho\widehat{C}_P \frac{\partial T}{\partial t} = \frac{1}{r}\lim_{\Delta r \to 0}\frac{(rq_r)|_r - (rq_r)|_{r+\Delta r}}{\Delta r} + \lim_{\Delta z \to 0}\frac{(q_z)|_z - (q_z)|_{z+\Delta z}}{\Delta z} \tag{10.2-90}$$

or,

$$\rho\widehat{C}_P \frac{\partial T}{\partial t} = -\frac{1}{r}\frac{\partial(rq_r)}{\partial r} - \frac{\partial q_z}{\partial z} \tag{10.2-91}$$

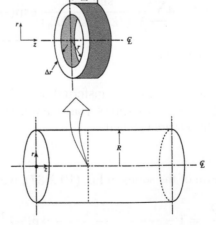

Figure 10.4. Heating of a solid cylinder.

Substitution of Eq. (10.2-88) into Eq. (10.2-91) leads to the following governing equation for temperature

$$\rho \widehat{C}_P \frac{\partial T}{\partial t} = \frac{k}{r} \frac{\partial}{\partial r}\left(r \frac{\partial T}{\partial r} \right) + k \frac{\partial^2 T}{\partial z^2} \tag{10.2-92}$$

The physical significance and the order of magnitude of the terms in Eq. (10.2-92) are given in Table 10.4. Note that the ratio of the orders of magnitude of the two conduction terms in Eq. (10.2-92) is expressed as

$$\frac{\text{Rate of conduction in } z\text{-direction}}{\text{Rate of conduction in } r\text{-direction}} = \left(\frac{R}{L} \right)^2 \tag{10.2-93}$$

Let us restrict our analysis to cases in which $R/L \ll 1$ so that the conduction in the z-direction can be neglected in favor of that in the r-direction[5]. Under these circumstances, Eq. (10.2-92) simplifies to

$$\boxed{\rho \widehat{C}_P \frac{\partial T}{\partial t} = \frac{k}{r} \frac{\partial}{\partial r}\left(r \frac{\partial T}{\partial r} \right)} \tag{10.2-94}$$

The initial and boundary conditions associated with Eq. (10.2-94) are given by

$$\text{at} \quad t = 0 \qquad T = T_o \tag{10.2-95}$$

$$\text{at} \quad r = 0 \qquad \frac{\partial T}{\partial r} = 0 \tag{10.2-96}$$

$$\text{at} \quad r = R \qquad k \frac{\partial T}{\partial r} = \langle h \rangle (T_\infty - T) \tag{10.2-97}$$

Introduction of the dimensionless quantities

$$\theta = \frac{T_\infty - T}{T_\infty - T_o} \qquad \tau = \frac{\alpha t}{R^2} \qquad \xi = \frac{r}{R} \qquad \text{Bi}_\text{H} = \frac{\langle h \rangle R}{k} \tag{10.2-98}$$

Table 10.4. The physical significance and the order of magnitude of the terms in Eq. (10.2-92)

Term	Physical Significance	Order of Magnitude
$\rho \widehat{C}_P \dfrac{\partial T}{\partial t}$	Rate of energy accumulation	$\dfrac{\rho \widehat{C}_P (T_\infty - T_o)}{t}$
$\dfrac{k}{r} \dfrac{\partial}{\partial r}\left(r \dfrac{\partial T}{\partial r} \right)$	Rate of conduction in r-direction	$\dfrac{k(T_\infty - T_o)}{R^2}$
$k \dfrac{\partial^2 T}{\partial z^2}$	Rate of conduction in z-direction	$\dfrac{k(T_\infty - T_o)}{L^2}$

[5]This is known as the *infinite cylinder* assumption.

reduces Eqs. (10.2-94)–(10.2-97) to

$$\frac{\partial \theta}{\partial \tau} = \frac{1}{\xi} \frac{\partial}{\partial \xi} \left(\xi \frac{\partial \theta}{\partial \xi} \right) \tag{10.2-99}$$

$$\text{at} \quad \tau = 0 \qquad \theta = 1 \tag{10.2-100}$$

$$\text{at} \quad \xi = 0 \qquad \frac{\partial \theta}{\partial \xi} = 0 \tag{10.2-101}$$

$$\text{at} \quad \xi = 1 \qquad -\frac{\partial \theta}{\partial \xi} = \text{Bi}_\text{H}\, \theta \tag{10.2-102}$$

Since the boundary conditions are homogeneous, the method of separation of variables can be used to solve Eq. (10.2-99). Representing the solution as a product of two functions of the form

$$\theta_t(\tau, \xi) = F(\tau) G(\xi) \tag{10.2-103}$$

reduces Eq. (10.2-99) to

$$\frac{1}{F} \frac{dF}{d\tau} = \frac{1}{G\xi} \frac{d}{d\xi} \left(\xi \frac{dG}{d\xi} \right) \tag{10.2-104}$$

While the left-hand side of Eq. (10.2-104) is a function of τ only, the right-hand side is dependent only on ξ. This is possible if both sides of Eq. (10.2-104) are equal to a constant, say $-\lambda^2$, i.e.,

$$\frac{1}{F} \frac{dF}{d\tau} = \frac{1}{G\xi} \frac{d}{d\xi} \left(\xi \frac{dG}{d\xi} \right) = -\lambda^2 \tag{10.2-105}$$

Equation (10.2-105) results in two ordinary differential equations. The equation for F is given by

$$\frac{dF}{d\tau} + \lambda^2 F = 0 \quad \Rightarrow \quad F(\tau) = e^{-\lambda^2 \tau} \tag{10.2-106}$$

On the other hand, the equation for G is

$$\frac{d}{d\xi} \left(\xi \frac{dG}{d\xi} \right) + \lambda^2 \xi G = 0 \tag{10.2-107}$$

which is subject to the following boundary conditions

$$\text{at} \quad \xi = 0 \qquad \frac{dG}{d\xi} = 0 \tag{10.2-108}$$

$$\text{at} \quad \xi = 1 \qquad -\frac{dG}{d\xi} = \text{Bi}_\text{H}\, G \tag{10.2-109}$$

Note that Eq. (10.2-107) is a Sturm-Liouville equation with a weight function of ξ. Comparison of Eq. (10.2-107) with Eq. (B.2-16) in Appendix B indicates that $p = 1$, $j = 1$, $a = \lambda^2$, and $b = 0$. Therefore, Eq. (10.2-107) is Bessel's equation and the use of Eqs. (B.2-17)–(B.2-19) gives $\alpha = 1$, $\beta = 0$, and $n = 0$. Equation (B.2-21) gives the solution as

$$G(\xi) = C_1 J_o(\lambda \xi) + C_2 Y_o(\lambda \xi) \tag{10.2-110}$$

where C_1 and C_2 are constants. Since $Y_o(0) = -\infty$, $C_2 = 0$. Application of Eq. (10.2-109) gives

$$\lambda J_1(\lambda) = \text{Bi}_\text{H} J_o(\lambda) \tag{10.2-111}$$

The transcendental equation given by Eq. (10.2-111) determines an infinite number of eigenvalues for a particular value of Bi_H. Designating any particular value of an eigenvalue by λ_n, Eq. (10.2-111) takes the form

$$\boxed{\lambda_n J_1(\lambda_n) = \text{Bi}_\text{H} J_o(\lambda_n)} \qquad n = 1, 2, 3, \ldots \tag{10.2-112}$$

The first five roots of Eq. (10.2-112) are given as a function of Bi_H in Table 10.5.

The general solution is the summation of all possible solutions, i.e.,

$$\theta = \sum_{n=1}^{\infty} A_n \exp\left(-\lambda_n^2 \tau\right) J_o(\lambda_n \xi) \tag{10.2-113}$$

The unknown coefficients A_n can be determined by using the initial condition given by Eq. (10.2-100). The result is

$$1 = \sum_{n=1}^{\infty} A_n J_o(\lambda_n \xi) \tag{10.2-114}$$

Since the eigenfunctions are orthogonal to each other with respect to the weight function, multiplication of Eq. (10.2-114) by $\xi J_o(\lambda_m \xi) \, d\xi$ and integration from $\xi = 0$ to $\xi = 1$ give

$$\int_0^1 \xi J_o(\lambda_m \xi) \, d\xi = \sum_{n=1}^{\infty} A_n \int_0^1 \xi J_o(\lambda_n \xi) J_o(\lambda_m \xi) \, d\xi \tag{10.2-115}$$

The integral on the right-hand side of Eq. (10.2-115) is zero when $n \neq m$ and nonzero when $n = m$. Therefore, the summation drops out when $n = m$, and Eq. (10.2-115) reduces to

$$\int_0^1 \xi J_o(\lambda_n \xi) \, d\xi = A_n \int_0^1 \xi J_o^2(\lambda_n \xi) \, d\xi \tag{10.2-116}$$

Table 10.5. The roots of Eq. (10.2-112)

Bi_H	λ_1	λ_2	λ_3	λ_4	λ_5
0	0.000	3.832	7.016	10.173	13.324
0.1	0.442	3.858	7.030	10.183	13.331
0.5	0.941	3.959	7.086	10.222	13.361
1.0	1.256	4.079	7.156	10.271	13.398
2.0	1.599	4.291	7.288	10.366	13.472
10.0	2.179	5.033	7.957	10.936	13.958

Evaluation of the integrals yields

$$A_n = \frac{2}{\lambda_n}\left[\frac{J_1(\lambda_n)}{J_o^2(\lambda_n) + J_1^2(\lambda_n)}\right] \tag{10.2-117}$$

which can be further simplified with the help of Eq. (10.2-112) to the form

$$A_n = \frac{2\,\mathrm{Bi_H}}{(\lambda_n^2 + \mathrm{Bi_H^2})J_o(\lambda_n)} \tag{10.2-118}$$

Substitution of Eq. (10.2-118) into Eq. (10.2-113) gives the dimensionless temperature distribution as

$$\boxed{\theta = 2\,\mathrm{Bi_H}\sum_{n=1}^{\infty}\frac{1}{(\lambda_n^2 + \mathrm{Bi_H^2})J_o(\lambda_n)}\exp(-\lambda_n^2\tau)J_o(\lambda_n\xi)} \tag{10.2-119}$$

When $\tau \geqslant 0.2$, the series solution given by Eq. (10.2-119) can be approximated by the first term of the series.

Example 10.10 A red oak log of 20 cm diameter is initially at a temperature of 20 °C. Estimate the maximum exposure time of the lumber to hot gases at 400 °C before ignition starts. The average heat transfer coefficient between the surface of the oak and the gases is 15 W/m^2·K and the ignition temperature of oak is 275 °C. Take $k = 0.15$ W/m·K and $\alpha = 1.6 \times 10^{-7}$ m^2/s.

Solution

The ignition starts when the surface temperature reaches 275 °C. The temperature at the surface, T_s, can be found by evaluating Eq. (10.2-119) at $\xi = 1$. The result is

$$\frac{T_\infty - T_s}{T_\infty - T_o} = 2\,\mathrm{Bi_H}\sum_{n=1}^{\infty}\frac{1}{(\lambda_n^2 + \mathrm{Bi_H^2})}\exp(-\lambda_n^2\tau) \tag{1}$$

The Biot number is

$$\mathrm{Bi_H} = \frac{\langle h\rangle R}{k} = \frac{(15)(0.1)}{(0.15)} = 10$$

If we consider the first five terms of the series in Eq. (1), the corresponding eigenvalues can be found from Table 10.5 as

$$\lambda_1 = 2.179 \qquad \lambda_2 = 5.033 \qquad \lambda_3 = 7.957 \qquad \lambda_4 = 10.936 \qquad \lambda_5 = 13.958$$

Substitution of the numerical values into Eq. (1) leads to

$$\frac{400 - 275}{400 - 20} = 2(10)\sum_{n=1}^{5}\frac{1}{(\lambda_n^2 + \mathrm{Bi_H^2})}\exp(-\lambda_n^2\tau) \quad\Rightarrow\quad \tau = 0.018 \tag{2}$$

Thus, the time of ignition is

$$t = \frac{\tau R^2}{\alpha} = \frac{(0.018)(0.1)^2}{1.6 \times 10^{-7}} = 1125 \text{ s} \simeq 19 \text{ min}$$

Comment: It should be kept in mind that the solutions expressed in series cannot be approximated by the first term in each problem. For example, the use of only the first term of the series in Eq. (1) leads to a negative time value!

10.2.3.1 *Macroscopic equation* Integration of the governing equation for temperature, Eq. (10.2-94), over the volume of the system gives

$$\int_0^L \int_0^{2\pi} \int_0^R \rho \widehat{C}_P \frac{\partial T}{\partial t} r \, dr \, d\theta \, dz = \int_0^L \int_0^{2\pi} \int_0^R \frac{k}{r} \frac{\partial}{\partial r}\left(r \frac{\partial T}{\partial r}\right) r \, dr \, d\theta \, dz \qquad (10.2\text{-}120)$$

or,

$$\underbrace{\frac{d}{dt}\left[\int_0^L \int_0^{2\pi} \int_0^R \rho \widehat{C}_P (T - T_o) r \, dr \, d\theta \, dz\right]}_{\text{Rate of accumulation of energy}} = \underbrace{2\pi RL\left(k \frac{\partial T}{\partial r}\bigg|_{r=R}\right)}_{\substack{\text{Rate of energy entering} \\ \text{from the lateral surface}}} \qquad (10.2\text{-}121)$$

which is the macroscopic energy balance by considering the solid cylinder as a system. The rate of energy entering the cylinder, $\dot{Q}$, is given by

$$\dot{Q} = 2\pi RL\left(k \frac{\partial T}{\partial r}\bigg|_{r=R}\right) = -2\pi Lk(T_\infty - T_o) \frac{\partial \theta}{\partial \xi}\bigg|_{\xi=1} \qquad (10.2\text{-}122)$$

The use of Eq. (10.2-119) in Eq. (10.2-122) results in

$$\dot{Q} = 4\pi Lk(T_\infty - T_o) \, \text{Bi}_\text{H} \sum_{n=1}^{\infty} \frac{\lambda_n J_1(\lambda_n)}{(\lambda_n^2 + \text{Bi}_\text{H}^2) J_o(\lambda_n)} \exp(-\lambda_n^2 \tau) \qquad (10.2\text{-}123)$$

The amount of heat transferred can be calculated from

$$Q = \int_0^t \dot{Q} \, dt = \frac{R^2}{\alpha} \int_0^{\tau} \dot{Q} \, d\tau \qquad (10.2\text{-}124)$$

Substitution of Eq. (10.2-123) into Eq. (10.2-124) yields

$$\frac{Q}{Q_\infty} = 1 - 4\,\text{Bi}_\text{H} \sum_{n=1}^{\infty} \frac{J_1(\lambda_n)}{\lambda_n (\lambda_n^2 + \text{Bi}_\text{H}^2) J_o(\lambda_n)} \exp(-\lambda_n^2 \tau) \qquad (10.2\text{-}125)$$

where Q_∞ is the amount of heat transferred to the cylinder when the driving force is constant and equal to its maximum value, i.e.,

$$Q_\infty = \pi R^2 L \rho \widehat{C}_P (T_\infty - T_o) \qquad (10.2\text{-}126)$$

10.2.4 Heating of a Spherical Particle

A spherical particle of radius R is initially at a uniform temperature of T_o. At $t = 0$, it is exposed to a fluid at constant temperature T_∞ ($T_\infty > T_o$). It is required to determine the amount of heat transferred to the spherical particle.

Since the heat transfer takes place in the r-direction, Table C.6 in Appendix C indicates that the only nonzero energy flux component is e_r, and it is given by

$$e_r = q_r = -k \frac{\partial T}{\partial r} \tag{10.2-127}$$

For a spherical differential volume of thickness Δr, as shown in Figure 10.5, Eq. (10.2-1) is expressed as

$$q_r|_r 4\pi r^2 - q_r|_{r+\Delta r} 4\pi (r + \Delta r)^2 = \frac{\partial}{\partial t}\left[4\pi r^2 \Delta r \rho \widehat{C}_P(T - T_o)\right] \tag{10.2-128}$$

Dividing Eq. (10.2-128) by $4\pi \Delta r$ and letting $\Delta r \to 0$ give

$$\rho \widehat{C}_P \frac{\partial T}{\partial t} = \frac{1}{r^2} \lim_{\Delta r \to 0} \frac{(r^2 q_r)|_r - (r^2 q_r)|_{r+\Delta r}}{\Delta r} \tag{10.2-129}$$

or,

$$\rho \widehat{C}_P \frac{\partial T}{\partial t} = -\frac{1}{r^2} \frac{\partial (r^2 q_r)}{\partial r} \tag{10.2-130}$$

Substitution of Eq. (10.2-127) into Eq. (10.2-130) gives the governing differential equation for temperature as

$$\boxed{\rho \widehat{C}_P \frac{\partial T}{\partial t} = \frac{k}{r^2} \frac{\partial}{\partial r}\left(r^2 \frac{\partial T}{\partial r}\right)} \tag{10.2-131}$$

The initial and boundary conditions associated with Eq. (10.2-131) are

$$\text{at} \quad t = 0 \qquad T = T_o \tag{10.2-132}$$

$$\text{at} \quad r = 0 \qquad \frac{\partial T}{\partial r} = 0 \tag{10.2-133}$$

$$\text{at} \quad r = R \qquad k \frac{\partial T}{\partial r} = \langle h \rangle (T_\infty - T) \tag{10.2-134}$$

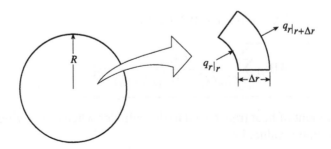

Figure 10.5. Heating of a spherical particle.

Introduction of the dimensionless quantities

$$\theta = \frac{T_\infty - T}{T_\infty - T_o} \qquad \tau = \frac{\alpha t}{R^2} \qquad \xi = \frac{r}{R} \qquad Bi_H = \frac{\langle h \rangle R}{k} \tag{10.2-135}$$

reduces Eqs. (10.2-131)–(10.2-134) to

$$\frac{\partial \theta}{\partial \tau} = \frac{1}{\xi^2} \frac{\partial}{\partial \xi} \left(\xi^2 \frac{\partial \theta}{\partial \xi} \right) \tag{10.2-136}$$

$$\text{at} \quad \tau = 0 \qquad \theta = 1 \tag{10.2-137}$$

$$\text{at} \quad \xi = 0 \qquad \frac{\partial \theta}{\partial \xi} = 0 \tag{10.2-138}$$

$$\text{at} \quad \xi = 1 \qquad -\frac{\partial \theta}{\partial \xi} = Bi_H \theta \tag{10.2-139}$$

Since the governing equation and the boundary conditions are homogeneous, the use of the method of separation of variables in which the solution is sought in the form

$$\theta(\tau, \xi) = F(\tau) G(\xi) \tag{10.2-140}$$

reduces Eq. (10.2-136) to

$$\frac{1}{F} \frac{dF}{d\tau} = \frac{1}{G\xi^2} \frac{d}{d\xi} \left(\xi^2 \frac{dG}{d\xi} \right) = -\lambda^2 \tag{10.2-141}$$

The equation for F is given by

$$\frac{dF}{d\tau} + \lambda^2 F = 0 \quad \Rightarrow \quad F(\tau) = e^{-\lambda^2 \tau} \tag{10.2-142}$$

The equation for G is

$$\frac{1}{\xi^2} \frac{d}{d\xi} \left(\xi^2 \frac{dG}{d\xi} \right) + \lambda^2 G = 0 \tag{10.2-143}$$

and it is subject to the boundary conditions

$$\text{at} \quad \xi = 0 \qquad \frac{dG}{d\xi} = 0 \tag{10.2-144}$$

$$\text{at} \quad \xi = 1 \qquad -\frac{dG}{d\xi} = Bi_H G \tag{10.2-145}$$

The transformation[6] $G = u(\xi)/\xi$ converts Eq. (10.2-143) to

$$\frac{d^2 u}{d\xi^2} + \lambda^2 u = 0 \tag{10.2-146}$$

---

[6] See Section 9.4.2.

which has the solution

$$u = C_1 \sin(\lambda \xi) + C_2 \cos(\lambda \xi) \tag{10.2-147}$$

or,

$$G = C_1 \frac{\sin(\lambda \xi)}{\xi} + C_2 \frac{\cos(\lambda \xi)}{\xi} \tag{10.2-148}$$

While the application of Eq. (10.2-144) gives $C_2 = 0$, the use of Eq. (10.2-145) results in

$$\boxed{\lambda_n \cot \lambda_n = 1 - \mathrm{Bi_H}} \qquad n = 1, 2, 3, \ldots \tag{10.2-149}$$

The first five roots of Eq. (10.2-149) are given as a function of $\mathrm{Bi_H}$ in Table 10.6.

The general solution is the summation of all possible solutions, i.e.,

$$\theta = \sum_{n=1}^{\infty} A_n \exp\left(-\lambda_n^2 \tau\right) \frac{\sin(\lambda_n \xi)}{\xi} \tag{10.2-150}$$

The unknown coefficients A_n can be determined by using the initial condition given by Eq. (10.2-137). The result is

$$A_n = \frac{\displaystyle\int_0^1 \xi \sin(\lambda_n \xi) \, d\xi}{\displaystyle\int_0^1 \sin^2(\lambda_n \xi) \, d\xi} = \frac{2}{\lambda_n} \left(\frac{\sin \lambda_n - \lambda_n \cos \lambda_n}{\lambda_n - \sin \lambda_n \cos \lambda_n} \right) \tag{10.2-151}$$

Equation (10.2-151) can be further simplified with the help of Eq. (10.2-149) to

$$A_n = 4\,\mathrm{Bi_H} \frac{\sin \lambda_n}{\lambda_n (2\lambda_n - \sin 2\lambda_n)} \tag{10.2-152}$$

Therefore, the dimensionless temperature distribution is given by

$$\boxed{\theta = 4\,\mathrm{Bi_H} \sum_{n=1}^{\infty} \frac{\sin \lambda_n}{\lambda_n (2\lambda_n - \sin 2\lambda_n)} \exp\left(-\lambda_n^2 \tau\right) \frac{\sin(\lambda_n \xi)}{\xi}} \tag{10.2-153}$$

Table 10.6. The roots of Eq. (10.2-149)

$\mathrm{Bi_H}$	λ_1	λ_2	λ_3	λ_4	λ_5
0	0.000	4.493	7.725	10.904	14.066
0.1	0.542	4.516	7.738	10.913	14.073
0.5	1.166	4.604	7.790	10.950	14.102
1.0	1.571	4.712	7.854	10.996	14.137
2.0	2.029	4.913	7.979	11.086	14.207
10.0	2.836	5.717	8.659	11.653	14.687

Example 10.11 Due to an unexpected cold spell, the air temperature drops to $-3\,°C$ accompanied by a wind blowing at a velocity of 3 m/s in Florida. Farmers have to take precautions in order to avoid frost in their orange orchards. If frost formation starts when the surface temperature of the orange reaches $0\,°C$, use your engineering judgement to estimate how much time the farmers have to take precautions. Assume the oranges are spherical with a diameter of 10 cm and are at an initial uniform temperature of $10\,°C$. The thermal conductivity and the thermal diffusivity of an orange are 0.51 W/m·K and 1.25×10^{-7} m²/s, respectively.

Solution

Physical properties

Initially the film temperature is $(-3 + 10)/2 = 3.5\,°C$.

For air at $3.5\,°C$ (276.5 K): $\begin{cases} \nu = 13.61 \times 10^{-6} \text{ m}^2/\text{s} \\ k = 24.37 \times 10^{-3} \text{ W/m·K} \\ Pr = 0.716 \end{cases}$

Analysis

It is first necessary to calculate the average heat transfer coefficient. The Reynolds number is

$$\text{Re}_P = \frac{D_P v_\infty}{\nu} = \frac{(10 \times 10^{-2})(3)}{13.61 \times 10^{-6}} = 22{,}043 \tag{1}$$

The use of the Ranz-Marshall correlation, Eq. (4.3-29), gives

$$\text{Nu} = 2 + 0.6\,\text{Re}_P^{1/2}\,\text{Pr}^{1/3} = 2 + 0.6(22{,}043)^{1/2}(0.716)^{1/3} = 81.7 \tag{2}$$

The average heat transfer coefficient is

$$\langle h \rangle = \text{Nu}\left(\frac{k}{D_P}\right) = (81.7)\left(\frac{24.37 \times 10^{-3}}{10 \times 10^{-2}}\right) = 19.9 \text{ W/m}^2\text{·K} \tag{3}$$

The Biot number is

$$\text{Bi}_H = \frac{\langle h \rangle R}{k} = \frac{(19.9)(5 \times 10^{-2})}{0.51} = 1.95 \tag{4}$$

The solution of Eq. (10.2-149) gives the first eigenvalue, λ_1, as 2.012. Considering only the first term of the series in Eq. (10.2-153), the temperature distribution becomes

$$\theta = \frac{T_\infty - T}{T_\infty - T_o} = \frac{4\,\text{Bi}_H}{\lambda_1}\left(\frac{\sin\lambda_1}{2\lambda_1 - \sin 2\lambda_1}\right)\exp\left(-\lambda_1^2 \tau\right)\frac{\sin(\lambda_1 \xi)}{\xi} \tag{5}$$

Evaluation of Eq. (5) at $\xi = 1$ gives the temperature at the surface, T_R, as

$$\frac{T_\infty - T_R}{T_\infty - T_o} = \frac{4\,\text{Bi}_H}{\lambda_1}\left(\frac{\sin^2\lambda_1}{2\lambda_1 - \sin 2\lambda_1}\right)\exp\left(-\lambda_1^2 \tau\right) \tag{6}$$

Substitution of the numerical values into Eq. (6) gives

$$\frac{-3-0}{-3-10} = \frac{4(1.95)}{2.012}\left[\frac{\sin^2 115.3}{2(2.012) - \sin 230.6}\right]\exp\left[-(2.012)^2\tau\right] \tag{7}$$

in which 2.012 rad $= 115.3°$. Solving Eq. (7) for τ yields

$$\tau = 0.26 \quad \Rightarrow \quad t = \frac{\tau R^2}{\alpha} = \frac{(0.26)(5 \times 10^{-2})^2}{1.25 \times 10^{-7}} = 5200\text{ s} \simeq 1\text{ h }27\text{ min}$$

Example 10.12 A 2-kg spherical rump roast is placed into a 175 °C oven. How long does it take for the center to reach 80 °C if the initial temperature is 5 °C? The average heat transfer coefficient in the oven is 15 W/m²·K and the physical properties of the meat are given as

$$\rho = 1076\text{ kg/m}^3 \qquad k = 0.514\text{ W/m·K} \qquad \widehat{C}_P = 3.431\text{ kJ/kg·K}$$

Solution

The diameter of the roast is

$$D = \left(\frac{6M}{\pi\rho}\right)^{1/3} = \left[\frac{6(2)}{\pi(1076)}\right]^{1/3} = 0.153\text{ m} \tag{1}$$

The Biot number is

$$\text{Bi}_\text{H} = \frac{\langle h\rangle R}{k} = \frac{(15)(0.153/2)}{0.514} = 2.23 \tag{2}$$

From Eq. (10.2-149), the first eigenvalue, λ_1, is calculated as 2.101. Considering only the first term of the series in Eq. (10.2-153), the temperature distribution becomes

$$\theta = \frac{T_\infty - T}{T_\infty - T_o} = \frac{4\,\text{Bi}_\text{H}}{\lambda_1}\left(\frac{\sin\lambda_1}{2\lambda_1 - \sin 2\lambda_1}\right)\exp\left(-\lambda_1^2\tau\right)\frac{\sin(\lambda_1\xi)}{\xi} \tag{3}$$

Noting that

$$\lim_{\xi\to 0}\frac{\sin(\lambda_1\xi)}{\xi} = \lambda_1 \tag{4}$$

evaluation of Eq. (3) at the center, i.e., $\xi = 0$, yields

$$\frac{T_\infty - T_c}{T_\infty - T_o} = 4\,\text{Bi}_\text{H}\left(\frac{\sin\lambda_1}{2\lambda_1 - \sin 2\lambda_1}\right)\exp\left(-\lambda_1^2\tau\right) \tag{5}$$

where T_c represents the center temperature. Substitution of the numerical values into Eq. (5) gives

$$\frac{175 - 80}{175 - 5} = (4)(2.23)\left[\frac{\sin 120.4}{2(2.101) - \sin 240.8}\right]\exp\left[-(2.101)^2\tau\right] \tag{6}$$

in which 2.101 rad $= 120.4°$. Solving Eq. (6) for τ yields

$$\tau = 0.226 \quad \Rightarrow \quad t = \frac{\tau R^2}{\alpha} = \frac{(0.226)(0.153/2)^2}{0.514/[(1076)(3431)]} = 9500\text{ s} \simeq 2\text{ h }38\text{ min}$$

10.2.4.1 *Macroscopic equation* Integration of the governing equation for temperature, Eq. (10.2-131), over the volume of the system gives

$$\int_0^{2\pi} \int_0^{\pi} \int_0^R \rho \widehat{C}_P \frac{\partial T}{\partial t} r^2 \sin\theta \, dr \, d\theta \, d\phi = \int_0^{2\pi} \int_0^{\pi} \int_0^R \frac{k}{r^2} \frac{\partial}{\partial r}\left(r^2 \frac{\partial T}{\partial r}\right) r^2 \sin\theta \, dr \, d\theta \, d\phi$$

(10.2-154)

or,

$$\underbrace{\frac{d}{dt}\left[\int_0^{2\pi} \int_0^{\pi} \int_0^R \rho \widehat{C}_P (T - T_o) r^2 \sin\theta \, dr \, d\theta \, d\phi\right]}_{\text{Rate of accumulation of energy}} = \underbrace{4\pi R^2 \left(k \frac{\partial T}{\partial r}\bigg|_{r=R}\right)}_{\substack{\text{Rate of energy entering} \\ \text{from the surface}}}$$

(10.2-155)

which is the macroscopic energy balance by considering the spherical particle as a system. The rate of energy entering the sphere, $\dot{Q}$, is given by

$$\dot{Q} = 4\pi R^2 \left(k \frac{\partial T}{\partial r}\bigg|_{r=R}\right) = -4\pi Rk(T_\infty - T_o) \frac{\partial \theta}{\partial \xi}\bigg|_{\xi=1}$$

(10.2-156)

The use of Eq. (10.2-153) in Eq. (10.2-156) results in

$$\dot{Q} = 16\pi Rk(T_\infty - T_o) \, \text{Bi}_\text{H}^2 \sum_{n=1}^{\infty} \frac{\sin^2 \lambda_n}{\lambda_n(2\lambda_n - \sin 2\lambda_n)} \exp\left(-\lambda_n^2 \tau\right)$$

(10.2-157)

The amount of heat transferred can be calculated from

$$Q = \int_0^t \dot{Q} \, dt = \frac{R^2}{\alpha} \int_0^{\tau} \dot{Q} \, d\tau$$

(10.2-158)

Substitution of Eq. (10.2-157) into Eq. (10.2-158) yields

$$\frac{Q}{Q_\infty} = 1 - 12\,\text{Bi}_\text{H}^2 \sum_{n=1}^{\infty} \frac{\sin^2 \lambda_n}{\lambda_n^3(2\lambda_n - \sin 2\lambda_n)} \exp\left(-\lambda_n^2 \tau\right)$$

(10.2-159)

where Q_∞ is defined by

$$Q_\infty = \frac{4}{3}\pi R^3 \rho \widehat{C}_P (T_\infty - T_o)$$

(10.2-160)

Example 10.13[7] A hen's egg of mass 50 grams requires 5 minutes to hard boil. How long will it take to hard boil an ostrich's egg of mass 3 kg?

---

[7]This problem is taken from Konak (1994).

Solution

If an egg is assumed to be spherical, then Eq. (10.2-131) represents the governing equation for temperature. Since this equation contains the terms representing the rate of accumulation and the rate of conduction, then the orders of magnitude of these terms must be, more or less, equal to each other. In other words, the Fourier number is in the order of unity. Let subscripts h and o represent hen and ostrich, respectively. Then it is possible to equate the Fourier numbers as we did in Example 10.3:

$$\left(\frac{\alpha t}{R^2}\right)_h = \left(\frac{\alpha t}{R^2}\right)_o \tag{1}$$

If the eggs are chemically similar, then $\alpha_h = \alpha_o$. Since volume and hence mass, M, is proportional to R^3, Eq. (1) reduces to

$$t_o = t_h \left(\frac{M_o}{M_h}\right)^{2/3} \tag{2}$$

Substitution of the numerical values into Eq. (2) gives the time required to hard boil an ostrich's egg as

$$t_o = (5)\left(\frac{3000}{50}\right)^{2/3} = 76.6 \text{ min} \tag{3}$$

10.2.5 Lumped-Parameter Analysis

In Sections 10.2.2, 10.2.3, and 10.2.4, we have considered the cases in which Bi_H varies between 0.1 and 40. The lumped-parameter analysis used in Chapter 7 can only be applied to problems when $Bi_H < 0.1$, i.e., internal resistance to heat transfer is negligible. Consider, for example, heating of a spherical particle as described in Section 10.2.4. The lumped-parameter analysis leads to

$$4\pi R^2 \langle h \rangle (T_\infty - T) = \frac{d}{dt}\left[\frac{4}{3}\pi R^3 \rho \widehat{C}_P (T - T_o)\right] \tag{10.2-161}$$

Rearrangement of Eq. (10.2-161) gives

$$\int_{T_o}^{T} \frac{dT}{T_\infty - T} = \frac{3\langle h \rangle}{\rho \widehat{C}_P R} \int_0^t dt \tag{10.2-162}$$

Evaluation of the integrations leads to

$$\boxed{T_\infty - T = (T_\infty - T_o)\exp\left(-\frac{3\langle h \rangle t}{\rho \widehat{C}_P R}\right)} \tag{10.2-163}$$

The amount of heat transferred to the sphere can be calculated as

$$Q = 4\pi R^2 \langle h \rangle \int_0^t (T_\infty - T)\, dt \tag{10.2-164}$$

Table 10.7. Comparison of Q/Q_∞ values obtained from Eqs. (10.2-159) and (10.2-166)

					Q/Q_∞			
τ	$Bi_H = 0.1$		τ	$Bi_H = 1$		τ	$Bi_H = 10$	
	Exact	Approx.		Exact	Approx.		Exact	Approx.
0	0.000	0.000	0	0.000	0.000	0	0.000	0.000
1	0.255	0.259	0.1	0.229	0.259	0.01	0.157	0.259
2	0.444	0.451	0.2	0.398	0.451	0.02	0.259	0.451
3	0.586	0.593	0.3	0.530	0.593	0.03	0.337	0.593
4	0.691	0.699	0.4	0.633	0.699	0.04	0.402	0.699
5	0.770	0.777	0.5	0.713	0.777	0.05	0.457	0.777
6	0.828	0.835	0.6	0.776	0.835	0.06	0.505	0.835
7	0.872	0.878	0.7	0.825	0.878	0.07	0.548	0.878
8	0.905	0.909	0.8	0.863	0.909	0.08	0.586	0.909
9	0.929	0.933	0.9	0.893	0.933	0.09	0.620	0.933
10	0.947	0.950	1.0	0.916	0.950	0.10	0.650	0.950

The use of Eq. (10.2-163) in Eq. (10.2-164) gives

$$Q = \frac{4}{3}\pi R^3 \rho \widehat{C}_P (T_\infty - T_o)\left[1 - \exp\left(-\frac{3\langle h\rangle t}{\rho \widehat{C}_P R}\right)\right] \tag{10.2-165}$$

or,

$$\boxed{\frac{Q}{Q_\infty} = 1 - \exp(-3\,Bi_H\,\tau)} \tag{10.2-166}$$

The exact values of Q/Q_∞ obtained from Eq. (10.2-159) are compared with the approximate results obtained from Eq. (10.2-166) for different values of Bi_H in Table 10.7. As expected, when $Bi_H = 0.1$, the approximate values are almost equal to the exact ones. For $Bi_H > 0.1$, the use of Eq. (10.2-166) overestimates the exact values.

10.3 MASS TRANSPORT

The conservation statement for species $\mathcal{A}$ is expressed as

$$\begin{pmatrix} \text{Rate of} \\ \text{species } \mathcal{A} \text{ in} \end{pmatrix} - \begin{pmatrix} \text{Rate of} \\ \text{species } \mathcal{A} \text{ out} \end{pmatrix} = \begin{pmatrix} \text{Rate of species } \mathcal{A} \\ \text{accumulation} \end{pmatrix} \tag{10.3-1}$$

As in Section 8.4, our analysis will be restricted to the application of Eq. (10.3-1) to diffusion in solids and stationary liquids. The solutions of almost all imaginable diffusion problems in different coordinate systems with various initial and boundary conditions are given by Crank (1956). As will be shown later, conduction and diffusion problems become analogous in dimensionless form. Therefore, the solutions given by Carslaw and Jaeger (1959) can also be used for diffusion problems.

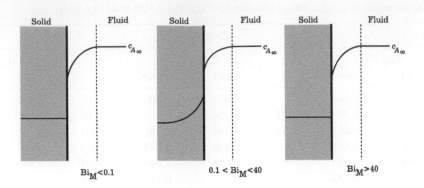

Figure 10.6. Effect of Bi_M on the concentration distribution.

Using Eq. (7.1-14), the Biot number for mass transfer is expressed in the form

$$Bi_M = \frac{\text{Concentration difference in solid}}{\text{Concentration difference in fluid}} \qquad (10.3\text{-}2)$$

When $Bi_M \ll 1$, the internal resistance to mass transfer is negligible and the concentration distribution is considered uniform within the solid phase. In this case, lumped-parameter analysis can be used as in Chapter 7. When $Bi_M \gg 1$, however, the concentration profile within the solid is obtained from the solution of a partial differential equation. Representative concentration profiles for species $\mathcal{A}$ within the solid and fluid phases depending on the value of the Biot number are shown in Figure 10.6.

10.3.1 Diffusion into a Rectangular Slab

Consider a rectangular slab (species $\mathcal{B}$) of thickness $2L$ as shown in Figure 10.7. Initially the concentration of species $\mathcal{A}$ within the slab is uniform at a value of c_{A_o}. At $t = 0$ the surfaces at $z = \pm L$ are exposed to a fluid having a constant concentration of c_{A_∞}. Let us assume $Bi_M > 40$ so that the resistance to mass transfer in the fluid phase is negligible. Under equilibrium conditions, a partition coefficient $\mathcal{H}$ relates the concentration of species $\mathcal{A}$ at the solid-fluid interface as $\mathcal{H} = c_A^{solid}/c_A^{fluid}$. Thus, the term $\mathcal{H}c_{A_\infty}$ represents the concentration of species $\mathcal{A}$ in the solid phase at the fluid-solid interface. As engineers, we are interested in the amount of species $\mathcal{A}$ transferred into the slab as a function of time. For this purpose, it is first necessary to determine the concentration distribution of species $\mathcal{A}$ within the slab as a function of position and time.

If $L/H \ll 1$ and $L/W \ll 1$, then it is possible to assume that the diffusion is one-dimensional and to postulate that $c_A = c_A(t, z)$. In that case, Table C.7 in Appendix C indicates that the only nonzero molar flux component is N_{A_z}, and it is given by

$$N_{A_z} = J^*_{A_z} = -\mathcal{D}_{AB}\frac{\partial c_A}{\partial z} \qquad (10.3\text{-}3)$$

For a rectangular differential volume element of thickness Δz, as shown in Figure 10.7, Eq. (10.3-1) is expressed as

$$N_{A_z}|_z WH - N_{A_z}|_{z+\Delta z} WH = \frac{\partial}{\partial t}\big[WH\Delta z(c_A - c_{A_o})\big] \qquad (10.3\text{-}4)$$

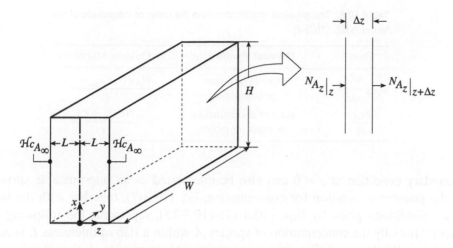

Figure 10.7. Mass transfer into a rectangular slab.

Dividing Eq. (10.3-4) by $W H \, \Delta z$ and letting $\Delta z \to 0$ give

$$\frac{\partial c_A}{\partial t} = \lim_{\Delta z \to 0} \frac{N_{A_z}|_z - N_{A_z}|_{z+\Delta z}}{\Delta z} \tag{10.3-5}$$

or,

$$\frac{\partial c_A}{\partial t} = -\frac{\partial N_{A_z}}{\partial z} \tag{10.3-6}$$

Substitution of Eq. (10.3-3) into Eq. (10.3-6) gives the governing equation for the concentration of species $\mathcal{A}$ as

$$\boxed{\frac{\partial c_A}{\partial t} = \mathcal{D}_{AB} \frac{\partial^2 c_A}{\partial z^2}} \tag{10.3-7}$$

in which the diffusion coefficient is considered constant. In the literature, Eq. (10.3-7) is also known as *Fick's second law of diffusion*. The initial and the boundary conditions associated with Eq. (10.3-7) are

$$\text{at} \quad t = 0 \qquad c_A = c_{A_o} \tag{10.3-8}$$

$$\text{at} \quad z = L \qquad c_A = \mathcal{H}c_{A_\infty} \tag{10.3-9}$$

$$\text{at} \quad z = -L \qquad c_A = \mathcal{H}c_{A_\infty} \tag{10.3-10}$$

Note that $z = 0$ represents a plane of symmetry across which there is no net flux, i.e., $\partial c_A / \partial z = 0$. Therefore, it is also possible to express the initial and boundary conditions as

$$\text{at} \quad t = 0 \qquad c_A = c_{A_o} \tag{10.3-11}$$

$$\text{at} \quad z = 0 \qquad \frac{\partial c_A}{\partial z} = 0 \tag{10.3-12}$$

$$\text{at} \quad z = L \qquad c_A = \mathcal{H}c_{A_\infty} \tag{10.3-13}$$

Table 10.8. The physical significance and the order of magnitude of the terms in Eq. (10.3-7)

Term	Physical Significance	Order of Magnitude
$\mathcal{D}_{AB}\dfrac{\partial^2 c_A}{\partial z^2}$	Rate of diffusion in z-direction	$\dfrac{\mathcal{D}_{AB}(\mathcal{H}c_{A_\infty} - c_{A_o})}{L^2}$
$\dfrac{\partial c_A}{\partial t}$	Rate of accumulation of mass (or mole)	$\dfrac{\mathcal{H}c_{A_\infty} - c_{A_o}}{t}$

The boundary condition at $z = 0$ can also be interpreted as an impermeable surface. As a result, the governing equation for concentration, Eq. (10.3-7), together with the initial and boundary conditions given by Eqs. (10.3-11)–(10.3-13), represents the following problem statement: "Initially the concentration of species A within a slab of thickness L is uniform at a value of c_{A_o}. While one of the surfaces is impermeable to species A, the other is exposed to a fluid having constant concentration c_{A_∞} with $c_{A_\infty} > c_{A_o}$ for $t > 0$."

The physical significance and the order of magnitude of the terms in Eq. (10.3-7) are given in Table 10.8. Note that the ratio of the rate of diffusion to the rate of accumulation of mass is given by

$$\frac{\text{Rate of diffusion}}{\text{Rate of accumulation of mass}} = \frac{\mathcal{D}_{AB}(\mathcal{H}c_{A_\infty} - c_{A_o})/L^2}{(\mathcal{H}c_{A_\infty} - c_{A_o})/t} = \frac{\mathcal{D}_{AB}t}{L^2} \tag{10.3-14}$$

which is completely analogous to the *Fourier number*.

Introduction of the dimensionless quantities

$$\theta = \frac{\mathcal{H}c_{A_\infty} - c_A}{\mathcal{H}c_{A_\infty} - c_{A_o}} \qquad \xi = \frac{z}{L} \qquad \tau = \frac{\mathcal{D}_{AB}t}{L^2} \tag{10.3-15}$$

reduces Eqs. (10.3-7)–(10.3-10) to

$$\frac{\partial \theta}{\partial \tau} = \frac{\partial^2 \theta}{\partial \xi^2} \tag{10.3-16}$$

$$\text{at} \quad \tau = 0 \qquad \theta = 1 \tag{10.3-17}$$

$$\text{at} \quad \xi = 1 \qquad \theta = 0 \tag{10.3-18}$$

$$\text{at} \quad \xi = -1 \qquad \theta = 0 \tag{10.3-19}$$

Since Eqs. (10.3-16)–(10.3-19) are exactly the same as Eqs. (10.2-16)–(10.2-19), the solution given by Eq. (10.2-36) is also valid for this case, i.e.,

$$\boxed{\theta = \frac{2}{\pi} \sum_{n=0}^{\infty} \frac{(-1)^n}{\left(n + \frac{1}{2}\right)} \exp\left[-\left(n + \tfrac{1}{2}\right)^2 \pi^2 \tau\right] \cos\left[\left(n + \tfrac{1}{2}\right)\pi\xi\right]} \tag{10.3-20}$$

When $\tau \geqslant 0.2$, the series solution given by Eq. (10.3-20) can be approximated by the first term of the series.

The average concentration of species $\mathcal{A}$ within the slab, $\langle c_A \rangle$, is defined by

$$\langle c_A \rangle = \frac{1}{L} \int_0^L c_A \, dz = \int_0^1 c_A \, d\xi \tag{10.3-21}$$

Therefore, the average dimensionless concentration, $\langle \theta \rangle$, becomes

$$\langle \theta \rangle = \frac{\mathcal{H}c_{A_\infty} - \langle c_A \rangle}{\mathcal{H}c_{A_\infty} - c_{A_o}} = \int_0^1 \theta \, d\xi \tag{10.3-22}$$

The use of Eq. (10.3-20) in Eq. (10.3-22) leads to

$$\boxed{\langle \theta \rangle = \frac{2}{\pi^2} \sum_{n=0}^{\infty} \frac{1}{\left(n + \frac{1}{2}\right)^2} \exp\left[-\left(n + \tfrac{1}{2}\right)^2 \pi^2 \tau\right]} \tag{10.3-23}$$

Example 10.14 A 1 mm thick membrane ($\mathcal{B}$) in the form of a flat sheet is immersed in a well-stirred 0.15 M solution of species $\mathcal{A}$. The diffusion coefficient of species $\mathcal{A}$ in $\mathcal{B}$ is 6.5×10^{-10} m^2/s.

a) Determine the concentration distribution as a function of position and time if species $\mathcal{A}$ has a partition coefficient of 0.4.
b) Calculate the time to reach steady-state conditions.

Solution

a) Since the membrane is initially $\mathcal{A}$-free, i.e., $c_{A_o} = 0$, Eq. (10.3-20) takes the form

$$\frac{(0.4)(0.15) - c_A}{(0.4)(0.15)} = \frac{2}{\pi} \sum_{n=0}^{\infty} \frac{(-1)^n}{\left(n + \frac{1}{2}\right)} \exp\left[-\left(n + \tfrac{1}{2}\right)^2 \frac{\pi^2 (6.5 \times 10^{-10})}{(0.5 \times 10^{-3})^2} t\right] \cos\left[\left(n + \tfrac{1}{2}\right)\pi\xi\right] \tag{1}$$

The calculated values of c_A as a function of the dimensionless distance, ξ, at four different times are given in the table below. Note that $\xi = 0$ and $\xi = 1$ represent the center and the surface of the membrane sheet, respectively.

ξ	$c_A \times 10^2$ (M)			
	$t = 1$ min	$t = 2$ min	$t = 5$ min	$t = 10$ min
0	0.881	2.465	4.885	5.837
0.1	0.936	2.510	4.889	5.839
0.2	1.103	2.637	4.940	5.845
0.3	1.381	2.848	5.007	5.855
0.4	1.770	3.137	5.098	5.868
0.5	2.268	3.497	5.212	5.885
0.6	2.869	3.918	5.345	5.904
0.7	3.561	4.391	5.494	5.926
0.8	4.329	4.905	5.656	5.950
0.9	5.151	5.445	5.826	5.975
1.0	6.000	6.000	6.000	6.000

b) If the time to reach steady-state, t_∞, is defined as the time for the center concentration to reach 99% of the surface concentration, Eq. (1) takes the form

$$\frac{(0.4)(0.15) - (0.99)(0.4)(0.15)}{(0.4)(0.15)} = \frac{4}{\pi} \exp\left(-\frac{\pi^2 \tau_\infty}{4}\right) \tag{2}$$

in which only the first term of the series is considered. The solution of Eq. (2) gives

$$\tau_\infty = 1.964 \quad \Rightarrow \quad t_\infty = \frac{\tau_\infty L^2}{\mathcal{D}_{AB}} = \frac{(1.964)(0.5 \times 10^{-3})^2}{6.5 \times 10^{-10}} = 755 \text{ s} = 12.6 \text{ min}$$

Comment: In Section 3.4.1, the time it takes for a given process to reach steady-state is defined by

$$t = \frac{L_{ch}^2}{\mathcal{D}_{AB}} \tag{3}$$

where L_{ch} is the half-thickness of the membrane. Substitution of the values into Eq. (3) yields

$$t = \frac{(0.5 \times 10^{-3})^2}{6.5 \times 10^{-10}} = 385 \text{ s} = 6.4 \text{ min}$$

which is quite satisfactory as far as the orders of magnitude are concerned.

10.3.1.1 *Macroscopic equation*   Integration of the governing equation, Eq. (10.3-7), over the volume of the system gives

$$\int_{-L}^{L} \int_{0}^{W} \int_{0}^{H} \frac{\partial c_A}{\partial t} \, dx \, dy \, dz = \int_{-L}^{L} \int_{0}^{W} \int_{0}^{H} \mathcal{D}_{AB} \frac{\partial^2 c_A}{\partial z^2} \, dx \, dy \, dz \tag{10.3-24}$$

or,

$$\underbrace{\frac{d}{dt}\left[\int_{-L}^{L} \int_{0}^{W} \int_{0}^{H} (c_A - c_{A_o}) \, dx \, dy \, dz\right]}_{\text{Rate of accumulation of species } \mathcal{A}} = \underbrace{2WH\left(\mathcal{D}_{AB} \frac{\partial c_A}{\partial z}\bigg|_{z=L}\right)}_{\substack{\text{Rate of species } \mathcal{A} \text{ entering} \\ \text{from surfaces at } z = \pm L}} \tag{10.3-25}$$

which is the macroscopic mass balance for species $\mathcal{A}$ by considering the rectangular slab as a system. The molar rate of species $\mathcal{A}$ entering the slab, $\dot{n}_A$, can be calculated from Eq. (10.3-25) as

$$\dot{n}_A = 2WH\left(\mathcal{D}_{AB} \frac{\partial c_A}{\partial z}\bigg|_{z=L}\right) = -\frac{2WH\mathcal{D}_{AB}(\mathcal{H}c_{A_\infty} - c_{A_o})}{L} \frac{\partial \theta}{\partial \xi}\bigg|_{\xi=1} \tag{10.3-26}$$

The use of Eq. (10.3-20) in Eq. (10.3-26) gives

$$\boxed{\dot{n}_A = \frac{4WH\mathcal{D}_{AB}(\mathcal{H}c_{A_\infty} - c_{A_o})}{L} \sum_{n=0}^{\infty} \exp\left[-\left(n + \tfrac{1}{2}\right)^2 \pi^2 \tau\right]} \tag{10.3-27}$$

The number of moles of species $\mathcal{A}$ transferred can be calculated from

$$n_A = \int_0^t \dot{n}_A \, dt = \frac{L^2}{\mathcal{D}_{AB}} \int_0^\tau \dot{n}_A \, d\tau \tag{10.3-28}$$

Substitution of Eq. (10.3-27) into Eq. (10.3-28) yields

$$\frac{M_A}{M_{A_\infty}} = 1 - \frac{2}{\pi^2} \sum_{n=0}^\infty \frac{1}{\left(n + \frac{1}{2}\right)^2} \exp\left[-\left(n + \tfrac{1}{2}\right)^2 \pi^2 \tau\right] \tag{10.3-29}$$

where M_A is the mass of species $\mathcal{A}$ transferred into the slab and M_{A_∞} is the maximum amount of species $\mathcal{A}$ transferred into the slab, i.e.,

$$M_{A_\infty} = 2LWH(\mathcal{H}c_{A_\infty} - c_{A_o})\mathcal{M}_A \tag{10.3-30}$$

10.3.1.2   Solution for short times   Let s be the distance measured from the surface of the slab, i.e.,

$$s = L - z \tag{10.3-31}$$

so that Eq. (10.3-7) reduces to

$$\frac{\partial c_A}{\partial t} = \mathcal{D}_{AB} \frac{\partial^2 c_A}{\partial s^2} \tag{10.3-32}$$

At small values of time, species $\mathcal{A}$ does not penetrate very far into the slab. Under these circumstances, it is possible to consider the slab a semi-infinite medium in the s-direction. The initial and boundary conditions associated with Eq. (10.3-32) become

$$\text{at} \quad t = 0 \qquad c_A = c_{A_o} \tag{10.3-33}$$

$$\text{at} \quad s = 0 \qquad c_A = \mathcal{H}c_{A_\infty} \tag{10.3-34}$$

$$\text{at} \quad s = \infty \qquad c_A = c_{A_o} \tag{10.3-35}$$

Introduction of the dimensionless concentration

$$\phi = \frac{c_A - c_{A_o}}{\mathcal{H}c_{A_\infty} - c_{A_o}} \tag{10.3-36}$$

reduces Eqs. (10.3-32)–(10.3-35) to

$$\frac{\partial \phi}{\partial t} = \mathcal{D}_{AB} \frac{\partial^2 \phi}{\partial s^2} \tag{10.3-37}$$

$$\text{at} \quad t = 0 \qquad \phi = 0 \tag{10.3-38}$$

$$\text{at} \quad s = 0 \qquad \phi = 1 \tag{10.3-39}$$

$$\text{at} \quad s = \infty \qquad \phi = 0 \tag{10.3-40}$$

Since Eqs. (10.3-37)–(10.3-40) are identical to Eqs. (10.2-50)–(10.2-53) with the exception that α is replaced by $\mathcal{D}_{AB}$, the solution is given by Eq. (10.2-65), i.e.,

$$\boxed{\frac{\mathcal{H}c_{A_\infty} - c_A}{\mathcal{H}c_{A_\infty} - c_{A_o}} = \mathrm{erf}\left(\frac{s}{\sqrt{4\mathcal{D}_{AB}t}}\right)} \tag{10.3-41}$$

Note that Eq. (10.3-41) is identical to Eq. (9.5-128) when $z/v_{\max}$ and c_A^* are replaced by t and $\mathcal{H}c_{A_\infty}$, respectively, in the latter equation.

Since $\mathrm{erf}(x) \simeq (2/\sqrt{\pi})x$ for small values of x, Eq. (10.3-41) reduces to

$$c_A = \mathcal{H}c_{A_\infty} - (\mathcal{H}c_{A_\infty} - c_{A_o})\frac{s}{\sqrt{\pi \mathcal{D}_{AB}t}} \tag{10.3-42}$$

indicating a linear distribution of concentration with position when t is large.

The molar rate of transfer of species $\mathcal{A}$ into the semi-infinite slab of cross-sectional area A is

$$\dot{n}_A = A\left(-\mathcal{D}_{AB}\left.\frac{\partial c_A}{\partial s}\right|_{s=0}\right) = A(\mathcal{H}c_{A_\infty} - c_{A_o})\sqrt{\frac{\mathcal{D}_{AB}}{\pi t}} \tag{10.3-43}$$

The number of moles of species $\mathcal{A}$ transferred is

$$n_A = \int_0^t \dot{n}_A \, dt = 2A(\mathcal{H}c_{A_\infty} - c_{A_o})\sqrt{\frac{\mathcal{D}_{AB}t}{\pi}} \tag{10.3-44}$$

The maximum amount of species $\mathcal{A}$ transferred to the slab is

$$M_{A_\infty} = AL(\mathcal{H}c_{A_\infty} - c_{A_o})M_A \tag{10.3-45}$$

Hence, the ratio of the uptake of species $\mathcal{A}$ relative to the maximum is given by

$$\boxed{\frac{M_A}{M_{A_\infty}} = \frac{2}{\sqrt{\pi}}\sqrt{\frac{\mathcal{D}_{AB}t}{L^2}}} \tag{10.3-46}$$

One should be careful in the interpretation of the term L in Eq. (10.3-46). If mass transfer takes place only from one surface, then L is the total thickness of the slab. On the other hand, if mass transfer takes place from both surfaces, then L is the half-thickness of the slab.

The values of M_A/M_{A_∞} calculated from Eqs. (10.3-29) and (10.3-46) are compared in Table 10.9. Note that the values obtained from the short time solution are almost equal to the exact values up to $\sqrt{\mathcal{D}_{AB}t/L^2} = 0.6$.

When $s/\sqrt{4\mathcal{D}_{AB}t} = 2$, Eq. (10.3-41) becomes

$$\frac{\mathcal{H}c_{A_\infty} - c_A}{\mathcal{H}c_{A_\infty} - c_{A_o}} = \mathrm{erf}(2) = 0.995 \tag{10.3-47}$$

indicating that $c_A \simeq c_{A_o}$. Therefore, the *diffusion penetration depth*, δ_c, is given by

$$\boxed{\delta_c = 4\sqrt{\mathcal{D}_{AB}t}} \tag{10.3-48}$$

Table 10.9. Comparison of the exact fractional uptake values with a short time solution

$\sqrt{\dfrac{\mathcal{D}_{AB}t}{L^2}}$	M_A/M_{A_∞} Exact Eq. (10.3-29)	Approx. Eq. (10.3-46)
0.1	0.113	0.113
0.2	0.226	0.226
0.3	0.339	0.339
0.4	0.451	0.451
0.5	0.562	0.564
0.6	0.667	0.677
0.7	0.758	0.790
0.8	0.833	0.903
0.9	0.890	1.016
1.0	0.931	1.128

The assumption of a semi-infinite medium (or short time solution) is no longer valid when the concentration at $s = L$ becomes equal to or greater than c_{A_o}. Therefore, the solution given by Eq. (10.3-41) holds as long as

$$1 \leqslant \text{erf}\left(\frac{L}{\sqrt{4\mathcal{D}_{AB}t}}\right) \tag{10.3-49}$$

Since $\text{erf}(2) \simeq 1$, Eq. (10.3-49) simplifies to

$$\boxed{t \leqslant \frac{L^2}{16\mathcal{D}_{AB}}} \quad \text{Criterion for semi-infinite medium assumption} \tag{10.3-50}$$

Example 10.15 Once the membrane in Example 10.14 is immersed in a solution of species $\mathcal{A}$, estimate the time it takes for the center concentration to start to rise.

Solution

The center concentration will start to rise when the diffusion penetration depth reaches half the thickness of the membrane. Thus, from Eq. (10.3-48)

$$t = \frac{L^2}{16\mathcal{D}_{AB}} = \frac{(0.5 \times 10^{-3})^2}{16(6.5 \times 10^{-10})} = 24 \text{ s}$$

Example 10.16 Carburization is the process of introducing carbon into a metal by diffusion. A steel sheet ($\mathcal{B}$) of thickness 1 cm initially has a uniform carbon (species $\mathcal{A}$) concentration of 0.15 wt%.

a) If the sheet is held in an atmosphere containing 1.25 wt% carbon at 1000 °C for an hour, estimate the concentration of carbon at a depth of 1 mm below the surface. Take $\mathcal{D}_{AB} = 2.8 \times 10^{-11} \text{ m}^2/\text{s}$.

b) Calculate the amount of carbon deposited in the steel sheet in one hour. Express your result as a fraction of the maximum amount.

Solution

Assumption

1. External resistance to mass transfer is negligible, i.e., $Bi_M > 40$, and the concentration of carbon on the surface of the steel remains constant at 1.25 wt%.

Analysis

a) First, it is necessary to check whether the steel sheet may be considered a semi-infinite medium. From Eq. (10.3-50)

$$\frac{L^2}{16 \mathcal{D}_{AB}} = \frac{(0.5 \times 10^{-2})^2}{(16)(2.8 \times 10^{-11})} = 55,804 \text{ s} \simeq 15 \text{ h } 30 \text{ min}$$

Since the time in question is one hour, Eq. (10.3-41) can be used to estimate the concentration. Substitution of the numerical values gives

$$\frac{1.25 - c_A}{1.25 - 0.15} = \underbrace{\text{erf} \left[\frac{1 \times 10^{-3}}{\sqrt{(4)(2.8 \times 10^{-11})(3600)}} \right]}_{0.974} \quad \Rightarrow \quad c_A = 0.179 \text{ wt%}$$

b) From Eq. (10.3-46)

$$\frac{M_A}{M_{A_\infty}} = \frac{2}{\sqrt{\pi}} \sqrt{\frac{\mathcal{D}_{AB} t}{L^2}} = \frac{2}{\sqrt{\pi}} \sqrt{\frac{(2.8 \times 10^{-11})(3600)}{(0.5 \times 10^{-2})^2}} = 0.07$$

10.3.2 Diffusion into a Rectangular Slab: Revisited

The equations developed in Section 10.3.1 are based on the assumption that the external resistance to mass transfer is negligible, i.e., $Bi_M > 40$. When $0.1 < Bi_M < 40$, one has to consider the external resistance, and the previously defined boundary condition at the fluid-solid interface, Eq. (10.3-13), has to be replaced by

$$\text{at} \quad z = L \qquad \mathcal{D}_{AB} \frac{\partial c_A}{\partial z} = \langle k_c \rangle \left(c_{A_\infty} - c_A^f \right) \tag{10.3-51}$$

The terms c_A and c_A^f represent the concentrations of species $\mathcal{A}$ in the solid phase and in the fluid phase, respectively, both at the fluid-solid interface. Since these concentrations are related by the partition coefficient, $\mathcal{H}$, Eq. (10.3-51) takes the form

$$\text{at} \quad z = L \qquad \mathcal{D}_{AB} \frac{\partial c_A}{\partial z} = \frac{\langle k_c \rangle}{\mathcal{H}} (\mathcal{H} c_{A_\infty} - c_A) \tag{10.3-52}$$

In terms of the dimensionless quantities defined by Eq. (10.3-15), Eq. (10.3-52) becomes

$$\text{at} \quad \xi = 1 \qquad -\frac{\partial \theta}{\partial \xi} = Bi_M \, \theta \tag{10.3-53}$$

where the Biot number for mass transfer, Bi_M, is defined by

$$Bi_M = \frac{\langle k_c \rangle L}{\mathcal{H} \mathcal{D}_{AB}} \tag{10.3-54}$$

In dimensionless form, this problem is similar to that described in Section 10.2.2. Therefore, the solution is given by Eq. (10.2-84), i.e.,

$$\theta = 4 \sum_{n=1}^{\infty} \frac{\sin \lambda_n}{2\lambda_n + \sin 2\lambda_n} \exp(-\lambda_n^2 \tau) \cos(\lambda_n \xi) \tag{10.3-55}$$

where the eigenvalues are the roots of the transcendental equation given by Eq. (10.2-81), i.e.,

$$\lambda_n \tan \lambda_n = Bi_M \qquad n = 1, 2, 3, \ldots \tag{10.3-56}$$

Table 10.3 gives the first five roots of Eq. (10.3-56) as a function of the Biot number. The use of Eq. (10.3-55) in Eq. (10.3-22) gives the average dimensionless concentration as

$$\langle \theta \rangle = 4 \sum_{n=1}^{\infty} \frac{\sin^2 \lambda_n}{\lambda_n (2\lambda_n + \sin 2\lambda_n)} \exp(-\lambda_n^2 \tau) \tag{10.3-57}$$

Example 10.17 Estimate the concentration at the center of the membrane in Example 10.14 after 5 min if the external mass transfer coefficient is 3.5×10^{-6} m/s.

Solution

The Biot number is

$$Bi_M = \frac{\langle k_c \rangle L}{\mathcal{H} \mathcal{D}_{AB}} = \frac{(3.5 \times 10^{-6})(0.5 \times 10^{-3})}{(0.4)(6.5 \times 10^{-10})} = 6.73$$

Therefore, the resistance of the solution to mass transfer should be taken into consideration. The concentration at the center of the membrane, $(c_A)_c$, can be found by evaluating Eq. (10.3-55) at $\xi = 0$. Since the Fourier number is

$$\tau = \frac{\mathcal{D}_{AB} t}{L^2} = \frac{(6.5 \times 10^{-10})(5 \times 60)}{(0.5 \times 10^{-3})^2} = 0.78$$

then it is possible to consider only the first term of the series. The result is

$$\frac{\mathcal{H} c_{A_\infty} - (c_A)_c}{\mathcal{H} c_{A_\infty} - c_{A_o}} = 4 \left(\frac{\sin \lambda_1}{2\lambda_1 + \sin 2\lambda_1} \right) \exp(-\lambda_1^2 \tau) \tag{1}$$

For $Bi_M = 6.73$, the solution of Eq. (10.3-56) gives $\lambda_1 = 1.37$. Substitution of the numerical values into Eq. (1) gives

$$\frac{(0.4)(0.15) - (c_A)_c}{(0.4)(0.15)} = 4 \left[\frac{\sin 78.5}{2(1.37) + \sin 157} \right] \exp\left[-(1.37)^2 (0.78)\right] \tag{2}$$

in which 1.37 rad = 78.5°. The solution of Eq. (2) yields

$$(c_A)_c = 4.26 \times 10^{-2} \, \text{M}$$

Comment: In Example 10.14, the concentration at the center after 5 min is calculated as 4.885×10^{-2} M. The existence of the external resistance obviously reduces the rate of mass transfer into the membrane.

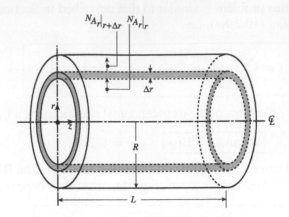

Figure 10.8. Unsteady diffusion into a solid cylinder.

10.3.3   Diffusion into a Cylinder

Consider a cylinder (species $\mathcal{B}$) of radius R and length L as shown in Figure 10.8. Initially the concentration of species $\mathcal{A}$ within the cylinder is uniform at a value of c_{A_o}. At $t = 0$, the cylinder is exposed to a fluid with a constant concentration of c_{A_∞}. Let us assume $\mathrm{Bi_M} > 40$ so that the resistance to mass transfer in the fluid phase is negligible. Under equilibrium conditions, a partition coefficient $\mathcal{H}$ relates the concentration of species $\mathcal{A}$ at the solid-fluid interface as $\mathcal{H} = c_A^{solid}/c_A^{fluid}$. As engineers, we are interested in the amount of species $\mathcal{A}$ transferred into the cylinder as a function of time. For this purpose, it is first necessary to determine the concentration distribution of species $\mathcal{A}$ within the cylinder as a function of position and time.

When $R/L \ll 1$, it is possible to assume that the diffusion is one-dimensional and to postulate that $c_A = c_A(t, r)$. In that case, Table C.8 in Appendix C indicates that the only nonzero molar flux component is N_{A_r}, and it is given by

$$N_{A_r} = J_{A_r}^* = -\mathcal{D}_{AB}\,\frac{\partial c_A}{\partial r} \tag{10.3-58}$$

For a cylindrical differential volume element of thickness Δr, as shown in Figure 10.8, Eq. (10.3-1) is expressed in the form

$$N_{A_r}|_r\, 2\pi r L - N_{A_r}|_{r+\Delta r}\, 2\pi(r + \Delta r)L = \frac{\partial}{\partial t}\big[2\pi r \Delta r L(c_A - c_{A_o})\big] \tag{10.3-59}$$

Dividing Eq. (10.3-59) by $2\pi L\, \Delta r$ and letting $\Delta r \to 0$ give

$$\frac{\partial c_A}{\partial t} = \frac{1}{r}\,\lim_{\Delta r \to 0}\,\frac{(rN_{A_r})|_r - (rN_{A_r})|_{r+\Delta r}}{\Delta r} \tag{10.3-60}$$

or,

$$\frac{\partial c_A}{\partial t} = -\frac{1}{r}\,\frac{\partial}{\partial r}(rN_{A_r}) \tag{10.3-61}$$

Substitution of Eq. (10.3-58) into Eq. (10.3-61) gives the governing equation for the concentration of species $\mathcal{A}$, i.e., Fick's second law of diffusion, as

$$\boxed{\frac{\partial c_A}{\partial t} = \frac{\mathcal{D}_{AB}}{r} \frac{\partial}{\partial r}\left(r\frac{\partial c_A}{\partial r}\right)}$$

(10.3-62)

in which the diffusion coefficient is considered constant.

The initial and boundary conditions associated with Eq. (10.3-62) are given by

$$\text{at} \quad t=0 \qquad c_A = c_{A_o}$$

(10.3-63)

$$\text{at} \quad r=0 \qquad \frac{\partial c_A}{\partial r} = 0$$

(10.3-64)

$$\text{at} \quad r=R \qquad c_A = \mathcal{H}c_{A_\infty}$$

(10.3-65)

Introduction of the dimensionless quantities

$$\theta = \frac{\mathcal{H}c_{A_\infty} - c_A}{\mathcal{H}c_{A_\infty} - c_{A_o}} \qquad \xi = \frac{r}{R} \qquad \tau = \frac{\mathcal{D}_{AB}t}{R^2}$$

(10.3-66)

reduces Eqs. (10.3-62)–(10.3-65) to

$$\frac{\partial\theta}{\partial\tau} = \frac{1}{\xi}\frac{\partial}{\partial\xi}\left(\xi\frac{\partial\theta}{\partial\xi}\right)$$

(10.3-67)

$$\text{at} \quad \tau=0 \qquad \theta=1$$

(10.3-68)

$$\text{at} \quad \xi=0 \qquad \frac{\partial\theta}{\partial\xi} = 0$$

(10.3-69)

$$\text{at} \quad \xi=1 \qquad \theta=0$$

(10.3-70)

The use of the method of separation of variables in which the solution is sought in the form

$$\theta(\tau,\xi) = F(\tau)G(\xi)$$

(10.3-71)

reduces Eq. (10.3-67) to

$$\frac{1}{F}\frac{dF}{d\tau} = \frac{1}{G\xi}\frac{d}{d\xi}\left(\xi\frac{dG}{d\xi}\right) = -\lambda^2$$

(10.3-72)

which results in two ordinary differential equations:

$$\frac{dF}{d\tau} + \lambda^2 F = 0 \quad \Rightarrow \quad F(\tau) = e^{-\lambda^2\tau}$$

(10.3-73)

$$\frac{d}{d\xi}\left(\xi\frac{dG}{d\xi}\right) + \lambda^2\xi G = 0 \quad \Rightarrow \quad G(\xi) = C_1 J_o(\lambda\xi) + C_2 Y_o(\lambda\xi)$$

(10.3-74)

The boundary conditions for $G(\xi)$ are

$$\text{at} \quad \xi = 0 \qquad \frac{dG}{d\xi} = 0 \tag{10.3-75}$$

$$\text{at} \quad \xi = 1 \qquad G = 0 \tag{10.3-76}$$

Since $Y_o(0) = -\infty$, $C_2 = 0$. Application of Eq. (10.3-76) yields

$$C_1 J_o(\lambda) = 0 \tag{10.3-77}$$

For a nontrivial solution, the eigenvalues are given by

$$\boxed{J_o(\lambda_n) = 0} \qquad n = 1, 2, 3, \ldots \tag{10.3-78}$$

The zeros of J_o are given as 2.405, 5.520, 8.654, 11.792, etc.

The general solution is the summation of all possible solutions, i.e.,

$$\theta = \sum_{n=1}^{\infty} A_n \exp\left(-\lambda_n^2 \tau\right) J_o(\lambda_n \xi) \tag{10.3-79}$$

The unknown coefficients A_n can be determined by using the initial condition given by Eq. (10.3-68). The result is

$$\int_0^1 \xi J_o(\lambda_n \xi) \, d\xi = A_n \int_0^1 \xi J_o^2(\lambda_n \xi) \, d\xi \tag{10.3-80}$$

Evaluation of the integrals yields

$$A_n = \frac{2}{\lambda_n J_1(\lambda_n)} \tag{10.3-81}$$

Substitution of Eq. (10.3-81) into Eq. (10.3-79) leads to the following expression for the dimensionless concentration profile

$$\boxed{\theta = 2 \sum_{n=1}^{\infty} \frac{1}{\lambda_n J_1(\lambda_n)} \exp\left(-\lambda_n^2 \tau\right) J_o(\lambda_n \xi)} \tag{10.3-82}$$

The average concentration of species $\mathcal{A}$ within the cylinder, $\langle c_A \rangle$, is defined by

$$\langle c_A \rangle = \frac{2}{R^2} \int_0^R c_A r \, dr = 2 \int_0^1 c_A \xi \, d\xi \tag{10.3-83}$$

Therefore, the average dimensionless concentration, $\langle \theta \rangle$, becomes

$$\langle \theta \rangle = \frac{\mathcal{H} c_{A_\infty} - \langle c_A \rangle}{\mathcal{H} c_{A_\infty} - c_{A_o}} = 2 \int_0^1 \theta \xi \, d\xi \tag{10.3-84}$$

The use of Eq. (10.3-82) in Eq. (10.3-84) leads to

$$\boxed{\langle \theta \rangle = 4 \sum_{n=1}^{\infty} \frac{1}{\lambda_n^2} \exp\left(-\lambda_n^2 \tau\right)} \tag{10.3-85}$$

10.3.3.1 *Macroscopic equation* Integration of the governing equation for the concentration of species $\mathcal{A}$, Eq. (10.3-62), over the volume of the system gives

$$\int_0^L \int_0^{2\pi} \int_0^R \frac{\partial c_A}{\partial t} r \, dr \, d\theta \, dz = \int_0^L \int_0^{2\pi} \int_0^R \frac{\mathcal{D}_{AB}}{r} \frac{\partial}{\partial r}\left(r \frac{\partial c_A}{\partial r}\right) r \, dr \, d\theta \, dz \qquad (10.3\text{-}86)$$

or,

$$\underbrace{\frac{d}{dt}\left[\int_0^L \int_0^{2\pi} \int_0^R (c_A - c_{A_o}) r \, dr \, d\theta \, dz\right]}_{\text{Rate of accumulation of species } \mathcal{A}} = \underbrace{2\pi R L \left(\mathcal{D}_{AB} \left.\frac{\partial c_A}{\partial r}\right|_{r=R}\right)}_{\substack{\text{Rate of species } \mathcal{A} \text{ entering} \\ \text{from the lateral surface}}} \qquad (10.3\text{-}87)$$

which is the macroscopic mass balance for species $\mathcal{A}$ by considering the cylinder as a system. The molar rate of species $\mathcal{A}$ entering the cylinder, $\dot{n}_A$, is given by

$$\dot{n}_A = 2\pi R L \left(\mathcal{D}_{AB} \left.\frac{\partial c_A}{\partial r}\right|_{r=R}\right) = -2\pi L \, \mathcal{D}_{AB}(\mathcal{H} c_{A_\infty} - c_{A_o}) \left.\frac{\partial \theta}{\partial \xi}\right|_{\xi=1} \qquad (10.3\text{-}88)$$

The use of Eq. (10.3-82) in Eq. (10.3-88) results in

$$\boxed{\dot{n}_A = 4\pi L \, \mathcal{D}_{AB}(\mathcal{H} c_{A_\infty} - c_{A_o}) \sum_{n=1}^{\infty} \exp(-\lambda_n^2 \tau)} \qquad (10.3\text{-}89)$$

The number of moles of species $\mathcal{A}$ transferred can be calculated from

$$n_A = \int_0^t \dot{n}_A \, dt = \frac{R^2}{\mathcal{D}_{AB}} \int_0^\tau \dot{n}_A \, d\tau \qquad (10.3\text{-}90)$$

Substitution of Eq. (10.3-89) into Eq. (10.3-90) yields

$$\boxed{\frac{M_A}{M_{A_\infty}} = 1 - 4 \sum_{n=1}^{\infty} \frac{1}{\lambda_n^2} \exp(-\lambda_n^2 \tau)} \qquad (10.3\text{-}91)$$

where M_{A_∞} is defined by

$$M_{A_\infty} = \pi R^2 L (\mathcal{H} c_{A_\infty} - c_{A_o}) \mathcal{M}_A \qquad (10.3\text{-}92)$$

10.3.3.2 *Solution for* $0.1 < \text{Bi}_\text{M} < 40$ In this case, the boundary condition at the fluid-solid interface, Eq. (10.3-65), has to be replaced by

$$\text{at} \quad r = R \qquad \mathcal{D}_{AB} \frac{\partial c_A}{\partial r} = \langle k_c \rangle \left(c_{A_\infty} - c_A^f\right) \qquad (10.3\text{-}93)$$

or,

$$\text{at} \quad r = R \qquad \mathcal{D}_{AB} \frac{\partial c_A}{\partial r} = \frac{\langle k_c \rangle}{\mathcal{H}} (\mathcal{H} c_{A_\infty} - c_A) \qquad (10.3\text{-}94)$$

In terms of the dimensionless quantities defined by Eq. (10.3-66), Eq. (10.3-94) becomes

$$\text{at} \quad \xi = 1 \qquad -\frac{\partial \theta}{\partial \xi} = \text{Bi}_\text{M}\, \theta \tag{10.3-95}$$

where the Biot number for mass transfer, Bi_M, is defined by

$$\text{Bi}_\text{M} = \frac{\langle k_c \rangle\, R}{\mathcal{H}\, \mathcal{D}_{AB}} \tag{10.3-96}$$

Note that this problem is similar to that described in Section 10.2.3. Therefore, the solution is given by Eq. (10.2-119), i.e.,

$$\boxed{\theta = 2\,\text{Bi}_\text{M} \sum_{n=1}^{\infty} \frac{1}{(\lambda_n^2 + \text{Bi}_\text{M}^2)\, J_o(\lambda_n)}\, \exp\!\left(-\lambda_n^2 \tau\right) J_o(\lambda_n \xi)} \tag{10.3-97}$$

where the eigenvalues are the roots of the transcendental equation given by Eq. (10.2-112), i.e.,

$$\boxed{\lambda_n J_1(\lambda_n) = \text{Bi}_\text{M}\, J_o(\lambda_n)} \qquad n = 1, 2, 3, \ldots \tag{10.3-98}$$

Table 10.5 gives the first five roots of Eq. (10.3-98) as a function of the Biot number. The use of Eq. (10.3-97) in Eq. (10.3-84) gives the average dimensionless concentration as

$$\boxed{\langle \theta \rangle = 4\,\text{Bi}_\text{M}^2 \sum_{n=1}^{\infty} \frac{1}{\lambda_n^2 (\lambda_n^2 + \text{Bi}_\text{M}^2)}\, \exp\!\left(-\lambda_n^2 \tau\right)} \tag{10.3-99}$$

Example 10.18 Cylindrical polymeric materials with $R/L \ll 1$ are soaked in a large volume of well-mixed solvent to remove the monomer impurity (species $\mathcal{A}$) left during their manufacture. The diameter of the cylinder is 1 cm, the diffusion coefficient of monomer in the polymer is 1.8×10^{-10} m^2/s, and the average mass transfer coefficient between the cylindrical surface and the solvent is 3.5×10^{-6} m/s. If the (polymer/solvent) partition coefficient of the monomer is 12, estimate the reduction in the monomer concentration at the center of the polymeric material after 10 h.

Solution

The Biot number for mass transfer is calculated from Eq. (10.3-96) as

$$\text{Bi}_\text{M} = \frac{(3.5 \times 10^{-6})(0.5 \times 10^{-2})}{(12)(1.8 \times 10^{-10})} = 8.1$$

The Fourier number is

$$\tau = \frac{\mathcal{D}_{AB}\, t}{R^2} = \frac{(1.8 \times 10^{-10})(10 \times 3600)}{(0.5 \times 10^{-2})^2} = 0.26$$

The centerline concentration can be found by evaluating Eq. (10.3-97) at $\xi = 0$. Considering only the first term of the series, the result is

$$\frac{c_A}{c_{A_o}} = \frac{2\,\mathrm{Bi_M}}{(\lambda_1^2 + \mathrm{Bi_M^2})J_o(\lambda_1)}\exp\left(-\lambda_1^2\tau\right) \tag{1}$$

in which c_{A_∞} is considered zero. The solution of Eq. (10.3-98) gives $\lambda_1 = 2.132$. Therefore, substitution of the numerical values into Eq. (1) yields

$$\frac{c_A}{c_{A_o}} = \frac{2(8.1)}{(2.132^2 + 8.1^2)J_o(2.132)}\exp\left[-(2.132)^2(0.26)\right] = 0.48$$

indicating approximately a two-fold decrease in the centerline concentration.

10.3.4 Gas Absorption Into a Spherical Liquid Droplet

Consider a liquid droplet ($\mathcal{B}$) of radius R surrounded by gas $\mathcal{A}$ as shown in Figure 10.9. We are interested in the rate of absorption of species $\mathcal{A}$ into the liquid. The problem will be analyzed with the following assumptions:

1. Convective flux is negligible with respect to the molecular flux.
2. The total concentration is constant.

Since $c_A = c_A(r)$, Table C.9 in Appendix C indicates that the only nonzero molar flux component is N_{A_r}, and it is given by

$$N_{A_r} = J_{A_r}^* = -\mathcal{D}_{AB}\frac{\partial c_A}{\partial r} \tag{10.3-100}$$

For a spherical differential volume element of thickness Δr, as shown in Figure 10.9, Eq. (10.3-1) is expressed in the form

$$N_{A_r}|_r 4\pi r^2 - N_{A_r}|_{r+\Delta r}4\pi(r+\Delta r)^2 = \frac{\partial}{\partial t}\left[4\pi r^2 \Delta r (c_A - c_{A_o})\right] \tag{10.3-101}$$

Dividing Eq. (10.3-101) by $4\pi\,\Delta r$ and taking the limit as $\Delta r \to 0$ give

$$\frac{\partial c_A}{\partial t} = \frac{1}{r^2}\lim_{\Delta r \to 0}\frac{(r^2 N_{A_r})|_r - (r^2 N_{A_r})|_{r+\Delta r}}{\Delta r} \tag{10.3-102}$$

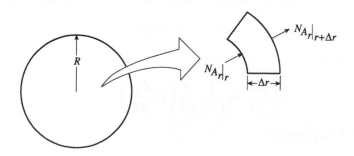

Figure 10.9. Gas absorption into a droplet.

or,

$$\frac{\partial c_A}{\partial t} = -\frac{1}{r^2}\frac{\partial(r^2 N_{A_r})}{\partial r} \qquad (10.3\text{-}103)$$

Substitution of Eq. (10.3-100) into Eq. (10.3-103) gives the governing differential equation for the concentration of species $\mathcal{A}$, i.e., Fick's second law of diffusion, as

$$\boxed{\frac{\partial c_A}{\partial t} = \frac{\mathcal{D}_{AB}}{r^2}\frac{\partial}{\partial r}\left(r^2\frac{\partial c_A}{\partial r}\right)} \qquad (10.3\text{-}104)$$

in which the diffusion coefficient is considered constant.

The initial and boundary conditions associated with Eq. (10.3-104) are

$$\text{at}\quad t = 0 \qquad c_A = c_{A_o} \qquad (10.3\text{-}105)$$

$$\text{at}\quad r = 0 \qquad \frac{\partial c_A}{\partial r} = 0 \qquad (10.3\text{-}106)$$

$$\text{at}\quad r = R \qquad c_A = c_A^* \qquad (10.3\text{-}107)$$

where c_A^* is the equilibrium solubility of species $\mathcal{A}$ in liquid $\mathcal{B}$.

Introduction of the dimensionless quantities

$$\theta = \frac{c_A^* - c_A}{c_A^* - c_{A_o}} \qquad \xi = \frac{r}{R} \qquad \tau = \frac{\mathcal{D}_{AB}t}{R^2} \qquad (10.3\text{-}108)$$

reduces Eqs. (10.3-104)–(10.3-107) to

$$\frac{\partial\theta}{\partial\tau} = \frac{1}{\xi^2}\frac{\partial}{\partial\xi}\left(\xi^2\frac{\partial\theta}{\partial\xi}\right) \qquad (10.3\text{-}109)$$

$$\text{at}\quad \tau = 0 \qquad \theta = 1 \qquad (10.3\text{-}110)$$

$$\text{at}\quad \xi = 0 \qquad \frac{\partial\theta}{\partial\xi} = 0 \qquad (10.3\text{-}111)$$

$$\text{at}\quad \xi = 1 \qquad \theta = 0 \qquad (10.3\text{-}112)$$

Since the governing equation and the boundary conditions are homogeneous, the use of the method of separation of variables in which the solution is sought in the form

$$\theta(\tau, \xi) = F(\tau)G(\xi) \qquad (10.3\text{-}113)$$

reduces Eq. (10.3-109) to

$$\frac{1}{F}\frac{dF}{d\tau} = \frac{1}{G\xi^2}\frac{d}{d\xi}\left(\xi^2\frac{dG}{d\xi}\right) = -\lambda^2 \qquad (10.3\text{-}114)$$

The equation for F is given by

$$\frac{dF}{d\tau} + \lambda^2 F = 0 \quad\Rightarrow\quad F(\tau) = e^{-\lambda^2\tau} \qquad (10.3\text{-}115)$$

The equation for G is

$$\frac{1}{\xi^2}\frac{d}{d\xi}\left(\xi^2\frac{dG}{d\xi}\right)+\lambda^2 G=0 \tag{10.3-116}$$

and it is subject to the boundary conditions

$$\text{at}\quad \xi=0\qquad \frac{dG}{d\xi}=0 \tag{10.3-117}$$

$$\text{at}\quad \xi=1\qquad G=0 \tag{10.3-118}$$

The transformation[8] $G=u(\xi)/\xi$ converts Eq. (10.3-116) to

$$\frac{d^2u}{d\xi^2}+\lambda^2 u=0 \tag{10.3-119}$$

which has the solution

$$u=C_1\sin(\lambda\xi)+C_2\cos(\lambda\xi) \tag{10.3-120}$$

or,

$$G=C_1\frac{\sin(\lambda\xi)}{\xi}+C_2\frac{\cos(\lambda\xi)}{\xi} \tag{10.3-121}$$

While the application of Eq. (10.3-117) gives $C_2=0$, the use of Eq. (10.3-118) results in

$$\sin\lambda=0\quad\Rightarrow\quad \lambda_n=n\pi\qquad n=1,2,3,\dots \tag{10.3-122}$$

The general solution is the summation of all possible solutions, i.e.,

$$\theta=\sum_{n=1}^{\infty}A_n\exp(-n^2\pi^2\tau)\frac{\sin(n\pi\xi)}{\xi} \tag{10.3-123}$$

The unknown coefficients A_n can be determined from the initial condition defined by Eq. (10.3-110). The result is

$$A_n=\frac{\displaystyle\int_0^1\xi\sin(n\pi\xi)\,d\xi}{\displaystyle\int_0^1\sin^2(n\pi\xi)\,d\xi}=\frac{2(-1)^{n+1}}{n\pi} \tag{10.3-124}$$

Hence, the dimensionless concentration distribution is expressed as

$$\theta=\frac{2}{\pi}\sum_{n=1}^{\infty}\frac{(-1)^{n+1}}{n}\exp(-n^2\pi^2\tau)\frac{\sin(n\pi\xi)}{\xi} \tag{10.3-125}$$

---

[8]See Section 9.4.2.

The average concentration of species A within the sphere, $\langle c_A \rangle$, is defined by

$$\langle c_A \rangle = \frac{3}{R^3} \int_0^R c_A r^2 \, dr = 3 \int_0^1 c_A \xi^2 \, d\xi \tag{10.3-126}$$

Therefore, the average dimensionless concentration, $\langle \theta \rangle$, becomes

$$\langle \theta \rangle = \frac{c_A^* - \langle c_A \rangle}{c_A^* - c_{A_o}} = 3 \int_0^1 \theta \xi^2 \, d\xi \tag{10.3-127}$$

The use of Eq. (10.3-125) in Eq. (10.3-127) leads to

$$\boxed{\langle \theta \rangle = \frac{6}{\pi^2} \sum_{n=1}^{\infty} \frac{1}{n^2} \exp(-n^2 \pi^2 \tau)} \tag{10.3-128}$$

10.3.4.1 *Macroscopic equation*   Integration of the governing equation for the concentration of species A, Eq. (10.3-104), over the volume of the system gives

$$\int_0^{2\pi} \int_0^{\pi} \int_0^R \frac{\partial c_A}{\partial t} r^2 \sin\theta \, dr \, d\theta \, d\phi = \int_0^{2\pi} \int_0^{\pi} \int_0^R \frac{\mathcal{D}_{AB}}{r^2} \frac{\partial}{\partial r} \left(r^2 \frac{\partial c_A}{\partial r} \right) r^2 \sin\theta \, dr \, d\theta \, d\phi \tag{10.3-129}$$

or,

$$\underbrace{\frac{d}{dt}\left[\int_0^{2\pi} \int_0^{\pi} \int_0^R (c_A - c_{A_o}) r^2 \sin\theta \, dr \, d\theta \, d\phi \right]}_{\text{Rate of accumulation of species } A} = \underbrace{4\pi R^2 \left(\mathcal{D}_{AB} \frac{\partial c_A}{\partial r} \bigg|_{r=R} \right)}_{\substack{\text{Rate of species } A \text{ entering} \\ \text{from the surface}}} \tag{10.3-130}$$

which is the macroscopic mass balance for species A by considering the liquid droplet as a system. The molar rate of absorption of species A is given by

$$\dot{n}_A = 4\pi R^2 \left(\mathcal{D}_{AB} \frac{\partial c_A}{\partial r} \bigg|_{r=R} \right) = -4\pi R \, \mathcal{D}_{AB} (c_A^* - c_{A_o}) \frac{\partial \theta}{\partial \xi} \bigg|_{\xi=1} \tag{10.3-131}$$

The use of Eq. (10.3-125) in Eq. (10.3-131) results in

$$\boxed{\dot{n}_A = 8\pi R \, \mathcal{D}_{AB} (c_A^* - c_{A_o}) \sum_{n=1}^{\infty} \exp(-n^2 \pi^2 \tau)} \tag{10.3-132}$$

The moles of species A absorbed can be calculated from

$$n_A = \int_0^t \dot{n}_A \, dt = \frac{R^2}{\mathcal{D}_{AB}} \int_0^{\tau} \dot{n}_A \, d\tau \tag{10.3-133}$$

Substitution of Eq. (10.3-132) into Eq. (10.3-133) yields

$$\boxed{n_A = \frac{8R^3 (c_A^* - c_{A_o})}{\pi} \sum_{n=1}^{\infty} \frac{1}{n^2} \left[1 - \exp(-n^2 \pi^2 \tau) \right]} \tag{10.3-134}$$

The maximum amount of species $\mathcal{A}$ absorbed by the droplet is given by

$$M_{A_\infty} = \frac{4}{3}\pi R^3 (c_A^* - c_{A_o})\mathcal{M}_A \qquad (10.3\text{-}135)$$

Therefore, the mass of species $\mathcal{A}$ absorbed by the droplet relative to the maximum is

$$\boxed{\frac{M_A}{M_{A_\infty}} = 1 - \frac{6}{\pi^2}\sum_{n=1}^{\infty}\frac{1}{n^2}\exp(-n^2\pi^2\tau)} \qquad (10.3\text{-}136)$$

10.3.4.2 *Solution for* $0.1 < \mathrm{Bi_M} < 40$ In this case, the boundary condition at the fluid-solid interface, Eq. (10.3-107), has to be replaced by

$$\text{at} \quad r = R \qquad \mathcal{D}_{AB}\frac{\partial c_A}{\partial r} = \langle k_c \rangle \left(c_{A_\infty} - c_A^f \right) \qquad (10.3\text{-}137)$$

or,

$$\text{at} \quad r = R \qquad \mathcal{D}_{AB}\frac{\partial c_A}{\partial r} = \frac{\langle k_c \rangle}{\mathcal{H}}(\mathcal{H}c_{A_\infty} - c_A) \qquad (10.3\text{-}138)$$

In terms of the dimensionless quantities defined by Eq. (10.3-108), Eq. (10.3-138) becomes

$$\text{at} \quad \xi = 1 \qquad -\frac{\partial\theta}{\partial\xi} = \mathrm{Bi_M}\,\theta \qquad (10.3\text{-}139)$$

where the Biot number for mass transfer, $\mathrm{Bi_M}$, is defined by

$$\mathrm{Bi_M} = \frac{\langle k_c \rangle R}{\mathcal{H}\mathcal{D}_{AB}} \qquad (10.3\text{-}140)$$

Note that this problem is similar to that described in Section 10.2.4. Therefore, the solution is given by Eq. (10.2-153), i.e.,

$$\boxed{\theta = 4\,\mathrm{Bi_M}\sum_{n=1}^{\infty}\frac{\sin\lambda_n}{\lambda_n(2\lambda_n - \sin 2\lambda_n)}\exp\left(-\lambda_n^2\tau\right)\frac{\sin(\lambda_n\xi)}{\xi}} \qquad (10.3\text{-}141)$$

where the eigenvalues are the roots of the transcendental equation given by Eq. (10.2-149), i.e.,

$$\boxed{\lambda_n \cot\lambda_n = 1 - \mathrm{Bi_M}} \qquad n = 1, 2, 3, \ldots \qquad (10.3\text{-}142)$$

Table 10.6 gives the first five roots of Eq. (10.3-142) as a function of the Biot number. The use of Eq. (10.3-141) in Eq. (10.3-127) gives the average dimensionless concentration as

$$\boxed{\langle\theta\rangle = 12\,\mathrm{Bi_M^2}\sum_{n=1}^{\infty}\frac{\sin^2\lambda_n}{\lambda_n^3(2\lambda_n - \sin 2\lambda_n)}\exp\left(-\lambda_n^2\tau\right)} \qquad (10.3\text{-}143)$$

NOTATION

A	area, m^2
$\widehat{C}_P$	heat capacity at constant pressure, kJ/kg·K
c_i	concentration of species i, kmol/m^3
D_P	particle diameter, m
$\mathcal{D}_{AB}$	diffusion coefficient for system $\mathcal{A}$-$\mathcal{B}$, m^2/s
e	total energy flux, W/m^2
F_D	drag force, N
$\mathcal{H}$	partition coefficient
h	heat transfer coefficient, W/m^2·K
J^*	molecular molar flux, kmol/m^2·s
k_c	mass transfer coefficient, m/s
L	length, m
M	mass, kg
$\dot{m}$	mass flow rate, kg/s
$\mathcal{M}$	molecular weight, kg/kmol
N	total molar flux, kmol/m^2·s
$\dot{n}$	molar flow rate, kmol/s
$\dot{Q}$	heat transfer rate, W
Q	volumetric flow rate, m^3/s
q	heat flux, W/m^2
R	radius, m
T	temperature, °C or K
t	time, s
V	velocity of the plate in Couette flow, m/s; volume, m^3
v	velocity, m/s
W	width, m
α	thermal diffusivity, m^2/s
δ	penetration distance for momentum, m
δ_c	penetration distance for mass, m
δ_t	penetration distance for heat, m
μ	viscosity, kg/m·s
ν	kinematic viscosity, m^2/s
ρ	density, kg/m^3
π	total momentum flux, N/m^2
τ	dimensionless time
τ_{ij}	shear stress (flux of j-momentum in the i-direction), N/m^2

Bracket

$\langle a \rangle$	average value of a

Subscripts

A, B	species in binary systems
c	center

ch   characteristic
ref   reference

Dimensionless Numbers

Bi_H Biot number for heat transfer
Bi_M Biot number for mass transfer
Fo   Fourier number
Nu   Nusselt number
Pr   Prandtl number
Re   Reynolds number

REFERENCES

Carslaw, H.S. and J.C. Jaeger, 1959, Conduction of Heat in Solids, 2nd Ed., Oxford University Press, London.
Crank, J., 1956, The Mathematics of Diffusion, Oxford University Press, London.
Konak, A.R., 1994, Magic unveiled through the concept of heat and its transfer, Chem. Eng. Ed. (Summer), 180.
Siegel, R.A., 2000, Theoretical analysis of inward hemispheric release above and below drug solubility, J. Control. Rel. 69, 109.

SUGGESTED REFERENCES FOR FURTHER STUDY

Bird, R.B., W.E. Stewart and E.N. Lightfoot, 2002, Transport Phenomena, 2nd Ed., Wiley, New York.
Deen, W.M., 1998, Analysis of Transport Phenomena, Oxford University Press, New York.
Middleman, S., 1998, An Introduction to Mass and Heat Transfer – Principles of Analysis and Design, Wiley, New York.
Slattery, J.C., 1999, Advanced Transport Phenomena, Cambridge University Press, Cambridge.

PROBLEMS

10.1 Consider the heating of a rectangular slab of thickness $2L$ as described in Section 10.2.1.

a) Show that the general governing equation for temperature is given by

$$\rho \widehat{C}_P \frac{\partial T}{\partial t} = k \frac{\partial^2 T}{\partial x^2} + k \frac{\partial^2 T}{\partial y^2} + k \frac{\partial^2 T}{\partial z^2} \tag{1}$$

b) Use order of magnitude analysis and show that Eq. (1) reduces to Eq. (10.2-7) when $L/H \ll 1$ and $L/W \ll 1$.

10.2 Consider one-dimensional temperature distribution in a slab of thickness L. The temperatures of the surfaces located at $z = 0$ and $z = L$ are held at T_o and T_L, respectively, until steady-state conditions prevail. Then, at $t = 0$, the temperatures of the surfaces are interchanged.

a) Consider a differential volume element of thickness Δz within the slab and in terms of the following dimensionless quantities

$$\theta = \frac{T - T_o}{T_L - T_o} \qquad \tau = \frac{\alpha t}{L^2} \qquad \xi = \frac{z}{L} \tag{1}$$

show that the governing differential equation for temperature is given by

$$\frac{\partial \theta}{\partial \tau} = \frac{\partial^2 \theta}{\partial \xi^2} \tag{2}$$

subject to the following initial and boundary conditions

$$\text{at} \quad \tau = 0 \qquad \theta = \xi \tag{3}$$

$$\text{at} \quad \xi = 0 \qquad \theta = 1 \tag{4}$$

$$\text{at} \quad \xi = 1 \qquad \theta = 0 \tag{5}$$

b) Since the boundary condition at $\xi = 0$ is not homogeneous, propose a solution in the form

$$\theta(\tau, \xi) = \theta_\infty(\xi) - \theta_t(\tau, \xi) \tag{6}$$

in which $\theta_\infty(\xi)$ is the steady-state solution, i.e.,

$$\frac{d^2 \theta_\infty}{d\xi^2} = 0 \tag{7}$$

with the following boundary conditions

$$\text{at} \quad \xi = 0 \qquad \theta_\infty = 1 \tag{8}$$

$$\text{at} \quad \xi = 1 \qquad \theta_\infty = 0 \tag{9}$$

Show that the steady-state solution is given by

$$\theta_\infty = 1 - \xi \tag{10}$$

c) Show that the governing equation for the transient contribution $\theta_t(\tau, \xi)$ is given by

$$\frac{\partial \theta_t}{\partial \tau} = \frac{\partial^2 \theta_t}{\partial \xi^2} \tag{11}$$

subject to the following initial and boundary conditions

$$\text{at} \quad \tau = 0 \qquad \theta_t = 1 - 2\xi \tag{12}$$

$$\text{at} \quad \xi = 0 \qquad \theta_t = 0 \tag{13}$$

$$\text{at} \quad \xi = 1 \qquad \theta_t = 0 \tag{14}$$

Use the method of separation of variables and show that the transient solution is given in the form

$$\theta_t = \frac{2}{\pi} \sum_{n=1}^{\infty} \frac{1}{n} e^{-4n^2\pi^2\tau} \sin(2n\pi\xi) \tag{15}$$

10.3 A circular rod of radius R and length L is insulated on the lateral surface and the steady-state temperature distribution within the rod is given by

$$\frac{T - T_o}{T_L - T_o} = \left(\frac{z}{L}\right)^2$$

where T_o and T_L are the temperatures at $z = 0$ and $z = L$, respectively. At time $t = 0$, both ends of the rod are also insulated.

a) Since there is no variation in temperature in the radial direction, consider a differential volume element of thickness Δz within the rod and, in terms of the following dimensionless quantities

$$\theta = \frac{T - T_o}{T_L - T_o} \qquad \tau = \frac{\alpha t}{L^2} \qquad \xi = \frac{z}{L} \tag{1}$$

show that the governing differential equation is given by

$$\frac{\partial\theta}{\partial\tau} = \frac{\partial^2\theta}{\partial\xi^2} \tag{2}$$

subject to the following initial and boundary conditions

$$\text{at} \quad \tau = 0 \qquad \theta = \xi^2 \tag{3}$$

$$\text{at} \quad \xi = 0 \qquad \frac{\partial\theta}{\partial\xi} = 0 \tag{4}$$

$$\text{at} \quad \xi = 1 \qquad \frac{\partial\theta}{\partial\xi} = 0 \tag{5}$$

b) Use the method of separation of variables and show that

$$\theta = \frac{1}{3} + \frac{4}{\pi^2} \sum_{n=1}^{\infty} \frac{(-1)^n}{n^2} e^{-n^2\pi^2\tau} \cos(n\pi\xi) \tag{6}$$

Note that the term $1/3$ comes from the fact that $\lambda = 0$ does not yield a trivial solution. It is simply the final steady-state temperature of the rod.

10.4 Two semi-infinite solids A and B, initially at T_{A_o} and T_{B_o} with $T_{A_o} > T_{B_o}$, are suddenly brought into contact at $t = 0$. The contact resistance between the metals is negligible.

a) Equating the heat fluxes at the interface, show that the interface temperature, T_i, is given by

$$\frac{T_i - T_{B_o}}{T_{A_o} - T_{B_o}} = \frac{\sqrt{\alpha_B}k_A}{\sqrt{\alpha_B}k_A + \sqrt{\alpha_A}k_B}$$

b) Consider two slabs, one made of copper and the other of wood, at a temperature of $80\,°\text{C}$. You want to check if they are hot by touching them with your finger. Explain why you think the copper slab feels hotter. The physical properties are given as follows:

	k (W/m·K)	α (m^2/s)
Skin	0.3	1.5×10^{-7}
Copper	401	117×10^{-6}
Wood	0.15	1.2×10^{-7}

10.5 The fuel oil pipe that supplies the heating system of a house is laid 1 m below the ground. Around a temperature of $2\,°\text{C}$ the viscosity of the fuel oil increases to a point at which pumping becomes almost impossible. When the air temperature drops to $-15\,°\text{C}$, how long is it before there are problems in the heating system? Assume that the initial ground temperature is $10\,°\text{C}$ and the physical properties are: $k = 0.38$ W/m·K and $\alpha = 4 \times 10^{-7}$ m^2/s.

(**Answer:** 351.3 h)

10.6 A long slab of thickness L is initially at a uniform temperature of T_o. At $t = 0$, both sides of the slab are exposed to the same fluid at a temperature of T_∞ ($T_\infty < T_o$). The fluids have different velocities and, as a result, unequal cooling conditions are applied at the surfaces, i.e., the average heat transfer coefficients between the surfaces and the fluid are different from each other. The geometry of the system is shown in the figure below.

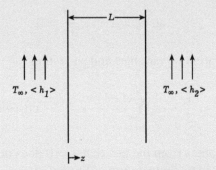

a) Consider a differential volume of thickness Δz within the slab and show that

$$\frac{\partial T}{\partial t} = \alpha \frac{\partial^2 T}{\partial z^2} \tag{1}$$

with the following initial and boundary conditions

$$\text{at} \quad t = 0 \qquad T = T_o \tag{2}$$

$$\text{at} \quad z = 0 \qquad k\frac{\partial T}{\partial z} = \langle h_1 \rangle (T - T_\infty) \tag{3}$$

$$\text{at} \quad z = L \qquad -k\frac{\partial T}{\partial z} = \langle h_2 \rangle (T - T_\infty) \tag{4}$$

b) In terms of the following dimensionless quantities

$$\theta = \frac{T - T_\infty}{T_o - T_\infty} \qquad \tau = \frac{\alpha t}{L^2} \qquad \xi = \frac{z}{L} \qquad (Bi_H)_1 = \frac{\langle h_1 \rangle L}{k} \qquad (Bi_H)_2 = \frac{\langle h_2 \rangle L}{k}$$

show that Eqs. (1)–(4) reduce to

$$\frac{\partial \theta}{\partial \tau} = \frac{\partial^2 \theta}{\partial \xi^2} \tag{5}$$

$$\text{at} \quad \tau = 0 \qquad \theta = 1 \tag{6}$$

$$\text{at} \quad \xi = 0 \qquad \frac{\partial \theta}{\partial \xi} = (Bi_H)_1 \theta \tag{7}$$

$$\text{at} \quad \xi = 1 \qquad -\frac{\partial \theta}{\partial \xi} = (Bi_H)_2 \theta \tag{8}$$

c) Use the method of separation of variables and obtain the solution in the form

$$\theta = \sum_{n=1}^{\infty} A_n \exp\left(-\lambda_n^2 \tau\right) X_n(\lambda_n \xi) \tag{9}$$

where the eigenvalues λ_n are the positive roots of

$$\frac{\lambda_n [(Bi_H)_1 + (Bi_H)_2]}{\lambda_n^2 - (Bi_H)_1 (Bi_H)_2} = \tan \lambda_n \tag{10}$$

and the eigenfunctions X_n are defined by

$$X_n(\lambda_n \xi) = \cos(\lambda_n \xi) + \frac{(Bi_H)_1}{\lambda_n} \sin(\lambda_n \xi) \tag{11}$$

d) Noting that the eigenfunctions satisfy the equation

$$\frac{d^2 X_n}{d\xi^2} + \lambda_n^2 X_n = 0 \tag{12}$$

subject to the following boundary conditions

$$\text{at} \quad \xi = 0 \qquad \frac{dX_n}{d\xi} = (Bi_H)_1 X_n \tag{13}$$

$$\text{at} \quad \xi = 1 \qquad -\frac{dX_n}{d\xi} = (Bi_H)_2 X_n \tag{14}$$

show that

$$\int_0^1 X_n(\lambda_n\xi)X_m(\lambda_n\xi)\,d\xi = 0 \tag{15}$$

$$\int_0^1 X_n^2(\lambda_n\xi)\,d\xi = \frac{\lambda_n^2(B_1 - B_2)^2 + (\lambda_n^2 + B_1 B_2)(\lambda_n^2 + B_1 B_2 + B_1 + B_2)}{2\lambda_n^2(\lambda_n^2 + B_2^2)} \tag{16}$$

in which $(Bi_H)_1$ and $(Bi_H)_2$ are designated as B_1 and B_2, respectively.

e) With the help of Eqs. (15) and (16), show that the coefficients A_n are given by

$$A_n = \frac{2(\lambda_n^2 + B_2^2)(\lambda_n \sin\lambda_n - B_1 \cos\lambda_n + B_1)}{\lambda_n^2(B_1 - B_2)^2 + (\lambda_n^2 + B_1 B_2)(\lambda_n^2 + B_1 B_2 + B_1 + B_2)} \tag{17}$$

f) Show that the average temperature is given by

$$\langle\theta\rangle = \int_0^1 \theta\,d\xi = 2\sum_{n=1}^{\infty}\left(1 + \frac{B_2^2}{\lambda_n^2}\right)\frac{[\lambda_n\sin\lambda_n - B_1(\cos\lambda_n - 1)]^2\exp(-\lambda_n^2\tau)}{\lambda_n^2(B_1 - B_2)^2 + (\lambda_n^2 + B_1 B_2)(\lambda_n^2 + B_1 B_2 + B_1 + B_2)}$$

Plot $\langle\theta\rangle$ versus τ when *i*) $(Bi_H)_1 = 0.1$ and $(Bi_H)_2 = 0.2$, *ii*) $(Bi_H)_1 = 1$ and $(Bi_H)_2 = 10$.

10.7 One side of a concrete wall ($\rho = 2300\text{ kg/m}^3$, $k = 1.9\text{ W/m·K}$, $\widehat{C}_P = 840\text{ J/m·K}$) of thickness 20 cm is insulated. The initial temperature of the wall is $15\,°C$. At $t = 0$, the other side of the wall is exposed to a hot gas at $600\,°C$. Estimate the time for the insulated surface to reach $500\,°C$ if the average heat transfer coefficient between the concrete surface and the hot gas is:

a) $40\text{ W/m}^2\text{·K}$
b) $400\text{ W/m}^2\text{·K}$

(**Answer:** a) 13.7 h   b) 9.2 h)

10.8 A solid cylinder of radius R is initially at a temperature of T_o. At $t = 0$, the surface temperature is increased to T_1.

a) Show that the governing equation and the associated initial and boundary conditions are given in dimensionless form as

$$\frac{\partial\theta}{\partial\tau} = \frac{1}{\xi}\frac{\partial}{\partial\xi}\left(\xi\frac{\partial\theta}{\partial\xi}\right) \tag{1}$$

$$\text{at}\quad \tau = 0 \quad \theta = 1 \tag{2}$$

$$\text{at}\quad \xi = 0 \quad \frac{\partial\theta}{\partial\xi} = 0 \tag{3}$$

$$\text{at}\quad \xi = 1 \quad \theta = 0 \tag{4}$$

where the dimensionless variables are defined by

$$\theta = \frac{T_1 - T}{T_1 - T_o} \qquad \xi = \frac{r}{R} \qquad \tau = \frac{\alpha t}{R^2} \tag{5}$$

b) Use the method of separation of variables and show that the dimensionless temperature distribution is given by

$$\theta = 2 \sum_{n=1}^{\infty} \frac{1}{\lambda_n J_1(\lambda_n)} \exp\left(-\lambda_n^2 \tau\right) J_o(\lambda_n \xi) \tag{6}$$

where the eigenvalues, λ_n, are the roots of

$$J_o(\lambda_n) = 0 \tag{7}$$

c) Show that Eqs. (6) and (7) can also be obtained from Eqs. (10.2-119) and (10.2-112), respectively, by letting $Bi_H \rightarrow \infty$.

10.9   A hot dog is composed of 15% fat, 18% carbohydrates, 11% protein, and 56% water, and has the following physical properties:

$$\rho = 1200 \text{ kg/m}^3 \qquad \widehat{C}_P = 3300 \text{ J/kg·K} \qquad k = 0.4 \text{ W/m·K}$$

A hot dog is considered cooked when its center temperature reaches $60\,^\circ$C. Consider a hot dog taken out of a refrigerator at $2\,^\circ$C and dropped into boiling water. If the average surface heat transfer coefficient is 120 W/m^2·K, estimate the time it takes to cook a hot dog. You can model the hot dog as an infinite cylinder with a diameter of 2 cm.

(**Answer:** 6.4 min)

10.10   A spherical material 15 cm in radius is initially at a uniform temperature of $60\,^\circ$C. It is placed in a room where the temperature is $23\,^\circ$C. Estimate the average heat transfer coefficient if it takes 42 min for the center temperature to reach $30\,^\circ$C. Take $k = 0.12 \text{ W/m·K}$ and $\alpha = 2.7 \times 10^{-6} \text{ m}^2/\text{s}$.

(**Answer:** 6.3 W/m^2·K)

10.11   A solid sphere ($k = 180 \text{ W/m·K}$, $\alpha = 8 \times 10^{-5} \text{ m}^2/\text{s}$) of diameter 20 cm is initially at a temperature of $150\,^\circ$C. Estimate the time required for the center of the sphere to cool to $50\,^\circ$C for the following two cases:

a) The sphere is exposed to an air stream at $40\,^\circ$C having an average heat transfer coefficient of 100 W/m^2·K.

b) The sphere is immersed in a well-mixed large bath at $40\,^\circ$C. The average heat transfer coefficient is 850 W/m^2·K.

(**Answer:** a) 30 min b) 69 s)

10.12 Consider mass transfer into a rectangular slab as described in Section 10.3.1. If the slab is initially A-free, show that the time required for the center concentration to reach 99% of the final concentration is given by

$$t \simeq \frac{2L^2}{\mathcal{D}_{AB}}$$

10.13 A slab of thickness 4 cm contains drug A with a uniform concentration c_{A_o}. If it is immersed in a large bath of pure liquid B, how long does it take for half of the drug to be released into the liquid? Take $\mathcal{D}_{AB} = 3 \times 10^{-8}$ m^2/s.

(**Answer:** 43.5 min)

10.14 Consider an unsteady-state diffusion of species A through a plane slab with the following initial and boundary conditions:

$$\frac{\partial c_A}{\partial t} = \mathcal{D}_{AB} \frac{\partial^2 c_A}{\partial z^2} \tag{1}$$

$$\text{at} \quad t = 0 \qquad c_A = 0 \tag{2}$$

$$\text{at} \quad z = 0 \qquad c_A = c_{A_o} \tag{3}$$

$$\text{at} \quad z = L \qquad c_A = 0 \tag{4}$$

a) In terms of the dimensionless quantities

$$\theta = \frac{c_{A_o} - c_A}{c_{A_o}} \qquad \xi = \frac{z}{L} \qquad \tau = \frac{\mathcal{D}_{AB}t}{L^2} \tag{5}$$

show that Eqs. (1)–(4) become

$$\frac{\partial \theta}{\partial \tau} = \frac{\partial^2 \theta}{\partial \xi^2} \tag{6}$$

$$\text{at} \quad \tau = 0 \qquad \theta = 1 \tag{7}$$

$$\text{at} \quad \xi = 0 \qquad \theta = 0 \tag{8}$$

$$\text{at} \quad \xi = 1 \qquad \theta = 1 \tag{9}$$

b) Since the boundary condition at $\xi = 1$ is not homogeneous, propose a solution in the form

$$\theta(\tau, \xi) = \theta_\infty(\xi) - \theta_t(\tau, \xi) \tag{10}$$

in which $\theta_\infty(\xi)$ is the steady-state solution, i.e.,

$$\frac{d^2 \theta_\infty}{d\xi^2} = 0 \tag{11}$$

with the following boundary conditions

$$\text{at} \quad \xi = 0 \qquad \theta_\infty = 0 \tag{12}$$

$$\text{at} \quad \xi = 1 \qquad \theta_\infty = 1 \tag{13}$$

Show that the steady-state solution is

$$\theta_\infty = \xi \tag{14}$$

c) Show that the governing equation for the transient contribution $\theta_t(\tau, \xi)$ is given by

$$\frac{\partial \theta_t}{\partial \tau} = \frac{\partial^2 \theta_t}{\partial \xi^2} \tag{15}$$

with the following initial and boundary conditions

$$\text{at} \quad \tau = 0 \qquad \theta_t = \xi - 1 \tag{16}$$

$$\text{at} \quad \xi = 0 \qquad \theta_t = 0 \tag{17}$$

$$\text{at} \quad \xi = 1 \qquad \theta_t = 0 \tag{18}$$

d) Use the method of separation of variables and obtain the transient solution as

$$\theta_t = -\frac{2}{\pi} \sum_{n=1}^{\infty} \frac{1}{n} \exp(-n^2 \pi^2 \tau) \sin(n \pi \xi) \tag{19}$$

10.15 In Section 10.3.1.2, the number of moles of species A transferred into the semi-infinite medium, n_A, is determined by integrating the molar transfer rate over time, i.e., Eq. (10.3-44). It is also possible to determine n_A from

$$n_A = A \int_0^\infty (c_A - c_{A_o}) \, ds \tag{1}$$

Show that the substitution of Eq. (10.3-41) into Eq. (1) and integration lead to Eq. (10.3-44).

10.16 Consider mass transfer into a rectangular slab for short times as described in Section 10.3.1.2. Start with Eq. (10.3-32) and show that the order of magnitude of the diffusion penetration depth is given by $\sqrt{\mathcal{D}_{AB} t}$.

10.17 A polymer sheet with the dimensions of $2 \times 50 \times 50$ mm is exposed to chloroform vapor at $20\,°C$ and 5 mmHg. The weight of the polymer sheet is recorded with the help of a sensitive electrobalance and the following data are obtained:

Time (h)	Weight of polymer sheet (g)
0	6.0000
54	6.0600
∞	6.1200

Assuming that the mass transport of chloroform in the polymer sheet is described by a Fickian-type diffusion process, estimate the diffusion coefficient of chloroform in the polymer sheet.

(**Answer:** 1.01×10^{-12} m^2/s)

10.18   Decarburization is the reversal of the process of carburization, i.e., removal of carbon (species $\mathcal{A}$) from a metal by diffusion. A thick steel plate (species $\mathcal{B}$) having a uniform carbon concentration of 0.6 wt% is decarburized in a vacuum at 950 °C. Estimate the time it takes for the carbon concentration at a depth of 1 mm below the surface to decrease to 0.3 wt%. Take $\mathcal{D}_{AB} = 2.1 \times 10^{-11}$ m^2/s.

(**Answer:** 14.5 h)

10.19   A dopant is an impurity added to a semiconductor in trace amounts to alter its electrical properties, producing n-type (negative) or p-type (positive) semiconductors. Boron is a common dopant for producing p-type semiconductors. For this purpose, one side of a silicon wafer is exposed to hot boron gas and boron atoms diffuse into the silicon. The process is stopped when the atoms reach a specified depth.

Boron (species $\mathcal{A}$) is to be diffused into a 1 mm thick silicon wafer (species $\mathcal{B}$) from one side for five hours at a temperature of 1000 °C. The surface concentration is constant at 3×10^{20} atoms/cm^3, and $\mathcal{D}_{AB} = 0.5 \times 10^{-18}$ m^2/s.

a) What is the diffusion penetration depth?
b) Estimate the boron concentration at a depth of 0.3 μm below the silicon surface.

(**Answer:** a) 0.38 μm   b) 7.6×10^{18} atoms/cm^3)

10.20   Consider gas absorption into a spherical liquid droplet as described in Section 10.3.4. If the liquid droplet is initially $\mathcal{A}$-free, show that the time required for the center concentration to reach 99% of the final concentration is given by

$$t \simeq \frac{0.42 R^2}{\mathcal{D}_{AB}}$$

10.21   The bottom of a large cylindrical tank is completely covered with a salt layer (species $\mathcal{A}$) of thickness L_o. At time $t = 0$, the tank is filled with pure water (species $\mathcal{B}$). The height of the water layer, H, is very large compared to the salt layer thickness, i.e., $H \gg L_o$.

a) Let z be the distance measured from the surface of the salt layer into the liquid phase and consider a differential volume element of thickness Δz in the liquid phase. If the dissolution of salt is diffusion-controlled, show that the conservation statement for salt is given by

$$\frac{\partial c_A}{\partial t} = \mathcal{D}_{AB} \frac{\partial^2 c_A}{\partial z^2} \tag{1}$$

with the following initial and boundary conditions

$$at \quad t=0 \qquad c_A = 0 \tag{2}$$

$$at \quad z=0 \qquad c_A = c_A^* \tag{3}$$

$$at \quad z \to \infty \qquad c_A = 0 \tag{4}$$

What is the physical significance of c_A^*? Note that in writing Eqs. (1)–(4) it is implicitly assumed that the dissolution process is quasi-steady, i.e., variation in the salt layer thickness as a result of dissolution is negligible.

b) Propose a solution of the form

$$\frac{c_A}{c_A^*} = f(\eta) \qquad \text{where} \qquad \eta = \frac{z}{\sqrt{4\mathcal{D}_{AB}t}} \tag{5}$$

and show that the concentration distribution is expressed as

$$\frac{c_A}{c_A^*} = 1 - \text{erf}\left(\frac{z}{\sqrt{4\mathcal{D}_{AB}t}}\right) \tag{6}$$

c) Show that the molar flux of salt from the surface is given by

$$N_{A_z}|_{z=0} = c_A^* \sqrt{\frac{\mathcal{D}_{AB}}{\pi t}} \tag{7}$$

d) Consider the salt layer as a system and show that the conservation statement for salt leads to the following expression for the thickness of the salt layer, $L(t)$, as a function of time

$$L(t) = L_o - \frac{2c_A^* \mathcal{M}_A}{\rho_A^s}\sqrt{\frac{\mathcal{D}_{AB}t}{\pi}} \tag{8}$$

where $\mathcal{M}_A$ and ρ_A^s are the molecular weight of salt and density of solid salt, respectively.

10.22 A solid sphere (species $\mathcal{A}$) of radius R is immersed in a stagnant liquid $\mathcal{B}$ of composition c_{A_o}. The solid is assumed to dissolve uniformly under isothermal conditions.

a) Consider a spherical differential volume element of thickness Δr in the liquid phase surrounding the particle. If the dissolution process is diffusion-controlled, show that the conservation statement for species $\mathcal{A}$ gives

$$\frac{\partial c_A}{\partial t} = \frac{\mathcal{D}_{AB}}{r^2}\frac{\partial}{\partial r}\left(r^2 \frac{\partial c_A}{\partial r}\right) \tag{1}$$

with the following initial and boundary conditions

$$at \quad t=0 \qquad c_A = c_{A_o} \tag{2}$$

$$at \quad r=R \qquad c_A = c_A^* \tag{3}$$

$$at \quad r=\infty \qquad c_A = c_{A_o} \tag{4}$$

What is the physical significance of c_A^*?

b) Rewrite Eqs. (1)–(4) in terms of the dimensionless variable θ defined by

$$\theta = \frac{c_A - c_{A_o}}{c_A^* - c_{A_o}} \tag{5}$$

c) Convert the spherical geometry to rectangular geometry by introducing a new dependent variable as

$$\theta = \frac{u}{r} \tag{6}$$

and show that Eqs. (1)–(4) take the form

$$\frac{\partial u}{\partial t} = \mathcal{D}_{AB} \frac{\partial^2 u}{\partial r^2} \tag{7}$$

$$\text{at} \quad t = 0 \qquad u = 0 \tag{8}$$

$$\text{at} \quad r = R \qquad u = R \tag{9}$$

$$\text{at} \quad r \to \infty \qquad u = 0 \tag{10}$$

d) Although the particle is dissolving, assume that the process is quasi-steady, i.e., variation in particle radius with time is negligible. Show that the use of the similarity transformation

$$u = u(\eta) \qquad \text{where} \qquad \eta = \frac{r - R}{\sqrt{4\mathcal{D}_{AB}t}} \tag{11}$$

reduces Eq. (7) to

$$\frac{d^2 u}{d\eta^2} + 2\eta \frac{du}{d\eta} = 0 \tag{12}$$

Note that the boundary conditions are $u = R$ for $\eta = 0$ and $u = 0$ for $\eta = \infty$.

e) Solve Eq. (12) and show that the concentration distribution is given by

$$\frac{c_A - c_{A_o}}{c_A^* - c_{A_o}} = \frac{R}{r}\left[1 - \text{erf}\left(\frac{r - R}{\sqrt{4\mathcal{D}_{AB}t}}\right)\right] \tag{13}$$

f) Show that the flux at the solid-fluid interface is given by

$$N_{A_r}|_{r=R} = \frac{\mathcal{D}_{AB}(c_A^* - c_{A_o})}{R}\left(1 + \frac{R}{\sqrt{\pi \mathcal{D}_{AB}t}}\right) \tag{14}$$

g) Consider the spherical solid particle as the system and show that the macroscopic mass balance leads to

$$-N_{A_r}|_{r=R} = \frac{\rho_A^s}{\mathcal{M}_A}\frac{dR}{dt} \tag{15}$$

where ρ_A^s and $\mathcal{M}_A$ are the solid density and the molecular weight of species $\mathcal{A}$, respectively.

h) Combine Eqs. (14) and (15) to obtain

$$-\phi\left(\frac{1}{\xi} + \frac{1}{\sqrt{\pi}}\frac{1}{\sqrt{\tau}}\right) = \frac{d\xi}{d\tau} \tag{16}$$

where

$$\phi = \frac{(c_A^* - c_{A_o})\mathcal{M}_A}{\rho_A^s} \qquad \xi = \frac{R}{R_o} \qquad \tau = \frac{\mathcal{D}_{AB}t}{R_o^2} \tag{17}$$

in which R_o is the initial radius of the spherical solid particle.

i) Making use of the substitution

$$X = \frac{\xi^2}{\tau} \tag{18}$$

show that Eq. (16) reduces to a separable first-order differential equation. Carry out the integrations and show that the time required for the complete dissolution of the sphere is given by

$$t = \frac{R_o^2}{2\mathcal{D}_{AB}\,\phi}\exp\left\{\frac{2\phi}{\sqrt{2\pi\phi - \phi^2}}\left[\tan^{-1}\left(\frac{\phi}{\sqrt{2\pi\phi - \phi^2}}\right) - \frac{\pi}{2}\right]\right\} \tag{19}$$

10.23 Microorganisms adhere to inert and/or living surfaces in moist environments by secreting a slimy, glue-like substance known as a *biofilm*. A typical example of a biofilm is the plaque that forms on your teeth.

a) The hypertextbook "Biofilms" (http://www.erc.montana.edu/biofilmbook) indicates that the time required for a solute to reach 90% of the bulk fluid concentration at the base of a flat slab biofilm is given by

$$t = 1.03\frac{L^2}{\mathcal{D}_{eff}} \tag{1}$$

where $\mathcal{D}_{eff}$ is the effective diffusion coefficient of solute in the biofilm. On the other hand, the time required for a solute to reach 90% of the bulk fluid concentration at the center of a spherical biofilm is given by

$$t = 0.31\frac{R^2}{\mathcal{D}_{eff}} \tag{2}$$

How can one derive these equations? What does L represent in Eq. (1)?

b) Suppose that there is hemispherical dental plaque with a radius of 250 μm on the surface of the tooth. How long must one rinse with mouthwash for the antimicrobial agent to penetrate to the tooth surface. Take $\mathcal{D}_{eff} = 0.75 \times 10^{-10}$ m^2/s.

(**Answer:** b) 258 s)

10.24 The effective use of a drug is achieved by its controlled release, and film-coated pellets, tablets, or capsules are used for this purpose. Consider the case in which a drug (species $\mathcal{A}$) is dissolved uniformly in a semispherical matrix with radii R_i and R_o as shown in the figure below.

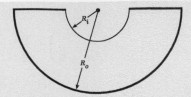

The thick lines in the figure represent the impermeable coating, and transfer of the drug takes place only through the surface of the tablet located at $r = R_i$. Let us assume that the drug concentration on this surface is zero. This is known as the *perfect sink* condition.

a) If the initial concentration of species $\mathcal{A}$, c_{A_o}, is below solubility, then the release of the drug is governed by diffusion. In terms of the following dimensionless quantities

$$\theta = \frac{c_A}{c_{A_o}} \qquad \kappa = \frac{R_i}{R_o} \qquad \xi = \frac{r}{R_o} \qquad \tau = \frac{\mathcal{D}_{AB} t}{R_o^2}$$

show that the governing equation, together with the initial and boundary conditions, takes the form

$$\frac{\partial \theta}{\partial \tau} = \frac{1}{\xi^2} \frac{\partial}{\partial \xi} \left(\xi^2 \frac{\partial \theta}{\partial \xi} \right) \tag{1}$$

$$\text{at} \quad \tau = 0 \qquad \theta = 1 \tag{2}$$

$$\text{at} \quad \xi = \kappa \qquad \theta = 0 \tag{3}$$

$$\text{at} \quad \xi = 1 \qquad \frac{\partial \theta}{\partial \xi} = 0 \tag{4}$$

b) To solve Eq. (1), use the transformation

$$\theta(\tau, \xi) = \frac{u(\tau, \xi)}{\xi} \tag{5}$$

and show that the solution is given by

$$\theta = \frac{2\kappa}{\xi} \sum_{n=1}^{\infty} \frac{1}{\lambda_n} \left[\frac{\lambda_n^2 + 1}{\lambda_n^2 (1 - \kappa) - \kappa} \right] \exp(-\lambda_n^2 \tau) \sin[\lambda_n(\xi - \kappa)] \tag{6}$$

where the eigenvalues are the positive roots of

$$\lambda_n = \tan[\lambda_n(1 - \kappa)] \tag{7}$$

c) Show that the molar flux of species on the tablet surface is given by

$$N_{A_r}\big|_{r=R_i} = \frac{2c_{A_o} \mathcal{D}_{AB}}{R_o} \sum_{n=1}^{\infty} \left[\frac{\lambda_n^2 + 1}{\lambda_n^2 (1 - \kappa) - \kappa} \right] \exp(-\lambda_n^2 \tau) \tag{8}$$

d) Show that the fractional release of the drug, F, is given by

$$F = 1 - \frac{6\kappa^2}{1-\kappa^3} \sum_{n=1}^{\infty} \left[\frac{\lambda_n^2 + 1}{\lambda_n^2(1-\kappa) - \kappa} \right] \frac{\exp(-\lambda_n^2 \tau)}{\lambda_n^2} \tag{9}$$

This problem was studied in detail by Siegel (2000).

10.25 Spherical particles of diameter 5 cm contain impurity $\mathcal{A}$ at a uniform concentration of c_{A_o}. Estimate the leaching time, i.e., the contact time of particles with a solvent, to reduce the species $\mathcal{A}$ content to 5% of its initial value. Take $\mathcal{D}_{AB} = 8.7 \times 10^{-9}$ m^2/s.

a) Assume that the external resistance to mass transfer is negligible, i.e., Bi$_M > 40$.
b) Assume that the average mass transfer coefficient at the surface of the particle is 6×10^{-6} m/s, and the (particle/solvent) partition coefficient of species $\mathcal{A}$ is 3.

(**Answer:** a) 5 h b) 8.1 h)

10.26 A solvent containing a small quantity of reactant $\mathcal{A}$ at concentration c_{A_o}, flows in plug flow through a tubular reactor of radius R. The inner surface of the tube is coated with a catalyst on which the reactant undergoes a first-order irreversible reaction.

a) Neglecting diffusion in the axial direction, show that the governing equation for the concentration of reactant is given by

$$v_o \frac{\partial c_A}{\partial z} = \frac{\mathcal{D}_{AB}}{r} \frac{\partial}{\partial r}\left(r \frac{\partial c_A}{\partial r} \right) \tag{1}$$

subject to the following boundary conditions

$$\text{at} \quad z = 0 \qquad c_A = c_{A_o} \tag{2}$$

$$\text{at} \quad r = 0 \qquad \frac{\partial c_A}{\partial r} = 0 \tag{3}$$

$$\text{at} \quad r = R \qquad -\mathcal{D}_{AB} \frac{\partial c_A}{\partial r} = k^s c_A \tag{4}$$

where v_o and k^s represent the plug flow velocity and the first-order surface reaction rate constant, respectively.
b) In terms of the following dimensionless quantities

$$\theta = \frac{c_A}{c_{A_o}} \qquad \tau = \frac{\mathcal{D}_{AB} z}{v_o R^2} \qquad \xi = \frac{r}{R} \qquad \Lambda = \frac{k^s R}{\mathcal{D}_{AB}} \tag{5}$$

show that Eqs. (1)–(4) take the form

$$\frac{\partial \theta}{\partial \tau} = \frac{1}{\xi} \frac{\partial}{\partial \xi}\left(\xi \frac{\partial \theta}{\partial \xi} \right) \tag{6}$$

$$\text{at} \quad \tau = 0 \qquad \theta = 1 \tag{7}$$

$$\text{at} \quad \xi = 0 \qquad \frac{\partial \theta}{\partial \xi} = 0 \tag{8}$$

$$\text{at} \quad \xi = 1 \qquad -\frac{\partial \theta}{\partial \xi} = \Lambda \theta \tag{9}$$

c) Note that Eqs. (6)–(9) are similar to Eqs. (10.2-99)–(10.2-102). Therefore, conclude that the solution is given by Eq. (10.2-119), i.e.,

$$\theta = 2\Lambda \sum_{n=1}^{\infty} \frac{1}{(\lambda_n^2 + \Lambda^2) J_o(\lambda_n)} \exp(-\lambda_n^2 \tau) J_o(\lambda_n \xi) \tag{10}$$

where the eigenvalues are the positive roots of

$$\lambda_n J_1(\lambda_n) = \Lambda J_o(\lambda_n) \tag{11}$$

11

UNSTEADY-STATE MICROSCOPIC BALANCES WITH GENERATION

This chapter briefly considers cases in which all the terms in the inventory rate equation are nonzero. The resulting governing equations for velocity, temperature, and concentration are non-homogeneous partial differential equations. Non-homogeneity may also be introduced by the initial and boundary conditions. Since the solutions are rather complicated, some representative examples will be included in this chapter.

11.1 MOMENTUM TRANSPORT

A horizontal tube of radius R is filled with a stationary incompressible Newtonian fluid as shown in Figure 11.1. At time $t = 0$, a constant pressure gradient is imposed and the fluid begins to flow. It is required to determine the development of velocity profile as a function of position and time.

Postulating $v_z = v_z(t, r)$ and $v_r = v_\theta = 0$, Table C.2 in Appendix C indicates that the only nonzero shear stress component is τ_{rz}, and the components of the total momentum flux are given by

$$\pi_{rz} = \tau_{rz} + (\rho v_z)v_r = \tau_{rz} = -\mu \frac{\partial v_z}{\partial r} \tag{11.1-1}$$

$$\pi_{\theta z} = \tau_{\theta z} + (\rho v_z)v_\theta = 0 \tag{11.1-2}$$

$$\pi_{zz} = \tau_{zz} + (\rho v_z)v_z = \rho v_z^2 \tag{11.1-3}$$

The conservation statement for momentum is expressed as

$$\begin{pmatrix} \text{Rate of} \\ \text{momentum in} \end{pmatrix} - \begin{pmatrix} \text{Rate of} \\ \text{momentum out} \end{pmatrix} + \begin{pmatrix} \text{Forces acting} \\ \text{on a system} \end{pmatrix} = \begin{pmatrix} \text{Rate of momentum} \\ \text{accumulation} \end{pmatrix} \tag{11.1-4}$$

Since the pressure in the pipe varies in the axial direction, it is necessary to consider only the z-component of the equation of motion. For a cylindrical differential volume element of thickness Δr and length Δz, as shown in Figure 11.1, Eq. (11.1-4) is expressed as

$$\left(\pi_{zz}|_z 2\pi r \Delta r + \pi_{rz}|_r 2\pi r \Delta z\right) - \left[\pi_{zz}|_{z+\Delta z} 2\pi r \Delta r + \pi_{rz}|_{r+\Delta r} 2\pi (r + \Delta r)\Delta z\right]$$

$$+ \left(P|_z - P|_{z+\Delta z}\right) 2\pi r \Delta r + 2\pi r \Delta r \Delta z \rho g = \frac{\partial}{\partial t}(2\pi r \Delta r \Delta z \rho v_z) \tag{11.1-5}$$

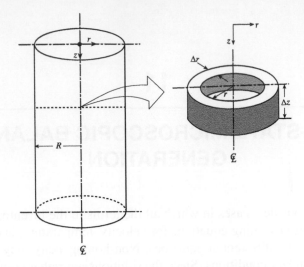

Figure 11.1. Unsteady-state flow in a circular pipe.

Dividing Eq. (11.1-5) by $2\pi\,\Delta r\,\Delta z$ and taking the limit as $\Delta r \to 0$ and $\Delta z \to 0$ give

$$\rho\,\frac{\partial v_z}{\partial t} = \lim_{\Delta z \to 0} \frac{P|_z - P|_{z+\Delta z}}{\Delta z} + \frac{1}{r}\lim_{\Delta r \to 0} \frac{(r\pi_{rz})|_r - (r\pi_{rz})|_{r+\Delta r}}{\Delta r}$$

$$+ \lim_{\Delta z \to 0} \frac{\pi_{zz}|_z - \pi_{zz}|_{z+\Delta z}}{\Delta z} + \rho g \tag{11.1-6}$$

or,

$$\rho\,\frac{\partial v_z}{\partial t} = -\frac{dP}{dz} - \frac{1}{r}\frac{\partial(r\pi_{rz})}{\partial r} - \frac{\partial \pi_{zz}}{\partial z} + \rho g \tag{11.1-7}$$

Substitution of Eqs. (11.1-1) and (11.1-3) into Eq. (11.1-7) and noting that $\partial v_z/\partial z = 0$ give

$$\rho\,\frac{\partial v_z}{\partial t} = -\frac{dP}{dz} + \frac{\mu}{r}\frac{\partial}{\partial r}\left(r\,\frac{\partial v_z}{\partial r}\right) + \rho g \tag{11.1-8}$$

The modified pressure is defined by

$$\mathcal{P} = P - \rho g z \tag{11.1-9}$$

so that

$$\frac{d\mathcal{P}}{dz} = \frac{dP}{dz} - \rho g \tag{11.1-10}$$

Substitution of Eq. (11.1-10) into Eq. (11.1-8) yields

$$\underbrace{\rho\,\frac{\partial v_z}{\partial t} - \frac{\mu}{r}\frac{\partial}{\partial r}\left(r\,\frac{\partial v_z}{\partial r}\right)}_{f(t,r)} = \underbrace{-\frac{d\mathcal{P}}{dz}}_{f(z)} \tag{11.1-11}$$

While the right-hand side of Eq. (11.1-11) is a function of z only, the left-hand side is dependent on r and t. This is possible if and only if both sides of Eq. (11.1-11) are equal to a constant, say λ. Hence,

$$-\frac{d\mathcal{P}}{dz} = \lambda \quad \Rightarrow \quad \lambda = \frac{\mathcal{P}_o - \mathcal{P}_L}{L} \tag{11.1-12}$$

where $\mathcal{P}_o$ and $\mathcal{P}_L$ are the values of $\mathcal{P}$ at $z = 0$ and $z = L$, respectively. Substitution of Eq. (11.1-12) into Eq. (11.1-11) gives the governing equation for velocity as

$$\boxed{\rho \frac{\partial v_z}{\partial t} = \frac{\mathcal{P}_o - \mathcal{P}_L}{L} + \frac{\mu}{r} \frac{\partial}{\partial r}\left(r \frac{\partial v_z}{\partial r} \right)} \tag{11.1-13}$$

The initial and the boundary conditions associated with Eq. (11.1-13) are

$$\text{at} \quad t = 0 \qquad v_z = 0 \tag{11.1-14}$$

$$\text{at} \quad r = 0 \qquad \frac{\partial v_z}{\partial r} = 0 \tag{11.1-15}$$

$$\text{at} \quad r = R \qquad v_z = 0 \tag{11.1-16}$$

11.1.1 Exact Solution

Introduction of the following dimensionless quantities

$$\theta = \frac{v_z}{\left(\dfrac{\mathcal{P}_o - \mathcal{P}_L}{4\mu L}\right)R^2} \qquad \xi = \frac{r}{R} \qquad \tau = \frac{\nu t}{R^2} \tag{11.1-17}$$

reduces Eqs. (11.1-13)–(11.1-16) to the form

$$\frac{\partial \theta}{\partial \tau} = 4 + \frac{1}{\xi} \frac{\partial}{\partial \xi}\left(\xi \frac{\partial \theta}{\partial \xi} \right) \tag{11.1-18}$$

$$\text{at} \quad \tau = 0 \qquad \theta = 0 \tag{11.1-19}$$

$$\text{at} \quad \xi = 0 \qquad \frac{\partial \theta}{\partial \xi} = 0 \tag{11.1-20}$$

$$\text{at} \quad \xi = 1 \qquad \theta = 0 \tag{11.1-21}$$

Since Eq. (11.1-18) is not homogeneous, the solution is proposed in the form

$$\theta(\tau, \xi) = \theta_\infty(\xi) - \theta_t(\tau, \xi) \tag{11.1-22}$$

in which θ_∞ is the steady-state solution, i.e.,

$$0 = 4 + \frac{1}{\xi} \frac{d}{d\xi}\left(\xi \frac{d\theta_\infty}{d\xi} \right) \tag{11.1-23}$$

with the following boundary conditions

$$
\text{at} \quad \xi = 0 \qquad \frac{d\theta_\infty}{d\xi} = 0 \tag{11.1-24}
$$

$$
\text{at} \quad \xi = 1 \qquad \theta_\infty = 0 \tag{11.1-25}
$$

The solution of Eq. (11.1-23) is

$$
\theta_\infty = 1 - \xi^2 \tag{11.1-26}
$$

which is identical to Eq. (9.1-79).

The use of Eq. (11.1-26) in Eq. (11.1-22) gives

$$
\theta(\tau, \xi) = 1 - \xi^2 - \theta_t(\tau, \xi) \tag{11.1-27}
$$

Substitution of Eq. (11.1-27) into Eqs. (11.1-18)–(11.1-21) leads to the following governing equation for the transient problem, together with the initial and boundary conditions

$$
\frac{\partial \theta_t}{\partial \tau} = \frac{1}{\xi} \frac{\partial}{\partial \xi} \left(\xi \frac{\partial \theta_t}{\partial \xi} \right) \tag{11.1-28}
$$

$$
\text{at} \quad \tau = 0 \qquad \theta_t = 1 - \xi^2 \tag{11.1-29}
$$

$$
\text{at} \quad \xi = 0 \qquad \frac{\partial \theta_t}{\partial \xi} = 0 \tag{11.1-30}
$$

$$
\text{at} \quad \xi = 1 \qquad \theta_t = 0 \tag{11.1-31}
$$

Representing the solution as a product of two functions of the form

$$
\theta_t(\tau, \xi) = F(\tau) G(\xi) \tag{11.1-32}
$$

reduces Eq. (11.1-28) to

$$
\frac{1}{F} \frac{dF}{d\tau} = \frac{1}{G\xi} \frac{d}{d\xi} \left(\xi \frac{dG}{d\xi} \right) = -\lambda^2 \tag{11.1-33}
$$

which results in two ordinary differential equations:

$$
\frac{dF}{d\tau} + \lambda^2 F = 0 \quad \Rightarrow \quad F(\tau) = e^{-\lambda^2 \tau} \tag{11.1-34}
$$

$$
\frac{d}{d\xi} \left(\xi \frac{dG}{d\xi} \right) + \lambda^2 \xi G = 0 \quad \Rightarrow \quad G(\xi) = C_1 J_o(\lambda \xi) + C_2 Y_o(\lambda \xi) \tag{11.1-35}
$$

The boundary conditions for $G(\xi)$ are

$$
\text{at} \quad \xi = 0 \qquad \frac{dG}{d\xi} = 0 \tag{11.1-36}
$$

$$
\text{at} \quad \xi = 1 \qquad G = 0 \tag{11.1-37}
$$

Since $Y_o(0) = -\infty$, $C_2 = 0$. Application of Eq. (11.1-37) gives

$$\boxed{J_o(\lambda_n) = 0} \qquad n = 1, 2, 3, \ldots \tag{11.1-38}$$

Therefore, the transient solution is

$$\theta_t = \sum_{n=1}^{\infty} A_n \exp\left(-\lambda_n^2 \tau\right) J_o(\lambda_n \xi) \tag{11.1-39}$$

The unknown coefficients A_n can be determined by using the initial condition given by Eq. (11.1-29). The result is

$$\int_0^1 \xi J_o(\lambda_n \xi) \, d\xi - \int_0^1 \xi^3 J_o(\lambda_n \xi) \, d\xi = A_n \int_0^1 \xi J_o^2(\lambda_n \xi) \, d\xi \tag{11.1-40}$$

Evaluation of the integrals with the help of Eqs. (B.2-30)–(B.2-32) in Appendix B gives

$$\int_0^1 \xi J_o(\lambda_n \xi) \, d\xi = \frac{J_1(\lambda_n)}{\lambda_n} \tag{11.1-41}$$

$$\int_0^1 \xi^3 J_o(\lambda_n \xi) \, d\xi = \left(\frac{1}{\lambda_n} - \frac{4}{\lambda_n^3} \right) J_1(\lambda_n) \tag{11.1-42}$$

$$\int_0^1 \xi J_o^2(\lambda_n \xi) \, d\xi = \frac{1}{2} J_1^2(\lambda_n) \tag{11.1-43}$$

Substitution of Eqs. (11.1-41)–(11.1-43) into Eq. (11.1-40) leads to

$$A_n = \frac{8}{\lambda_n^3 J_1(\lambda_n)} \tag{11.1-44}$$

Hence, the solution is expressed as

$$\boxed{\theta = 1 - \xi^2 - 8 \sum_{n=1}^{\infty} \frac{1}{\lambda_n^3 J_1(\lambda_n)} \exp\left(-\lambda_n^2 \tau\right) J_o(\lambda_n \xi)} \tag{11.1-45}$$

The volumetric flow rate can be determined by integrating the velocity distribution over the cross-sectional area of the tube, i.e.,

$$\mathcal{Q} = \int_0^{2\pi} \int_0^R v_z \, r \, dr \, d\theta \tag{11.1-46}$$

Substitution of Eq. (11.1-45) into Eq. (11.1-46) gives

$$\boxed{\mathcal{Q} = \frac{\pi (\mathcal{P}_o - \mathcal{P}_L) R^4}{8 \mu L} \left[1 - 32 \sum_{n=1}^{\infty} \frac{1}{\lambda_n^4} \exp\left(-\lambda_n^2 \tau\right) \right]} \tag{11.1-47}$$

Note that, when $\tau \to \infty$, $\mathcal{Q} \to \pi(\mathcal{P}_o - \mathcal{P}_L)R^4/8\mu L$, which is identical to Eq. (9.1-83).

11.1.2 Approximate Solution by the Area Averaging Technique

It should be kept in mind that the purpose of obtaining the velocity distribution is to establish a relationship between the volumetric flow rate and the pressure drop in order to estimate the power required to pump the fluid.

The area averaging technique[1] enables one to calculate the average velocity, and hence the volumetric flow rate, without determining the velocity distribution. Multiplication of Eq. (11.1-13) by $r \, dr \, d\theta$ and integration over the cross-sectional area of the pipe give

$$\rho \int_0^{2\pi} \int_0^R \frac{\partial v_z}{\partial t} r \, dr \, d\theta = \int_0^{2\pi} \int_0^R \left(\frac{\mathcal{P}_o - \mathcal{P}_L}{L} \right) r \, dr \, d\theta + \int_0^{2\pi} \int_0^R \frac{\mu}{r} \frac{\partial}{\partial r} \left(r \frac{\partial v_z}{\partial r} \right) r \, dr \, d\theta \tag{11.1-48}$$

The term on the left-hand side of Eq. (11.1-48) can be rearranged in the form

$$\rho \int_0^{2\pi} \int_0^R \frac{\partial v_z}{\partial t} r \, dr \, d\theta = \rho \left(\frac{d}{dt} \underbrace{\int_0^{2\pi} \int_0^R v_z r \, dr \, d\theta}_{\pi R^2 \langle v_z \rangle} \right) = \rho \pi R^2 \frac{d \langle v_z \rangle}{dt} \tag{11.1-49}$$

Therefore, Eq. (11.1-48) becomes

$$\rho \pi R^2 \frac{d \langle v_z \rangle}{dt} = \pi R^2 \left(\frac{\mathcal{P}_o - \mathcal{P}_L}{L} \right) + 2\pi \mu R \frac{\partial v_z}{\partial r} \bigg|_{r=R} \tag{11.1-50}$$

Note that the area averaging technique transforms a partial differential equation into an ordinary differential equation. However, one has to pay the price for this simplification. That is, to proceed further, it is necessary to express the velocity gradient at the wall, $(\partial v_z / \partial r)_{r=R}$, in terms of the average velocity, $\langle v_z \rangle$. If it is assumed that the velocity gradient at the wall is approximately equal to that for the steady-state case, from Eqs. (9.1-79) and (9.1-84)

$$\frac{\partial v_z}{\partial r} \bigg|_{r=R} = -\frac{4 \langle v_z \rangle}{R} \tag{11.1-51}$$

Substitution of Eq. (11.1-51) into Eq. (11.1-50) yields the following linear ordinary differential equation

$$\frac{d \langle v_z \rangle}{dt} + \frac{8\nu}{R^2} \langle v_z \rangle = \frac{1}{\rho} \left(\frac{\mathcal{P}_o - \mathcal{P}_L}{L} \right) \tag{11.1-52}$$

The initial condition associated with Eq. (11.1-52) is

$$\text{at} \quad t = 0 \qquad \langle v_z \rangle = 0 \tag{11.1-53}$$

The integrating factor is

$$\text{Integrating factor} = \exp \left(\frac{8\nu t}{R^2} \right) \tag{11.1-54}$$

---

[1]This development is taken from Slattery (1972).

Multiplication of Eq. (11.1-52) by the integrating factor and rearrangement give

$$\frac{d}{dt}\left[\langle v_z\rangle \exp\left(\frac{8vt}{R^2}\right)\right] = \frac{1}{\rho}\left(\frac{\mathcal{P}_o - \mathcal{P}_L}{L}\right)\exp\left(\frac{8vt}{R^2}\right) \tag{11.1-55}$$

Integration of Eq. (11.1-55) leads to

$$\langle v_z\rangle = \frac{(\mathcal{P}_o - \mathcal{P}_L)R^2}{8\mu L}\left[1 - \exp\left(\frac{8vt}{R^2}\right)\right] \tag{11.1-56}$$

Therefore, the volumetric flow rate is

$$\boxed{\mathcal{Q} = \frac{\pi(\mathcal{P}_o - \mathcal{P}_L)R^4}{8\mu L}\left[1 - \exp(8\tau)\right]} \tag{11.1-57}$$

Slattery (1972) compared Eq. (11.1-57) with the exact solution, Eq. (11.1-47), and concluded that the error introduced is less than 20% when $\tau > 0.05$.

11.2 ENERGY TRANSPORT

The conservation statement for energy is expressed as

$$\begin{pmatrix} \text{Rate of} \\ \text{energy in} \end{pmatrix} - \begin{pmatrix} \text{Rate of} \\ \text{energy in} \end{pmatrix} + \begin{pmatrix} \text{Rate of energy} \\ \text{generation} \end{pmatrix} = \begin{pmatrix} \text{Rate of energy} \\ \text{accumulation} \end{pmatrix} \tag{11.2-1}$$

11.2.1 Rectangular Geometry

Consider a slab of thickness $2L$ with a uniform initial temperature of T_o. At $t = 0$, heat starts to generate within the slab at a uniform rate of $\Re$ (W/m^3) and, to avoid excessive heating of the slab, the surfaces at $z = \pm L$ are exposed to a fluid at constant temperature T_∞ ($T_\infty < T_o$) as shown in Figure 11.2. Let us assume Bi$_\text{H}$ > 40 so that the slab surfaces are

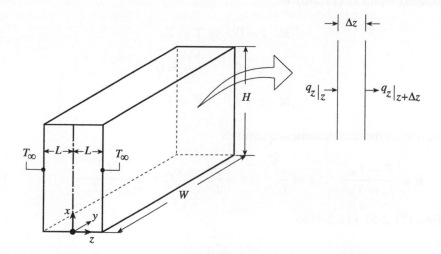

Figure 11.2. Unsteady-state conduction in a slab with generation.

also at temperature T_∞. We are interested in the temperature distribution within the slab as a function of position and time.

If $L/H \ll 1$ and $L/W \ll 1$, then it is possible to assume that the conduction is one-dimensional and to postulate that $T = T(t, z)$. In that case, Table C.4 in Appendix C indicates that the only nonzero energy flux component is e_z, and it is given by

$$e_z = q_z = -k\frac{\partial T}{\partial z} \qquad (11.2\text{-}2)$$

For a rectangular differential volume element of thickness Δz, as shown in Figure 11.2, Eq. (11.2-1) is expressed as

$$q_z|_z A - q_z|_{z+\Delta z} A + A\Delta z \Re = \frac{\partial}{\partial t}\left[A\Delta z \rho \widehat{C}_P (T - T_o) \right] \qquad (11.2\text{-}3)$$

Dividing Eq. (11.2-3) by $A\Delta z$ and taking the limit as $\Delta z \to 0$ give

$$\rho \widehat{C}_P \frac{\partial T}{\partial t} = \lim_{\Delta z \to 0} \frac{q_z|_z - q_z|_{z+\Delta z}}{\Delta z} + \Re \qquad (11.2\text{-}4)$$

or,

$$\rho \widehat{C}_P \frac{\partial T}{\partial t} = -\frac{\partial q_z}{\partial z} + \Re \qquad (11.2\text{-}5)$$

Substitution of Eq. (11.2-2) into Eq. (11.2-5) gives the governing equation for temperature as

$$\boxed{\rho \, \widehat{C}_P \frac{\partial T}{\partial t} = k\frac{\partial^2 T}{\partial z^2} + \Re} \qquad (11.2\text{-}6)$$

in which all physical properties are assumed to be constant. The initial and boundary conditions associated with Eq. (11.2-6) are

$$\text{at} \quad t = 0 \qquad T = T_o \qquad (11.2\text{-}7)$$

$$\text{at} \quad z = 0 \qquad \frac{\partial T}{\partial z} = 0 \qquad (11.2\text{-}8)$$

$$\text{at} \quad z = L \qquad T = T_\infty \qquad (11.2\text{-}9)$$

Introduction of the dimensionless quantities

$$\theta = \frac{T - T_\infty}{T_o - T_\infty} \qquad \xi = \frac{z}{L} \qquad \tau = \frac{\alpha t}{L^2} \qquad \Omega = \frac{\Re L^2}{k(T_o - T_\infty)} \qquad (11.2\text{-}10)$$

reduces Eqs. (11.2-6)–(11.2-9) to

$$\frac{\partial \theta}{\partial \tau} = \frac{\partial^2 \theta}{\partial \xi^2} + \Omega \qquad (11.2\text{-}11)$$

$$\text{at} \quad \tau = 0 \qquad \theta = 1 \tag{11.2-12}$$

$$\text{at} \quad \xi = 0 \qquad \frac{\partial \theta}{\partial \xi} = 0 \tag{11.2-13}$$

$$\text{at} \quad \xi = 1 \qquad \theta = 0 \tag{11.2-14}$$

Since Eq. (11.2-11) is not homogeneous, the solution is proposed in the form

$$\theta(\tau, \xi) = \theta_\infty(\xi) - \theta_t(\tau, \xi) \tag{11.2-15}$$

in which θ_∞ is the steady-state solution, i.e.,

$$\frac{d^2 \theta_\infty}{d\xi^2} + \Omega = 0 \tag{11.2-16}$$

with the following boundary conditions

$$\text{at} \quad \xi = 0 \qquad \frac{d\theta_\infty}{d\xi} = 0 \tag{11.2-17}$$

$$\text{at} \quad \xi = 1 \qquad \theta_\infty = 0 \tag{11.2-18}$$

The solution of Eq. (11.2-16) is

$$\theta_\infty = \frac{\Omega}{2}(1 - \xi^2) \tag{11.2-19}$$

The use of Eq. (11.2-19) in Eq. (11.2-15) gives

$$\theta(\tau, \xi) = \frac{\Omega}{2}(1 - \xi^2) - \theta_t(\tau, \xi) \tag{11.2-20}$$

Substitution of Eq. (11.2-20) into Eqs. (11.2-11)–(11.2-14) leads to the following governing equation for the transient problem, together with the initial and boundary conditions

$$\frac{\partial \theta_t}{\partial \tau} = \frac{\partial^2 \theta_t}{\partial \xi^2} \tag{11.2-21}$$

$$\text{at} \quad \tau = 0 \qquad \theta_t = \frac{\Omega}{2}(1 - \xi^2) - 1 \tag{11.2-22}$$

$$\text{at} \quad \xi = 0 \qquad \frac{\partial \theta_t}{\partial \xi} = 0 \tag{11.2-23}$$

$$\text{at} \quad \xi = 1 \qquad \theta_t = 0 \tag{11.2-24}$$

The solution of Eq. (11.2-21) by the method of separation of variables gives

$$\theta_t = \sum_{n=0}^{\infty} A_n \exp\left[-\left(n + \tfrac{1}{2}\right)^2 \pi^2 \tau\right] \cos\left[\left(n + \tfrac{1}{2}\right)\pi\xi\right] \tag{11.2-25}$$

The unknown coefficients A_n can be determined by using the initial condition, Eq. (11.2-22), with the result

$$A_n = \frac{\int_0^1 \left[\frac{\Omega}{2}(1 - \xi^2) - 1 \right] \cos\left[(n + \frac{1}{2})\pi\xi\right] d\xi}{\int_0^1 \cos^2\left[(n + \frac{1}{2})\pi\xi\right] d\xi} \tag{11.2-26}$$

Evaluation of the integrals gives

$$A_n = \frac{2(-1)^n}{(n + \frac{1}{2})\pi} \left[\frac{\Omega}{(n + \frac{1}{2})^2 \pi^2} - 1 \right] \tag{11.2-27}$$

Therefore, the solution is given by

$$\theta = \frac{\Omega}{2}(1 - \xi^2) + \frac{2}{\pi} \sum_{n=0}^{\infty} \frac{(-1)^n}{(n + \frac{1}{2})} \left[1 - \frac{\Omega}{(n + \frac{1}{2})^2 \pi^2} \right] \exp\left[-(n + \frac{1}{2})^2 \pi^2 \tau\right] \cos\left[(n + \frac{1}{2})\pi\xi\right] \tag{11.2-28}$$

When there is no internal generation, i.e., $\Omega = 0$, Eq. (11.2-28) reduces to Eq. (10.2-36).

11.2.1.1 *Macroscopic equation* Integration of the governing equation for temperature, Eq. (11.2-6), over the volume of the system gives

$$\int_{-L}^{L} \int_0^W \int_0^H \rho \widehat{C}_P \frac{\partial T}{\partial t} \, dx \, dy \, dz = \int_{-L}^{L} \int_0^W \int_0^H k \frac{\partial^2 T}{\partial z^2} \, dx \, dy \, dz + \int_{-L}^{L} \int_0^W \int_0^H \Re \, dx \, dy \, dz \tag{11.2-29}$$

or,

$$\underbrace{\frac{d}{dt}\left[\int_{-L}^{L} \int_0^W \int_0^H \rho \widehat{C}_P (T - T_o) \, dx \, dy \, dz \right]}_{\text{Rate of accumulation of energy}} = \underbrace{-2WH\left(-k \frac{\partial T}{\partial z}\Big|_{z=L} \right)}_{\substack{\text{Rate of energy leaving} \\ \text{from surfaces at } z = \pm L}} + \underbrace{\Re(2WHL)}_{\substack{\text{Rate of energy} \\ \text{generation}}} \tag{11.2-30}$$

which is the macroscopic energy balance by considering the rectangular slab as a system. The rate of energy leaving the slab, $\dot{Q}$, is given by

$$\dot{Q} = 2WH\left(-k \frac{\partial T}{\partial z}\Big|_{z=L} \right) = -\frac{2WHk(T_o - T_\infty)}{L} \frac{\partial \theta}{\partial \xi}\Big|_{\xi=1} \tag{11.2-31}$$

The use of Eq. (11.2-28) in Eq. (11.2-31) gives

$$\dot{Q} = \frac{2WHk(T_o - T_\infty)}{L} \left\{ \Omega + 2\sum_{n=0}^{\infty} \left[1 - \frac{\Omega}{(n + \frac{1}{2})^2 \pi^2} \right] \exp\left[-(n + \frac{1}{2})^2 \pi^2 \tau\right] \right\} \tag{11.2-32}$$

When $\Omega = 0$, Eq. (11.2-32) reduces to Eq. (10.2-40). On the other hand, under steady conditions, i.e., $\tau \to \infty$, $\dot{Q} \to 2WHL\Re$, indicating that the rate of heat loss from the system equals the rate of internal generation of heat.

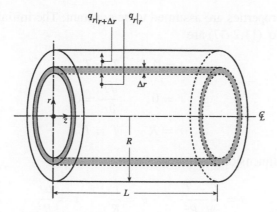

Figure 11.3. Unsteady-state conduction in a cylinder with generation.

11.2.2 Cylindrical Geometry

A cylindrical rod of radius R is initially at a uniform temperature of T_o. At time $t = 0$, a switch is turned on and heat starts to generate uniformly at a rate $\Re_e$ (W/m^3) as a result of the electric current passing through the rod. The outer surface of the rod is maintained constant at T_o to avoid excessive heating. The geometry of the system is shown in Figure 11.3 and we are interested in obtaining the temperature distribution within the rod as a function of position and time.

If $R/L \ll 1$, then it is possible to assume that the conduction is one-dimensional and to postulate that $T = T(t, r)$. In that case, Table C.5 in Appendix C indicates that the only nonzero energy flux component is e_r, and it is given by

$$e_r = q_r = -k \frac{\partial T}{\partial r} \tag{11.2-33}$$

For a cylindrical differential volume element of thickness Δr, as shown in Figure 11.3, the conservation statement given by Eq. (11.2-1) is expressed as

$$q_r|_r 2\pi r L - q_r|_{r+\Delta r} 2\pi (r + \Delta r) L + 2\pi r L \Delta r \Re_e = \frac{\partial}{\partial t} \left[2\pi r L \Delta r \rho \widehat{C}_P (T - T_o) \right] \tag{11.2-34}$$

Dividing Eq. (11.2-34) by $2\pi L \Delta r$ and letting $\Delta r \to 0$ give

$$\rho \widehat{C}_P \frac{\partial T}{\partial t} = \frac{1}{r} \lim_{\Delta r \to 0} \frac{(rq_r)|_r - (rq_r)|_{r+\Delta r}}{\Delta r} + \Re_e \tag{11.2-35}$$

or,

$$\rho \widehat{C}_P \frac{\partial T}{\partial t} = -\frac{1}{r} \frac{\partial (rq_r)}{\partial r} + \Re_e \tag{11.2-36}$$

Substitution of Eq. (11.2-33) into Eq. (11.2-36) gives the governing equation for temperature as

$$\boxed{\rho \widehat{C}_P \frac{\partial T}{\partial t} = \frac{k}{r} \frac{\partial}{\partial r} \left(r \frac{\partial T}{\partial r} \right) + \Re_e} \tag{11.2-37}$$

in which all physical properties are assumed to be constant. The initial and boundary conditions associated with Eq. (11.2-37) are

$$\text{at} \quad t = 0 \qquad T = T_o \tag{11.2-38}$$

$$\text{at} \quad r = 0 \qquad \frac{\partial T}{\partial r} = 0 \tag{11.2-39}$$

$$\text{at} \quad r = R \qquad T = T_o \tag{11.2-40}$$

Introduction of the dimensionless quantities

$$\theta = \frac{T - T_o}{\dfrac{\Re_e R^2}{4k}} \qquad \xi = \frac{r}{R} \qquad \tau = \frac{\alpha t}{R^2} \tag{11.2-41}$$

reduces Eqs. (11.2-37)–(11.2-40) to

$$\frac{\partial \theta}{\partial \tau} = \frac{1}{\xi} \frac{\partial}{\partial \xi} \left(\xi \frac{\partial \theta}{\partial \xi} \right) + 4 \tag{11.2-42}$$

$$\text{at} \quad \tau = 0 \qquad \theta = 0 \tag{11.2-43}$$

$$\text{at} \quad \xi = 0 \qquad \frac{\partial \theta}{\partial \xi} = 0 \tag{11.2-44}$$

$$\text{at} \quad \xi = 1 \qquad \theta = 0 \tag{11.2-45}$$

Note that Eqs. (11.2-42)–(11.2-45) are similar to Eqs. (11.1-18)–(11.1-21). Therefore, the solution is given by Eq. (11.1-45), i.e.,

$$\boxed{\theta = 1 - \xi^2 - 8 \sum_{n=1}^{\infty} \frac{1}{\lambda_n^3 J_1(\lambda_n)} \exp\left(-\lambda_n^2 \tau\right) J_o(\lambda_n \xi)} \tag{11.2-46}$$

When $\tau \to \infty$, Eq. (11.2-46) reduces to

$$\theta = 1 - \xi^2 \quad \Rightarrow \quad T = T_o + \frac{\Re_e R^2}{4k} \left[1 - \left(\frac{r}{R} \right)^2 \right] \tag{11.2-47}$$

which is identical to Eq. (9.2-45) when $T_R = T_o$ and $\Re = \Re_e$.

11.2.2.1 *Macroscopic equation* Integration of the governing equation for temperature, Eq. (11.2-37), over the volume of the system gives

$$\int_0^L \int_0^{2\pi} \int_0^R \rho \widehat{C}_P \frac{\partial T}{\partial t} r \, dr \, d\theta \, dz = \int_0^L \int_0^{2\pi} \int_0^R \frac{k}{r} \frac{\partial}{\partial r} \left(r \frac{\partial T}{\partial r} \right) r \, dr \, d\theta \, dz$$

$$+ \int_0^L \int_0^{2\pi} \int_0^R \Re_e \, r \, dr \, d\theta \, dz \tag{11.2-48}$$

or,

$$\frac{d}{dt}\left[\int_0^L \int_0^{2\pi} \int_0^R \rho \widehat{C}_P (T - T_o) r\, dr\, d\theta\, dz\right] = \underbrace{-2\pi RL\left(-k\frac{\partial T}{\partial r}\bigg|_{r=R}\right)}_{\substack{\text{Rate of energy leaving} \\ \text{from the lateral surface}}} + \underbrace{\Re_e \pi R^2 L}_{\substack{\text{Rate of energy} \\ \text{generation}}}$$

$$\underbrace{\phantom{\frac{d}{dt}\left[\int_0^L \int_0^{2\pi} \int_0^R \rho \widehat{C}_P (T - T_o) r\, dr\, d\theta\, dz\right]}}_{\text{Rate of accumulation of energy}}$$

(11.2-49)

which is the macroscopic energy balance by considering the rod as a system. The rate of energy leaving from the lateral surface, $\dot{Q}$, is given by

$$\dot{Q} = 2\pi RL\left(-k\frac{\partial T}{\partial r}\bigg|_{r=R}\right) = -\frac{\pi R^2 L \Re_e}{2}\frac{\partial \theta}{\partial \xi}\bigg|_{\xi=1} \tag{11.2-50}$$

The use of Eq. (11.2-46) in Eq. (11.2-50) gives

$$\dot{Q} = \pi R^2 L \Re_e\left[1 - 4\sum_{n=1}^{\infty}\frac{1}{\lambda_n^2}\exp(-\lambda_n^2 \tau)\right] \tag{11.2-51}$$

Under steady conditions, i.e., $\tau \to \infty$, $\dot{Q} \to \pi R^2 L \Re_e$, indicating that the rate of heat loss from the system equals the rate of internal generation of heat.

11.2.3 Spherical Geometry

A spherical nuclear fuel element of radius R is initially at a uniform temperature of T_o. At $t = 0$, energy is generated within the sphere at a uniform rate of $\Re$ (W/m^3). The outside surface temperature is kept constant at T_∞ by a coolant ($T_\infty < T_o$). We are interested in the temperature distribution within the sphere as a function of position and time.

Since heat transfer takes place in the r-direction, Table C.6 in Appendix C indicates that the only nonzero energy flux component is e_r, and it is given by

$$e_r = q_r = -k\frac{\partial T}{\partial r} \tag{11.2-52}$$

For a spherical differential volume of thickness Δr, as shown in Figure 11.4, Eq. (11.2-1) is expressed as

$$q_r|_r 4\pi r^2 - q_r|_{r+\Delta r} 4\pi (r + \Delta r)^2 + 4\pi r^2 \Delta r \, \Re = \frac{\partial}{\partial t}\left[4\pi r^2 \Delta r \rho \widehat{C}_P (T - T_o)\right] \tag{11.2-53}$$

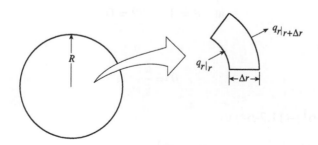

Figure 11.4. Unsteady-state conduction in a sphere with generation.

Dividing Eq. (11.2-53) by $4\pi\,\Delta r$ and letting $\Delta r \to 0$ give

$$\rho\widehat{C}_P \frac{\partial T}{\partial t} = \frac{1}{r^2}\lim_{\Delta r \to 0}\frac{(r^2 q_r)|_r - (r^2 q_r)|_{r+\Delta r}}{\Delta r} + \Re \tag{11.2-54}$$

or,

$$\rho\widehat{C}_P \frac{\partial T}{\partial t} = -\frac{1}{r^2}\frac{\partial(r^2 q_r)}{\partial r} + \Re \tag{11.2-55}$$

Substitution of Eq. (11.2-52) into Eq. (11.2-55) gives the governing differential equation for temperature as

$$\boxed{\rho\widehat{C}_P \frac{\partial T}{\partial t} = \frac{k}{r^2}\frac{\partial}{\partial r}\left(r^2 \frac{\partial T}{\partial r}\right) + \Re} \tag{11.2-56}$$

The initial and boundary conditions associated with Eq. (11.2-56) are

$$\text{at} \quad t = 0 \qquad T = T_o \tag{11.2-57}$$

$$\text{at} \quad r = 0 \qquad \frac{\partial T}{\partial r} = 0 \tag{11.2-58}$$

$$\text{at} \quad r = R \qquad T = T_\infty \tag{11.2-59}$$

Introduction of the dimensionless quantities

$$\theta = \frac{T - T_\infty}{T_o - T_\infty} \qquad \xi = \frac{r}{R} \qquad \tau = \frac{\alpha t}{R^2} \qquad \Omega = \frac{\Re R^2}{k(T_o - T_\infty)} \tag{11.2-60}$$

reduces Eqs. (11.2-56)–(11.2-59) to

$$\frac{\partial\theta}{\partial\tau} = \frac{1}{\xi^2}\frac{\partial}{\partial\xi}\left(\xi^2\frac{\partial\theta}{\partial\xi}\right) + \Omega \tag{11.2-61}$$

$$\text{at} \quad \tau = 0 \qquad \theta = 1 \tag{11.2-62}$$

$$\text{at} \quad \xi = 0 \qquad \frac{\partial\theta}{\partial\xi} = 0 \tag{11.2-63}$$

$$\text{at} \quad \xi = 1 \qquad \theta = 0 \tag{11.2-64}$$

The transformation

$$\theta = \frac{u}{\xi} \tag{11.2-65}$$

reduces Eqs. (11.2-61)–(11.2-64) to

$$\frac{\partial u}{\partial\tau} = \frac{\partial^2 u}{\partial\xi^2} + \Omega\xi \tag{11.2-66}$$

$$\text{at} \quad \tau = 0 \qquad u = \xi \tag{11.2-67}$$

$$\text{at} \quad \xi = 0 \qquad u = 0 \tag{11.2-68}$$

$$\text{at} \quad \xi = 1 \qquad u = 0 \tag{11.2-69}$$

Since Eq. (11.2-66) is not homogeneous, the solution is proposed in the form

$$u(\tau, \xi) = u_\infty(\xi) - u_t(\tau, \xi) \tag{11.2-70}$$

in which u_∞ is the steady-state solution, i.e.,

$$\frac{d^2 u_\infty}{d\xi^2} + \Omega\xi = 0 \tag{11.2-71}$$

with the following boundary conditions

$$\text{at} \quad \xi = 0 \qquad u_\infty = 0 \tag{11.2-72}$$

$$\text{at} \quad \xi = 1 \qquad u_\infty = 0 \tag{11.2-73}$$

The solution of Eq. (11.2-71) is

$$u_\infty = \frac{\Omega}{6}(\xi - \xi^3) \tag{11.2-74}$$

The use of Eq. (11.2-74) in Eq. (11.2-70) gives

$$u(\tau, \xi) = \frac{\Omega}{6}(\xi - \xi^3) - u_t(\tau, \xi) \tag{11.2-75}$$

Substitution of Eq. (11.2-75) into Eqs. (11.2-66)–(11.2-69) leads to the following governing equation for the transient problem, together with the initial and boundary conditions

$$\frac{\partial u_t}{\partial \tau} = \frac{\partial^2 u_t}{\partial \xi^2} \tag{11.2-76}$$

$$\text{at} \quad \tau = 0 \qquad u_t = \frac{\Omega}{6}(\xi - \xi^3) - \xi \tag{11.2-77}$$

$$\text{at} \quad \xi = 0 \qquad u_t = 0 \tag{11.2-78}$$

$$\text{at} \quad \xi = 1 \qquad u_t = 0 \tag{11.2-79}$$

The solution of Eq. (11.2-76) by the method of separation of variables is straightforward and given by

$$u_t = \sum_{n=1}^{\infty} A_n \exp(-n^2\pi^2\tau) \sin(n\pi\xi) \tag{11.2-80}$$

The unknown coefficients A_n can be determined by using the initial condition given by Eq. (11.2-77). The result is

$$\int_0^1 \left[\frac{\Omega}{6}(\xi - \xi^3) - \xi\right] \sin(n\pi\xi)\,d\xi = A_n \int_0^1 \sin^2(n\pi\xi)\,d\xi \tag{11.2-81}$$

Evaluation of the integrals yields

$$A_n = \frac{2(-1)^n}{n\pi}\left(1 - \frac{\Omega}{n^2\pi^2}\right) \tag{11.2-82}$$

Therefore, the solution becomes

$$\theta = \frac{\Omega}{6}(1 - \xi^2) - \frac{2}{\pi}\sum_{n=1}^{\infty}\frac{(-1)^n}{n}\left(1 - \frac{\Omega}{n^2\pi^2}\right)\exp(-n^2\pi^2\tau)\frac{\sin(n\pi\xi)}{\xi} \tag{11.2-83}$$

When $\Omega = 0$, Eq. (11.2-83) reduces to Eq. (10.3-125).

11.2.3.1 *Macroscopic equation* Integration of the governing equation for temperature, Eq. (11.2-56), over the volume of the system gives

$$\int_0^{2\pi}\int_0^{\pi}\int_0^R \rho\widehat{C}_P\frac{\partial T}{\partial t}r^2\sin\theta\,dr\,d\theta\,d\phi = \int_0^{2\pi}\int_0^{\pi}\int_0^R \frac{k}{r^2}\frac{\partial}{\partial r}\left(r^2\frac{\partial T}{\partial r}\right)r^2\sin\theta\,dr\,d\theta\,d\phi$$

$$+ \int_0^{2\pi}\int_0^{\pi}\int_0^R \Re\,r^2\sin\theta\,dr\,d\theta\,d\phi \tag{11.2-84}$$

or,

$$\underbrace{\frac{d}{dt}\left[\int_0^{2\pi}\int_0^{\pi}\int_0^R \rho\widehat{C}_P(T - T_o)r^2\sin\theta\,dr\,d\theta\,d\phi\right]}_{\text{Rate of accumulation of energy}} = \underbrace{-4\pi R^2\left(-k\frac{\partial T}{\partial r}\bigg|_{r=R}\right)}_{\substack{\text{Rate of energy leaving} \\ \text{from the surface}}} + \underbrace{\frac{4}{3}\pi R^3\Re}_{\substack{\text{Rate of energy} \\ \text{generation}}}$$

$$\tag{11.2-85}$$

which is the macroscopic energy balance by considering the sphere as a system. The rate of energy leaving from the surface, $\dot{Q}$, is given by

$$\dot{Q} = 4\pi R^2\left(-k\frac{\partial T}{\partial r}\bigg|_{r=R}\right) = -4\pi Rk(T_o - T_\infty)\frac{\partial\theta}{\partial\xi}\bigg|_{\xi=1} \tag{11.2-86}$$

The use of Eq. (11.2-83) in Eq. (11.2-86) leads to

$$\dot{Q} = 4\pi Rk(T_o - T_\infty)\left[\frac{\Omega}{3} + 2\sum_{n=1}^{\infty}\left(1 - \frac{\Omega}{n^2\pi^2}\right)\exp(-n^2\pi^2\tau)\right] \tag{11.2-87}$$

Under steady conditions, i.e., $\tau \to \infty$, $\dot{Q} \to (4/3)\pi R^3\Re$, indicating that the rate of heat loss from the system equals the rate of internal generation of heat.

11.3 MASS TRANSPORT

The conservation statement for species $\mathcal{A}$ is expressed as

$$\begin{pmatrix}\text{Rate of} \\ \text{species } \mathcal{A} \text{ in}\end{pmatrix} - \begin{pmatrix}\text{Rate of} \\ \text{species } \mathcal{A} \text{ out}\end{pmatrix} + \begin{pmatrix}\text{Rate of species } \mathcal{A} \\ \text{generation}\end{pmatrix} = \begin{pmatrix}\text{Rate of species } \mathcal{A} \\ \text{accumulation}\end{pmatrix}$$

$$\tag{11.3-1}$$

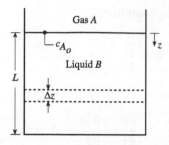

Figure 11.5. Unsteady diffusion with a homogeneous reaction.

11.3.1 Rectangular Geometry

Steady diffusion of species $\mathcal{A}$ in a liquid with a homogeneous reaction is described in Section 9.4.1. Now let us consider the unsteady-state version of the same problem.

For a differential volume element of thickness Δz, as shown in Figure 11.5, Eq. (11.3-1) is expressed as

$$N_{A_z}|_z A - N_{A_z}|_{z+\Delta z} A + \Re_A A \Delta z = \frac{\partial}{\partial t}(A \Delta z c_A) \tag{11.3-2}$$

Dividing Eq. (11.3-2) by $A \Delta z$ and taking the limit as $\Delta z \to 0$ give

$$\frac{\partial c_A}{\partial t} = \lim_{\Delta z \to 0} \frac{N_{A_z}|_z - N_{A_z}|_{z+\Delta z}}{\Delta z} + \Re_A \tag{11.3-3}$$

or,

$$\frac{\partial c_A}{\partial t} = -\frac{\partial N_{A_z}}{\partial z} + \Re_A \tag{11.3-4}$$

The molar flux of species $\mathcal{A}$ in the z-direction, N_{A_z}, and the rate of depletion of species $\mathcal{A}$ per unit volume, $\Re_A$, are given by

$$N_{A_z} = -\mathcal{D}_{AB} \frac{dc_A}{dz} \quad \text{and} \quad \Re_A = -kc_A \tag{11.3-5}$$

Substitution of Eq. (11.3-5) into Eq. (11.3-4) gives the governing equation for the concentration of species $\mathcal{A}$ as

$$\boxed{\frac{\partial c_A}{\partial t} = \mathcal{D}_{AB} \frac{\partial^2 c_A}{\partial z^2} - kc_A} \tag{11.3-6}$$

The initial and boundary conditions associated with Eq. (11.3-6) are

$$\text{at} \quad t = 0 \qquad c_A = 0 \tag{11.3-7}$$

$$\text{at} \quad z = 0 \qquad c_A = c_{A_o} \tag{11.3-8}$$

$$\text{at} \quad z = L \qquad \frac{\partial c_A}{\partial z} = 0 \tag{11.3-9}$$

Introduction of the dimensionless quantities

$$\theta = \frac{c_A}{c_{A_o}} \qquad \xi = \frac{z}{L} \qquad \tau = \frac{\mathcal{D}_{AB}t}{L^2} \qquad \Lambda = \sqrt{\frac{kL^2}{\mathcal{D}_{AB}}} \qquad (11.3\text{-}10)$$

reduces Eqs. (11.3-6)–(11.3-9) to

$$\frac{\partial \theta}{\partial \tau} = \frac{\partial^2 \theta}{\partial \xi^2} - \Lambda^2 \theta \qquad (11.3\text{-}11)$$

$$\text{at} \quad \tau = 0 \qquad \theta = 0 \qquad (11.3\text{-}12)$$

$$\text{at} \quad \xi = 0 \qquad \theta = 1 \qquad (11.3\text{-}13)$$

$$\text{at} \quad \xi = 1 \qquad \frac{\partial \theta}{\partial \xi} = 0 \qquad (11.3\text{-}14)$$

The solution is proposed in the form

$$\theta(\tau, \xi) = \theta_\infty(\xi) - \theta_t(\tau, \xi) \qquad (11.3\text{-}15)$$

in which θ_∞ is the steady-state solution, i.e.,

$$\frac{d^2 \theta_\infty}{d\xi^2} - \Lambda^2 \theta_\infty = 0 \qquad (11.3\text{-}16)$$

with the following boundary conditions

$$\text{at} \quad \xi = 0 \qquad \theta_\infty = 1 \qquad (11.3\text{-}17)$$

$$\text{at} \quad \xi = 1 \qquad \frac{d\theta_\infty}{d\xi} = 0 \qquad (11.3\text{-}18)$$

The solution of Eq. (11.3-16) is given by Eq. (9.4-16), i.e.,

$$\theta_\infty = \frac{\cosh[\Lambda(1 - \xi)]}{\cosh \Lambda} \qquad (11.3\text{-}19)$$

The use of Eq. (11.3-19) in Eq. (11.3-15) gives

$$\theta(\tau, \xi) = \frac{\cosh[\Lambda(1 - \xi)]}{\cosh \Lambda} - \theta_t(\tau, \xi) \qquad (11.3\text{-}20)$$

Substitution of Eq. (11.3-20) into Eqs. (11.3-11)–(11.3-14) leads to the following governing equation for the transient problem, together with the initial and the boundary conditions

$$\frac{\partial \theta_t}{\partial \tau} = \frac{\partial^2 \theta_t}{\partial \xi^2} - \Lambda^2 \theta_t \qquad (11.3\text{-}21)$$

$$\text{at} \quad \tau = 0 \qquad \theta_t = \frac{\cosh[\Lambda(1 - \xi)]}{\cosh \Lambda} \qquad (11.3\text{-}22)$$

$$\text{at} \quad \xi = 0 \qquad \theta_t = 0 \qquad (11.3\text{-}23)$$

$$\text{at} \quad \xi = 1 \qquad \frac{\partial \theta_t}{\partial \xi} = 0 \qquad (11.3\text{-}24)$$

The separation of variables method assumes that the solution can be represented as a product of two functions of the form

$$\theta_t(\tau, \xi) = F(\tau)G(\xi) \tag{11.3-25}$$

Substitution of Eq. (11.3-25) into Eq. (11.3-21) and rearrangement give

$$\frac{1}{F}\frac{dF}{d\tau} + \Lambda^2 = \frac{1}{G}\frac{d^2G}{d\xi^2} = -\lambda^2 \tag{11.3-26}$$

Equation (11.3-26) results in two ordinary differential equations. The equation for F is given by

$$\frac{dF}{d\tau} + (\lambda^2 + \Lambda^2)F = 0 \quad \Rightarrow \quad F(\tau) = e^{-(\lambda^2 + \Lambda^2)\tau} \tag{11.3-27}$$

On the other hand, the equation for G is

$$\frac{d^2G}{d\xi^2} + \lambda^2 G = 0 \quad \Rightarrow \quad G(\xi) = C_1 \sin(\lambda\xi) + C_2 \cos(\lambda\xi) \tag{11.3-28}$$

and it is subject to the following boundary conditions

$$\text{at} \quad \xi = 0 \qquad G = 0 \tag{11.3-29}$$

$$\text{at} \quad \xi = 1 \qquad \frac{dG}{d\xi} = 0 \tag{11.3-30}$$

While the application of Eq. (11.3-29) gives $C_2 = 0$, the use of Eq. (11.3-30) results in

$$\cos\lambda = 0 \quad \Rightarrow \quad \lambda_n = (n + \tfrac{1}{2})\pi \qquad n = 0, 1, 2, \ldots \tag{11.3-31}$$

Therefore, the transient solution is expressed as

$$\theta_t = \sum_{n=0}^{\infty} A_n \exp\left[-(\lambda_n^2 + \Lambda^2)\tau\right]\sin(\lambda_n\xi) \tag{11.3-32}$$

The unknown coefficients A_n can be determined by using the initial condition, i.e., Eq. (11.3-22). The result is

$$\int_0^1 \frac{\cosh[\Lambda(1 - \xi)]}{\cosh\Lambda}\sin(\lambda_n\xi)\,d\xi = A_n \int_0^1 \sin^2(\lambda_n\xi)\,d\xi \tag{11.3-33}$$

Evaluation of the integrals yields

$$A_n = \frac{2\lambda_n}{\lambda_n^2 + \Lambda^2} \tag{11.3-34}$$

Therefore, the solution becomes

$$\boxed{\theta = \frac{\cosh[\Lambda(1 - \xi)]}{\cosh\Lambda} - 2\sum_{n=0}^{\infty}\frac{\lambda_n}{\lambda_n^2 + \Lambda^2}\exp\left[-(\lambda_n^2 + \Lambda^2)\tau\right]\sin(\lambda_n\xi)} \tag{11.3-35}$$

11.3.1.1 *Macroscopic Balance* Integration of the governing equation, Eq. (11.3-6), over the volume of the liquid in the tank gives

$$A \int_0^L \frac{\partial c_A}{\partial t}\, dz = A \int_0^L \mathcal{D}_{AB} \frac{\partial^2 c_A}{\partial z^2}\, dz - A \int_0^L k c_A\, dz \tag{11.3-36}$$

or,

$$A \frac{d}{dt}\left(\int_0^L c_A\, dz \right) = A \left(-\mathcal{D}_{AB} \frac{\partial c_A}{\partial z}\bigg|_{z=0} \right) - A \int_0^L k c_A\, dz \tag{11.3-37}$$

$$\underbrace{\phantom{A \frac{d}{dt}\left(\int_0^L c_A\, dz \right)}}_{\substack{\text{Rate of accumulation}\\ \text{of species } \mathcal{A}}} \qquad \underbrace{\phantom{A \left(-\mathcal{D}_{AB} \frac{\partial c_A}{\partial z}\bigg|_{z=0} \right)}}_{\substack{\text{Rate of species } \mathcal{A}\\ \text{entering the liquid}}} \qquad \underbrace{}_{\substack{\text{Rate of depletion}\\ \text{of species } \mathcal{A}}}$$

which is the macroscopic mass balance for species $\mathcal{A}$ by considering the liquid in the tank as a system. The molar rate of species $\mathcal{A}$ entering the liquid, $\dot{n}_A$, is given by

$$\dot{n}_A = A \left(-\mathcal{D}_{AB} \frac{\partial c_A}{\partial z}\bigg|_{z=0} \right) = -\frac{A c_{A_o} \mathcal{D}_{AB}}{L} \frac{\partial \theta}{\partial \xi}\bigg|_{\xi=0} \tag{11.3-38}$$

The use of Eq. (11.3-35) in Eq. (11.3-38) leads to

$$\dot{n}_A = \frac{A c_{A_o} \mathcal{D}_{AB}}{L} \left\{ \Lambda \tanh \Lambda + 2 \sum_{n=0}^{\infty} \frac{\lambda_n^2}{\lambda_n^2 + \Lambda^2} \exp\left[-(\lambda_n^2 + \Lambda^2)\tau \right] \right\} \tag{11.3-39}$$

Under steady conditions, i.e., $\tau \to \infty$, Eq. (11.3-39) reduces to Eq. (9.4-19).

11.3.2 Cylindrical Geometry

Consider unsteady-state diffusion in a long cylinder of radius R with a homogeneous chemical reaction. Initially the reactant (species $\mathcal{A}$) concentration is zero within the cylinder. For $t > 0$, the reactant concentration at the lateral surface of the cylinder is kept constant at c_{A_R}. The reaction is first-order and irreversible.

Assuming one-dimensional diffusion, we postulate that $c_A = c_A(t, r)$. From Table C.8 in Appendix C, the nonzero molar flux component is given by

$$N_{A_r} = J_{A_r}^* = -\mathcal{D}_{AB} \frac{\partial c_A}{\partial r} \tag{11.3-40}$$

For a cylindrical differential volume element of thickness, as shown in Figure 11.6, Eq. (11.3-1) takes the form

$$N_{A_r}|_r 2\pi r L - N_{A_r}|_{r+\Delta r} 2\pi (r + \Delta r) L - (k c_A) 2\pi r \Delta r L = \frac{\partial}{\partial t}(2\pi r \Delta r L c_A) \tag{11.3-41}$$

Dividing Eq. (11.3-41) by $2\pi L \Delta r$ and taking the limit as $\Delta r \to 0$ give

$$\frac{\partial c_A}{\partial t} = \frac{1}{r} \lim_{\Delta r \to 0} \frac{(r N_{A_r})|_r - (r N_{A_r})|_{r+\Delta r}}{\Delta r} - k c_A \tag{11.3-42}$$

or,

$$\frac{\partial c_A}{\partial t} = -\frac{1}{r} \frac{\partial (r N_{A_r})}{\partial r} - k c_A \tag{11.3-43}$$

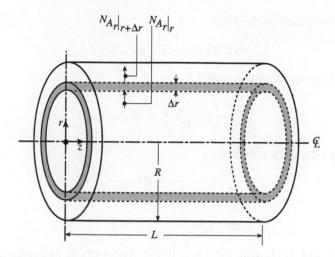

Figure 11.6. Unsteady diffusion in a cylinder with a homogeneous reaction.

Substitution of Eq. (11.3-40) into Eq. (11.3-43) gives the governing differential equation for the concentration of species $\mathcal{A}$ as

$$\frac{\partial c_A}{\partial t} = \frac{\mathcal{D}_{AB}}{r} \frac{\partial}{\partial r}\left(r \frac{\partial c_A}{\partial r} \right) - k c_A \qquad (11.3\text{-}44)$$

The initial and boundary conditions associated with Eq. (11.3-44) are

$$\text{at} \quad t = 0 \qquad c_A = 0 \qquad\qquad (11.3\text{-}45)$$

$$\text{at} \quad r = 0 \qquad \frac{\partial c_A}{\partial r} = 0 \qquad\qquad (11.3\text{-}46)$$

$$\text{at} \quad r = R \qquad c_A = c_{A_R} \qquad\qquad (11.3\text{-}47)$$

Introduction of the dimensionless quantities

$$\theta = \frac{c_A}{c_{A_R}} \qquad \xi = \frac{r}{R} \qquad \tau = \frac{\alpha t}{R^2} \qquad \Lambda = \sqrt{\frac{kR^2}{\mathcal{D}_{AB}}} \qquad (11.3\text{-}48)$$

reduces Eqs. (11.3-44)–(11.3-47) to

$$\frac{\partial \theta}{\partial \tau} = \frac{1}{\xi} \frac{\partial}{\partial \xi}\left(\xi \frac{\partial \theta}{\partial \xi} \right) - \Lambda^2 \theta \qquad (11.3\text{-}49)$$

$$\text{at} \quad \tau = 0 \qquad \theta = 0 \qquad\qquad (11.3\text{-}50)$$

$$\text{at} \quad \xi = 0 \qquad \frac{\partial \theta}{\partial \xi} = 0 \qquad\qquad (11.3\text{-}51)$$

$$\text{at} \quad \xi = 1 \qquad \theta = 1 \qquad\qquad (11.3\text{-}52)$$

The solution is proposed in the form

$$\theta(\tau, \xi) = \theta_\infty(\xi) - \theta_t(\tau, \xi) \tag{11.3-53}$$

in which θ_∞ is the steady-state solution, i.e.,

$$\frac{1}{\xi} \frac{d}{d\xi} \left(\xi \frac{d\theta_\infty}{d\xi} \right) - \Lambda^2 \theta_\infty = 0 \tag{11.3-54}$$

with the following boundary conditions

$$\text{at} \quad \xi = 0 \qquad \frac{d\theta_\infty}{d\xi} = 0 \tag{11.3-55}$$

$$\text{at} \quad \xi = 1 \qquad \theta_\infty = 1 \tag{11.3-56}$$

Comparison of Eq. (11.3-54) with Eq. (B.2-16) in Appendix B indicates that $p = 1$, $j = 1$, $a = -\Lambda^2$, and $b = 0$. Therefore, Eq. (11.3-54) is Bessel's equation and the use of Eqs. (B.2-17)–(B.2-19) gives $\alpha = 1$, $\beta = 0$, and $n = 0$. Equation (B.2-26) gives the solution as

$$\theta_\infty = C_1 I_0(\Lambda \xi) + C_2 K_0(\Lambda \xi) \tag{11.3-57}$$

Since $K_0(0) = \infty$, $C_2 = 0$. Application of Eq. (11.3-56) gives $C_1 = 1/I_0(\Lambda)$. Thus, the steady-state solution becomes

$$\theta_\infty = \frac{I_0(\Lambda \xi)}{I_0(\Lambda)} \tag{11.3-58}$$

The use of Eq. (11.3-58) in Eq. (11.3-53) gives

$$\theta(\tau, \xi) = \frac{I_0(\Lambda \xi)}{I_0(\Lambda)} - \theta_t(\tau, \xi) \tag{11.3-59}$$

Substitution of Eq. (11.3-59) into Eqs. (11.3-49)–(11.3-52) leads to the following governing equation for the transient problem together with the initial and the boundary conditions

$$\frac{\partial \theta_t}{\partial \tau} = \frac{1}{\xi} \frac{\partial}{\partial \xi} \left(\xi \frac{\partial \theta_t}{\partial \xi} \right) - \Lambda^2 \theta_t \tag{11.3-60}$$

$$\text{at} \quad \tau = 0 \qquad \theta_t = \frac{I_0(\Lambda \xi)}{I_0(\Lambda)} \tag{11.3-61}$$

$$\text{at} \quad \xi = 0 \qquad \theta_t = 0 \tag{11.3-62}$$

$$\text{at} \quad \xi = 1 \qquad \theta_t = 0 \tag{11.3-63}$$

Representing the solution as a product of two functions of the form

$$\theta_t(\tau, \xi) = F(\tau) G(\xi) \tag{11.3-64}$$

reduces Eq. (11.3-60) to

$$\frac{1}{F} \frac{dF}{d\tau} + \Lambda^2 = \frac{1}{G\xi} \frac{d}{d\xi} \left(\xi \frac{dG}{d\xi} \right) = -\lambda^2 \tag{11.3-65}$$

which results in two ordinary differential equations:

$$\frac{dF}{d\tau} + (\lambda^2 + \Lambda^2)F = 0 \quad \Rightarrow \quad F(\tau) = e^{-(\lambda^2 + \Lambda^2)\tau} \tag{11.3-66}$$

$$\frac{d^2G}{d\xi^2} + \lambda^2 G = 0 \quad \Rightarrow \quad G(\xi) = C_3 J_o(\lambda\xi) + C_4 Y_o(\lambda\xi) \tag{11.3-67}$$

The boundary conditions for $G(\xi)$ are

$$\text{at} \quad \xi = 0 \quad G = 0 \tag{11.3-68}$$

$$\text{at} \quad \xi = 1 \quad G = 0 \tag{11.3-69}$$

Since $Y_o(0) = -\infty$, $C_4 = 0$. Application of Eq. (11.3-69) yields

$$C_3 J_o(\lambda) = 0 \tag{11.3-70}$$

For a nontrivial solution, the eigenvalues are given by

$$\boxed{J_o(\lambda_n) = 0} \quad \lambda_n = 1, 2, 3, \ldots \tag{11.3-71}$$

The general solution is the summation of all possible solutions, i.e.,

$$\theta_t = \sum_{n=1}^{\infty} A_n \exp\left[-(\lambda_n^2 + \Lambda^2)\tau\right] J_o(\lambda_n\xi) \tag{11.3-72}$$

The unknown coefficients A_n can be determined by using the initial condition given by Eq. (11.3-61). The result is

$$\frac{1}{I_o(\Lambda)} \int_0^1 \xi J_o(\lambda_n\xi) I_o(\Lambda\xi)\, d\xi = A_n \int_0^1 \xi J_o^2(\lambda_n\xi)\, d\xi \tag{11.3-73}$$

Evaluation of the integral with the help of

$$\int x J_o(ax) I_o(bx)\, dx = \frac{x}{a^2 + b^2}\left[a J_1(ax) I_o(bx) + b J_o(ax) I_1(bx)\right] \tag{11.3-74}$$

gives the coefficients A_n in the form

$$A_n = \frac{2\lambda_n}{(\lambda_n^2 + \Lambda^2) J_1(\lambda_n)} \tag{11.3-75}$$

Thus, the solution becomes

$$\boxed{\theta = \frac{I_o(\Lambda\xi)}{I_o(\Lambda)} - 2\sum_{n=1}^{\infty} \frac{\lambda_n}{(\lambda_n^2 + \Lambda^2) J_1(\lambda_n)} \exp\left[-(\lambda_n^2 + \Lambda^2)\tau\right] J_o(\lambda_n\xi)} \tag{11.3-76}$$

11.3.2.1 *Macroscopic Balance* Integration of the governing differential equation, Eq. (11.3-44), over the volume of the cylinder gives

$$
\int_0^L \int_0^{2\pi} \int_0^R \frac{\partial c_A}{\partial t}\, r\, dr\, d\theta\, dz = \int_0^L \int_0^{2\pi} \int_0^R \frac{\mathcal{D}_{AB}}{r} \frac{\partial}{\partial r}\left(r\, \frac{\partial c_A}{\partial r} \right) r\, dr\, d\theta\, dz
$$

$$
- \int_0^L \int_0^{2\pi} \int_0^R k c_A\, r\, dr\, d\theta\, dz \qquad (11.3\text{-}77)
$$

or,

$$
\underbrace{\frac{d}{dt}\left[\int_0^L \int_0^{2\pi} \int_0^R c_A\, r\, dr\, d\theta\, dz \right]}_{\text{Rate of accumulation of species } \mathcal{A}} = \underbrace{2\pi RL\left(\mathcal{D}_{AB} \frac{\partial c_A}{\partial r}\Big|_{r=R} \right)}_{\substack{\text{Rate of species } \mathcal{A} \text{ entering} \\ \text{from the lateral surface}}} - \underbrace{\int_0^L \int_0^{2\pi} \int_0^R k c_A\, r\, dr\, d\theta\, dz}_{\text{Rate of depletion of species } \mathcal{A}}
$$

$$(11.3\text{-}78)$$

which is the macroscopic mass balance for species $\mathcal{A}$ by considering the cylinder as a system. The molar rate of species $\mathcal{A}$ entering the cylinder, $\dot{n}_A$, is given by

$$
\dot{n}_A = 2\pi RL\left(\mathcal{D}_{AB} \frac{\partial c_A}{\partial r}\Big|_{r=R} \right) = 2\pi L c_{A_R} \mathcal{D}_{AB} \frac{\partial \theta}{\partial \xi}\Big|_{\xi=1} \qquad (11.3\text{-}79)
$$

The use of Eq. (11.3-76) in Eq. (11.3-79) leads to

$$
\dot{n}_A = 2\pi L c_{A_R} \mathcal{D}_{AB} \left\{ \Lambda \frac{I_1(\Lambda)}{I_o(\Lambda)} + 2\sum_{n=1}^{\infty} \frac{\lambda_n^2}{\lambda_n^2 + \Lambda^2} \exp\left[-(\lambda_n^2 + \Lambda^2)\tau \right] \right\} \qquad (11.3\text{-}80)
$$

11.3.3 Spherical Geometry

A liquid droplet ($\mathcal{B}$) of radius R is initially $\mathcal{A}$-free. At $t = 0$, it is surrounded by gas $\mathcal{A}$ as shown in Figure 11.7. As species $\mathcal{A}$ diffuses into $\mathcal{B}$, it undergoes an irreversible chemical reaction with $\mathcal{B}$ to form $\mathcal{AB}$, i.e.,

$$
A + B \rightarrow AB
$$

The rate of reaction is expressed by

$$
r = k c_A
$$

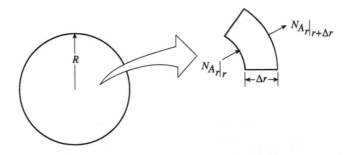

Figure 11.7. Unsteady-state absorption with a chemical reaction.

We are interested in the rate of absorption of species $\mathcal{A}$ into the liquid during the unsteady-state period. The problem will be analyzed with the following assumptions:

1. Convective flux is negligible with respect to the molecular flux.
2. The total concentration is constant.
3. Pseudo-binary behavior.

Since $c_A = c_A(t, r)$, Table C.9 in Appendix C indicates that the only nonzero molar flux component is N_{A_r}, and it is given by

$$N_{A_r} = J_{A_r}^* = -\mathcal{D}_{AB} \frac{\partial c_A}{\partial r} \tag{11.3-81}$$

For a spherical differential volume element of thickness Δr, as shown in Figure 11.7, Eq. (11.3-1) is expressed in the form

$$N_{A_r}|_r 4\pi r^2 - N_{A_r}|_{r+\Delta r} 4\pi (r + \Delta r)^2 - (kc_A)4\pi r^2 \Delta r = \frac{\partial}{\partial t}(4\pi r^2 \Delta r c_A) \tag{11.3-82}$$

Dividing Eq. (11.3-82) by $4\pi \Delta r$ and taking the limit as $\Delta r \to 0$ give

$$\frac{\partial c_A}{\partial t} = \frac{1}{r^2} \lim_{\Delta r \to 0} \frac{(r^2 N_{A_r})|_r - (r^2 N_{A_r})|_{r+\Delta r}}{\Delta r} - kc_A \tag{11.3-83}$$

or,

$$\frac{\partial c_A}{\partial t} = -\frac{1}{r^2} \frac{\partial (r^2 N_{A_r})}{\partial r} - kc_A \tag{11.3-84}$$

Substitution of Eq. (11.3-81) into Eq. (11.3-84) gives the governing differential equation for the concentration of species $\mathcal{A}$ as

$$\boxed{\frac{\partial c_A}{\partial t} = \frac{\mathcal{D}_{AB}}{r^2} \frac{\partial}{\partial r}\left(r^2 \frac{\partial c_A}{\partial r}\right) - kc_A} \tag{11.3-85}$$

The initial and the boundary conditions associated with Eq. (11.3-85) are

$$\text{at} \quad t = 0 \qquad c_A = 0 \tag{11.3-86}$$

$$\text{at} \quad r = 0 \qquad \frac{\partial c_A}{\partial r} = 0 \tag{11.3-87}$$

$$\text{at} \quad r = R \qquad c_A = c_A^* \tag{11.3-88}$$

where c_A^* is the equilibrium solubility of species $\mathcal{A}$ in liquid $\mathcal{B}$.

Danckwerts (1951) showed that the partial differential equation of the form

$$\frac{\partial c}{\partial t} = \mathcal{D} \frac{\partial^2 c}{\partial x^2} - kc \tag{11.3-89}$$

with the initial and the boundary conditions of the form

$$\text{at} \quad t = 0 \qquad c = 0 \tag{11.3-90}$$

and either

$$\text{at all points on the surface} \quad c = c^* \tag{11.3-91}$$

or

$$\text{at all points on the surface} \quad \mathcal{D}\frac{\partial c}{\partial x} = k_c(c^* - c) \tag{11.3-92}$$

has the solution

$$c = k\int_0^t \phi(\eta, x)e^{-k\eta}\,d\eta + \phi(t, x)e^{-kt} \tag{11.3-93}$$

where $\phi(t, x)$ is the solution of Eq. (11.3-89) without the chemical reaction, i.e.,

$$\frac{\partial\phi}{\partial t} = \mathcal{D}\frac{\partial^2\phi}{\partial x^2} \tag{11.3-94}$$

and is subject to the same initial and boundary conditions given by Eqs. (11.3-90)–(11.3-92). Note that η is a dummy variable of the integration in Eq. (11.3-93).

The solution of Eq. (11.3-85) without the chemical reaction is given by Eq. (10.3-125), i.e.,

$$\frac{c_A}{c_A^*} = 1 + \frac{2R}{\pi r}\sum_{n=1}^{\infty}\frac{(-1)^n}{n}\exp\left(-\frac{n^2\pi^2\mathcal{D}_{AB}t}{R^2}\right)\sin\left(\frac{n\pi r}{R}\right) \tag{11.3-95}$$

Substitution of Eq. (11.3-95) into Eq. (11.3-93) gives

$$\frac{c_A}{c_A^*} = k\int_0^t\left[1 + \frac{2R}{\pi r}\sum_{n=1}^{\infty}\frac{(-1)^n}{n}\exp\left(-\frac{n^2\pi^2\mathcal{D}_{AB}\eta}{R^2}\right)\sin\left(\frac{n\pi r}{R}\right)\right]e^{-k\eta}\,d\eta$$
$$+ \left[1 + \frac{2R}{\pi r}\sum_{n=1}^{\infty}\frac{(-1)^n}{n}\exp\left(-\frac{n^2\pi^2\mathcal{D}_{AB}t}{R^2}\right)\sin\left(\frac{n\pi r}{R}\right)\right]e^{-kt} \tag{11.3-96}$$

Carrying out the integration gives the solution as

$$\boxed{\frac{c_A}{c_A^*} = 1 + \frac{2R}{\pi r}\sum_{n=1}^{\infty}\frac{(-1)^n}{n}\left\{\frac{1 + \Omega\exp[-(1 + \Omega)kt]}{1 + \Omega}\right\}\sin\left(\frac{n\pi r}{R}\right)} \tag{11.3-97}$$

where

$$\Omega = \frac{n^2\pi^2\mathcal{D}_{AB}}{kR^2} \tag{11.3-98}$$

11.3.3.1 *Macroscopic equation* Integration of the governing equation for the concentration of species $\mathcal{A}$, Eq. (11.3-85), over the volume of the system gives

$$
\int_0^{2\pi} \int_0^{\pi} \int_0^R \frac{\partial c_A}{\partial t} r^2 \sin\theta \, dr \, d\theta \, d\phi = \int_0^{2\pi} \int_0^{\pi} \int_0^R \frac{\mathcal{D}_{AB}}{r^2} \frac{\partial}{\partial r} \left(r^2 \frac{\partial c_A}{\partial r} \right) r^2 \sin\theta \, dr \, d\theta \, d\phi
$$

$$
- \int_0^{2\pi} \int_0^{\pi} \int_0^R k c_A r^2 \sin\theta \, dr \, d\theta \, d\phi \qquad (11.3\text{-}99)
$$

or,

$$
\underbrace{\frac{d}{dt} \left[\int_0^{2\pi} \int_0^{\pi} \int_0^R c_A r^2 \sin\theta \, dr \, d\theta \, d\phi \right]}_{\text{Rate of accumulation of species } \mathcal{A}} = \underbrace{4\pi R^2 \left(\mathcal{D}_{AB} \frac{\partial c_A}{\partial r} \bigg|_{r=R} \right)}_{\substack{\text{Rate of species } \mathcal{A} \text{ entering} \\ \text{from the surface}}}
$$

$$
\underbrace{- \int_0^{2\pi} \int_0^{\pi} \int_0^R k c_A r^2 \sin\theta \, dr \, d\theta \, d\phi}_{\text{Rate of depletion of species } \mathcal{A}} \qquad (11.3\text{-}100)
$$

which is the macroscopic mass balance for species $\mathcal{A}$ by considering the liquid droplet as a system. The molar rate of absorption of species $\mathcal{A}$, $\dot{n}_A$, is given by

$$
\dot{n}_A = 4\pi R^2 \left(\mathcal{D}_{AB} \frac{\partial c_A}{\partial r} \bigg|_{r=R} \right) \qquad (11.3\text{-}101)
$$

The use of Eq. (11.3-97) in Eq. (11.3-101) leads to

$$
\dot{n}_A = 8\pi R \mathcal{D}_{AB} c_A^* \sum_{n=1}^{\infty} \left\{ \frac{1 + \Omega \exp[-(1+\Omega)kt]}{1+\Omega} \right\} \qquad (11.3\text{-}102)
$$

The moles of species $\mathcal{A}$ absorbed can be calculated from

$$
n_A = \int_0^t \dot{n}_A \, dt \qquad (11.3\text{-}103)
$$

Substitution of Eq. (11.3-102) into Eq. (11.3-103) and integration yield

$$
n_A = 8\pi R \mathcal{D}_{AB} c_A^* \sum_{n=1}^{\infty} \frac{t}{1+\Omega} \left(1 + \frac{\Omega}{(1+\Omega)kt} \{ 1 - \exp[-(1+\Omega)kt] \} \right) \qquad (11.3\text{-}104)
$$

Example 11.1 Show that the solution given by Eq. (11.3-93) satisfies Eq. (11.3-89).

Solution

Differentiation of Eq. (11.3-93) with respect to t by using Leibnitz's rule gives

$$
\frac{\partial c}{\partial t} = k\phi(t,x)e^{-kt} - k\phi(t,x)e^{-kt} + \frac{\partial \phi}{\partial t} e^{-kt} = \frac{\partial \phi}{\partial t} e^{-kt} \qquad (1)
$$

Differentiation of Eq. (11.3-93) twice with respect to x yields

$$\frac{\partial^2 c}{\partial x^2} = k \int_0^t \frac{\partial^2 \phi(\eta, x)}{\partial x^2} e^{-k\eta} d\eta + \frac{\partial^2 \phi(t, x)}{\partial x^2} e^{-kt} \tag{2}$$

The use of Eq. (11.3-94) in Eq. (2) leads to

$$\mathcal{D} \frac{\partial^2 c}{\partial x^2} = k \int_0^t \frac{\partial \phi(\eta, x)}{\partial \eta} e^{-k\eta} d\eta + \frac{\partial \phi(t, x)}{\partial t} e^{-kt} \tag{3}$$

Substitution of Eq. (1) into Eq. (3) yields

$$\mathcal{D} \frac{\partial^2 c}{\partial x^2} = k \int_0^t \frac{\partial c}{\partial \eta} d\eta + \frac{\partial c}{\partial t} \tag{4}$$

or,

$$\mathcal{D} \frac{\partial^2 c}{\partial x^2} = kc + \frac{\partial c}{\partial t} \tag{5}$$

which is identical to Eq. (11.3-89).

NOTATION

A	area, m^2
$\widehat{C}_P$	heat capacity at constant pressure, kJ/kg·K
c_i	concentration of species i, kmol/m^3
$\mathcal{D}_{AB}$	diffusion coefficient for system $\mathcal{A}$-$\mathcal{B}$, m^2/s
e	total energy flux, W/m^2
$\mathcal{H}$	partition coefficient
J^*	molecular molar flux, kmol/m^2·s
k	thermal conductivity, W/m·K
L	length, m
$\dot{m}$	mass flow rate, kg/s
$\mathcal{M}$	molecular weight, kg/kmol
N	total molar flux, kmol/m^2·s
n	number of moles, kmol
$\dot{n}$	molar flow rate, kmol/s
$\dot{Q}$	heat transfer rate, W
Q	volumetric flow rate, m^3/s
q	heat flux, W/m^2
R	radius, m
$\mathfrak{R}$	rate of generation (momentum, energy, mass) per unit volume
T	temperature, °C or K
t	time, s
V	volume, m^3
v	velocity, m/s
α	thermal diffusivity, m^2/s
μ	viscosity, kg/m·s

ν kinematic viscosity, m^2/s

ρ density, kg/m^3

π total momentum flux, N/m^2

τ_{ij} shear stress (flux of j-momentum in the i-direction), N/m^2

Bracket

$\langle a \rangle$ average value of a

Subscripts

A, B species in binary systems

i species in multicomponent systems

REFERENCES

Danckwerts, P.V., 1951, Absorption by simultaneous diffusion and chemical reaction into particles of various shapes and into falling drops, J. Chem. Soc. Faraday Trans. 47, 1014.

Slattery, J.C., 1972, Momentum, Energy, and Mass Transfer in Continua, McGraw-Hill, New York.

SUGGESTED REFERENCES FOR FURTHER STUDY

Astarita, G., 1967, Mass Transfer With Chemical Reaction, Elsevier, Amsterdam.

Bird, R.B., W.E. Stewart and E.N. Lightfoot, 2002, Transport Phenomena, 2nd Ed., Wiley, New York.

Carslaw, H.S. and J.C. Jaeger, 1959, Conduction of Heat in Solids, 2nd Ed., Oxford University Press, London.

Crank, J., 1956, The Mathematics of Diffusion, Oxford University Press, London.

Deen, W.M., 1998, Analysis of Transport Phenomena, Oxford University Press, New York.

Middleman, S., 1998, An Introduction to Mass and Heat Transfer – Principles of Analysis and Design, Wiley, New York.

Slattery, J.C., 1999, Advanced Transport Phenomena, Cambridge University Press, Cambridge.

PROBLEMS

11.1 A stationary incompressible Newtonian fluid is contained between two parallel plates. At time $t = 0$, a constant pressure gradient is imposed and the fluid begins to flow. Repeat the analysis given in Section 11.1 as follows:

a) Considering the flow geometry shown in Figure 9.1, write the governing differential equation, and initial and boundary conditions in terms of the following dimensionless variables

$$\theta = \frac{v_z}{\left(\dfrac{\mathcal{P}_o - \mathcal{P}_L}{2\mu L}\right) B^2} \qquad \xi = \frac{x}{B} \qquad \tau = \frac{\nu t}{B^2}$$

and show that

$$\frac{\partial \theta}{\partial \tau} = 2 + \frac{\partial^2 \theta}{\partial \xi^2} \tag{1}$$

$$\text{at} \quad \tau = 0 \qquad \theta = 0 \tag{2}$$

$$\text{at} \quad \xi = 0 \qquad \theta = 0 \tag{3}$$

$$\text{at} \quad \xi = 1 \qquad \theta = 0 \tag{4}$$

b) Since Eq. (1) is not homogeneous, propose a solution in the form

$$\theta(\tau, \xi) = \theta_\infty(\xi) - \theta_t(\tau, \xi) \tag{5}$$

in which θ_∞ is the steady-state solution, i.e.,

$$\frac{d^2\theta_\infty}{d\xi^2} + 2 = 0 \tag{6}$$

with the following boundary conditions

$$\text{at} \quad \xi = 0 \qquad \theta_\infty = 0 \tag{7}$$

$$\text{at} \quad \xi = 1 \qquad \theta_\infty = 0 \tag{8}$$

Obtain the steady-state solution as

$$\theta_\infty = \xi - \xi^2 \tag{9}$$

c) Show that the governing equation for the transient contribution $\theta_t(\tau, \xi)$ is given by

$$\frac{\partial \theta_t}{\partial \tau} = \frac{\partial^2 \theta_t}{\partial \xi^2} \tag{10}$$

with the following initial and boundary conditions

$$\text{at} \quad \tau = 0 \qquad \theta_t = \xi - \xi^2 \tag{11}$$

$$\text{at} \quad \xi = 0 \qquad \theta_t = 0 \tag{12}$$

$$\text{at} \quad \xi = 1 \qquad \theta_t = 0 \tag{13}$$

Use the method of separation of variables and obtain the solution in the form

$$\theta_t = \frac{8}{\pi^3} \sum_{n=0}^{\infty} \frac{1}{(2n+1)^3} \exp\left[-(2n+1)^2\pi^2\tau\right] \sin\left[(2n+1)\pi\xi\right] \tag{14}$$

d) Integrate the velocity distribution over the flow area and show that the volumetric flow rate is given by

$$Q = \frac{(\mathcal{P}_o - \mathcal{P}_L)WB^3}{12\mu L} \left\{ 1 - \frac{96}{\pi^4} \sum_{n=0}^{\infty} \frac{1}{(2n+1)^4} \exp\left[-(2n+1)^2\pi^2\tau\right] \right\} \tag{15}$$

11.2 A stationary incompressible Newtonian fluid is contained between two concentric cylinders of radii κR and R. At time $t = 0$, a constant pressure gradient is imposed and the fluid begins to flow. Repeat the analysis given in Section 11.1 as follows:

a) Considering the flow geometry shown in Figure 9.4, write the governing differential equation, and initial and boundary conditions in terms of the following dimensionless variables

$$\theta = \frac{v_z}{\left(\dfrac{P_o - P_L}{4\mu L}\right) R^2} \qquad \xi = \frac{r}{R} \qquad \tau = \frac{vt}{R^2}$$

and show that

$$\frac{\partial \theta}{\partial \tau} = 4 + \frac{1}{\xi} \frac{\partial}{\partial \xi} \left(\xi \frac{\partial \theta}{\partial \xi} \right) \tag{1}$$

$$\text{at} \quad \tau = 0 \qquad \theta = 0 \tag{2}$$

$$\text{at} \quad \xi = \kappa \qquad \theta = 0 \tag{3}$$

$$\text{at} \quad \xi = 1 \qquad \theta = 0 \tag{4}$$

b) Since Eq. (1) is not homogeneous, propose a solution in the form

$$\theta(\tau, \xi) = \theta_\infty(\xi) - \theta_t(\tau, \xi) \tag{5}$$

in which θ_∞ is the steady-state solution, i.e.,

$$\frac{1}{\xi} \frac{d}{d\xi} \left(\xi \frac{d\theta_\infty}{d\xi} \right) + 4 = 0 \tag{6}$$

with the following boundary conditions

$$\text{at} \quad \xi = \kappa \qquad \theta_\infty = 0 \tag{7}$$

$$\text{at} \quad \xi = 1 \qquad \theta_\infty = 0 \tag{8}$$

Obtain the steady-state solution as

$$\theta_\infty = 1 - \xi^2 - \left(\frac{1 - \kappa^2}{\ln \kappa} \right) \ln \xi \tag{9}$$

c) Show that the governing equation for the transient contribution $\theta_t(\tau, \xi)$ is given by

$$\frac{\partial \theta_t}{\partial \tau} = \frac{1}{\xi} \frac{\partial}{\partial \xi} \left(\xi \frac{\partial \theta_t}{\partial \xi} \right) \tag{10}$$

with the following initial and boundary conditions

$$\text{at} \quad \tau = 0 \qquad \theta_t = 1 - \xi^2 - \left(\frac{1 - \kappa^2}{\ln \kappa} \right) \ln \xi \tag{11}$$

$$\text{at} \quad \xi = \kappa \qquad \theta_t = 0 \tag{12}$$

$$\text{at} \quad \xi = 1 \qquad \theta_t = 0 \tag{13}$$

Use the method of separation of variables and obtain the solution in the form

$$\theta_t = \sum_{n=1}^{\infty} A_n \exp\left(-\lambda_n^2 \tau\right) Z_o(\lambda_n \xi) \tag{14}$$

where

$$Z_n(\lambda_n \xi) = \frac{Y_o(\lambda_n \kappa) J_n(\lambda_n \xi) - J_o(\lambda_n \kappa) Y_n(\lambda_n \xi)}{J_o(\lambda_n \kappa) Y_o(\lambda_n \kappa)} \tag{15}$$

and the eigenvalues λ_n are the roots of

$$Z_o(\lambda_n) = 0 \tag{16}$$

d) The unknown coefficients A_n in Eq. (14) can be determined by using the initial condition given by Eq. (11). Note that Eqs. (B.2-30)–(B.2-32) in Appendix B are also applicable for Z, i.e.,

$$\int x Z_o(\lambda x)\, dx = \frac{x}{\lambda} Z_1(\lambda x) \tag{17}$$

$$\int x^3 Z_o(\lambda x)\, dx = \left(\frac{x^3}{\lambda} - \frac{4x}{\lambda^3}\right) Z_1(\lambda x) + \frac{2x^2}{\lambda^2} Z_o(\lambda x) \tag{18}$$

$$\int x Z_o^2(\lambda x)\, dx = \frac{x^2}{2}\left[Z_o^2(\lambda x) + Z_1^2(\lambda x)\right] \tag{19}$$

and show that

$$A_n = \frac{8}{\lambda_n^3}\left[\frac{1}{Z_1(\lambda_n) + \kappa Z_1(\lambda_n \kappa)}\right] \tag{20}$$

e) Integrate the velocity distribution over the flow area and show that the volumetric flow rate is given by

$$Q = \frac{\pi(\mathcal{P}_o - \mathcal{P}_L) R^4}{8\mu L}$$

$$\times \left\{ 1 - \kappa^4 + \frac{(1 - \kappa^2)^2}{\ln \kappa} - 32 \sum_{n=1}^{\infty} \frac{1}{\lambda_n^4}\left[\frac{Z_1(\lambda_n) - \kappa Z_1(\lambda_n \kappa)}{Z_1(\lambda_n) + \kappa Z_1(\lambda_n \kappa)}\right] \exp\left(-\lambda_n^2 \tau\right) \right\} \tag{21}$$

f) Use Eq. (16) together with the identity

$$J_1(x) Y_o(x) - J_o(x) Y_1(x) = \frac{2}{\pi x} \tag{22}$$

to simplify Eq. (21) to

$$Q = \frac{\pi(\mathcal{P}_o - \mathcal{P}_L) R^4}{8\mu L}\left\{ 1 - \kappa^4 + \frac{(1 - \kappa^2)^2}{\ln \kappa} - 32 \sum_{n=1}^{\infty} \frac{1}{\lambda_n^4}\left[\frac{J_o(\lambda_n \kappa) - J_o(\lambda_n)}{J_o(\lambda_n \kappa) + J_o(\lambda_n)}\right] \exp\left(-\lambda_n^2 \tau\right) \right\}$$
$$\tag{23}$$

g) When $\tau \to \infty$, show that Eq. (23) reduces to Eq. (9.1-99). Also show that Eq. (23) reduces to Eq. (11.1-47) when $\kappa \to 0$.

11.3 Repeat the analysis given in Section 11.2.1 for the geometry shown in the figure below and show that the dimensionless temperature distribution is given by

$$\theta = \frac{\Omega}{2}(\xi - \xi^2) + \frac{4}{\pi} \sum_{n=1,3,5}^{\infty} \frac{1}{n}\left(1 - \frac{\Omega}{n^2\pi^2}\right) \exp(-n^2\pi^2\tau)\sin(n\pi\xi)$$

where

$$\theta = \frac{T - T_\infty}{T_o - T_\infty} \qquad \xi = \frac{z}{L} \qquad \tau = \frac{\alpha t}{L^2} \qquad \Omega = \frac{\Re L^2}{k(T_o - T_\infty)}$$

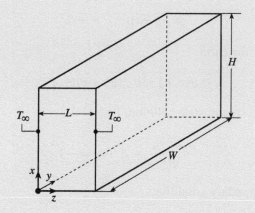

11.4 A rectangular slab of thickness $2L$ is initially at a uniform temperature of T_o. At $t = 0$, heat starts to generate within the slab with a volumetric generation rate of

$$\Re = a + bT$$

where a and b are known constants. To avoid excessive heating of the slab, both surfaces are kept constant at temperature T_o.

a) Show that the governing equation for temperature together with the initial and boundary conditions are given by

$$\rho \widehat{C}_P \frac{\partial T}{\partial t} = k\frac{\partial^2 T}{\partial z^2} + a + bT \tag{1}$$

$$\text{at} \quad t = 0 \qquad T = T_o \tag{2}$$

$$\text{at} \quad z = 0 \qquad \frac{\partial T}{\partial z} = 0 \tag{3}$$

$$\text{at} \quad z = L \qquad T = T_o \tag{4}$$

b) In terms of the following variables

$$\theta = \frac{T - T_o}{\dfrac{(a + bT_o)L^2}{k}} \qquad \xi = \frac{z}{L} \qquad \tau = \frac{\alpha t}{L^2} \qquad \Lambda = \sqrt{\frac{bL^2}{k}} \qquad (5)$$

show that Eqs. (1)–(4) reduce to

$$\frac{\partial \theta}{\partial \tau} = \frac{\partial^2 \theta}{\partial \xi^2} + 1 + \Lambda^2 \theta \qquad (6)$$

$$\text{at} \quad \tau = 0 \qquad \theta = 0 \qquad (7)$$

$$\text{at} \quad \xi = 0 \qquad \frac{\partial \theta}{\partial \xi} = 0 \qquad (8)$$

$$\text{at} \quad \xi = 1 \qquad \theta = 0 \qquad (9)$$

c) Propose a solution of the form

$$\theta(\tau, \xi) = \theta_\infty(\xi) - \theta_t(\tau, \xi) \qquad (10)$$

in which $\theta_\infty(\xi)$ is the steady-state solution, i.e.,

$$\frac{d^2 \theta_\infty}{d\xi^2} + \Lambda^2 \theta_\infty = -1 \qquad (11)$$

with the following boundary conditions

$$\text{at} \quad \xi = 0 \qquad \frac{d\theta_\infty}{d\xi} = 0 \qquad (12)$$

$$\text{at} \quad \xi = 1 \qquad \theta_\infty = 0 \qquad (13)$$

Show that the steady-state solution is given by

$$\theta_\infty = \frac{1}{\Lambda^2} \left[\frac{\cos(\Lambda \xi)}{\cos \Lambda} - 1 \right] \qquad (14)$$

d) Show that the governing equation for the transient contribution $\theta_t(\tau, \xi)$ is given by

$$\frac{\partial \theta_t}{\partial \tau} = \frac{\partial^2 \theta_t}{\partial \xi^2} + \Lambda^2 \theta_t \qquad (15)$$

$$\text{at} \quad \tau = 0 \qquad \theta_t = \frac{1}{\Lambda^2} \left[\frac{\cos(\Lambda \xi)}{\cos \Lambda} - 1 \right] \qquad (16)$$

$$\text{at} \quad \xi = 0 \qquad \frac{\partial \theta_t}{\partial \xi} = 0 \qquad (17)$$

$$\text{at} \quad \xi = 1 \qquad \theta_t = 0 \qquad (18)$$

e) Solve Eq. (15) and show that the solution is given by

$$\theta_t = \sum_{n=0}^{\infty} A_n \exp\left[-(\lambda_n^2 - \Lambda^2)\tau\right]\cos(\lambda_n\xi) \tag{19}$$

where

$$\lambda_n = \left(n + \frac{1}{2}\right)\pi \qquad n = 0, 1, 2, \ldots \tag{20}$$

and the coefficients A_n are given by

$$A_n = \frac{2(-1)^n}{\lambda_n(\lambda_n^2 - \Lambda^2)} \tag{21}$$

f) Show that the solution is stable as long as

$$\frac{\pi}{2} > \sqrt{\frac{bL^2}{k}} \tag{22}$$

11.5   A long cylindrical rod of radius R is initially at a uniform temperature of T_o. At $t = 0$, heat starts to generate within the rod with a volumetric generation rate of

$$\Re = a + bT$$

where a and b are known constants. To avoid excessive heating of the rod, the outer surface is kept constant at temperature T_o.

a) In terms of the following variables

$$\theta = \frac{T - T_o}{\dfrac{(a + bT_o)R^2}{k}} \qquad \xi = \frac{r}{R} \qquad \tau = \frac{\alpha t}{R^2} \qquad \Lambda = \sqrt{\frac{bR^2}{k}} \tag{1}$$

show that the governing equation, together with the initial and boundary conditions, takes the form

$$\frac{\partial\theta}{\partial\tau} = \frac{1}{\xi}\frac{\partial}{\partial\xi}\left(\xi\frac{\partial\theta}{\partial\xi}\right) + 1 + \Lambda^2\theta \tag{2}$$

$$\text{at} \quad \tau = 0 \qquad \theta = 0 \tag{3}$$

$$\text{at} \quad \xi = 0 \qquad \frac{\partial\theta}{\partial\xi} = 0 \tag{4}$$

$$\text{at} \quad \xi = 1 \qquad \theta = 0 \tag{5}$$

b) Follow the procedure outlined in Problem 11.4 and show that the solution is given by

$$\theta = \frac{1}{\Lambda^2}\left[\frac{J_o(\Lambda\xi)}{J_o(\Lambda)} - 1\right] - 2\sum_{n=1}^{\infty}\frac{1}{\lambda_n(\lambda_n^2 - \Lambda^2)}\exp\left[-(\lambda_n^2 - \Lambda^2)\tau\right]J_o(\lambda_n\xi) \tag{6}$$

where the eigenvalues are the positive roots of the equation

$$J_o(\lambda_n) = 0 \tag{7}$$

Also conclude that the solution is stable as long as

$$\lambda_1^2 > \frac{bR^2}{k} \tag{8}$$

11.6 A cylindrical rod of radius R is initially at a uniform temperature of T_o. At $t = 0$, heat starts to generate within the rod and the rate of heat generation per unit volume is given by

$$\Re = \Re_o r^2 \tag{1}$$

where $\Re_o$ is a known constant. The outer surface of the rod is maintained constant at T_R to avoid excessive heating of the rod.

a) Consider a cylindrical differential volume element of thickness Δr and length L within the rod and show that the conservation statement for energy leads to

$$\rho \widehat{C}_P \frac{\partial T}{\partial t} = \frac{k}{r} \frac{\partial}{\partial r} \left(r \frac{\partial T}{\partial r} \right) + \Re_o r^2 \tag{2}$$

with the following initial and boundary conditions

$$\text{at} \quad t = 0 \qquad T = T_o \tag{3}$$

$$\text{at} \quad r = 0 \qquad \frac{\partial T}{\partial r} = 0 \tag{4}$$

$$\text{at} \quad r = R \qquad T = T_R \tag{5}$$

b) In terms of the dimensionless quantities

$$\theta = \frac{T - T_R}{T_o - T_R} \qquad \xi = \frac{r}{R} \qquad \tau = \frac{\alpha t}{R^2} \qquad \Omega = \frac{\Re_o R^4}{k(T_o - T_R)}$$

show that Eqs. (2)–(5) become

$$\frac{\partial \theta}{\partial \tau} = \frac{1}{\xi} \frac{\partial}{\partial \xi} \left(\xi \frac{\partial \theta}{\partial \xi} \right) + \Omega \xi^2 \tag{6}$$

$$\text{at} \quad \tau = 0 \qquad \theta = 1 \tag{7}$$

$$\text{at} \quad \xi = 0 \qquad \frac{\partial \theta}{\partial \xi} = 0 \tag{8}$$

$$\text{at} \quad \xi = 1 \qquad \theta = 0 \tag{9}$$

c) Since Eq. (6) is not homogeneous, propose a solution in the form

$$\theta(\tau, \xi) = \theta_\infty(\xi) - \theta_t(\tau, \xi) \tag{10}$$

in which $\theta_\infty(\xi)$ is the steady-state solution, i.e.,

$$\frac{1}{\xi}\frac{d}{d\xi}\left(\xi\frac{d\theta_\infty}{d\xi}\right) + \Omega\xi^2 = 0 \tag{11}$$

with the following boundary conditions

$$\text{at} \quad \xi = 0 \qquad \frac{d\theta_\infty}{d\xi} = 0 \tag{12}$$

$$\text{at} \quad \xi = 1 \qquad \theta_\infty = 0 \tag{13}$$

Obtain the steady-state solution as

$$\theta_\infty = \frac{\Omega}{16}(1 - \xi^4) \tag{14}$$

d) Show that the governing equation for the transient contribution $\theta_t(\tau, \xi)$ is given by

$$\frac{\partial\theta_t}{\partial\tau} = \frac{1}{\xi}\frac{\partial}{\partial\xi}\left(\xi\frac{\partial\theta_t}{\partial\xi}\right) \tag{15}$$

with the following initial and boundary conditions

$$\text{at} \quad \tau = 0 \qquad \theta_t = \frac{\Omega}{16}(1 - \xi^4) - 1 \tag{16}$$

$$\text{at} \quad \xi = 0 \qquad \frac{\partial\theta_t}{\partial\xi} = 0 \tag{17}$$

$$\text{at} \quad \xi = 1 \qquad \theta_t = 0 \tag{18}$$

Use the method of separation of variables and show that the solution of Eq. (15) is given by

$$\theta_t = \sum_{n=1}^{\infty} A_n \exp\left(-\lambda_n^2\tau\right)J_o(\lambda_n\xi) \tag{19}$$

where the eigenvalues λ_n are the roots of

$$J_o(\lambda_n) = 0 \tag{20}$$

and the coefficients are given by

$$A_n = \frac{\displaystyle\int_0^1 \left[\frac{\Omega}{16}(1 - \xi^4) - 1\right]\xi J_o(\lambda_n\xi)\, d\xi}{\displaystyle\int_0^1 \xi J_o^2(\lambda_n\xi)\, d\xi} \tag{21}$$

e) Evaluate the integrals in Eq. (21) and show that

$$A_n = \frac{2}{\lambda_n}\left[\frac{\Omega}{\lambda_n^2}\left(1 - \frac{4}{\lambda_n^2}\right) - 1\right]\frac{1}{J_1(\lambda_n)} \tag{22}$$

11.7   A solid sphere of radius R is initially at a temperature of T_o. At $t = 0$, the solid sphere experiences a uniform internal heat generation rate per unit volume, $\Re$, and heat is dissipated from the surface to the surrounding fluid at a temperature of T_∞ with an average heat transfer coefficient of $\langle h \rangle$.

a) Consider a spherical differential volume element of thickness Δr within the solid and show that the conservation statement for energy leads to

$$\rho \widehat{C}_P \frac{\partial T}{\partial t} = \frac{k}{r^2} \frac{\partial}{\partial r}\left(r^2 \frac{\partial T}{\partial r}\right) + \Re \tag{1}$$

with the following initial and boundary conditions

$$\text{at} \quad t = 0 \qquad T = T_o \tag{2}$$

$$\text{at} \quad r = 0 \qquad \frac{\partial T}{\partial r} = 0 \tag{3}$$

$$\text{at} \quad r = R \qquad -k\frac{\partial T}{\partial r} = \langle h \rangle (T - T_\infty) \tag{4}$$

b) In terms of the dimensionless quantities

$$\theta = \frac{T - T_\infty}{T_o - T_\infty} \qquad \xi = \frac{r}{R} \qquad \tau = \frac{\alpha t}{R^2} \qquad \Omega = \frac{\Re R^2}{k(T_o - T_\infty)} \qquad \text{Bi}_\text{H} = \frac{\langle h \rangle R}{k}$$

show that Eqs. (1)–(4) become

$$\frac{\partial \theta}{\partial \tau} = \frac{1}{\xi^2} \frac{\partial}{\partial \xi}\left(\xi^2 \frac{\partial \theta}{\partial \xi}\right) + \Omega \tag{5}$$

$$\text{at} \quad \tau = 0 \qquad \theta = 1 \tag{6}$$

$$\text{at} \quad \xi = 0 \qquad \frac{\partial \theta}{\partial \xi} = 0 \tag{7}$$

$$\text{at} \quad \xi = 1 \qquad -\frac{\partial \theta}{\partial \xi} = \text{Bi}_\text{H}\, \theta \tag{8}$$

c) Since Eq. (5) is not homogeneous, propose a solution in the form

$$\theta(\tau, \xi) = \theta_\infty(\xi) - \theta_t(\tau, \xi) \tag{9}$$

in which $\theta_\infty(\xi)$ is the steady-state solution, i.e.,

$$\frac{1}{\xi^2} \frac{d}{d\xi}\left(\xi^2 \frac{d\theta_\infty}{d\xi}\right) + \Omega = 0 \tag{10}$$

with the following boundary conditions

$$\text{at} \quad \xi = 0 \qquad \frac{d\theta_\infty}{d\xi} = 0 \tag{11}$$

$$\text{at} \quad \xi = 1 \qquad -\frac{d\theta_\infty}{d\xi} = \text{Bi}_\text{H}\, \theta_\infty \tag{12}$$

Obtain the steady-state solution as

$$\theta_\infty = \frac{\Omega}{6}(1 - \xi^2) + \frac{\Omega}{3\,\mathrm{Bi_H}} \tag{13}$$

d) Show that the governing equation for the transient contribution $\theta_t(\tau, \xi)$ is given by

$$\frac{\partial \theta_t}{\partial \tau} = \frac{1}{\xi^2} \frac{\partial}{\partial \xi}\left(\xi^2 \frac{\partial \theta_t}{\partial \xi}\right) \tag{14}$$

with the following initial and boundary conditions

$$\text{at} \quad \tau = 0 \qquad \theta_t = \frac{\Omega}{6}(1 - \xi^2) + \frac{\Omega}{3\,\mathrm{Bi_H}} - 1 \tag{15}$$

$$\text{at} \quad \xi = 0 \qquad \frac{\partial \theta_t}{\partial \xi} = 0 \tag{16}$$

$$\text{at} \quad \xi = 1 \qquad -\frac{\partial \theta_t}{\partial \xi} = \mathrm{Bi_H}\,\theta_t \tag{17}$$

First convert the spherical geometry to rectangular geometry by introducing a new dependent variable as

$$\theta_t = \frac{u}{\xi} \tag{18}$$

then use the method of separation of variables and show that the solution of Eq. (14) is given by

$$\theta_t = \sum_{n=1}^{\infty} A_n\, e^{-\lambda_n^2 \tau} \frac{\sin(\lambda_n \xi)}{\xi} \tag{19}$$

where the eigenvalues λ_n are the roots of

$$\lambda_n \cot \lambda_n = 1 - \mathrm{Bi_H} \tag{20}$$

and the coefficients A_n are given by

$$A_n = 2\,\mathrm{Bi_H}\left(\frac{\Omega}{\lambda_n^2} - 1\right) \frac{\cos \lambda_n}{\lambda_n(1 - \mathrm{Bi_H} - \cos^2 \lambda_n)} \tag{21}$$

11.8 In Section 10.3, when the initial concentration is zero, solutions to diffusion problems without a homogeneous reaction, i.e., Eqs. (10.3-20), (10.3-55), (10.3-82), (10.3-97), (10.3-125), and (10.3-141), are expressed in the form

$$\frac{c_A^{\text{no rxn}}}{c_A^*} = 1 - \sum \varphi \exp(-\beta t) \tag{1}$$

where c_A^* is the concentration on the surface and φ is a position-dependent function. Show that the use of Eq. (1) in Eq. (11.3-93) leads to the following concentration distribution in the case of diffusion with a homogeneous reaction

$$\frac{c_A}{c_A^*} = 1 - \sum \varphi \left\{ \frac{1 + \Omega \exp[-(1 + \Omega)kt]}{1 + \Omega} \right\} \tag{2}$$

where

$$\Omega = \frac{\beta}{k} \tag{3}$$

with k being the first-order reaction rate constant.

11.9 For diffusion accompanied by a first-order homogeneous reaction, Danckwerts (1951) showed that the molar rate of absorption of species $\mathcal{A}$ can be calculated by the following equation

$$\dot{n}_A = k \int_0^t \dot{n}_A^{\text{no rxn}}(\eta) e^{-k\eta} \, d\eta + \dot{n}_A^{\text{no rxn}}(t) e^{-kt} \tag{1}$$

where $\dot{n}_A^{\text{no rxn}}$ is the molar rate of absorption of species $\mathcal{A}$ without a chemical reaction. In Chapter 10, the molar rate of absorption of species $\mathcal{A}$, i.e., Eqs. (10.3-27), (10.3-89), and (10.3-132), is expressed in the form

$$\dot{n}_A^{\text{no rxn}} = \sum \chi \exp(-\beta t) \tag{2}$$

where the function χ does not depend on position or time.

a) Show that the use of Eq. (2) in Eq. (1) leads to

$$\dot{n}_A = \sum \chi \left\{ \frac{1 + \Omega \exp[-(1 + \Omega)kt]}{1 + \Omega} \right\} \tag{3}$$

where $\Omega = \beta/k$.

b) Consider the unsteady diffusion of species $\mathcal{A}$ into a cylinder of radius R. If the cylinder is initially $\mathcal{A}$-free and $0.1 < \text{Bi}_M < 40$, start with Eq. (10.3-97) and show that

$$\dot{n}_A^{\text{no rxn}} = 4\pi L \mathcal{D}_{AB} \mathcal{H} c_{A_\infty} \, \text{Bi}_M^2 \sum_{n=1}^{\infty} \frac{1}{(\lambda_n^2 + \text{Bi}_M^2)} \exp(-\lambda_n^2 \tau) \tag{4}$$

c) Show that the molar rate of absorption with a first-order homogeneous chemical reaction is given by

$$\dot{n}_A = 4\pi L \mathcal{D}_{AB} \mathcal{H} c_{A_\infty} \, \text{Bi}_M^2 \sum_{n=1}^{\infty} \frac{1}{(\lambda_n^2 + \text{Bi}_M^2)} \left\{ \frac{1 + \Omega \exp[-(1 + \Omega)kt]}{1 + \Omega} \right\} \tag{5}$$

where

$$\Omega = \frac{\lambda_n^2 \mathcal{D}_{AB}}{kR^2} \tag{6}$$

Appendix A

MATHEMATICAL PRELIMINARIES

A.1 CYLINDRICAL AND SPHERICAL COORDINATE SYSTEMS

For cylindrical coordinates, the variables (r, θ, z) are related to the rectangular coordinates (x, y, z) as follows:

$$x = r\cos\theta \qquad r = \sqrt{x^2 + y^2} \tag{A.1-1}$$

$$y = r\sin\theta \qquad \theta = \arctan(y/x) \tag{A.1-2}$$

$$z = z \qquad z = z \tag{A.1-3}$$

The ranges of the variables (r, θ, z) are

$$0 \leqslant r \leqslant \infty \qquad 0 \leqslant \theta \leqslant 2\pi \qquad -\infty \leqslant z \leqslant \infty$$

For spherical coordinates, the variables (r, θ, ϕ) are related to the rectangular coordinates (x, y, z) as follows:

$$x = r\sin\theta\cos\phi \qquad r = \sqrt{x^2 + y^2 + z^2} \tag{A.1-4}$$

$$y = r\sin\theta\sin\phi \qquad \theta = \arctan\left(\sqrt{x^2 + y^2}/z\right) \tag{A.1-5}$$

$$z = r\cos\theta \qquad \phi = \arctan(y/x) \tag{A.1-6}$$

The ranges of the variables (r, θ, ϕ) are

$$0 \leqslant r \leqslant \infty \qquad 0 \leqslant \theta \leqslant \pi \qquad 0 \leqslant \phi \leqslant 2\pi$$

The cylindrical and spherical coordinate systems are shown in Figure A.1. The differential volumes in these coordinate systems are given by

$$dV = \begin{cases} r\,dr\,d\theta\,dz & \text{cylindrical} \\ r^2\sin\theta\,dr\,d\theta\,d\phi & \text{spherical} \end{cases} \tag{A.1-7}$$

The application of Eq. (1.3-1) to determine the rate of a quantity requires the integration of the flux of a quantity over a differential area. The differential areas in the cylindrical and spherical coordinate systems are given as follows:

$$dA_{cylindrical} = \begin{cases} R\,d\theta\,dz & \text{flux is in the } r\text{-direction} \\ dr\,dz & \text{flux is in the } \theta\text{-direction} \\ r\,dr\,d\theta & \text{flux is in the } z\text{-direction} \end{cases} \tag{A.1-8}$$

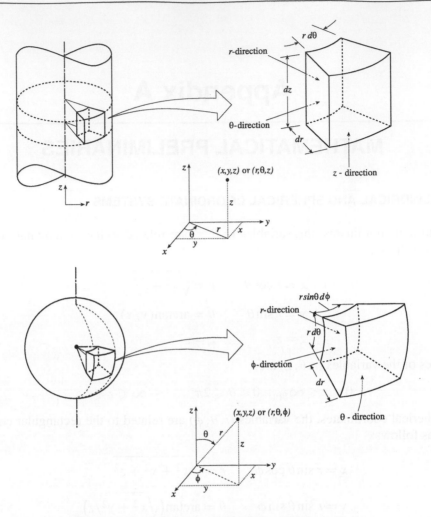

Figure A.1. The cylindrical and spherical coordinate systems.

$$dA_{spherical} = \begin{cases} R^2 \sin\theta \, d\theta \, d\phi & \text{flux is in the } r\text{-direction} \\ r\sin\theta \, dr \, d\phi & \text{flux is in the } \theta\text{-direction} \\ r \, dr \, d\theta & \text{flux is in the } \phi\text{-direction} \end{cases} \tag{A.1-9}$$

A.2   MEAN VALUE THEOREM

If $f(x)$ is continuous in the interval $a \leqslant x \leqslant b$, then the value of the integration of $f(x)$ over an interval $x = a$ to $x = b$ is

$$I = \int_a^b f(x) \, dx = \langle f \rangle \int_a^b dx = \langle f \rangle (b - a) \tag{A.2-1}$$

where $\langle f \rangle$ is the average value of f in the interval $a \leqslant x \leqslant b$.

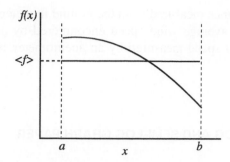

Figure A.2. The mean value of the function $f(x)$.

In Figure A.2 note that $\int_a^b f(x)\,dx$ is the area under the curve between a and b. On the other hand, $\langle f \rangle (b - a)$ is the area under the rectangle of height $\langle f \rangle$ and width $(b - a)$. The average value of f, $\langle f \rangle$, is defined such that these two areas are equal to each other.

It is possible to extend the definition of the mean value to two- and three-dimensional cases as

$$\langle f \rangle = \frac{\displaystyle\iint_A f(x, y)\,dx\,dy}{\displaystyle\iint_A dx\,dy} \quad \text{and} \quad \langle f \rangle = \frac{\displaystyle\iiint_V f(x, y, z)\,dx\,dy\,dz}{\displaystyle\iiint_V dx\,dy\,dz} \tag{A.2-2}$$

PROBLEMS

A.1 Two rooms have the same average temperature, $\langle T \rangle$, defined by

$$\langle T \rangle = \frac{\displaystyle\iiint_V T(x, y, z)\,dx\,dy\,dz}{\displaystyle\iiint_V dx\,dy\,dz}$$

However, while one of the rooms is very comfortable, the other is very uncomfortable. With the mean value theorem in mind, how would you explain the difference in comfort levels between the two rooms? What design alterations would you suggest to make the uncomfortable room comfortable?

A.2 Wind speed is measured by anemometers placed at an altitude of 10 m from the ground. Buckler (1969) carried out a series of experiments to determine the effect of height above ground level on wind speed and proposed the following equation for the winter months

$$v = v_{10} \left(\frac{z}{10} \right)^{0.21}$$

where z is the vertical distance measured from the ground in meters and v_{10} is the measured wind speed. Estimate the average wind speed encountered by a person of height 1.7 m at ground level if the wind speed measured by an anemometer 10 m above the ground is 30 km/h.

(**Answer:** 17.1 km/h)

A.3 SLOPES ON LOG-LOG AND SEMI-LOG GRAPH PAPER

A mathematical transformation that converts the logarithm of a number to a length in the x-direction is given by

$$\ell_x = L_x \log x \tag{A.3-1}$$

where ℓ_x is the distance in the x-direction and L_x is the cycle length for the x-coordinate. Therefore, if the cycle length is taken as 10 cm, the distances in the x-direction for various values of x are given in Table A.1.

The slope of a straight line, m, on log-log graph paper is

$$m = \frac{\log y_2 - \log y_1}{\log x_2 - \log x_1} = \left(\frac{\ell_{y_2} - \ell_{y_1}}{\ell_{x_2} - \ell_{x_1}}\right)\frac{L_x}{L_y} \tag{A.3-2}$$

On the other hand, the slope of a straight line, m, on semi-log graph paper (y-axis is logarithmic) is

$$m = \frac{\log y_2 - \log y_1}{x_2 - x_1} = \left(\frac{\ell_{y_2} - \ell_{y_1}}{x_2 - x_1}\right)\frac{1}{L_y} \tag{A.3-3}$$

A.4 LEIBNITZ'S RULE FOR DIFFERENTIATION OF INTEGRALS

Let $f(x, t)$ be continuous and have a continuous derivative $\partial f/\partial t$ in a domain of the xt plane, which includes the rectangle $a \leqslant x \leqslant b$, $t_1 \leqslant t \leqslant t_2$. Then for $t_1 \leqslant t \leqslant t_2$

$$\frac{d}{dt} \int_a^b f(x, t)\, dx = \int_a^b \frac{\partial f}{\partial t}\, dx \tag{A.4-1}$$

In other words, differentiation and integration can be interchanged if the limits of the integration are fixed.

On the other hand, if the limits of the integral in Eq. (A.4-1) are dependent on time, then

$$\frac{d}{dt} \int_{a(t)}^{b(t)} f(x, t)\, dx = \int_{a(t)}^{b(t)} \frac{\partial f}{\partial t}\, dx + f\big[b(t), t\big]\frac{db}{dt} - f\big[a(t), t\big]\frac{da}{dt} \tag{A.4-2}$$

Table A.1. Distances in the x-direction for a logarithmic x-axis

x	1	2	3	4	5	6	7	8	9	10
ℓ_x	0.00	3.01	4.77	6.02	6.99	7.78	8.45	9.03	9.54	10.00

If $f = f(x)$ only, then Eq. (A.4-2) reduces to

$$\frac{d}{dt}\int_{a(t)}^{b(t)} f(x)\,dx = f[b(t)]\frac{db}{dt} - f[a(t)]\frac{da}{dt} \tag{A.4-3}$$

A.5 NUMERICAL DIFFERENTIATION OF EXPERIMENTAL DATA

The determination of a rate requires the differentiation of the original experimental data. As explained by De Nevers (1966), given a table of $x - y$ data, the value of dy/dx can be calculated by:

1. Plotting the data on graph paper, drawing a smooth curve through the points with the help of a French curve, and then drawing a tangent to this curve.
2. Fitting the entire set of data with an empirical equation, such as a polynomial, and then differentiating the empirical equation.
3. Fitting short sections of the data by using arbitrary functions.
4. Using the difference table method, i.e., plotting the differences and smoothing them graphically.

De Nevers also points out the fact that although the value of dy/dx obtained by any of the above four methods is approximately equal to each other, the value of d^2y/dx^2 is extremely sensitive to the method used.

In the case of the graphical method, there are an infinite number of ways of drawing the curve through the data points. As a result, the slope of the tangent will be affected by the mechanics of drawing the curved line and the tangent.

The availability of computer programs makes the second and third methods very attractive. However, since the choice of the functional form of the equation is highly arbitrary, the final result is almost as subjective and biased as that obtained using a French curve.

Two methods, namely the Douglass-Avakian (1933) and Whitaker-Pigford (1960) methods, are worth mentioning as part of the third approach. Both methods require the values of the independent variable, x, be *equally* spaced by an amount Δx.

A.5.1 Douglass-Avakian Method

In this method, the value of dy/dx is determined by fitting a fourth-degree polynomial to seven consecutive data points, with the point in question as the mid-point, by least squares. If the mid-point is designated by x_c, then the value of dy/dx at this particular location is given by

$$\frac{dy}{dx} = \frac{397(\sum Xy) - 49(\sum X^3 y)}{1512\Delta x} \tag{A.5-1}$$

where

$$X = \frac{x - x_c}{\Delta x} \tag{A.5-2}$$

A.5.2 Whitaker-Pigford Method

In this case, a parabola is fitted to five consecutive data points, with the point in question as the mid-point, by least squares. The value of dy/dx at x_c is given by

$$\frac{dy}{dx} = \frac{\sum Xy}{10\Delta x} \tag{A.5-3}$$

where X is defined by Eq. (A.5-2).

Example A.1 Given the enthalpy of steam at $P = 0.01$ MPa as a function of temperature as follows, determine the heat capacity at constant pressure at 500 °C.

T (°C)	$\widehat{H}$ (J/g)	T (°C)	$\widehat{H}$ (J/g)
100	2687.5	700	3928.7
200	2879.5	800	4159.0
300	3076.5	900	4396.4
400	3279.6	1000	4640.0
500	3489.1	1100	4891.2
600	3705.4		

Solution

The heat capacity at constant pressure, $\widehat{C}_P$, is defined as $(\partial \widehat{H}/\partial T)_P$. Therefore, determination of $\widehat{C}_P$ requires numerical differentiation of the $\widehat{H}$ versus the T data.

Graphical method

The plot of $\widehat{H}$ versus T is given in the figure below. The slope of the tangent to the curve at $T = 500$ °C gives $\widehat{C}_P = 2.12$ J/g·K.

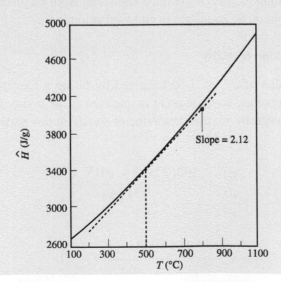

Douglass-Avakian method

The values required to use Eq. (A.5-1) are given in the table below:

$x = T$	$y = \widehat{H}$	X	Xy	$X^3 y$
200	2879.5	−3	−8638.5	−77,746.5
300	3076.5	−2	−6153	−24,612
400	3279.6	−1	−3279.6	−3279.6
500	3489.1	0	0	0
600	3705.4	1	3705.4	3705.4
700	3928.7	2	7857.4	31,429.6
800	4159.0	3	12,477	112,293
			$\sum = 5968.7$	$\sum = 41,789.9$

Therefore, the heat capacity at constant pressure at 500 °C is given by

$$\widehat{C}_P = \frac{397(\sum Xy) - 49(\sum X^3 y)}{1512 \Delta x} = \frac{(397)(5968.7) - (49)(41,789.9)}{(1512)(100)} = 2.13 \text{ J/g·K}$$

Whitaker-Pigford method

By taking $X = T$ and $y = \widehat{H}$, the parameters in Eq. (A.5-3) are given in the following table:

$X = T$	$y = \widehat{H}$	X	Xy
300	3076.5	−2	−6153
400	3279.6	−1	−3279.6
500	3489.1	0	0
600	3705.4	1	3705.4
700	3928.7	2	7857.4
			$\sum = 2130.2$

Therefore, the use of Eq. (A.5-3) gives the heat capacity at constant pressure as

$$\widehat{C}_P = \frac{\sum Xy}{10 \Delta x} = \frac{2130.2}{(10)(100)} = 2.13 \text{ J/g·K}$$

The difference table method

The use of the difference table method is explained in detail by Churchill (1974). To smooth the data by using this method, the divided differences $\Delta \widehat{H} / \Delta T$ shown in the table below are plotted versus temperature in the figure.

T	$\widehat{H}$	ΔT	$\Delta \widehat{H}$	$\Delta \widehat{H}/\Delta T$
100	2687.5			
		100	192	1.92
200	2879.5			
		100	197	1.97
300	3076.5			
		100	203.1	2.031
400	3279.6			
		100	209.5	2.095
500	3489.1			
		100	216.3	2.163
600	3705.4			
		100	223.3	2.233
700	3928.7			
		100	230.3	2.303
800	4159.0			
		100	237.4	2.374
900	4396.4			
		100	243.6	2.436
1000	4640.0			
		100	251.2	2.512
1100	4891.2			

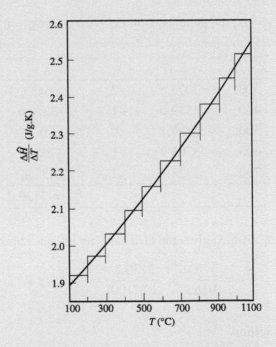

Each line represents the average value of $d\widehat{H}/dT$ over the specified temperature range. The smooth curve should be drawn so as to equalize the area under the group of bars. From the figure, the heat capacity at constant pressure at 500 °C is 2.15 J/g·K.

A.6 REGRESSION AND CORRELATION

To predict the mechanism of a process, we often need to know the relationship of one process variable to another, i.e., how the reactor yield depends on pressure. A relationship between the two variables x and y, measured over a range of values, can be obtained by proposing linear relationships first, because they are the simplest. The analyses we use for this are *correlation*, which indicates whether there is indeed a linear relationship, and *regression*, which finds the equation of a straight line that best fits the observed x-y data.

A.6.1 Simple Linear Regression

The equation describing a straight line is

$$y = ax + b \qquad \text{(A.6-1)}$$

where a denotes the slope of the line and b denotes the y-axis intercept. Most of the time the variables x and y do not have a linear relationship. However, transformation of the variables may result in a linear relationship. Some examples of transformation are given in Table A.2. Thus, linear regression can be applied even to nonlinear data.

A.6.2 Sum of Squared Deviations

Suppose we have a set of observations x_1, x_2, x_3, ..., x_N. The sum of the squares of their deviations from some mean value, x_m, is

$$S = \sum_{i=1}^{N} (x_i - x_m)^2 \qquad \text{(A.6-2)}$$

Now suppose we wish to minimize S with respect to the mean value x_m, i.e.,

$$\frac{\partial S}{\partial x_m} = 0 = \sum_{i=1}^{N} -2(x_i - x_m) = 2\left(Nx_m - \sum_{i=1}^{N} x_i\right) \qquad \text{(A.6-3)}$$

or,

$$x_m = \frac{1}{N} \sum_i x_i = \overline{x} \qquad \text{(A.6-4)}$$

Therefore, the mean value that minimizes the sum of the squares of the deviations is the arithmetic mean, $\overline{x}$.

Table A.2. Transformation of nonlinear equations to linear forms

Nonlinear Form	Linear Form	
$y = \dfrac{ax}{b+cx}$	$\dfrac{x}{y} = \dfrac{c}{a}x + \dfrac{b}{a}$	$\dfrac{x}{y}$ vs x is linear
	$\dfrac{1}{y} = \dfrac{b}{a}\dfrac{1}{x} + \dfrac{c}{a}$	$\dfrac{1}{y}$ vs $\dfrac{1}{x}$ is linear
$y = ax^n$	$\log y = n \log x + \log a$	$\log y$ vs $\log x$ is linear

A.6.3 The Method of Least Squares

The parameters a and b in Eq. (A.6-1) are estimated by the method of least squares. These values have to be chosen such that the sum of the squares of the deviations

$$S = \sum_{i=1}^{N}\left[y_i - (ax_i + b)\right]^2 \tag{A.6-5}$$

is minimum. This is accomplished by differentiating the function S with respect to a and b, and setting these derivatives equal to zero:

$$\frac{\partial S}{\partial a} = 0 = -2\sum_i (y_i - ax_i - b)x_i \tag{A.6-6}$$

$$\frac{\partial S}{\partial b} = 0 = -2\sum_i (y_i - ax_i - b) \tag{A.6-7}$$

Equations (A.6-6) and (A.6-7) can be simplified as

$$a\sum_i x_i^2 + b\sum_i x_i = \sum_i x_i y_i \tag{A.6-8}$$

$$a\sum_i x_i + Nb = \sum_i y_i \tag{A.6-9}$$

Simultaneous solution of Eqs. (A.6-8) and (A.6-9) gives

$$a = \frac{N(\sum_i x_i y_i) - (\sum_i x_i)(\sum_i y_i)}{N(\sum_i x_i^2) - (\sum_i x_i^2)} \tag{A.6-10}$$

$$b = \frac{(\sum_i y_i)(\sum_i x_i^2) - (\sum_i x_i)(\sum_i x_i y_i)}{N(\sum_i x_i^2) - (\sum_i x_i)^2} \tag{A.6-11}$$

Example A.2 Experimental measurements of the density of benzene vapor at 563 K are given as follows:

P (atm)	$\widetilde{V}$ (cm^3/mol)	P (atm)	$\widetilde{V}$ (cm^3/mol)
30.64	1164	40.04	707
31.60	1067	41.79	646
32.60	1013	43.59	591
33.89	956	45.48	506
35.17	900	47.07	443
36.63	842	48.07	386
38.39	771		

Assume that the data obey the virial equation of state, i.e.,

$$Z = \frac{P\widetilde{V}}{\mathcal{R}T} = 1 + \frac{B}{\widetilde{V}} + \frac{C}{\widetilde{V}^2}$$

and determine the virial coefficients B and C.

Solution

The equation of state can be rearranged as

$$\left(\frac{P\widetilde{V}}{\mathcal{R}T} - 1\right)\widetilde{V} = B + \frac{C}{\widetilde{V}}$$

Note that this equation has the form

$$y = B + C x$$

where

$$y = \left(\frac{P\widetilde{V}}{\mathcal{R}T} - 1\right)\widetilde{V} \quad \text{and} \quad x = \frac{1}{\widetilde{V}}$$

Taking $\mathcal{R} = 82.06$ cm$^3 \cdot$atm/mol$\cdot$K, the required values are calculated as follows:

y_i	$x_i \times 10^3$	$x_i y_i$	$x_i^2 \times 10^6$
−265.4	0.859	−0.2280	0.738
−288.3	0.937	−0.2702	0.878
−288.9	0.987	−0.2852	0.975
−285.6	1.046	−0.2987	1.094
−283.4	1.111	−0.3149	1.235
−279.9	1.188	−0.3324	1.411
−277	1.297	−0.3593	1.682
−273.8	1.414	−0.3873	2.001
−268.5	1.548	−0.4157	2.396
−261.4	1.692	−0.4424	2.863
−254	1.976	−0.5019	3.906
−243.1	2.257	−0.5487	5.096
−231	2.591	−0.5984	6.712
$\sum y_i = -3500.3$	$\sum x_i = 0.0189$	$\sum x_i y_i = -4.9831$	$\sum x_i^2 = 30.99 \times 10^{-6}$

The values of B and C are

$$B = \frac{(\sum_i y_i)(\sum_i x_i^2) - (\sum_i x_i)(\sum_i x_i y_i)}{N(\sum_i x_i^2) - (\sum_i x_i)^2}$$

$$= \frac{(-3500.3)(30.99 \times 10^{-6}) - (0.0189)(-4.9831)}{(13)(30.99 \times 10^{-6}) - (0.0189)^2} = -313 \text{ cm}^3/\text{mol}$$

$$C = \frac{N(\sum_i x_i y_i) - (\sum_i x_i)(\sum_i y_i)}{N(\sum_i x_i^2) - (\sum_i x_i)^2}$$

$$= \frac{(13)(-4.9831) - (0.0189)(-3500.3)}{(13)(30.99 \times 10^{-6}) - (0.0189)^2} = 30,122 \; (\text{cm}^3/\text{mol})^2$$

The method of least squares can also be applied to higher order polynomials. For example, consider a second-order polynomial

$$y = ax^2 + bx + c \tag{A.6-12}$$

To find the constants a, b, and c, the sum of the squared deviations

$$S = \sum_{i=1}^{N} [y_i - (ax_i^2 + bx_i + c)]^2 \tag{A.6-13}$$

must be minimum. Hence,

$$\frac{\partial S}{\partial a} = \frac{\partial S}{\partial b} = \frac{\partial S}{\partial c} = 0 \tag{A.6-14}$$

Partial differentiation of Eq. (A.6-13) gives

$$a \sum_i x_i^4 + b \sum_i x_i^3 + c \sum_i x_i^2 = \sum_i x_i^2 y_i \tag{A.6-15}$$

$$a \sum_i x_i^3 + b \sum_i x_i^2 + c \sum_i x_i = \sum_i x_i y_i \tag{A.6-16}$$

$$a \sum_i x_i^2 + b \sum_i x_i + c N = \sum_i y_i \tag{A.6-17}$$

These equations may then be solved for the constants a, b, and c.

If the equation is of the form

$$y = ax^n + b \tag{A.6-18}$$

then the parameters a, b, and n can be determined as follows:

1. Least squares values of a and b can be found for a series of chosen values of n.
2. The sum of the squares of the deviations can then be calculated and plotted versus n to find the minimum and, hence, the best value of n. The corresponding values of a and b are readily found by plotting the calculated values versus n and interpolating.

Alternatively, Eq. (A.6-18) might first be arranged as

$$\log(y - b) = n \log x + \log a \tag{A.6-19}$$

and the least squares values of n and $\log a$ are determined for a series of chosen values of b, etc.

Example A.3[1] It is proposed to correlate the data for forced convection heat transfer to a sphere in terms of the equation

$$Nu = 2 + a\,Re^n$$

The following values were obtained from McAdams (1954) for heat transfer from air to spheres by forced convection:

Re	10	100	1000
Nu	2.8	6.3	19.0

Solution

The equation can be rearranged as

$$\log(Nu - 2) = n\log Re + \log a$$

Note that this equation has the form

$$y = nx + b$$

where

$$y = \log(Nu - 2) \qquad x = \log Re \qquad b = \log a$$

y_i	x_i	$x_i y_i$	x_i^2
-0.09691	1	-0.09691	1
0.63347	2	1.26694	4
1.23045	3	3.69135	9
$\sum y_i = 1.76701$	$\sum x_i = 6$	$\sum x_i y_i = 4.86138$	$\sum x_i^2 = 14$

The values of n and b are

$$n = \frac{(3)(4.86138) - (6)(1.76701)}{(3)(14) - (6)^2} = 0.66368$$

$$b = \frac{(14)(1.76701) - (6)(4.86138)}{(3)(14) - (6)^2} = -0.73835 \quad \Rightarrow \quad a = 0.1827$$

A.6.4 Correlation Coefficient

If two variables, x and y, are related in such a way that the points of a scatter plot tend to fall in a straight line, then we say that there is an association between the variables and that they are linearly correlated. The most common measure of the strength of the association between the variables is the *Pearson correlation coefficient*, r. It is defined by

$$r = \frac{\sum x_i y_i - \dfrac{\sum x_i \sum y_i}{n}}{\sqrt{\left(\sum x_i^2 - \dfrac{(\sum x_i)^2}{n}\right)\left(\sum y_i^2 - \dfrac{(\sum y_i)^2}{n}\right)}} \tag{A.6-20}$$

---

[1]This problem is taken from Churchill (1974).

The value of r can range from -1 to $+1$. A value of -1 means a *perfect negative correlation*. A perfect negative correlation implies that $y = ax + b$ where $a < 0$. A *perfect positive correlation* $(r = +1)$ implies that $y = ax + b$ where $a > 0$. When $r = 0$, the variables are uncorrelated. This, however, does not imply that the variables are unrelated. It simply indicates that if a relationship exists, then it is not linear.

A.7   THE ROOT OF AN EQUATION

In engineering problems, we frequently encounter equations of the form

$$f(x) = 0 \tag{A.7-1}$$

and want to determine the values of x satisfying Eq. (A.7-1). These values are called the *roots* of $f(x)$ and may be real or imaginary. Since imaginary roots appear as complex conjugates, the number of imaginary roots must always be even.

The function $f(x)$ may be a polynomial in x or it may be a transcendental equation involving trigonometric and/or logarithmic terms.

A.7.1   Roots of a Polynomial

If $f(x)$ is a polynomial, then *Descartes' rule of sign* determines the maximum number of real roots:

- The maximum number of real positive roots is equal to the number of sign changes in $f(x) = 0$.
- The maximum number of real negative roots is equal to the number of sign changes in $f(-x) = 0$.

In applying the sign rule, zero coefficients are regarded as positive.

A.7.1.1 *Quadratic equation*   The roots of a quadratic equation

$$ax^2 + bx + c = 0 \tag{A.7-2}$$

are given as

$$x_{1,2} = \frac{-b \pm \sqrt{b^2 - 4ac}}{2a} \tag{A.7-3}$$

If a, b, and c are real and if $\Delta = b^2 - 4ac$ is the discriminant, then

- $\Delta > 0$; the roots are real and unequal,
- $\Delta = 0$; the roots are real and equal,
- $\Delta < 0$; the roots are complex conjugate.

A.7.1.2 *Cubic equation*   Consider the cubic equation

$$x^3 + px^2 + qx + r = 0 \tag{A.7-4}$$

Let us define the terms M and N as

$$M = \frac{3q - p^2}{9} \tag{A.7-5}$$

$$N = \frac{9pq - 27r - 2p^3}{54} \tag{A.7-6}$$

If p, q, and r are real and if $\Delta = M^3 + N^2$ is the discriminant, then

- $\Delta > 0$; one root is real and two complex conjugate,
- $\Delta = 0$; all roots are real and at least two are equal,
- $\Delta < 0$; all roots are real and unequal.

Case (i) **Solutions for** $\Delta \geqslant 0$

In this case, the roots are given by

$$x_1 = S + T - \frac{1}{3}p \tag{A.7-7}$$

$$x_2 = -\frac{1}{2}(S + T) - \frac{1}{3}p + \frac{1}{2}i\sqrt{3}(S - T) \tag{A.7-8}$$

$$x_3 = -\frac{1}{2}(S + T) - \frac{1}{3}p - \frac{1}{2}i\sqrt{3}(S - T) \tag{A.7-9}$$

where

$$S = \sqrt[3]{N + \sqrt{\Delta}} \tag{A.7-10}$$

$$T = \sqrt[3]{N - \sqrt{\Delta}} \tag{A.7-11}$$

Case (ii) **Solutions for** $\Delta < 0$

The roots are given by

$$x_1 = \pm 2\sqrt{-M} \cos\left(\frac{\theta}{3}\right) - \frac{1}{3}p \tag{A.7-12}$$

$$x_2 = \pm 2\sqrt{-M} \cos\left(\frac{\theta}{3} + 120°\right) - \frac{1}{3}p \tag{A.7-13}$$

$$x_3 = \pm 2\sqrt{-M} \cos\left(\frac{\theta}{3} + 240°\right) - \frac{1}{3}p \tag{A.7-14}$$

where

$$\theta = \arccos\sqrt{\frac{N^2}{(-M)^3}} \qquad (\theta \text{ is in degrees}) \tag{A.7-15}$$

In Eqs. (A.7-12)-(A.7-14) the upper sign applies if N is positive, and the lower sign applies if N is negative.

Example A.4 Cubic equations of state are frequently used in thermodynamics to describe the *PVT* behavior of liquids and vapors. These equations are expressed in the form

$$P = \frac{\mathcal{R}T}{\widetilde{V} - b} - \frac{a(T)}{\widetilde{V}^\alpha + \beta\widetilde{V} + \gamma} \tag{A.7-16}$$

where the terms α, β, γ, and $a(T)$ for different types of equations of state are given by

Eqn. of State	α	β	γ	$a(T)$
van der Waals	2	0	0	a
Redlich-Kwong	2	b	0	$a/\sqrt{T}$
Peng-Robinson	2	$2b$	$-b^2$	$a(T)$

When Eq. (A.7-16) has three real roots, the largest and the smallest roots correspond to the molar volumes of the vapor and liquid phases, respectively. The intermediate root has no physical meaning.

Predict the density of saturated methanol vapor at 10.84 atm and 140 °C using the van der Waals equation of state. The coefficients a and b are given as

$$a = 9.3424 \text{ m}^6\cdot\text{atm/kmol}^2 \quad \text{and} \quad b = 0.0658 \text{ m}^3/\text{kmol}$$

The experimental value of the density of saturated methanol vapor is 0.01216 g/cm³.

Solution

For the van der Waals equation of state, Eq. (A.7-16) takes the form

$$\widetilde{V}^3 - \left(b + \frac{\mathcal{R}T}{P}\right)\widetilde{V}^2 + \frac{a}{P}\widetilde{V} - \frac{ab}{P} = 0 \tag{1}$$

Substitution of the values of a, b, $\mathcal{R}$, and P into Eq. (1) gives

$$\widetilde{V}^3 - 3.1923\widetilde{V}^2 + 0.8618\widetilde{V} - 0.0567 = 0 \tag{2}$$

Application of the sign rule indicates that the maximum number of real positive roots is equal to three. The terms M and N are

$$M = \frac{3q - p^2}{9} = \frac{(3)(0.8618) - (3.1923)^2}{9} = -0.845 \tag{3}$$

$$N = \frac{9pq - 27r - 2p^3}{54} = \frac{(9)(-3.1923)(0.8618) - (27)(-0.0567) + (2)(3.1923)^3}{54} = 0.775 \tag{4}$$

The discriminant, Δ, is

$$\Delta = M^3 + N^2 = (-0.845)^3 + (0.775)^2 = -0.003 \tag{5}$$

Therefore, all the roots of Eq. (2) are real and unequal. Before calculating the roots by using Eqs. (A.7-12)-(A.7-14), θ must be determined. From Eq. (A.7-15)

$$\theta = \arccos \sqrt{\frac{N^2}{(-M)^3}} = \arccos \sqrt{\frac{(0.775)^2}{(0.845)^3}} = 3.85° \tag{6}$$

Hence, the roots are

$$\widetilde{V}_1 = (2)\sqrt{0.845}\cos\left(\frac{3.85}{3}\right) + \frac{3.1923}{3} = 2.902 \tag{7}$$

$$\widetilde{V}_2 = (2)\sqrt{0.845}\cos\left(\frac{3.85}{3} + 120\right) + \frac{3.1923}{3} = 0.109 \tag{8}$$

$$\widetilde{V}_3 = (2)\sqrt{0.845}\cos\left(\frac{3.85}{3} + 240\right) + \frac{3.1923}{3} = 0.181 \tag{9}$$

The molar volume of saturated vapor, $\widetilde{V}_g$, corresponds to the largest root, i.e., 2.902 m^3/kmol. Since the molecular weight, M, of methanol is 32, the density of saturated vapor, ρ_g, is given by

$$\rho_g = \frac{M}{\widetilde{V}_g} = \frac{32}{(2.902)(1 \times 10^3)} = 0.01103 \text{ g/cm}^3 \tag{10}$$

A.7.2   Numerical Methods

Numerical methods should be used when the equations to be solved are complex and do not have direct analytical solutions. Various numerical methods have been developed for solving Eq. (A.7-1). Some of the most convenient techniques to solve chemical engineering problems are summarized by Serghides (1982), Gjumbir and Olujic (1984), and Tao (1988).

One of the most important problems in the application of numerical techniques is *convergence*. It can be promoted by finding a good starting value and/or a suitable transformation of the variable or the equation.

When using numerical methods, it is always important to use engineering common sense. The following advice given by Tao (1989) should always be remembered in the application of numerical techniques:

- To err is digital, to catch the error is divine.
- An ounce of theory is worth 100 lb of computer output.
- Numerical methods are like political candidates: they'll tell you anything you want to hear.

A.7.2.1   *Newton-Raphson method*   The *Newton-Raphson method* is one of the most widely used techniques to solve an equation of the form $f(x) = 0$. It is based on the expansion of the function $f(x)$ by Taylor series around an estimate x_{k-1} as

$$f(x) = f(x_{k-1}) + (x - x_{k-1})\frac{df}{dx}\bigg|_{x_{k-1}} + \frac{(x - x_{k-1})^2}{2!}\frac{d^2 f}{dx^2}\bigg|_{x_{k-1}} + \cdots \tag{A.7-17}$$

If we neglect the derivatives higher than the first order and let $x = x_k$ be the value of x that makes $f(x) = 0$, then Eq. (A.7-17) becomes

$$x_k = x_{k-1} - \frac{f(x_{k-1})}{\left.\dfrac{df}{dx}\right|_{x_{k-1}}}$$ (A.7-18)

with $k > 0$.

Iterations start with an initial estimate x_o and the required number of iterations to get x_k is dependent on the following error control methods:

• **Absolute error control:** Convergence is achieved when

$$|x_k - x_{k-1}| < \varepsilon$$ (A.7-19)

where ε is a small positive number determined by the desired accuracy.

• **Relative error control:** Convergence is achieved when

$$\left| \frac{x_k - x_{k-1}}{x_k} \right| \times 100 < \varepsilon_s$$ (A.7-20)

where

$$\varepsilon_s = \frac{1}{2} 10^{2-n}$$ (A.7-21)

with n being the number of correct digits. The result, x_k, is correct to at least n significant digits.

A graphical representation of the Newton-Raphson method is shown in Figure A.3. Note that the slope of the tangent drawn to the curve at x_{k-1} is given by

$$\text{slope} = \tan \alpha = \left.\frac{df}{dx}\right|_{x_{k-1}} = \frac{f(x_{k-1})}{x_{k-1} - x_k}$$ (A.7-22)

which is identical to Eq. (A.7-18).

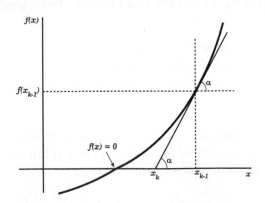

Figure A.3. The Newton-Raphson method.

The Newton-Raphson method has two main drawbacks: (i) the first derivative of the function is not always easy to evaluate, (ii) the method breaks down if $(df/dx)_{x_{k-1}} = 0$ at some point. To circumvent these disadvantages, the first derivative of the function at x_{k-1} is expressed by the central difference approximation as

$$\left.\frac{df}{dx}\right|_{x_{k-1}} = \frac{f(x_{k-1} + \Delta) - f(x_{k-1} - \Delta)}{2\Delta} \tag{A.7-23}$$

where

$$\Delta = \frac{x_{k-1}}{100} \tag{A.7-24}$$

Substitution of Eq. (A.7-23) into Eq. (A.7-18) leads to

$$x_k = x_{k-1} - \frac{0.02x_{k-1} \, f(x_{k-1})}{f(1.01x_{k-1}) - f(0.99x_{k-1})} \tag{A.7-25}$$

with $k > 0$. The main advantages of Eq. (A.7-25) over the numerical techniques proposed to replace the Newton-Raphson method, i.e., the secant method, are: (i) it requires only one initial guess, x_o, instead of two, (ii) the rate of convergence is faster.

PROBLEMS

A.3 Caffeine is extracted from coffee grains by means of a crossflow extractor. The standard error of the exit concentration versus time curve was found as $\sigma = 1.31$, where the standard deviation, σ^2, is given as

$$\sigma^2 = \frac{2}{\text{Pe}} + \frac{2}{\text{Pe}^2}\left(1 - e^{-\text{Pe}}\right)$$

Solve this equation and determine the Peclet number, Pe, which is a measure of axial dispersion in the extractor.

(**Answer:** 1.72)

A.4 The roof of a building absorbs energy at a rate of 225 kW due to solar radiation. The roof loses energy by radiation and convection. The loss of energy flux as a result of convection from the roof to the surrounding air at $25\,°C$ is expressed as

$$q = 2.5(T - T_\infty)^{1.25}$$

where T and T_∞ are the temperatures of the roof and the air in degrees Kelvin, respectively, and q is in W/m^2. Calculate the steady-state temperature of the roof if it has dimensions of $10\,m \times 30\,m$ and its emissivity is 0.9.

(**Answer:** 352 K)

A.8 METHODS OF INTEGRATION

Analytical evaluation of a definite integral

$$I = \int_a^b f(x)\,dx \qquad (A.8\text{-}1)$$

is possible only for limited cases. When analytical evaluation is impossible, then the following techniques can be used to estimate the value of the integral.

A.8.1 Mean Value Theorem

As stated in Section A.2, if $f(x)$ is continuous in the interval $a \leqslant x \leqslant b$, then the value of I is

$$I = \int_a^b f(x)\,dx = \langle f \rangle (b - a) \qquad (A.8\text{-}2)$$

where $\langle f \rangle$ is the average value of f in the interval $a \leqslant x \leqslant b$.

If $f(x)$ is a monotonic function, then the value of I is bounded by I_{min} and I_{max} such that

$$f(x) = \begin{cases} \text{Monotonically increasing function} \begin{cases} I_{min} = f(a)(b-a) \\ I_{max} = f(b)(b-a) \end{cases} \\ \text{Monotonically decreasing function} \begin{cases} I_{min} = f(b)(b-a) \\ I_{max} = f(a)(b-a) \end{cases} \end{cases} \qquad (A.8\text{-}3)$$

In some cases, only part of the integrand may be approximated to permit analytical integration, i.e.,

$$I = \int_a^b f(x)g(x)\,dx = \begin{cases} \langle f \rangle \int_a^b g(x)\,dx \\ \langle g \rangle \int_a^b f(x)\,dx \end{cases} \qquad (A.8\text{-}4)$$

Example A.5 Evaluate the integral

$$I = \int_0^{10} x^2 \sqrt{0.1x + 2}\,dx$$

Solution

Analytical evaluation of the integral is possible and the result is

$$I = \int_0^{10} x^2 \sqrt{0.1x + 2}\,dx = \frac{2(0.15x^2 - 2.4x + 32)}{0.105} \sqrt{(0.1x+2)^3}\Big|_{x=0}^{x=10} = 552.4$$

The same integral can be evaluated approximately as follows: Note that the integrand is the product of two terms and the integral can be written as

$$I = \int_a^b f(x)\,g(x)\,dx \qquad (1)$$

where

$$f(x) = x^2 \quad \text{and} \quad g(x) = \sqrt{0.1x + 2} \tag{2}$$

The value of $g(x)$ is 1.732 and 1.414 at $x = 10$ and $x = 0$, respectively. Since the value of $g(x)$ does not change drastically over the interval $0 \leqslant x \leqslant 10$, Eq. (1) can be expressed in the form

$$I = \langle g \rangle \int_0^{10} f(x)\, dx \tag{3}$$

As a rough approximation, the average value of the function g, $\langle g \rangle$, can be taken as the arithmetic average, i.e.,

$$\langle g \rangle = \frac{1.732 + 1.414}{2} = 1.573 \tag{4}$$

Therefore, Eq. (3) becomes

$$I = 1.573 \int_0^{10} x^2\, dx = \frac{1.573}{3} x^3 \Big|_{x=0}^{x=10} = 524.3 \tag{5}$$

with a percent error of approximately 5%.

A.8.2   Graphical Integration

In order to evaluate the integral given by Eq. (A.8-1) graphically, first $f(x)$ is plotted as a function of x. Then the area under this curve in the interval $[a, b]$ is determined.

A.8.3   Numerical Integration or Quadrature

Numerical integration or *quadrature*[2] is an alternative to graphical and analytical integration. In this method, the integrand is replaced by a polynomial and this polynomial is integrated to give a summation:

$$I = \int_a^b f(x)\, dx = \int_c^d F(u)\, du = \sum_{i=0}^n w_i F(u_i) \tag{A.8-5}$$

Numerical integration is preferred for the following cases:

- The function $f(x)$ is not known but the values of $f(x)$ are known at equally spaced discrete points.
- The function $f(x)$ is known, but is too difficult to integrate analytically.

A.8.3.1   *Numerical integration with equally spaced base points*   Consider Figure A.4 in which $f(x)$ is known only at five equally spaced base points. The two most frequently used numerical integration methods for this case are the *trapezoidal rule* and *Simpson's rule*.

Trapezoidal rule

In this method, the required area under the solid curve is approximated by the area under the dotted straight line (the shaded trapezoid) as shown in Figure A.5.

---

[2]The word *quadrature* is used for *approximate integration*.

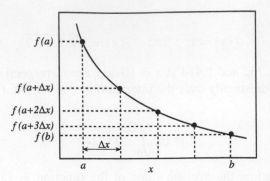

Figure A.4. Values of the function $f(x)$ at five equally spaced points.

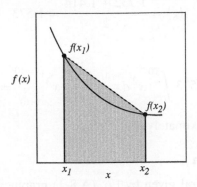

Figure A.5. The trapezoidal rule.

The area of the trapezoid is then

$$\text{Area} = \frac{[f(x_1) + f(x_2)](x_2 - x_1)}{2} \tag{A.8-6}$$

If this procedure is repeated at four equally spaced intervals given in Figure A.5, the value of the integral is

$$I = \int_a^b f(x)\,dx = \frac{[f(a) + f(a + \Delta x)]\Delta x}{2} + \frac{[f(a + \Delta x) + f(a + 2\Delta x)]\Delta x}{2}$$

$$+ \frac{[f(a + 2\Delta x) + f(a + 3\Delta x)]\Delta x}{2} + \frac{[f(a + 3\Delta x) + f(b)]\Delta x}{2} \tag{A.8-7}$$

or,

$$I = \int_a^b f(x)\,dx = \Delta x \left[\frac{f(a)}{2} + f(a + \Delta x) + f(a + 2\Delta x) + f(a + 3\Delta x) + \frac{f(b)}{2} \right] \tag{A.8-8}$$

This result can be generalized as

$$I = \int_a^b f(x)\,dx = \frac{\Delta x}{2} \left[f(a) + 2\sum_{i=1}^{n-2} f(a + i\Delta x) + f(b) \right] \tag{A.8-9}$$

where

$$n = 1 + \frac{b-a}{\Delta x} \tag{A.8-10}$$

Simpson's rule

The trapezoidal rule fits a straight line (first-order polynomial) between the two points. Simpson's rule, on the other hand, fits a second-order polynomial between the two points. In this case, the general formula is

$$I = \int_a^b f(x)\,dx = \frac{\Delta x}{3}\, f(a) + 4 \sum_{i=1,3,5}^{n-1} f(a+i\Delta x) + 2 \sum_{i=2,4,6}^{n-2} f(a+i\Delta x) + f(b)$$

$$\tag{A.8-11}$$

where

$$n = \frac{b-a}{\Delta x} \tag{A.8-12}$$

Note that this formula requires the division of the interval of integration into an even number of subdivisions.

Example A.6 Determine the heat required to increase the temperature of benzene vapor from 300 K to 1000 K at atmospheric pressure. The heat capacity of benzene vapor varies as a function of temperature as follows:

T (K)	300	400	500	600	700	800	900	1000
$\widetilde{C}_P$ (cal/mol·K)	19.65	26.74	32.80	37.74	41.75	45.06	47.83	50.16

Solution

The amount of heat necessary to increase the temperature of benzene vapor from 300 K to 1000 K under constant pressure is calculated from the formula

$$\widetilde{Q} = \Delta\widetilde{H} = \int_{300}^{1000} \widetilde{C}_P\,dT$$

The variation of $\widetilde{C}_P$ as a function of temperature is shown in the figure below:

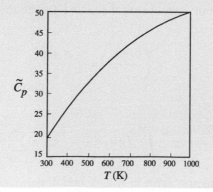

Since the function is monotonically increasing, the bounding values are

$$\widetilde{Q}_{\min} = (19.65)(1000 - 300) = 13,755 \text{ cal/mol}$$

$$\widetilde{Q}_{\max} = (50.16)(1000 - 300) = 35,112 \text{ cal/mol}$$

Trapezoidal rule with $n = 8$

From Eq. (A.8-10)

$$\Delta T = \frac{1000 - 300}{8 - 1} = 100$$

The value of the integral can be calculated from Eq. (A.8-9) as

$$\widetilde{Q} = \frac{100}{2}\big[19.65 + 2(26.74 + 32.80 + 37.74 + 41.75 + 45.06 + 47.83) + 50.16\big]$$

$$= 26,683 \text{ cal/mol}$$

Simpson's rule with $n = 4$

From Eq. (A.8-12)

$$\Delta T = \frac{1000 - 300}{4} = 175$$

Therefore, the values of $\widetilde{C}_P$ at five equally spaced points are given in the following table:

T (K)	300	475	650	825	1000
$\widetilde{C}_P$ (cal/mol·K)	19.65	31.50	39.50	45.75	50.16

The value of the integral using Eq. (A.8-11) is

$$\widetilde{Q} = \frac{175}{3}\big[19.65 + 4(31.50 + 45.75) + 2(39.50) + 50.16\big] = 26,706 \text{ cal/mol}$$

A.8.4 Numerical Integration when the Integrand Is a Continuous Function

A.8.4.1 *Gauss-Legendre quadrature*   The evaluation of an integral given by Eq. (A.8-1), where a and b are arbitrary but finite, using the Gauss-Legendre quadrature requires the following transformation:

$$x = \left(\frac{b - a}{2}\right)u + \frac{a + b}{2} \tag{A.8-13}$$

Then Eq. (A.8-1) becomes

$$I = \int_a^b f(x)\,dx = \frac{b - a}{2}\int_{-1}^1 F(u)\,du = \frac{b - a}{2}\sum_{i=0}^n w_i F(u_i) \tag{A.8-14}$$

where the roots and weight factors for $n = 1, 2, 3$, and 4 are given in Table A.3.

Table A.3. Roots and weight factors for the Gauss-Legendre quadrature
(Abramowitz and Stegun, 1970)

n	Roots (u_i)	Weight Factors (w_i)
1	±0.57735 02691 89626	1.00000 00000 00000
2	0.00000 00000 00000	0.88888 88888 88889
	±0.77459 66692 41483	0.55555 55555 55556
3	±0.33998 10435 84856	0.65214 51548 62546
	±0.86113 63115 94053	0.34785 48451 37454
4	0.00000 00000 00000	0.56888 88888 88889
	±0.53846 93101 05683	0.47862 86704 99366
	±0.90617 98459 38664	0.23692 68850 56189

Example A.7 Evaluate

$$I = \int_1^2 \frac{1}{x+2}\,dx$$

using the five-point ($n = 4$) Gauss-Legendre quadrature formula and compare it with the analytical solution.

Solution

Since $b = 2$ and $a = 1$, from Eq. (A.8-13)

$$x = \frac{u+3}{2}$$

Then

$$F(u) = \frac{1}{\left(\dfrac{u+3}{2}\right)+2} = \frac{2}{u+7}$$

The five-point quadrature is given by

$$I = \int_1^2 \frac{1}{x+2}\,dx = \frac{1}{2}\sum_{i=0}^{4} w_i\, F(u_i)$$

The values of w_i and $F(u_i)$ are given in the table below:

i	u_i	w_i	$F(u_i) = \dfrac{2}{u_i+7}$	$w_i\, F(u_i)$
0	0.00000000	0.56888889	0.28571429	0.16253969
1	+0.53846931	0.47862867	0.26530585	0.12698299
2	−0.53846931	0.47862867	0.30952418	0.14814715
3	+0.90617985	0.23692689	0.25296667	0.05993461
4	−0.90617985	0.23692689	0.32820135	0.07775973
			$\sum_{i=0}^{i=4} w_i\, F(u_i) =$	0.57536417

Therefore,

$$I = (0.5)(0.57536417) = 0.28768209$$

Analytically,

$$I = \ln(x+2)|_{x=1}^{x=2} = \ln\left(\frac{4}{3}\right) = 0.28768207$$

A.8.4.2 *Gauss-Laguerre quadrature*   The Gauss-Laguerre quadrature can be used to evaluate integrals of the form

$$I = \int_a^\infty e^{-x} f(x)\, dx \tag{A.8-15}$$

where a is arbitrary and finite. The transformation

$$x = u + a \tag{A.8-16}$$

reduces Eq. (A.8-15) to

$$I = \int_a^\infty e^{-x} f(x)\, dx = e^{-a} \int_0^\infty e^{-u} F(u)\, du = e^{-a} \sum_{i=0}^n w_i F(u_i) \tag{A.8-17}$$

where the w_i and u_i are given in Table A.4.

Table A.4. Roots and weight factors for the Gauss-Laguerre quadrature (Abramowitz and Stegun, 1970)

n	Roots (u_i)	Weight Factors (w_i)
1	0.58578 64376 27	0.85355 33905 93
	3.41421 35623 73	0.14644 66094 07
2	0.41577 45567 83	0.71109 30099 29
	2.29428 03602 79	0.27851 77335 69
	6.28994 50829 37	0.01038 92565 02
3	0.32254 76896 19	0.60315 41043 42
	1.74576 11011 58	0.35741 86924 38
	4.53662 02969 21	0.03888 79085 15
	9.39507 09123 01	0.00053 92947 06
4	0.26356 03197 18	0.52175 56105 83
	1.41340 30591 07	0.39866 68110 83
	3.59642 57710 41	0.07594 24496 82
	7.08581 00058 59	0.00361 17586 80
	12.64080 08442 76	0.00002 33699 72

Example A.8 The gamma function, $\Gamma(n)$, is defined by

$$\Gamma(n) = \int_0^\infty \beta^{n-1} e^{-\beta}\, d\beta$$

where the variable β in the integrand is the dummy variable of integration. Estimate $\Gamma(1.5)$ by using the Gauss-Laguerre quadrature with $n = 3$.

Solution

Since $a = 0$, then

$$\beta = u \quad \text{and} \quad F(u) = \sqrt{u}$$

The four-point quadrature is given by

$$\Gamma(1.5) = \int_0^\infty \sqrt{\beta}\, e^{-\beta}\, d\beta = \sum_{i=0}^3 w_i F(u_i)$$

The values of w_i and $F(u_i)$ are given in the table below:

i	u_i	w_i	$F(u_i) = \sqrt{u_i}$	$w_i F(u_i)$
0	0.32254769	0.60315410	0.56793282	0.34255101
1	1.74576110	0.35741869	1.32127253	0.47224750
2	4.53662030	0.03888791	2.12993434	0.08282869
3	9.39507091	0.00053929	3.06513799	0.00165300
			$\Gamma(1.5) = \sum_{i=0}^{i=3} w_i F(u_i) = 0.8992802$	

The exact value of $\Gamma(1.5)$ is 0.8862269255.

A.8.4.3 *Gauss-Hermite quadrature*   The Gauss-Hermite quadrature can be used to evaluate integrals of the form

$$I = \int_{-\infty}^\infty e^{-x^2} f(x)\, dx = \sum_{i=0}^n w_i f(x_i) \tag{A.8-18}$$

The weight factors and appropriate roots for the first few quadrature formulas are given in Table A.5.

Table A.5. Roots and weight factors for the Gauss-Hermite quadrature (Abramowitz and Stegun, 1970)

n	Roots (x_i)	Weight Factors (w_i)
1	$\pm 0.70710\ 67811$	$0.88622\ 69255$
2	$\pm 1.22474\ 48714$	$0.29540\ 89752$
	$0.00000\ 00000$	$1.18163\ 59006$
3	$\pm 1.65068\ 01239$	$0.08131\ 28354$
	$\pm 0.52464\ 76233$	$0.80491\ 40900$
4	$\pm 2.02018\ 28705$	$0.01995\ 32421$
	$\pm 0.95857\ 24646$	$0.39361\ 93232$
	$0.00000\ 00000$	$0.94530\ 87205$

A.9 MATRICES

A rectangular array of elements or functions is called a *matrix*. If the array has m rows and n columns, it is called an $m \times n$ matrix and expressed in the form

$$\mathbf{A} = \begin{bmatrix} a_{11} & a_{12} & a_{13} & \cdots & a_{1n} \\ a_{21} & a_{22} & a_{23} & \cdots & a_{2n} \\ \cdots & \cdots & \cdots & \cdots & \cdots \\ a_{m1} & a_{m2} & a_{m3} & \cdots & a_{mn} \end{bmatrix} \tag{A.9-1}$$

The numbers or functions a_{ij} are called the elements of a matrix. Equation (A.9-1) is also expressed as

$$\mathbf{A} = (a_{ij}) \tag{A.9-2}$$

in which the subscripts i and j represent the row and the column of the matrix, respectively.

A matrix having only one row is called a *row matrix* (or *row vector*), while a matrix having only one column is called a *column matrix* (or *column vector*). When the number of rows and the number of columns are the same, i.e., $m = n$, the matrix is called a *square matrix* or a matrix of order n.

A.9.1 Fundamental Algebraic Operations

1. Two matrices $\mathbf{A} = (a_{ij})$ and $\mathbf{B} = (b_{ij})$ of the same order are equal if and only if $a_{ij} = b_{ij}$.

2. If $\mathbf{A} = (a_{ij})$ and $\mathbf{B} = (b_{ij})$ have the same order, the sum of $\mathbf{A}$ and $\mathbf{B}$ is defined as

$$\mathbf{A} + \mathbf{B} = (a_{ij} + b_{ij}) \tag{A.9-3}$$

If $\mathbf{A}$, $\mathbf{B}$, and $\mathbf{C}$ are matrices of the same order, addition is commutative and associative, i.e.,

$$\mathbf{A} + \mathbf{B} = \mathbf{B} + \mathbf{A} \tag{A.9-4}$$

$$\mathbf{A} + (\mathbf{B} + \mathbf{C}) = (\mathbf{A} + \mathbf{B}) + \mathbf{C} \tag{A.9-5}$$

3. If $\mathbf{A} = (a_{ij})$ and $\mathbf{B} = (b_{ij})$ have the same order, the difference between $\mathbf{A}$ and $\mathbf{B}$ is defined as

$$\mathbf{A} - \mathbf{B} = (a_{ij} - b_{ij}) \tag{A.9-6}$$

Example A.9 If

$$\mathbf{A} = \begin{bmatrix} 2 & -1 \\ 1 & 0 \\ 3 & 5 \end{bmatrix} \quad \text{and} \quad \mathbf{B} = \begin{bmatrix} 2 & -4 \\ 3 & 0 \\ 0 & 1 \end{bmatrix}$$

determine $\mathbf{A} + \mathbf{B}$ and $\mathbf{A} - \mathbf{B}$.

Solution

$$\mathbf{A} + \mathbf{B} = \begin{bmatrix} 2+2 & -1-4 \\ 1+3 & 0+0 \\ 3+0 & 5+1 \end{bmatrix} = \begin{bmatrix} 4 & -5 \\ 4 & 0 \\ 3 & 6 \end{bmatrix}$$

$$\mathbf{A} - \mathbf{B} = \begin{bmatrix} 2-2 & -1+4 \\ 1-3 & 0-0 \\ 3-0 & 5-1 \end{bmatrix} = \begin{bmatrix} 0 & 3 \\ -2 & 0 \\ 3 & 4 \end{bmatrix}$$

4. If $\mathbf{A} = (a_{ij})$ and λ is any number, the product of $\mathbf{A}$ by λ is defined as

$$\lambda \mathbf{A} = \mathbf{A}\lambda = (\lambda a_{ij}) \tag{A.9-7}$$

5. The product of two matrices $\mathbf{A}$ and $\mathbf{B}$, $\mathbf{AB}$, is defined only if the number of columns in $\mathbf{A}$ is equal to the number of rows in $\mathbf{B}$. In this case, the two matrices are said to be *conformable* in the order stated. For example, if $\mathbf{A}$ is of order 4×2 and $\mathbf{B}$ is of order 2×3, then the product $\mathbf{AB}$ is

$$\mathbf{AB} = \begin{bmatrix} a_{11} & a_{12} \\ a_{21} & a_{22} \\ a_{31} & a_{32} \\ a_{41} & a_{42} \end{bmatrix} \begin{bmatrix} b_{11} & b_{12} & b_{13} \\ b_{21} & b_{22} & b_{23} \end{bmatrix}$$

$$= \begin{bmatrix} a_{11}b_{11} + a_{12}b_{21} & a_{11}b_{12} + a_{12}b_{22} & a_{11}b_{13} + a_{12}b_{23} \\ a_{21}b_{11} + a_{22}b_{21} & a_{21}b_{12} + a_{22}b_{22} & a_{21}b_{13} + a_{22}b_{23} \\ a_{31}b_{11} + a_{32}b_{21} & a_{31}b_{12} + a_{32}b_{22} & a_{31}b_{13} + a_{32}b_{23} \\ a_{41}b_{11} + a_{42}b_{21} & a_{41}b_{12} + a_{42}b_{22} & a_{41}b_{13} + a_{42}b_{23} \end{bmatrix} \tag{A.9-8}$$

In general, if a matrix of order (m, r) is multiplied by a matrix of order (r, n), the product is a matrix of order (m, n). Symbolically, this may be expressed as

$$(m, r) \times (r, n) = (m, n) \tag{A.9-9}$$

Example A.10 If

$$\mathbf{A} = \begin{bmatrix} 1 & -1 \\ 2 & 0 \\ -1 & 5 \end{bmatrix} \quad \text{and} \quad \mathbf{B} = \begin{bmatrix} 1 \\ 2 \end{bmatrix}$$

determine $\mathbf{AB}$.

Solution

$$\mathbf{AB} = \begin{bmatrix} 1 & -1 \\ 2 & 0 \\ -1 & 5 \end{bmatrix} \begin{bmatrix} 1 \\ 2 \end{bmatrix}$$

$$= \begin{bmatrix} (1)(1) + (-1)(2) \\ (2)(1) + (0)(2) \\ (-1)(1) + (5)(2) \end{bmatrix} = \begin{bmatrix} -1 \\ 2 \\ 9 \end{bmatrix}$$

6. A matrix $\mathbf{A}$ can be multiplied by itself if and only if it is a square matrix. The product $\mathbf{AA}$ can be expressed as $\mathbf{A}^2$. If the relevant products are defined, the multiplication of matrices is associative, i.e.,

$$\mathbf{A}(\mathbf{BC}) = (\mathbf{AB})\mathbf{C} \qquad (A.9\text{-}10)$$

and distributive, i.e.,

$$\mathbf{A}(\mathbf{B} + \mathbf{C}) = \mathbf{AB} + \mathbf{AC} \qquad (A.9\text{-}11)$$

$$(\mathbf{B} + \mathbf{C})\mathbf{A} = \mathbf{BA} + \mathbf{CA} \qquad (A.9\text{-}12)$$

but, in general, not commutative.

A.9.2 Determinants

For each square matrix $\mathbf{A}$, it is possible to associate a scalar quantity called the determinant of $\mathbf{A}$, $|\mathbf{A}|$. If the matrix $\mathbf{A}$ in Eq. (A.9-1) is a square matrix, then the determinant of $\mathbf{A}$ is given by

$$|\mathbf{A}| = \begin{vmatrix} a_{11} & a_{12} & a_{13} & \dots & a_{1n} \\ a_{21} & a_{22} & a_{23} & \dots & a_{2n} \\ \dots & \dots & \dots & \dots & \dots \\ a_{n1} & a_{n2} & a_{n3} & \dots & a_{nn} \end{vmatrix} \qquad (A.9\text{-}13)$$

If the row and column containing an element a_{ij} in a square matrix $\mathbf{A}$ are deleted, the determinant of the remaining square array is called the *minor* of a_{ij} and denoted by M_{ij}. The cofactor of a_{ij}, denoted by A_{ij}, is then defined by the relation

$$A_{ij} = (-1)^{i+j} M_{ij} \qquad (A.9\text{-}14)$$

Thus, if the sum of the row and column indices of an element is even, the cofactor and the minor of that element are identical; otherwise they differ in sign.

The determinant of a square matrix $\mathbf{A}$ can be calculated by the following formula:

$$|\mathbf{A}| = \sum_{k=1}^{n} a_{ik} A_{ik} = \sum_{k=1}^{n} a_{kj} A_{kj} \qquad (A.9\text{-}15)$$

where i and j may stand for any row and column, respectively. Therefore, the determinants of 2×2 and 3×3 matrices are

$$\begin{vmatrix} a_{11} & a_{12} \\ a_{21} & a_{22} \end{vmatrix} = a_{11}a_{22} - a_{12}a_{21} \qquad (A.9\text{-}16)$$

$$\begin{vmatrix} a_{11} & a_{12} & a_{13} \\ a_{21} & a_{22} & a_{23} \\ a_{31} & a_{32} & a_{33} \end{vmatrix} = a_{11}a_{22}a_{33} + a_{12}a_{23}a_{31} + a_{13}a_{21}a_{32}$$

$$- a_{11}a_{23}a_{32} - a_{12}a_{21}a_{33} - a_{13}a_{22}a_{31} \qquad (A.9\text{-}17)$$

Example A.11 Determine $|\mathbf{A}|$ if

$$\mathbf{A} = \begin{bmatrix} 1 & 0 & 1 \\ 3 & 2 & 1 \\ -1 & 1 & 0 \end{bmatrix}$$

Solution

Expanding on the first row, i.e., $i = 1$, gives

$$|\mathbf{A}| = 1 \begin{vmatrix} 2 & 1 \\ 1 & 0 \end{vmatrix} - 0 \begin{vmatrix} 3 & 1 \\ -1 & 0 \end{vmatrix} + 1 \begin{vmatrix} 3 & 2 \\ -1 & 1 \end{vmatrix} = -1 + 5 = 4$$

A.9.2.1 *Some properties of determinants*

1. If all elements in a row or column are zero, the determinant is zero, i.e.,

$$\begin{vmatrix} a_1 & b_1 & c_1 \\ a_2 & b_2 & c_2 \\ 0 & 0 & 0 \end{vmatrix} = 0 \qquad \begin{vmatrix} 0 & b_1 & c_1 \\ 0 & b_2 & c_2 \\ 0 & b_3 & c_3 \end{vmatrix} = 0 \tag{A.9-18}$$

2. The value of a determinant is not altered when the rows are changed to columns or the columns to rows, i.e., when the rows and columns are interchanged.
3. The interchange of any two columns or any two rows of a determinant changes the sign of the determinant.
4. If two columns or two rows of a determinant are identical, the determinant is equal to zero.
5. If each element in any column or row of a determinant is expressed as the sum of two quantities, the determinant can be expressed as the sum of two determinants of the same order, i.e.,

$$\begin{vmatrix} a_1 + d_1 & b_1 & c_1 \\ a_2 + d_2 & b_2 & c_2 \\ a_3 + d_3 & b_3 & c_3 \end{vmatrix} = \begin{vmatrix} a_1 & b_1 & c_1 \\ a_2 & b_2 & c_2 \\ a_3 & b_3 & c_3 \end{vmatrix} + \begin{vmatrix} d_1 & b_1 & c_1 \\ d_2 & b_2 & c_2 \\ d_3 & b_3 & c_3 \end{vmatrix} \tag{A.9-19}$$

6. Adding the same multiple of each element of one row to the corresponding element of another row does not change the value of the determinant. The same is true for the columns.

$$\begin{vmatrix} a_1 & b_1 & c_1 \\ a_2 & b_2 & c_2 \\ a_3 & b_3 & c_3 \end{vmatrix} = \begin{vmatrix} (a_1 + nb_1) & b_1 & c_1 \\ (a_2 + nb_2) & b_2 & c_2 \\ (a_3 + nb_3) & b_3 & c_3 \end{vmatrix} \tag{A.9-20}$$

This result follows immediately from Properties 4 and 5.

7. If all the elements in any column or row are multiplied by any factor, the determinant is multiplied by that factor, i.e.,

$$\begin{vmatrix} \lambda a_1 & b_1 & c_1 \\ \lambda a_2 & b_2 & c_2 \\ \lambda a_3 & b_3 & c_3 \end{vmatrix} = \lambda \begin{vmatrix} a_1 & b_1 & c_1 \\ a_2 & b_2 & c_2 \\ a_3 & b_3 & c_3 \end{vmatrix} \tag{A.9-21}$$

$$\begin{vmatrix} (1/\lambda)a_1 & b_1 & c_1 \\ (1/\lambda)a_2 & b_2 & c_2 \\ (1/\lambda)a_3 & b_3 & c_3 \end{vmatrix} = \frac{1}{\lambda} \begin{vmatrix} a_1 & b_1 & c_1 \\ a_2 & b_2 & c_2 \\ a_3 & b_3 & c_3 \end{vmatrix} \tag{A.9-22}$$

A.9.3 Types of Matrices

A.9.3.1 *The transpose of a matrix* The matrix obtained from **A** by interchanging rows and columns is called the *transpose* of **A** and denoted by $\mathbf{A}^T$.

The transpose of the product **AB** is the product of the transposes in the form

$$(\mathbf{AB})^T = \mathbf{B}^T \mathbf{A}^T \tag{A.9-23}$$

A.9.3.2 *Unit matrix* The *unit matrix* **I** of order n is the square $n \times n$ matrix having ones in its principal diagonal and zeros elsewhere, i.e.,

$$\mathbf{I} = \begin{pmatrix} 1 & 0 & \dots & 0 \\ 0 & 1 & \dots & 0 \\ \dots & \dots & \dots & \dots \\ 0 & 0 & \dots & 1 \end{pmatrix} \tag{A.9-24}$$

For any matrix

$$\mathbf{AI} = \mathbf{IA} = \mathbf{A} \tag{A.9-25}$$

A.9.3.3 *Symmetric and skew-symmetric matrices* A square matrix **A** is said to be *symmetric* if

$$\mathbf{A} = \mathbf{A}^T \quad \text{or} \quad a_{ij} = a_{ji} \tag{A.9-26}$$

A square matrix A is said to be *skew-symmetric* (or *antisymmetric*) if

$$\mathbf{A} = -\mathbf{A}^T \quad \text{or} \quad a_{ij} = -a_{ji} \tag{A.9-27}$$

Equation (A.9-27) implies that the diagonal elements of a skew-symmetric matrix are all zero.

A.9.3.4 *Singular matrix* A square matrix **A** for which the determinant $|\mathbf{A}|$ of its elements is zero is termed a *singular* matrix. If $|\mathbf{A}| \neq 0$, then **A** is *nonsingular*.

A.9.3.5 *The inverse matrix* If the determinant $|\mathbf{A}|$ of a square matrix **A** does not vanish, i.e., a nonsingular matrix, it then possesses an *inverse* (or *reciprocal*) matrix $\mathbf{A}^{-1}$ such that

$$\mathbf{AA}^{-1} = \mathbf{A}^{-1}\mathbf{A} = \mathbf{I} \tag{A.9-28}$$

The inverse of a matrix **A** is defined by

$$\mathbf{A}^{-1} = \frac{\text{Adj}\,\mathbf{A}}{|\mathbf{A}|} \tag{A.9-29}$$

where $\text{Adj}\,\mathbf{A}$ is called the *adjoint* of **A**. It is obtained from a square matrix **A** by replacing each element by its cofactor and then interchanging rows and columns.

Example A.12 Find the inverse of the matrix **A** given in Example A.11.

Solution

The minor of **A** is given by

$$
M_{ij} = \begin{bmatrix}
\begin{vmatrix} 2 & 1 \\ 1 & 0 \end{vmatrix} & \begin{vmatrix} 3 & 1 \\ -1 & 0 \end{vmatrix} & \begin{vmatrix} 3 & 2 \\ -1 & 1 \end{vmatrix} \\[2mm]
\begin{vmatrix} 0 & 1 \\ 1 & 0 \end{vmatrix} & \begin{vmatrix} 1 & 1 \\ -1 & 0 \end{vmatrix} & \begin{vmatrix} 1 & 0 \\ -1 & 1 \end{vmatrix} \\[2mm]
\begin{vmatrix} 0 & 1 \\ 2 & 1 \end{vmatrix} & \begin{vmatrix} 1 & 1 \\ 3 & 1 \end{vmatrix} & \begin{vmatrix} 1 & 0 \\ 3 & 2 \end{vmatrix}
\end{bmatrix}
= \begin{bmatrix}
-1 & 1 & 5 \\
-1 & 1 & 1 \\
-2 & -2 & 2
\end{bmatrix}
$$

The cofactor matrix is

$$
A_{ij} = \begin{bmatrix}
-1 & -1 & 5 \\
1 & 1 & -1 \\
-2 & 2 & 2
\end{bmatrix}
$$

The transpose of the cofactor matrix gives the adjoint of **A** as

$$
\text{Adj}\,\mathbf{A} = \begin{bmatrix}
-1 & 1 & -2 \\
-1 & 1 & 2 \\
5 & -1 & 2
\end{bmatrix}
$$

Since $|\mathbf{A}| = 4$, the use of Eq. (A.9-29) gives the inverse of **A** in the form

$$
\mathbf{A}^{-1} = \frac{\text{Adj}\,\mathbf{A}}{|\mathbf{A}|} = \begin{bmatrix}
-0.25 & 0.25 & -0.5 \\
-0.25 & 0.25 & 0.5 \\
1.25 & -0.25 & 0.5
\end{bmatrix}
$$

A.9.4 Solution of Simultaneous Algebraic Equations

Consider the system of n non-homogeneous algebraic equations

$$
\begin{aligned}
a_{11}x_1 + a_{12}x_2 + \cdots + a_{1n}x_n &= c_1 \\
a_{21}x_1 + a_{22}x_2 + \cdots + a_{2n}x_n &= c_2 \\
\cdots\cdots\cdots\cdots\cdots\cdots\cdots\cdots &= \ldots \\
a_{n1}x_1 + a_{n2}x_2 + \cdots + a_{nn}x_n &= c_n
\end{aligned}
\tag{A.9-30}
$$

in which the coefficients a_{ij} and the constants c_i are independent of $x_1, x_2, \ldots, x_n$ but are otherwise arbitrary. In matrix notation, Eq. (A.9-30) is expressed as

$$
\begin{bmatrix}
a_{11} & a_{12} & \ldots & a_{1n} \\
a_{21} & a_{22} & \ldots & a_{2n} \\
\ldots & \ldots & \ldots & \ldots \\
a_{n1} & a_{n2} & \ldots & a_{nn}
\end{bmatrix}
\begin{bmatrix}
x_1 \\ x_2 \\ \ldots \\ x_n
\end{bmatrix}
=
\begin{bmatrix}
c_1 \\ c_2 \\ \ldots \\ c_n
\end{bmatrix}
\tag{A.9-31}
$$

or,

$$
\mathbf{AX} = \mathbf{C}
\tag{A.9-32}
$$

Multiplication of Eq. (A.9-32) by the inverse of the coefficient matrix $\mathbf{A}$ gives

$$\mathbf{X} = \mathbf{A}^{-1}\mathbf{C} \tag{A.9-33}$$

A.9.4.1 *Cramer's rule* Cramer's rule states that, if the determinant of $\mathbf{A}$ is not equal to zero, the system of linear algebraic equations has a solution given by

$$x_j = \frac{|\mathbf{A}_j|}{|\mathbf{A}|} \tag{A.9-34}$$

where $|\mathbf{A}|$ and $|\mathbf{A}_j|$ are the determinants of the coefficient and substituted matrices, respectively. The *substituted matrix*, $\mathbf{A}_j$, is obtained by replacing the jth column of $\mathbf{A}$ by the column of c's, i.e.,

$$\mathbf{A}_j = \begin{bmatrix} a_{11} & a_{12} & \dots & c_1 & \dots & a_{1n} \\ a_{21} & a_{22} & \dots & c_2 & \dots & a_{2n} \\ \dots & \dots & \dots & \dots & \dots & \dots \\ a_{n1} & a_{n2} & \dots & c_n & \dots & a_{nn} \end{bmatrix} \tag{A.9-35}$$

REFERENCES

Abramowitz, M. and I.A. Stegun, 1970, Handbook of Mathematical Functions, Dover Publications, New York.

Buckler, S.J., 1969, The vertical wind profile of monthly mean winds over the prairies, Canada Department of Transport, Tech. Memo. TEC 718.

Churchill, S.W., 1974, The Interpretation and Use of Rate Data: The Rate Concept, Scripta Publishing Co., Washington, D.C.

De Nevers, N., 1966, Rate data and derivatives, AIChE Journal 12, 1110.

Gjumbir, M. and Z. Olujic, 1984, Effective ways to solve single nonlinear equations, Chem. Eng. 91 (July 23), 51.

McAdams, W.H., 1954, Heat Transmission, 3rd Ed., McGraw-Hill, New York.

Mickley, H.S., T.S. Sherwood and C.E. Reed, 1975, Applied Mathematics in Chemical Engineering, 2nd Ed., p. 25, Tata McGraw-Hill, New Delhi.

Whitaker, S. and R.L. Pigford, 1960, An approach to numerical differentiation of experimental data, Ind. Eng. Chem. 52, 185.

Serghides, T.K., 1982, Iterative solutions by direct substitution, Chem. Eng. 89 (Sept. 6), 107.

Tao, B.Y., 1988, Finding your roots, Chem. Eng. 95 (Apr. 25), 85.

Tao, B.Y., 1989, Linear programming, Chem. Eng. 96 (July), 146.

SUGGESTED REFERENCES FOR FURTHER STUDY

Amundson, N.R., 1966, Mathematical Methods in Chemical Engineering: Matrices and Their Applications, Prentice-Hall, Englewood Cliffs, New Jersey.

Hildebrand, F.B., 1965, Methods of Applied Mathematics, 2nd Ed., Prentice-Hall, Englewood Cliffs, New Jersey.

Appendix B

SOLUTIONS OF DIFFERENTIAL EQUATIONS

A *differential equation* is an equation involving derivatives or differentials of one or more dependent variables with respect to one or more independent variables.

The *order* of a differential equation is the order of the highest derivative in the equation.

The *degree* of a differential equation is the power of the highest derivative after the equation has been rationalized and cleared of fractions.

A differential equation is *linear* when: (*i*) every dependent variable and every derivative involved occurs to the first-degree only, (*ii*) neither products nor powers of dependent variables nor products of dependent variables with differentials exist.

B.1 TYPES OF FIRST-ORDER EQUATIONS WITH EXACT SOLUTIONS

There are five types of differential equations for which solutions may be obtained by exact methods. These are:

- Separable equations,
- Exact equations,
- Homogeneous equations,
- Linear equations,
- Bernoulli equations.

B.1.1 Separable Equations

An equation of the form

$$f_1(x)g_1(y)\, dx + f_2(x)g_2(y)\, dy = 0 \tag{B.1-1}$$

is called a *separable equation*. Division of Eq. (B.1-1) by $g_1(y)f_2(x)$ results in

$$\frac{f_1(x)}{f_2(x)}\, dx + \frac{g_2(y)}{g_1(y)}\, dy = 0 \tag{B.1-2}$$

Integration of Eq. (B.1-2) gives

$$\int \frac{f_1(x)}{f_2(x)}\, dx + \int \frac{g_2(y)}{g_1(y)}\, dy = C \tag{B.1-3}$$

where C is the integration constant.

Example B.1 Solve the following equation

$$(2x + xy^2)\, dx + (3y + x^2 y)\, dy = 0$$

Solution

The differential equation can be rewritten in the form

$$x(2 + y^2)\, dx + y(3 + x^2)\, dy = 0 \tag{1}$$

Note that Eq. (1) is a separable equation and can be expressed as

$$\frac{x}{3 + x^2}\, dx + \frac{y}{2 + y^2}\, dy = 0 \tag{2}$$

Integration of Eq. (2) gives

$$(3 + x^2)(2 + y^2) = C \tag{3}$$

B.1.2 Exact Equations

The expression $M\, dx + N\, dy$ is called an *exact differential*[1] if there exists some $\phi = \phi(x, y)$ for which this expression is the total differential $d\phi$, i.e.,

$$M\, dx + N\, dy = d\phi \tag{B.1-4}$$

A necessary and sufficient condition for the expression $M\, dx + N\, dy$ to be expressed as a total differential is that

$$\frac{\partial M}{\partial y} = \frac{\partial N}{\partial x} \tag{B.1-5}$$

If $M\, dx + N\, dy$ is an exact differential, then the differential equation

$$M\, dx + N\, dy = 0 \tag{B.1-6}$$

is called an *exact differential equation*. Since an exact differential can be expressed in the form of a total differential $d\phi$, then

$$M\, dx + N\, dy = d\phi = 0 \tag{B.1-7}$$

and the solution can easily be obtained as

$$\phi = C \tag{B.1-8}$$

where C is a constant.

Example B.2 Solve the following differential equation

$$(4x - 3y)\, dx + (1 - 3x)\, dy = 0$$

---

[1] In thermodynamics, an exact differential is called a state function.

Solution

Note that $M = 4x - 3y$ and $N = 1 - 3x$. Since

$$\frac{\partial M}{\partial y} = \frac{\partial N}{\partial x} = -3 \tag{1}$$

the differential equation is exact and can be expressed in the form of a total differential $d\phi$,

$$(4x - 3y)\,dx + (1 - 3x)\,dy = d\phi = \frac{\partial \phi}{\partial x}\,dx + \frac{\partial \phi}{\partial y}\,dy = 0 \tag{2}$$

From Eq. (2) we see that

$$\frac{\partial \phi}{\partial x} = 4x - 3y \tag{3}$$

$$\frac{\partial \phi}{\partial y} = 1 - 3x \tag{4}$$

Partial integration of Eq. (3) with respect to x gives

$$\phi = 2x^2 - 3xy + h(y) \tag{5}$$

Substitution of Eq. (5) into Eq. (4) yields

$$\frac{dh}{dy} = 1 \tag{6}$$

Integration of Eq. (6) gives the function h as

$$h = y + C \tag{7}$$

where C is a constant. Substitution of Eq. (7) into Eq. (5) gives the function ϕ as

$$\phi = 2x^2 - 3xy + y + C \tag{8}$$

Hence, the solution is

$$2x^2 - 3xy + y = C^* \tag{9}$$

where C^* is a constant.

If the equation $M\,dx + N\,dy$ is not exact, multiplication of it by some function μ, called an *integrating factor*, may make it an exact equation, i.e.,

$$\mu M\,dx + \mu N\,dy = 0 \quad \Rightarrow \quad \frac{\partial(\mu M)}{\partial y} = \frac{\partial(\mu N)}{\partial x} \tag{B.1-9}$$

For example, all thermodynamic functions except heat and work are state functions. Although dQ is a path function, dQ/T is a state function. Therefore, $1/T$ is an integrating factor in this case.

B.1.3 Homogeneous Equations

A function $f(x, y)$ is said to be *homogeneous of degree n* if

$$f(\lambda x, \lambda y) = \lambda^n f(x, y) \tag{B.1-10}$$

for all λ. For an equation

$$M\, dx + N\, dy = 0 \tag{B.1-11}$$

if M and N are homogeneous of the same degree, the transformation

$$y = ux \tag{B.1-12}$$

will make the equation separable.

For a homogeneous function of degree n, *Euler's theorem* states that

$$nf(x_1, x_2, \ldots, x_\alpha) = \sum_{i=1}^{\alpha} \left(\frac{\partial f}{\partial x_i} \right) x_i \tag{B.1-13}$$

Note that the *extensive properties* in thermodynamics can be regarded as homogeneous functions of order unity. Therefore, for every extensive property we can write

$$f(x_1, x_2, \ldots, x_\alpha) = \sum_{i=1}^{\alpha} \left(\frac{\partial f}{\partial x_i} \right) x_i \tag{B.1-14}$$

On the other hand, the *intensive properties* are homogeneous functions of order zero and can be expressed as

$$0 = \sum_{i=1}^{\alpha} \left(\frac{\partial f}{\partial x_i} \right) x_i \tag{B.1-15}$$

Example B.3 Solve the following differential equation

$$xy\, dx - (x^2 + y^2)\, dy = 0$$

Solution

Since both of the functions

$$M = xy \tag{1}$$

$$N = -(x^2 + y^2) \tag{2}$$

are homogeneous of degree 2, the transformation

$$y = ux \quad \text{and} \quad dy = u\, dx + x\, du \tag{3}$$

reduces the equation to the form

$$\frac{dx}{x} + \frac{1 + u^2}{u^3} du = 0 \tag{4}$$

Integration of Eq. (4) gives

$$x\,u = C \exp\left(\frac{1}{2u^2}\right) \tag{5}$$

where C is an integration constant. Substitution of $u = y/x$ into Eq. (5) gives the solution as

$$y = C \exp\left[\frac{1}{2}\left(\frac{x}{y}\right)^2\right] \tag{6}$$

B.1.4 Linear Equations

In order to solve an equation of the form

$$\frac{dy}{dx} + P(x)y = Q(x) \tag{B.1-16}$$

the first step is to find out an *integrating factor*, μ, which is defined by

$$\mu = \exp\left[\int P(x)\,dx\right] \tag{B.1-17}$$

Multiplication of Eq. (B.1-16) by the integrating factor gives

$$\frac{d(\mu y)}{dx} = Q\mu \tag{B.1-18}$$

Integration of Eq. (B.1-18) gives the solution as

$$y = \frac{1}{\mu}\int Q\mu\,dx + \frac{C}{\mu} \tag{B.1-19}$$

where C is an integration constant.

Example B.4 Solve the following differential equation

$$x\frac{dy}{dx} - 2y = x^3 \sin x$$

Solution

The differential equation can be rewritten as

$$\frac{dy}{dx} - \frac{2}{x}y = x^2 \sin x \tag{1}$$

The integrating factor, μ, is

$$\mu = \exp\left(-\int \frac{2}{x}\,dx\right) = x^{-2} \tag{2}$$

Multiplication of Eq. (1) by the integrating factor gives

$$\frac{1}{x^2}\frac{dy}{dx} - \frac{2}{x^3}y = \sin x \tag{3}$$

Note that Eq. (3) can also be expressed in the form

$$\frac{d}{dx}\left(\frac{y}{x^2}\right) = \sin x \tag{4}$$

Integration of Eq. (4) gives

$$y = -x^2 \cos x + Cx^2 \tag{5}$$

B.1.5 Bernoulli Equations

A *Bernoulli equation* has the form

$$\frac{dy}{dx} + P(x)y = Q(x)y^n \quad n \neq 0, 1 \tag{B.1-20}$$

The transformation

$$z = y^{1-n} \tag{B.1-21}$$

reduces the Bernoulli equation to a linear equation, Eq. (B.1-16).

B.2 SECOND-ORDER LINEAR DIFFERENTIAL EQUATIONS

A general second-order linear differential equation with constant coefficients is written as

$$a_o \frac{d^2 y}{dx^2} + a_1 \frac{dy}{dx} + a_2 y = R(x) \tag{B.2-1}$$

If $R(x) = 0$, the equation

$$a_o \frac{d^2 y}{dx^2} + a_1 \frac{dy}{dx} + a_2 y = 0 \tag{B.2-2}$$

is called a *homogeneous equation*.

The second-order homogeneous equation can be solved by proposing a solution of the form

$$y = e^{mx} \tag{B.2-3}$$

where m is a constant. Substitution of Eq. (B.2-3) into Eq. (B.2-2) gives

$$a_o m^2 + a_1 m + a_2 = 0 \tag{B.2-4}$$

which is known as the *characteristic* or *auxiliary equation*. Solution of the given differential equation depends on the roots of the characteristic equation.

Distinct real roots

When the roots of Eq. (B.2-4), m_1 and m_2, are real and distinct, then the solution is

$$y = C_1 e^{m_1 x} + C_2 e^{m_2 x} \tag{B.2-5}$$

Repeated real roots

When the roots of Eq. (B.2-4), m_1 and m_2, are real and equal to each other, i.e., $m_1 = m_2 = m$, then the solution is

$$y = (C_1 + C_2 x)e^{mx} \tag{B.2-6}$$

Conjugate complex roots

When the roots of Eq. (B.2-4), m_1 and m_2, are complex and conjugate, i.e., $m_{1,2} = a \pm ib$, then the solution is

$$y = e^{ax}(C_1 \cos bx + C_2 \sin bx) \tag{B.2-7}$$

B.2.1 Special Case of a Second-Order Equation

A second-order ordinary differential equation of the form

$$\frac{d^2 y}{dx^2} - \lambda^2 y = 0 \tag{B.2-8}$$

where λ is a constant, is frequently encountered in heat and mass transfer problems. Since the roots of the characteristic equation are

$$m_{1,2} = \pm \lambda \tag{B.2-9}$$

the solution becomes

$$y = C_1 e^{\lambda x} + C_2 e^{-\lambda x} \tag{B.2-10}$$

Using the identities

$$\cosh \lambda x = \frac{e^{\lambda x} + e^{-\lambda x}}{2} \quad \text{and} \quad \sinh \lambda x = \frac{e^{\lambda x} - e^{-\lambda x}}{2} \tag{B.2-11}$$

Eq. (B.2-10) can be rewritten as

$$y = C_1^* \sinh \lambda x + C_2^* \cosh \lambda x \tag{B.2-12}$$

B.2.2 Solution of a Non-Homogeneous Differential Equation

Consider the second-order differential equation

$$\frac{d^2 y}{dx^2} + P(x)\frac{dy}{dx} + Q(x)y = R(x) \tag{B.2-13}$$

If one solution of the homogeneous solution is known, i.e., say $y = y_1(x)$, then the complete solution is (Murray, 1924)

$$y = C_1 y_1(x) + C_2 y_1(x) \int \frac{\exp(-\int P(x)\,dx)}{y_1^2}\,dx + y_1 \int \frac{\exp(-\int P(x)\,dx)}{y_1^2}$$

$$\times \left[\int^x y_1(u) R(u) \exp\left(\int P(u)\,du \right) \right] dx \tag{B.2-14}$$

Example B.5 Obtain the complete solution of the following non-homogeneous differential equation if one of the solutions of the homogeneous part is $y_1 = e^{2x}$.

$$\frac{d^2y}{dx^2} - \frac{dy}{dx} - 2y = 3e^{-x} + 10\sin x - 4x$$

Solution

Comparison of the equation with Eq. (B.2-13) indicates that

$$P(x) = -1 \qquad Q(x) = -2 \qquad R(x) = 3e^{-x} + 10\sin x - 4x$$

Therefore, Eq. (B.2-14) takes the form

$$y = C_1 e^{2x} + C_2 e^{2x} \int e^{-3x} dx + e^{2x} \int e^{-3x}\left[\int^x e^u (3e^{-u} + 10\sin u - 4u)\, du\right] dx$$

The use of the integral formulas

$$\int xe^{ax}\, dx = \frac{e^{ax}}{a}\left(x - \frac{1}{a}\right)$$

$$\int e^{ax}\sin bx\, dx = e^{ax}\left(\frac{a\sin bx - b\cos bx}{a^2 + b^2}\right)$$

$$\int e^{ax}\cos bx\, dx = e^{ax}\left(\frac{a\cos bx + b\sin bx}{a^2 + b^2}\right)$$

gives the complete solution as

$$y = C_1 e^{2x} + C_2^* e^{-x} - xe^{-x} - \frac{1}{3}e^{-x} - 3\sin x + \cos x + 2x - 1$$

B.2.3 Bessel's Equation

There is a large class of ordinary differential equations that cannot be solved in closed form in terms of elementary functions. Over certain intervals, the differential equation may possess solutions in power series or *Frobenius series*.

An expression of the form

$$a_o + a_1(x - x_o) + a_2(x - x_o)^2 + \cdots + a_n(x - x_o)^n = \sum_{n=0}^{\infty} a_n(x - x_o)^n \qquad \text{(B.2-15)}$$

is called a power series in powers of $(x - x_o)$, with x_o being the center of expansion. Such a series is said to *converge* if it approaches a finite value as n approaches infinity.

An ordinary differential equation given in the general form

$$\frac{d}{dx}\left(x^p \frac{dy}{dx}\right) + (ax^j + bx^k)y = 0 \quad j > k \qquad \text{(B.2-16)}$$

with either $k = p - 2$ or $b = 0$, is known as *Bessel's equation*. Solutions to Bessel's equations are expressed in the form of power series.

Example B.6 Show that the equation

$$x^2 \frac{d^2 y}{dx^2} + x \frac{dy}{dx} - \left(x^2 + \frac{1}{4}\right) y = 0$$

is reducible to Bessel's equation.

Solution

A second-order differential equation

$$a_o(x) \frac{d^2 y}{dx^2} + a_1(x) \frac{dy}{dx} + a_2(x) y = 0 \tag{1}$$

can be expressed in the form of Eq. (B.2-16) as follows. Dividing each term in Eq. (1) by $a_o(x)$ gives

$$\frac{d^2 y}{dx^2} + \frac{a_1(x)}{a_o(x)} \frac{dy}{dx} + \frac{a_2(x)}{a_o(x)} y = 0 \tag{2}$$

The integrating factor, μ, is

$$\mu = \exp\left(\int \frac{a_1(x)}{a_o(x)} dx\right) \tag{3}$$

Multiplication of Eq. (2) by the integrating factor results in

$$\frac{d}{dx}\left(\mu \frac{dy}{dx}\right) + qy = 0 \tag{4}$$

where

$$q = \frac{a_2(x)}{a_o(x)} \mu \tag{5}$$

To express the given equation in the form of Eq. (B.2-16), the first step is to divide each term by x^2 to get

$$\frac{d^2 y}{dx^2} + \frac{1}{x} \frac{dy}{dx} - \left(1 + \frac{1}{4} x^{-2}\right) y = 0 \tag{6}$$

Note that the integrating factor is

$$\mu = \exp\left(\int \frac{1}{x} dx\right) = x \tag{7}$$

Multiplication of Eq. (6) by the integrating factor and rearrangement give

$$\frac{d}{dx}\left(x \frac{dy}{dx}\right) - \left(x + \frac{1}{4} x^{-1}\right) y = 0 \tag{8}$$

Comparison of Eq. (8) with Eq. (B.2-16) gives $p = 1$; $a = -1$; $b = -\frac{1}{4}$; $j = 1$; $k = -1$. Since $k = p - 2$, then Eq. (8) is Bessel's equation.

B.2.3.1 *Solution of Bessel's equation* If an ordinary differential equation is reducible to Bessel's equation, then the constants α, β, and n are defined by

$$\alpha = \frac{2 - p + j}{2} \tag{B.2-17}$$

$$\beta = \frac{1 - p}{2 - p + j} \tag{B.2-18}$$

$$n = \frac{\sqrt{(1 - p)^2 - 4b}}{2 - p + j} \tag{B.2-19}$$

The solution depends on whether the term a is positive or negative.

Case (i) $a > 0$

In this case, the solution is given by

$$y = x^{\alpha\beta}\left[C_1 J_n(\Omega x^\alpha) + C_2 J_{-n}(\Omega x^\alpha)\right] \quad n \neq \text{integer} \tag{B.2-20}$$

$$y = x^{\alpha\beta}\left[C_1 J_n(\Omega x^\alpha) + C_2 Y_n(\Omega x^\alpha)\right] \quad n = \text{integer} \tag{B.2-21}$$

where C_1 and C_2 are constants, and Ω is defined by

$$\Omega = \frac{\sqrt{a}}{\alpha} \tag{B.2-22}$$

The term $J_n(x)$ is known as the *Bessel function of the first kind of order n* and is given by

$$J_n(x) = \sum_{i=0}^{\infty} \frac{(-1)^i (x/2)^{2i+n}}{i!\,\Gamma(i + n + 1)} \tag{B.2-23}$$

$J_{-n}(x)$ is obtained by simply replacing n in Eq. (B.2-23) with $-n$. When n is not an integer, the functions $J_n(x)$ and $J_{-n}(x)$ are linearly independent solutions of Bessel's equation as given by Eq. (B.2-20). When n is an integer, however, these two functions are no longer linearly independent. In this case, the solution is given by Eq. (B.2-21) in which $Y_n(x)$ is known as *Weber's Bessel function of the second kind of order n* and is given by

$$Y_n(x) = \frac{(\cos n\pi)\,J_n(x) - J_{-n}(x)}{\sin n\pi} \tag{B.2-24}$$

Case (ii) $a < 0$

In this case, the solution is given by

$$y = x^{\alpha\beta}\left[C_1 I_n(\Omega x^\alpha) + C_2 I_{-n}(\Omega x^\alpha)\right] \quad n \neq \text{integer} \tag{B.2-25}$$

$$y = x^{\alpha\beta}\left[C_1 I_n(\Omega x^\alpha) + C_2 K_n(\Omega x^\alpha)\right] \quad n = \text{integer} \tag{B.2-26}$$

where C_1 and C_2 are constants, and Ω is defined by

$$\Omega = -i\,\frac{\sqrt{a}}{\alpha} \tag{B.2-27}$$

The term $I_n(x)$ is known as the *modified Bessel function of the first kind of order n* and is given by

$$I_n(x) = \sum_{i=0}^{\infty} \frac{(x/2)^{2i+n}}{i!\,\Gamma(i+n+1)} \tag{B.2-28}$$

$I_{-n}(x)$ is obtained by simply replacing n in Eq. (B.2-28) with $-n$. When n is not an integer, the functions $I_n(x)$ and $I_{-n}(x)$ are linearly independent solutions of Bessel's equation as given by Eq. (B.2-25). However, when n is an integer, the functions $I_n(x)$ and $I_{-n}(x)$ are linearly dependent. In this case, the solution is given by Eq. (B.2-26) in which $K_n(x)$ is known as the *modified Bessel function of the second kind of order n* and is given by

$$K_n(x) = \frac{\pi}{2}\,\frac{I_{-n}(x) - I_n(x)}{\sin n\pi} \tag{B.2-29}$$

Example B.7 Obtain the general solution of the following equations in terms of Bessel functions:

a) $x\,\dfrac{d^2 y}{dx^2} - 3\dfrac{dy}{dx} + xy = 0$

b) $\dfrac{d^2 y}{dx^2} - x^2 y = 0$

Solution

a) Note that the integrating factor is x^{-3} and the equation can be rewritten as

$$\frac{d}{dx}\left(x^{-3}\frac{dy}{dx}\right) + x^{-3}y = 0 \tag{1}$$

Therefore, $p = -3$; $a = 1$; $j = -3$; $b = 0$. Since $b = 0$, the equation is reducible to Bessel's equation. The terms α, β, and n are calculated from Eqs. (B.2-17)–(B.2-19) as

$$\alpha = \frac{2 - p + j}{2} = \frac{2 + 3 - 3}{2} = 1 \tag{2}$$

$$\beta = \frac{1 - p}{2 - p + j} = \frac{1 + 3}{2 + 3 - 3} = 2 \tag{3}$$

$$n = \frac{\sqrt{(1-p)^2 - 4b}}{2 - p + j} = \frac{\sqrt{(1+3)^2 - (4)(0)}}{2 + 3 - 3} = 2 \tag{4}$$

Note that $a > 0$ and Ω is calculated from Eq. (B.2-22) as

$$\Omega = \frac{\sqrt{a}}{\alpha} = \frac{\sqrt{1}}{1} = 1 \tag{5}$$

Since n is an integer, the solution is given in the form of Eq. (B.2-21)

$$y = x^2 \left[C_1 J_2(x) + C_2 Y_2(x) \right] \tag{6}$$

b) The equation can be rearranged in the form

$$\frac{d}{dx}\left(\frac{dy}{dx}\right) - x^2 y = 0 \tag{7}$$

Therefore, $p = 0$; $a = -1$; $j = 2$; $b = 0$. Since $b = 0$, the equation is reducible to Bessel's equation. The terms α, β, and n are calculated from Eqs. (B.2-17)–(B.2-19) as

$$\alpha = \frac{2 - p + j}{2} = \frac{2 - 0 + 2}{2} = 2 \tag{8}$$

$$\beta = \frac{1 - p}{2 - p + j} = \frac{1 - 0}{2 - 0 + 2} = \frac{1}{4} \tag{9}$$

$$n = \frac{\sqrt{(1 - p)^2 - 4b}}{2 - p + j} = \frac{\sqrt{(1 - 0)^2 - (4)(0)}}{2 - 0 + 2} = \frac{1}{4} \tag{10}$$

Note that $a < 0$ and Ω is calculated from Eq. (B.2-27) as

$$\Omega = -i\,\frac{\sqrt{a}}{\alpha} = -i\,\frac{\sqrt{-1}}{2} = \frac{1}{2} \tag{11}$$

Since n is not an integer, the solution is given in the form of Eq. (B.2-25)

$$y = \sqrt{x}\left[C_1\, I_{1/4}(x^2/2) + C_2\, I_{-1/4}(x^2/2) \right]$$

The properties of the Bessel functions are summarized in Table B.1.

B.2.3.2 *Useful integration formulas involving Bessel functions*   The following integration formulas are useful in the evaluation of Fourier coefficients (see Section B.3.5) appearing in the solution of partial differential equations:

$$\int x^{n+1} J_n(\lambda x)\, dx = \frac{x^{n+1}}{\lambda}\, J_{n+1}(\lambda x) \tag{B.2-30}$$

$$\int x J_n^2(\lambda x)\, dx = \frac{x^2}{2}\left[J_n^2(\lambda x) + J_{n+1}^2(\lambda x) \right] - \frac{nx}{\lambda}\, J_n(\lambda x) J_{n+1}(\lambda x) \tag{B.2-31}$$

$$\int x^{n+3} J_n(\lambda x)\, dx = \left[\frac{x^{n+3}}{\lambda} - \frac{4(n+1)x^{n+1}}{\lambda^3}\right] J_{n+1}(\lambda x) + \frac{2x^{n+2}}{\lambda^2}\, J_n(\lambda x) \tag{B.2-32}$$

B.2.4 Numerical Solution of Initial Value Problems

Consider an initial value problem of the type

$$\frac{dy}{dt} = f(t, y) \tag{B.2-33}$$

$$y(0) = a = \text{given} \tag{B.2-34}$$

Table B.1. Properties of the Bessel functions

BEHAVIOR NEAR THE ORIGIN

$$J_o(0) = I_o(0) = 1$$
$$-Y_n(0) = K_n(0) = \infty \quad \text{for all } n$$
$$J_n(0) = I_n(0) = 0 \quad \text{for } n > 0$$

Note that if the origin is a point in the calculation field, then $J_n(x)$ and $I_n(x)$ are the only physically permissible solutions.

BESSEL FUNCTIONS OF NEGATIVE ORDER
(n is an integer)

$$J_{-n}(\lambda x) = (-1)^n J_n(\lambda x) \quad Y_{-n}(\lambda x) = (-1)^n Y_n(\lambda x)$$
$$I_{-n}(\lambda x) = I_n(\lambda x) \quad K_{-n}(\lambda x) = K_n(\lambda x)$$

RECURRENCE FORMULAS

$$J_n(\lambda x) = \frac{\lambda x}{2n} \left[J_{n+1}(\lambda x) + J_{n-1}(\lambda x) \right]$$

$$Y_n(\lambda x) = \frac{\lambda x}{2n} \left[Y_{n+1}(\lambda x) + Y_{n-1}(\lambda x) \right]$$

$$I_n(\lambda x) = -\frac{\lambda x}{2n} \left[I_{n+1}(\lambda x) - I_{n-1}(\lambda x) \right]$$

$$K_n(\lambda x) = \frac{\lambda x}{2n} \left[K_{n+1}(\lambda x) - K_{n-1}(\lambda x) \right]$$

INTEGRAL PROPERTIES

$$\int \lambda x^n J_{n-1}(\lambda x)\, dx = x^n J_n(\lambda x) \quad \int \lambda x^n Y_{n-1}(\lambda x)\, dx = x^n Y_n(\lambda x)$$

$$\int \lambda x^n I_{n-1}(\lambda x)\, dx = x^n I_n(\lambda x) \quad \int \lambda x^n K_{n-1}(\lambda x)\, dx = -x^n K_n(\lambda x)$$

DIFFERENTIAL RELATIONS

$$\frac{d}{dx} J_n(\lambda x) = \lambda J_{n-1}(\lambda x) - \frac{n}{x} J_n(\lambda x) = -\lambda J_{n+1}(\lambda x) + \frac{n}{x} J_n(\lambda x)$$

$$\frac{d}{dx} Y_n(\lambda x) = \lambda Y_{n-1}(\lambda x) - \frac{n}{x} Y_n(\lambda x) = -\lambda Y_{n+1}(\lambda x) + \frac{n}{x} Y_n(\lambda x)$$

$$\frac{d}{dx} I_n(\lambda x) = \lambda I_{n-1}(\lambda x) - \frac{n}{x} I_n(\lambda x) = \lambda I_{n+1}(\lambda x) + \frac{n}{x} I_n(\lambda x)$$

$$\frac{d}{dx} K_n(\lambda x) = -\lambda K_{n-1}(\lambda x) - \frac{n}{x} K_n(\lambda x) = -\lambda K_{n+1}(\lambda x) + \frac{n}{x} K_n(\lambda x)$$

Among the various numerical methods available for the integration of Eq. (B.2-33), the *fourth-order Runge-Kutta method* is the most frequently used. It is expressed by the following algorithm:

$$y_{n+1} = y_n + \frac{1}{6}(k_1 + k_4) + \frac{1}{3}(k_2 + k_3) \tag{B.2-35}$$

The terms k_1, k_2, k_3, and k_4 in Eq. (B.2-35) are defined by

$$k_1 = hf(t_n, y_n) \tag{B.2-36}$$

$$k_2 = hf\left(t_n + \frac{1}{2}h, y_n + \frac{1}{2}k_1\right) \tag{B.2-37}$$

$$k_3 = hf\left(t_n + \frac{1}{2}h, y_n + \frac{1}{2}k_2\right) \tag{B.2-38}$$

$$k_4 = hf(t_n + h, y_n + k_3) \tag{B.2-39}$$

in which h is the time step used in the numerical solution of the differential equation.

Example B.8 An irreversible chemical reaction

$$A \rightarrow B$$

takes place in an isothermal batch reactor. The rate of reaction is given by

$$r = kc_A$$

with a rate constant of $k = 2 \text{ h}^{-1}$. If the initial number of moles of species A is 1.5 mol, determine the variation in the number of moles of A during the first hour of the reaction. Compare your results with the analytical solution.

Solution

The inventory rate equation based on the moles of species A is

$$-(kc_A)V = \frac{dn_A}{dt} \tag{1}$$

or,

$$\frac{dn_A}{dt} = -kn_A \tag{2}$$

Analytical solution

Equation (2) is a separable equation with the solution

$$n_A = n_{A_o} \exp(-kt) \tag{3}$$

in which n_{A_o} is the initial number of moles of species A.

Numerical solution

In terms of the notation of the Runge-Kutta method, Eq. (2) is expressed as

$$\frac{dy}{dt} = -2y \tag{4}$$

with an initial condition of

$$y(0) = 1.5 \tag{5}$$

Therefore,

$$f(t, y) = -2y \tag{6}$$

$$y_o = 1.5 \tag{7}$$

Integration of Eq. (4) from $t = 0$ to $t = 1$ by using the fourth-order Runge-Kutta method with a time step of $h = 0.1$ is given as follows:

Calculation of y at $t = 0.1$ h

First, it is necessary to determine k_1, k_2, k_3, and k_4:

$$k_1 = hf(y_o) = (0.1)(-2)(1.5) = -0.3000 \tag{8}$$

$$k_2 = hf\left(y_o + \frac{1}{2}k_1\right) = (0.1)(-2)\left(1.5 - \frac{0.3}{2}\right) = -0.2700 \tag{9}$$

$$k_3 = hf\left(y_o + \frac{1}{2}k_2\right) = (0.1)(-2)\left(1.5 - \frac{0.2700}{2}\right) = -0.2730 \tag{10}$$

$$k_4 = hf(y_o + k_3) = (0.1)(-2)(1.5 - 0.2730) = -0.2454 \tag{11}$$

Substitution of these values into Eq. (B.2-35) gives the value of y at $t = 0.1$ h as

$$y_1 = 1.5 - \frac{1}{6}(0.3 + 0.2454) - \frac{1}{3}(0.2700 + 0.2730) = 1.2281 \tag{12}$$

Calculation of y at $t = 0.2$ h

The constants k_1, k_2, k_3, and k_4 are calculated as

$$k_1 = hf(y_1) = (0.1)(-2)(1.2281) = -0.2456 \tag{13}$$

$$k_2 = hf\left(y_1 + \frac{1}{2}k_1\right) = (0.1)(-2)\left(1.2281 - \frac{0.2456}{2}\right) = -0.2211 \tag{14}$$

$$k_3 = hf\left(y_1 + \frac{1}{2}k_2\right) = (0.1)(-2)\left(1.2281 - \frac{0.2211}{2}\right) = -0.2235 \tag{15}$$

$$k_4 = hf(y_1 + k_3) = (0.1)(-2)(1.2281 - 0.2235) = -0.2009 \tag{16}$$

Substitution of these values into Eq. (B.2-35) gives the value of y at $t = 0.2$ h as

$$y_2 = 1.2281 - \frac{1}{6}(0.2456 + 0.2009) - \frac{1}{3}(0.2211 + 0.2235) = 1.0055 \tag{17}$$

Repeated application of this procedure gives the value of y at every 0.1 hour. The results of such calculations are given in Table 1. The last column of Table 1 gives the analytical results obtained from Eq. (3). In this case, the numerical and analytical results are equal to each other. However, this is not always the case. The accuracy of the numerical results depends on the time step chosen for the calculations. For example, for a time step of $h = 0.5$, the numerical results are slightly different from the exact ones as shown in Table 2.

Table 1. Comparison of numerical and exact values for $h = 0.1$

t (h)	k_1	k_2	k_3	k_4	y (num.)	y (exact)
0.1	−0.3000	−0.2700	−0.2730	−0.2454	1.2281	1.2281
0.2	−0.2456	−0.2211	−0.2235	−0.2009	1.0055	1.0055
0.3	−0.2011	−0.1810	−0.1830	−0.1645	0.8232	0.8232
0.4	−0.1646	−0.1482	−0.1498	−0.1347	0.6740	0.6740
0.5	−0.1348	−0.1213	−0.1227	−0.1103	0.5518	0.5518
0.6	−0.1104	−0.0993	−0.1004	−0.0903	0.4518	0.4518
0.7	−0.0904	−0.0813	−0.0822	−0.0739	0.3699	0.3699
0.8	−0.0740	−0.0666	−0.0673	−0.0605	0.3028	0.3028
0.9	−0.0606	−0.0545	−0.0551	−0.0495	0.2479	0.2479
1.0	−0.0496	−0.0446	−0.0451	−0.0406	0.2030	0.2030

Table 2. Comparison of numerical and exact values for $h = 0.5$

t (h)	k_1	k_2	k_3	k_4	y (num.)	y (exact)
0.5	−1.5000	−0.7500	−1.1250	−0.3750	0.5625	0.5518
1.0	−0.5625	−0.2813	−0.4219	−0.1406	0.2109	0.2030

B.2.5 Solution of Simultaneous Differential Equations

The solution procedure presented for a single ordinary differential equation can be easily extended to solve sets of simultaneous differential equations. For example, for the case of two simultaneous ordinary differential equations

$$\frac{dy}{dx} = f(t, y, z) \tag{B.2-40}$$

$$\frac{dz}{dt} = g(t, y, z) \tag{B.2-41}$$

the fourth-order Runge-Kutta solution algorithm is given by

$$y_{n+1} = y_n + \frac{1}{6}(k_1 + k_4) + \frac{1}{3}(k_2 + k_3) \tag{B.2-42}$$

and

$$z_{n+1} = z_n + \frac{1}{6}(\ell_1 + \ell_4) + \frac{1}{3}(\ell_2 + \ell_3) \tag{B.2-43}$$

The terms $k_1 \to k_4$ and $\ell_1 \to \ell_4$ are defined by

$$k_1 = hf(t_n, y_n, z_n) \tag{B.2-44}$$

$$\ell_1 = hg(t_n, y_n, z_n) \tag{B.2-45}$$

$$k_2 = hf\left(t_n + \frac{1}{2}h, y_n + \frac{1}{2}k_1, z_n + \frac{1}{2}\ell_1\right) \tag{B.2-46}$$

$$\ell_2 = hg\left(t_n + \frac{1}{2}h, y_n + \frac{1}{2}k_1, z_n + \frac{1}{2}\ell_1\right) \tag{B.2-47}$$

$$k_3 = hf\left(t_n + \frac{1}{2}h, y_n + \frac{1}{2}k_2, z_n + \frac{1}{2}\ell_2\right) \tag{B.2-48}$$

$$\ell_3 = hg\left(t_n + \frac{1}{2}h, y_n + \frac{1}{2}k_2, z_n + \frac{1}{2}\ell_2\right) \tag{B.2-49}$$

$$k_4 = hf(t_n + h, y_n + k_3, z_n + \ell_3) \tag{B.2-50}$$

$$\ell_4 = hg(t_n + h, y_n + k_3, z_n + \ell_3) \tag{B.2-51}$$

Example B.9 The following liquid phase reactions are carried out in a batch reactor under isothermal conditions:

$$A \rightarrow B \qquad r = k_1 c_A \qquad k_1 = 0.4 \text{ h}^{-1}$$
$$B + C \rightarrow D \qquad r = k_2 c_B c_C \qquad k_2 = 0.7 \text{ m}^3/\text{mol·h}$$

If the initial concentrations of species A and C are 1 mol/m^3, determine the concentration of species D after 18 min. Compare your results with the analytical solution.

Solution

The inventory rate expressions for species A and D are given by

$$\frac{dc_A}{dt} = -k_1 c_A \tag{1}$$

$$\frac{dc_D}{dt} = k_2 c_B c_C \tag{2}$$

From the stoichiometry of the reactions, the concentrations of B and C are expressed in terms of A and D as

$$c_B = c_{A_o} - c_A - c_D \tag{3}$$

$$c_C = c_{C_o} - c_D \tag{4}$$

Substitution of Eqs. (3) and (4) into Eq. (2) yields

$$\frac{dc_D}{dt} = k_2(c_{A_o} - c_A - c_D)(c_{C_o} - c_D) \tag{5}$$

Analytical solution

Equation (1) is a separable equation with the solution

$$c_A = c_{A_o} \exp(-k_1 t) \tag{6}$$

in which c_{A_o} is the initial concentration of species A. Substitution of Eq. (6) into Eq. (5) gives

$$\frac{dc_D}{dt} = k_2 c_D^2 + k_2\left(c_{A_o} e^{-k_1 t} - c_{A_o} - c_{C_o}\right)c_D + k_2 c_{A_o} c_{C_o}(1 - e^{-k_1 t}) \tag{7}$$

In terms of numerical values, Eq. (7) becomes

$$\frac{dc_D}{dt} = 0.7 c_D^2 + 0.7(e^{-0.4t} - 2)c_D + 0.7(1 - e^{-0.4t}) \tag{8}$$

The non-linear first-order differential equation

$$\frac{dy}{dx} = a(x)y^2 + b(x)y + c(x)$$

is called a Riccati equation. If $y_1(x)$ is any known solution of the given equation, then the transformation

$$y = y_1(x) + \frac{1}{u}$$

leads to a linear equation in u. Equation (8) is in the form of a Riccati equation and note that $c_D = 1$ is a solution. Therefore, the solution is

$$c_D = 1 - e^{-1.75\tau} \left(1.75 \int_\tau^1 \frac{e^{-1.75\eta}}{\eta} d\eta + e^{-1.75} \right)^{-1} \tag{9}$$

where

$$\tau = e^{-0.4t} \tag{10}$$

When $t = 0.3$ h, Eq. (9) gives $c_D = 0.0112$ mol/m^3.

Numerical solution

In terms of the notation of the Runge-Kutta method, Eqs. (1) and (5) are expressed in the form

$$\frac{dy}{dt} = -0.4y \tag{11}$$

$$\frac{dz}{dt} = 0.7(1 - y - z)(1 - z) \tag{12}$$

with initial conditions of

$$y(0) = 1 \quad \text{and} \quad z(0) = 0 \tag{13}$$

Therefore,

$$f(t, y, z) = -0.4y \tag{14}$$

$$g(t, y, z) = 0.7(1 - y - z)(1 - z) \tag{15}$$

with the initial conditions

$$y_o = 1 \quad and \quad z_o = 0 \tag{16}$$

Choosing $h = 0.05$, the values of y_1 and z_1 are calculated as follows:

$$k_1 = hf(y_o, z_o) = (0.05)(-0.4)(1) = -0.0200 \tag{17}$$

$$\ell_1 = hg(y_o, z_o) = (0.05)(0.7)(1 - 1 - 0)(1 - 0) = 0 \tag{18}$$

$$k_2 = hf\left(y_o + \frac{1}{2}k_1, z_o + \frac{1}{2}\ell_1\right) = (0.05)(-0.4)\left(1 - \frac{0.0200}{2}\right) = -0.0198 \tag{19}$$

$$\ell_2 = hg\left(y_o + \frac{1}{2}k_1, z_o + \frac{1}{2}\ell_1\right)$$

$$= (0.05)(0.7)\left[1 - \left(1 - \frac{0.0200}{2}\right) - 0\right](1 - 0) = 3.5 \times 10^{-4} \tag{20}$$

$$k_3 = hf\left(y_o + \frac{1}{2}k_2, z_o + \frac{1}{2}\ell_2\right) = (0.05)(-0.4)\left(1 - \frac{0.0198}{2}\right) = -0.0198 \tag{21}$$

$$\ell_3 = hg\left(y_o + \frac{1}{2}k_2, z_o + \frac{1}{2}\ell_2\right)$$

$$= (0.05)(0.7)\left[1 - \left(1 - \frac{0.0198}{2}\right) - \frac{3.5 \times 10^{-4}}{2}\right]\left(1 - \frac{3.5 \times 10^{-4}}{2}\right)$$

$$= 3.4032 \times 10^{-4} \tag{22}$$

$$k_4 = hf(y_o + k_3, z_o + \ell_3) = (0.05)(-0.4)(1 - 0.0198) = -0.0196 \tag{23}$$

$$\ell_4 = hg(y_o + k_3, z_o + \ell_3)$$

$$= (0.05)(0.7)\left[1 - (1 - 0.0198) - 3.4032 \times 10^{-4}\right](1 - 3.4032 \times 10^{-4})$$

$$= 6.8086 \times 10^{-4} \tag{24}$$

Substitution of $k_1 \rightarrow k_4$ and $\ell_1 \rightarrow \ell_4$ into Eqs. (B.2-42) and (B.2-43), respectively, gives the values of y_1 and z_1 as

$$y_1 = 1 - \frac{1}{6}(0.0200 + 0.0196) - \frac{1}{3}(0.0198 + 0.0198) = 0.9802 \tag{25}$$

$$z_1 = 0 + \frac{1}{6}(0 + 6.8086 \times 10^{-4}) + \frac{1}{3}(3.5 \times 10^{-4} + 3.4032 \times 10^{-4}) = 3.4358 \times 10^{-4} \tag{26}$$

Repeated application of this procedure gives the values of y and z at every 0.05 h. The results are given in Tables 1 and 2.

Table 1. Values of y as a function of time

t (h)	k_1	k_2	k_3	k_4	y
0.05	−0.0200	−0.0198	−0.0198	−0.0196	0.9802
0.10	−0.0196	−0.0194	−0.0194	−0.0192	0.9608
0.15	−0.0192	−0.0190	−0.0190	−0.0188	0.9418
0.20	−0.0188	−0.0186	−0.0186	−0.0185	0.9232
0.25	−0.0185	−0.0183	−0.0183	−0.0181	0.9049
0.30	−0.0181	−0.0179	−0.0179	−0.0177	0.8870

Table 2. Values of z as a function of time

t (h)	ℓ_1	ℓ_2	ℓ_3	ℓ_4	z
0.05	0.0000	0.0004	0.0003	0.0007	0.0003
0.10	0.0007	0.0010	0.0010	0.0013	0.0013
0.15	0.0013	0.0016	0.0016	0.0019	0.0029
0.20	0.0019	0.0022	0.0022	0.0025	0.0051
0.25	0.0025	0.0028	0.0028	0.0030	0.0079
0.30	0.0030	0.0033	0.0033	0.0035	0.0112

B.3 SECOND-ORDER PARTIAL DIFFERENTIAL EQUATIONS

B.3.1 Classification of Partial Differential Equations

As a function of two independent variables, x and y, the most general form of a second-order linear partial differential equation is as follows:

$$A(x, y)\frac{\partial^2 u}{\partial x^2} + 2B(x, y)\frac{\partial^2 u}{\partial x \partial y} + C(x, y)\frac{\partial^2 u}{\partial y^2} + D(x, y)\frac{\partial u}{\partial x}$$

$$+ E(x, y)\frac{\partial u}{\partial y} + F(x, y)u = G(x, y) \tag{B.3-1}$$

It is assumed that the coefficient functions and the given function G are real-valued and twice continuously differentiable on a region $\mathbb{R}$ of the x, y plane. When $G = 0$, the equation is *homogeneous*; otherwise the equation is *non-homogeneous*.

The criterion, $B^2 - AC$, that will indicate whether the second-order equation is a graph of a parabola, ellipse, or hyperbola is called the discriminant, Δ, i.e.,

$$\Delta = B^2 - AC \begin{cases} > 0 & \text{Hyperbolic} \\ = 0 & \text{Parabolic} \\ < 0 & \text{Elliptic} \end{cases}$$

B.3.2 Orthogonal Functions

Let $f(x)$ and $g(x)$ be real-valued functions defined on the interval $a \leqslant x \leqslant b$. The *inner product* of $f(x)$ and $g(x)$ with respect to $w(x)$ is defined by

$$\langle f, g \rangle = \int_a^b w(x)f(x)g(x)\,dx \tag{B.3-2}$$

in which the *weight function* $w(x)$ is considered positive on the interval (a, b).

Example B.10 Find the inner product of $f(x) = x$ and $g(x) = 1$ with respect to the weight function $w(x) = x^{1/2}$ on the interval $0 \leqslant x \leqslant 1$.

Solution

Application of Eq. (B.3-2) gives the inner product as

$$\langle f, g \rangle = \int_0^1 \sqrt{x}\, x\, dx = \frac{2}{5}x^{5/2}\Big|_0^1 = \frac{2}{5}$$

The inner product has the following properties:

$$\langle f, g \rangle = \langle g, f \rangle \tag{B.3-3}$$

$$\langle f, g + h \rangle = \langle f, g \rangle + \langle f, h \rangle \tag{B.3-4}$$

$$\langle \alpha f, g \rangle = \alpha \langle f, g \rangle \quad \alpha \text{ is a scalar} \tag{B.3-5}$$

The inner product of f with respect to itself is

$$\langle f, f \rangle = \int_a^b w(x) f^2(x) \, dx = \| f(x) \|^2 > 0 \tag{B.3-6}$$

in which the *norm* of $f(x)$ is defined as

$$\| f(x) \| = \sqrt{\langle f, f \rangle} \tag{B.3-7}$$

When $\langle f, g \rangle = 0$ on (a, b), then $f(x)$ is *orthogonal* to $g(x)$ with respect to the weight function $w(x)$ on (a, b), and, when $\langle f, f \rangle = 1$, then $f(x)$ is an *orthonormal* function. In the special case where $w(x) = 1$ for $a \leqslant x \leqslant b$, $f(x)$ and $g(x)$ are said to be *simply orthogonal*.

A sequence of functions $\{f_n\}_{n=0}^{\infty}$ is an *orthogonal set* of functions if

$$\langle f_n, f_m \rangle = 0 \quad n \neq m \tag{B.3-8}$$

The orthogonal set is a linearly independent set. If

$$\langle f_n, f_m \rangle = \begin{cases} 0 & \text{if } n \neq m \\ 1 & \text{if } n = m \end{cases} \tag{B.3-9}$$

such a set is called an *orthonormal set*. Note that an orthonormal set can be obtained from an orthogonal set by dividing each function by its norm on the interval under consideration.

Example B.11 Let $\phi_n(x) = \sin(n\pi x)$ for $n = 1, 2, 3, \ldots$ and for $0 < x < 1$. Show that the sequence $\{\phi_n\}_{n=1}^{\infty}$ is simply orthogonal on $(0, 1)$. Find the norms of the functions ϕ_n.

Solution

The inner product is

$$\langle \phi_n, \phi_m \rangle = \int_0^1 \sin(n\pi x) \sin(m\pi x) \, dx \tag{1}$$

The use of the identity

$$\sin A \sin B = \frac{1}{2} \left[\cos(A - B) - \cos(A + B) \right] \tag{2}$$

reduces Eq. (1) to the form

$$\langle \phi_n, \phi_m \rangle = \frac{1}{2} \int_0^1 \left\{ \cos[(n - m)\pi x] - \cos[(n + m)\pi x] \right\} dx$$

$$= \frac{1}{2} \left\{ \frac{\sin[(n - m)\pi x]}{(n - m)\pi} - \frac{\sin[(n + m)\pi x]}{(n - m)\pi} \right\} \Big|_0^1 = 0 \tag{3}$$

On the other hand,

$$\langle \phi_n, \phi_n \rangle = \int_0^1 \sin^2(n\pi x)\, dx$$

$$= \frac{1}{2} \int_0^1 \left[1 - \cos(2n\pi x) \right] dx = \frac{1}{2} \left[x - \frac{\sin(2n\pi x)}{2n\pi} \right]_0^1 = \frac{1}{2} \tag{4}$$

Therefore, the norm is

$$\|\phi_n\| = \sqrt{\langle \phi_n, \phi_n \rangle} = \frac{1}{\sqrt{2}} \tag{5}$$

Hence, the corresponding orthonormal set is $\{\sqrt{2}\sin(n\pi x)\}_{n=1}^{\infty}$.

B.3.3 Self-Adjoint Problems

Consider a second-order ordinary differential equation of the form

$$a_o(x)\frac{d^2y}{dx^2} + a_1(x)\frac{dy}{dx} + a_2(x)y = 0 \tag{B.3-10}$$

Multiplication of Eq. (B.3-10) by $p(x)/a_o(x)$ in which $p(x)$ is the integrating factor defined by

$$p(x) = \exp\left(\int \frac{a_1(x)}{a_o(x)}\, dx \right) \tag{B.3-11}$$

gives

$$p(x)\frac{d^2y}{dx^2} + \frac{a_1(x)}{a_o(x)}\, p(x)\frac{dy}{dx} + \frac{a_2(x)}{a_o(x)}\, p(x)y = 0 \tag{B.3-12}$$

Equation (B.3-12) can be rewritten as

$$p(x)\frac{d^2y}{dx^2} + \frac{dp(x)}{dx}\frac{dy}{dx} + q(x)y = 0 \tag{B.3-13}$$

where

$$q(x) = \frac{a_2(x)}{a_o(x)}\, p(x) \tag{B.3-14}$$

Rearrangement of Eq. (B.3-13) yields

$$\frac{d}{dx}\left[p(x)\frac{dy}{dx} \right] + q(x)y = 0 \tag{B.3-15}$$

A second-order differential equation in this form is said to be in *self-adjoint form*.

Example B.12 Write the following differential equation in self-adjoint form:

$$x^2 \frac{d^2 y}{dx^2} - x \frac{dy}{dx} + (x-3)y = 0$$

Solution

Dividing the given equation by x^2 gives

$$\frac{d^2 y}{dx^2} - \frac{1}{x}\frac{dy}{dx} + \left(\frac{1}{x} - \frac{3}{x^2}\right)y = 0 \tag{1}$$

Note that

$$p(x) = \exp\left(-\int \frac{dx}{x}\right) = \frac{1}{x} \tag{2}$$

Multiplication of Eq. (1) by $p(x)$ gives

$$\frac{1}{x}\frac{d^2 y}{dx^2} - \frac{1}{x^2}\frac{dy}{dx} + \left(\frac{1}{x^2} - \frac{3}{x^3}\right)y = 0 \tag{3}$$

Note that Eq. (3) can be rearranged as

$$\frac{d}{dx}\left(\frac{1}{x}\frac{dy}{dx}\right) + \left(\frac{1}{x^2} - \frac{3}{x^3}\right)y = 0 \tag{4}$$

B.3.4 The Sturm-Liouville Problem

The linear, homogeneous, second-order equation

$$\frac{1}{w(x)}\frac{d}{dx}\left[p(x)\frac{dy}{dx}\right] + q(x)y = -\lambda y \tag{B.3-16}$$

on some interval $a \leqslant x \leqslant b$ satisfying boundary conditions of the form

$$\alpha_1 y(a) + \alpha_2 \frac{dy}{dx}\bigg|_{x=a} = 0 \tag{B.3-17}$$

$$\beta_1 y(b) + \beta_2 \frac{dy}{dx}\bigg|_{x=b} = 0 \tag{B.3-18}$$

where $\alpha_1, \alpha_2, \beta_1, \beta_2$ are given constants; $p(x)$, $q(x)$, $w(x)$ are given functions that are differentiable; and λ is an unspecified parameter independent of x, is called the *Sturm-Liouville equation*.

The values of λ for which the problem given by Eqs. (B.3-16)–(B.3-18) has a nontrivial solution, i.e., a solution other than $y = 0$, are called the *eigenvalues*. The corresponding solutions are the *eigenfunctions*.

Eigenfunctions corresponding to different eigenvalues are orthogonal with respect to the weight function $w(x)$. All the eigenvalues are positive. In particular, $\lambda = 0$ is not an eigenvalue.

Example B.13 Solve

$$\frac{d^2y}{dx^2} + \lambda y = 0$$

subject to the boundary conditions

$$\begin{array}{lll} \text{at} & x = 0 & y = 0 \\ \text{at} & x = \pi & y = 0 \end{array}$$

Solution

The equation can be rewritten in the form

$$\frac{d}{dx}\left(\frac{dy}{dx}\right) = -\lambda y \tag{1}$$

Comparison of Eq. (1) with Eq. (B.3-16) indicates that this is a Sturm-Liouville problem with $p(x) = 1$, $q(x) = 0$, and $w(x) = 1$.

The solution of Eq. (1) is

$$y = A\sin\left(\sqrt{\lambda}x\right) + B\cos\left(\sqrt{\lambda}x\right) \tag{2}$$

Application of the boundary condition at $x = 0$ implies that $B = 0$. On the other hand, the use of the boundary condition at $x = \pi$ gives

$$A\sin\left(\sqrt{\lambda}\,\pi\right) = 0 \tag{3}$$

In order to have a nontrivial solution

$$\sin\left(\sqrt{\lambda}\pi\right) = 0 \quad \Rightarrow \quad \sqrt{\lambda}\,\pi = n\pi \quad n = 1, 2, 3, \ldots \tag{4}$$

or,

$$\sqrt{\lambda} = n \quad \Rightarrow \quad \lambda_n = n^2 \quad n = 1, 2, 3, \ldots \tag{5}$$

Equation (5) represents the eigenvalues of the problem. The corresponding eigenfunctions are

$$y_n = A_n\sin(nx) \quad n = 1, 2, 3, \ldots \tag{6}$$

where A_n is an arbitrary nonzero constant.

Since the eigenfunctions are orthogonal to each other with respect to the weight function $w(x)$, it is possible to write

$$\int_0^\pi \sin(nx)\sin(mx)\,dx = 0 \quad n \neq m \tag{7}$$

B.3.4.1 *The method of Stodola and Vianello*   The method of Stodola and Vianello (Bird *et al.*, 1987; Hildebrand, 1976) is an iterative procedure that makes use of successive approximation to estimate λ in the following differential equation

$$\frac{d}{dx}\left[p(x)\frac{dy}{dx}\right] = -\lambda w(x)y \tag{B.3-19}$$

with appropriate homogeneous boundary conditions at $x = a$ and $x = b$.

The procedure is as follows:

1. Assume a trial function for $y_1(x)$ that satisfies the boundary conditions $x = a$ and $x = b$.
2. On the right-hand side of Eq. (B.3-19), replace $y(x)$ with $y_1(x)$.
3. Solve the resulting differential equation and express the solution in the form

$$y(x) = \lambda f_1(x) \tag{B.3-20}$$

4. Repeat step (2) with a second trial function $y_2(x)$ defined by

$$y_2(x) = f_1(x) \tag{B.3-21}$$

5. Solve the resulting differential equation and express the solution in the form

$$y(x) = \lambda f_2(x) \tag{B.3-22}$$

6. Continue the process as long as desired. The nth approximation to the smallest permissible value of λ is given by

$$\lambda_1^n = \frac{\displaystyle\int_a^b w(x)f_n(x)y_n(x)\,dx}{\displaystyle\int_a^b w(x)[f_n(x)]^2\,dx} \tag{B.3-23}$$

B.3.5   Fourier Series

Let $f(x)$ be an arbitrary function defined on $a \leqslant x \leqslant b$, and let $\{\phi_n\}_{n=1}^\infty$ be an orthogonal set of functions over the same interval with weight function $w(x)$. Let us assume that $f(x)$ can be represented by an infinite series of the form

$$f(x) = \sum_{n=1}^\infty A_n\phi_n(x) \tag{B.3-24}$$

The series $\sum A_n\phi_n(x)$ is called the *Fourier series* of $f(x)$, and the coefficients A_n are called the *Fourier coefficients* of $f(x)$ with respect to the orthogonal functions $\phi_n(x)$.

To determine the Fourier coefficients, multiply both sides of Eq. (B.3-24) by $w(x)\phi_m(x)$ and integrate from $x = a$ to $x = b$,

$$\int_a^b f(x)w(x)\phi_m(x)\,dx = \sum_{n=1}^\infty A_n \int_a^b \phi_n(x)\phi_m(x)w(x)\,dx \tag{B.3-25}$$

Because of the orthogonality, all the integrals on the right-hand side of Eq. (B.3-25) are zero except when $n = m$. Therefore, the summation drops and Eq. (B.3-25) takes the form

$$\int_a^b f(x)w(x)\phi_n(x)\,dx = A_n \int_a^b \phi_n^2(x)w(x)\,dx \tag{B.3-26}$$

or,

$$A_n = \frac{\langle f, \phi_n \rangle}{\|\phi_n\|^2} \tag{B.3-27}$$

Example B.14 Let $f(x) = x$ for $0 \leqslant x \leqslant \pi$. Find the Fourier series of $f(x)$ with respect to the simply orthogonal set $\{\sin(nx)\}_{n=1}^\infty$.

Solution

The function $f(x) = x$ is represented in the form of a Fourier series

$$x = \sum_{n=1}^\infty A_n \sin(nx) \tag{1}$$

The Fourier coefficients can be calculated from Eq. (B.3-27) as

$$A_n = \frac{\displaystyle\int_0^\pi x\sin(nx)\,dx}{\displaystyle\int_0^\pi \sin^2(nx)\,dx} = -2\frac{\cos(n\pi)}{n} \tag{2}$$

Since

$$\cos(n\pi) = (-1)^n \tag{3}$$

the coefficients A_n become

$$A_n = -2\frac{(-1)^n}{n} = 2\frac{(-1)^{n+1}}{n} \tag{4}$$

Substitution of Eq. (4) into Eq. (1) yields

$$x = 2\sum_{n=1}^\infty \frac{(-1)^{n+1}}{n}\sin(nx) \tag{5}$$

B.3.6 Solution of Partial Differential Equations

Various analytical methods are available to solve partial differential equations. In the determination of the method to be used, the structure of the equation is not the only factor that should be taken into consideration as in the case for ordinary differential equations. The boundary conditions are almost as important as the equation itself.

B.3.6.1   *The method of separation of variables*   The method of separation of variables requires the partial differential equation to be homogeneous and the boundary conditions to be defined over a limited interval, i.e., semi-infinite and infinite domains do not permit the use of the separation of variables method. Moreover, boundary conditions must be homogeneous in at least one dimension.

Let us apply the method of separation of variables to an unsteady-state heat transfer problem. Consider a slab that is initially at temperature T_o. At time $t = 0$, both surfaces are suddenly exposed to a constant temperature T_∞ with $T_\infty > T_o$. The governing differential equation together with the initial and boundary conditions is

$$\frac{\partial T}{\partial t} = \alpha \frac{\partial^2 T}{\partial x^2} \tag{B.3-28}$$

$$\text{at} \quad t = 0 \qquad T = T_o \tag{B.3-29}$$

$$\text{at} \quad x = 0 \qquad T = T_\infty \tag{B.3-30}$$

$$\text{at} \quad x = L \qquad T = T_\infty \tag{B.3-31}$$

While the differential equation is linear and homogeneous, the boundary conditions, although linear, are not homogeneous[2]. The boundary conditions in the x-direction become homogeneous by the introduction of the dimensionless quantities

$$\theta = \frac{T_\infty - T}{T_\infty - T_o} \qquad \xi = \frac{x}{L} \qquad \tau = \frac{\alpha t}{L^2} \tag{B.3-32}$$

In dimensionless form, Eqs. (B.3-28)–(B.3-31) become

$$\frac{\partial \theta}{\partial \tau} = \frac{\partial^2 \theta}{\partial \xi^2} \tag{B.3-33}$$

$$\text{at} \quad \tau = 0 \qquad \theta = 1 \tag{B.3-34}$$

$$\text{at} \quad \xi = 0 \qquad \theta = 0 \tag{B.3-35}$$

$$\text{at} \quad \xi = 1 \qquad \theta = 0 \tag{B.3-36}$$

The separation of variables method assumes that the solution can be represented as a product of two functions of the form

$$\theta(\tau, \xi) = F(\tau) G(\xi) \tag{B.3-37}$$

Substitution of Eq. (B.3-37) into Eq. (B.3-33) and rearrangement give

$$\frac{1}{F} \frac{dF}{d\tau} = \frac{1}{G} \frac{d^2 G}{d\xi^2} \tag{B.3-38}$$

---

[2]A linear differential equation or a linear boundary condition is said to be *homogeneous* if, when satisfied by a function f, it is also satisfied by βf, where β is an arbitrary constant.

While the left-hand side of Eq. (B.3-38) is a function of τ only, the right-hand side is dependent only on ξ. This is possible only if both sides of Eq. (B.3-38) are equal to a constant, say $-\lambda^2$, i.e.,

$$\frac{1}{F}\frac{dF}{d\tau} = \frac{1}{G}\frac{d^2G}{d\xi^2} = -\lambda^2 \tag{B.3-39}$$

The choice of a negative constant is due to the fact that the solution will decay to zero as time increases. The choice of a positive constant would give a solution that becomes infinite as time increases.

Equation (B.3-39) results in two ordinary differential equations. The equation for F is given by

$$\frac{dF}{d\tau} + \lambda^2 F = 0 \tag{B.3-40}$$

The solution of Eq. (B.3-40) is

$$F(\tau) = e^{-\lambda^2\tau} \tag{B.3-41}$$

On the other hand, the equation for G is

$$\frac{d^2G}{d\xi^2} + \lambda^2 G = 0 \tag{B.3-42}$$

subject to the boundary conditions

$$\text{at} \quad \xi = 0 \quad G = 0 \tag{B.3-43}$$

$$\text{at} \quad \xi = 1 \quad G = 0 \tag{B.3-44}$$

Note that Eq. (B.3-42) is a Sturm-Liouville equation with a weight function of unity. The solution of Eq. (B.3-42) is

$$G(\xi) = C_1 \sin(\lambda\xi) + C_2 \cos(\lambda\xi) \tag{B.3-45}$$

where C_1 and C_2 are constants. The use of the boundary condition defined by Eq. (B.3-43) implies $C_2 = 0$. Application of the boundary condition defined by Eq. (B.3-44) gives

$$C_1 \sin\lambda = 0 \tag{B.3-46}$$

For a nontrivial solution, the eigenvalues are given by

$$\sin\lambda = 0 \quad \Rightarrow \quad \lambda_n = n\pi \quad n = 1, 2, 3, \ldots \tag{B.3-47}$$

The corresponding eigenfunctions are

$$G_n(\xi) = \sin(n\pi\xi) \tag{B.3-48}$$

Note that each of the product functions

$$\theta_n(\tau, \xi) = e^{-n^2\pi^2\tau} \sin(n\pi\xi) \quad n = 1, 2, 3, \ldots \tag{B.3-49}$$

is a solution of Eq. (B.3-33) and satisfies the initial and boundary conditions, Eqs. (B.3-34)–(B.3-36).

If θ_1 and θ_2 are the solutions satisfying the linear and homogeneous partial differential equation and the boundary conditions, then the linear combination of the solutions, i.e., $A_1\theta_1 + A_2\theta_2$, also satisfies the partial differential equation and the boundary conditions. Therefore, the complete solution is

$$\theta = \sum_{n=1}^{\infty} A_n \exp(-n^2\pi^2\tau) \sin(n\pi\xi) \tag{B.3-50}$$

The unknown coefficients A_n can be determined by using the initial condition. The use of Eq. (B.3-34) results in

$$1 = \sum_{n=1}^{\infty} A_n \sin(n\pi\xi) \tag{B.3-51}$$

Since the eigenfunctions are simply orthogonal, multiplication of Eq. (B.3-51) by $\sin m\pi\xi\, d\xi$ and integration from $\xi = 0$ to $\xi = 1$ give

$$\int_0^1 \sin(m\pi\xi)\, d\xi = \sum_{n=1}^{\infty} A_n \int_0^1 \sin(n\pi\xi) \sin(m\pi\xi)\, d\xi \tag{B.3-52}$$

The integral on the right-hand side of Eq. (B.3-52) is zero when $m \neq n$ and nonzero when $m = n$. Therefore, when $m = n$ the summation drops out and Eq. (B.3-52) reduces to the form

$$\int_0^1 \sin(n\pi\xi)\, d\xi = A_n \int_0^1 \sin^2(n\pi\xi)\, d\xi \tag{B.3-53}$$

Evaluation of the integrals shows that

$$A_n = \frac{2}{\pi n}\left[1 - (-1)^n\right] \tag{B.3-54}$$

The coefficients A_n take the following values depending on the value of n:

$$A_n = \begin{cases} 0 & n = 2, 4, 6, \ldots \\ \dfrac{4}{\pi n} & n = 1, 3, 5, \ldots \end{cases} \tag{B.3-55}$$

Therefore, the solution becomes

$$\theta = \frac{4}{\pi} \sum_{n=1,3,5}^{\infty} \frac{1}{n} \exp(-n^2\pi^2\tau) \sin(n\pi\xi) \tag{B.3-56}$$

Replacing n with $2k + 1$ gives

$$\theta = \frac{4}{\pi} \sum_{k=0}^{\infty} \frac{1}{2k+1} \exp\left[-(2k+1)^2\pi^2\tau\right] \sin\left[(2k+1)\pi\xi\right] \tag{B.3-57}$$

B.3.6.2 *Similarity solution* This is also known as the *method of combination of variables*. Similarity solutions are a special class of solutions used to solve parabolic second-order partial differential equations when there is no geometric length scale in the problem, i.e., the domain must be either semi-infinite or infinite. Furthermore, the initial condition should match the boundary condition at infinity.

The basis of this method is to combine the two independent variables in a single variable so as to transform the second-order partial differential equation into an ordinary differential equation.

Let us consider the following parabolic second-order partial differential equation together with the initial and boundary conditions:

$$\frac{\partial v_z}{\partial t} = v \frac{\partial^2 v_z}{\partial x^2} \tag{B.3-58}$$

$$\text{at} \quad t = 0 \qquad v_z = 0 \tag{B.3-59}$$

$$\text{at} \quad x = 0 \qquad v_z = V \tag{B.3-60}$$

$$\text{at} \quad x = \infty \qquad v_z = 0 \tag{B.3-61}$$

Such a problem represents the velocity profile in a fluid adjacent to a wall suddenly set in motion and is also known as *Stokes' first problem*.

The solution is sought in the form

$$\frac{v_z}{V} = f(\eta) \tag{B.3-62}$$

where

$$\eta = \beta t^m x^n \tag{B.3-63}$$

The term η is called the *similarity variable*. The proportionality constant β is included in Eq. (B.3-63) so as to make η dimensionless.

The chain rule of differentiation gives

$$\frac{\partial (v_z/V)}{\partial t} = \frac{df}{d\eta} \frac{\partial \eta}{\partial t} = \beta m t^{m-1} x^n \frac{df}{d\eta} \tag{B.3-64}$$

$$\frac{\partial^2 (v_z/V)}{\partial x^2} = \frac{d^2 f}{d\eta^2} \left(\frac{\partial \eta}{\partial x} \right)^2 + \frac{df}{d\eta} \frac{\partial^2 \eta}{\partial x^2} = \beta^2 n^2 t^{2m} x^{2(n-1)} \frac{d^2 f}{d\eta^2} + \beta n (n-1) t^m x^{n-2} \frac{df}{d\eta} \tag{B.3-65}$$

Substitution of Eqs. (B.3-64) and (B.3-65) into Eq. (B.3-58) gives

$$\left(\frac{v \beta n^2 t^{m+1}}{m x^{2-n}} \right) \frac{d^2 f}{d\eta^2} + \left[\frac{v n (n-1) t}{m x^2} - 1 \right] \frac{df}{d\eta} = 0 \tag{B.3-66}$$

or,

$$\left[\left(\frac{v n^2 \eta}{m} \right) t x^{-2} \right] \frac{d^2 f}{d\eta^2} + \left\{ \left[\frac{v n (n-1)}{m} \right] t x^{-2} - 1 \right\} \frac{df}{d\eta} = 0 \tag{B.3-67}$$

It should be kept in mind that the purpose of introducing the similarity variable is to reduce the order of the partial differential equation by one. Therefore, the coefficients of $d^2 f/d\eta^2$ and $df/d\eta$ in Eq. (B.3-67) must depend only on η. This can be achieved if

$$tx^{-2} \propto t^m x^n \tag{B.3-68}$$

which implies that

$$n = -2m \tag{B.3-69}$$

If $n = 1$, then $m = -1/2$ and the similarity variable defined by Eq. (B.3-63) becomes

$$\eta = \beta \frac{x}{\sqrt{t}} \tag{B.3-70}$$

Note that $x/\sqrt{t}$ has the units of $m/s^{1/2}$. Since the kinematic viscosity, ν, has the units of m^2/s, η becomes dimensionless if $\beta = 1/\sqrt{\nu}$. It is also convenient to introduce a factor 2 in the denominator so that the similarity variable takes the form

$$\beta = \frac{x}{2\sqrt{\nu t}} \tag{B.3-71}$$

Hence, Eq. (B.3-67) becomes

$$\frac{d^2 f}{d\eta^2} + 2\eta \frac{df}{d\eta} = 0 \tag{B.3-72}$$

The boundary conditions associated with Eq. (B.3-72) are

$$\text{at} \quad \eta = 0 \qquad f = 1 \tag{B.3-73}$$
$$\text{at} \quad \eta = \infty \qquad f = 0 \tag{B.3-74}$$

The integrating factor for Eq. (B.3-72) is $\exp(\eta^2)$. Multiplication of Eq. (B.3-72) by the integrating factor yields[3]

$$\frac{d}{d\eta}\left(e^{\eta^2} \frac{df}{d\eta}\right) = 0 \tag{B.3-75}$$

which implies that

$$\frac{df}{d\eta} = C_1 e^{-\eta^2} \tag{B.3-76}$$

Integration of Eq. (B.3-76) gives

$$f = C_1 \int_0^\eta e^{-u^2} du + C_2 \tag{B.3-77}$$

---

[3]The advantage of including the term 2 in the denominator of the similarity variable can be seen here. Without it, the result would have been

$$\frac{d}{d\eta}\left(e^{\eta^2/2} \frac{df}{d\eta}\right) = 0$$

where u is a dummy variable of integration. Application of the boundary condition defined by Eq. (B.3-73) gives $C_2 = 1$. On the other hand, the use of the boundary condition defined by Eq. (B.3-74) gives

$$C_1 = -\frac{1}{\displaystyle\int_0^\infty e^{-u^2}\, du} = -\frac{2}{\sqrt{\pi}} \tag{B.3-78}$$

Therefore, the solution becomes

$$f = 1 - \frac{2}{\sqrt{\pi}} \int_0^\eta e^{-u^2}\, du = 1 - \operatorname{erf}(\eta) \tag{B.3-79}$$

where $\operatorname{erf}(x)$ is the *error function* defined by

$$\operatorname{erf}(x) = \frac{2}{\sqrt{\pi}} \int_0^x e^{-u^2}\, du \tag{B.3-80}$$

Finally, the velocity distribution as a function of t and x is given by

$$\frac{v_z}{V} = 1 - \operatorname{erf}\left(\frac{x}{2\sqrt{vt}}\right) \tag{B.3-81}$$

REFERENCES

Bird, R.B., R.C. Armstrong and O. Hassager, 1987, Dynamics of Polymeric Liquids, Volume 1: Fluid Dynamics, 2nd Ed., Wiley, New York.

Hildebrand, F.B., 1976, Advanced Calculus for Applications, 2nd Ed., Prentice-Hall, Englewood Cliffs, New Jersey.

Murray, D.A., 1924, Introductory Course in Differential Equations, Longmans, Green and Co., London.

SUGGESTED REFERENCES FOR FURTHER STUDY

Ames, W.F., 1963, Similarity for the nonlinear diffusion equation, Ind. Eng. Chem. Fund. 4, 72.

Holland, C.D. and R.G. Anthony, 1979, Fundamentals of Chemical Reaction Engineering, Prentice-Hall, Englewood Cliffs, New Jersey.

Jenson, V.G. and G.V. Jeffreys, 1963, Mathematical Methods in Chemical Engineering, Academic Press, London.

Lee, S.Y. and W.F. Ames, 1966, Similarity solutions for non-Newtonian fluids, AIChE Journal 12, 700.

Mickley, H.S., T.S. Sherwood and C.E. Reed, 1975, Applied Mathematics in Chemical Engineering, 2nd Ed., Tata McGraw-Hill, New Delhi.

Rice, R.G. and D.D. Do, 1995, Applied Mathematics and Modeling for Chemical Engineers, Wiley, New York.

Suzuki, M., S. Matsumoto and S. Maeda, 1977, New analytical method for a nonlinear diffusion problem, Int. J. Heat Mass Transfer 20, 883.

Suzuki, M., 1979, Some solutions of a nonlinear diffusion problem, J. Chem. Eng. Japan 12, 400.

Appendix C

FLUX EXPRESSIONS FOR MASS, MOMENTUM, AND ENERGY

Table C.1. Components of the stress tensor for Newtonian fluids in rectangular coordinates

$$\tau_{xx} = -\mu\left[2\frac{\partial v_x}{\partial x} - \frac{2}{3}(\nabla \bullet \mathbf{v})\right] \tag{A}$$

$$\tau_{yy} = -\mu\left[2\frac{\partial v_y}{\partial y} - \frac{2}{3}(\nabla \bullet \mathbf{v})\right] \tag{B}$$

$$\tau_{zz} = -\mu\left[2\frac{\partial v_z}{\partial z} - \frac{2}{3}(\nabla \bullet \mathbf{v})\right] \tag{C}$$

$$\tau_{xy} = \tau_{yx} = -\mu\left(\frac{\partial v_x}{\partial y} + \frac{\partial v_y}{\partial x}\right) \tag{D}$$

$$\tau_{yz} = \tau_{zy} = -\mu\left(\frac{\partial v_y}{\partial z} + \frac{\partial v_z}{\partial y}\right) \tag{E}$$

$$\tau_{zx} = \tau_{xz} = -\mu\left(\frac{\partial v_z}{\partial x} + \frac{\partial v_x}{\partial z}\right) \tag{F}$$

$$(\nabla \bullet \mathbf{v}) = \frac{\partial v_x}{\partial x} + \frac{\partial v_y}{\partial y} + \frac{\partial v_z}{\partial z} \tag{G}$$

Table C.2. Components of the stress tensor for Newtonian fluids in cylindrical coordinates

$$\tau_{rr} = -\mu\left[2\frac{\partial v_r}{\partial r} - \frac{2}{3}(\nabla \bullet \mathbf{v})\right] \tag{A}$$

$$\tau_{\theta\theta} = -\mu\left[2\left(\frac{1}{r}\frac{\partial v_\theta}{\partial \theta} + \frac{v_r}{r}\right) - \frac{2}{3}(\nabla \bullet \mathbf{v})\right] \tag{B}$$

$$\tau_{zz} = -\mu\left[2\frac{\partial v_z}{\partial z} - \frac{2}{3}(\nabla \bullet \mathbf{v})\right] \tag{C}$$

$$\tau_{r\theta} = \tau_{\theta r} = -\mu\left[r\frac{\partial}{\partial r}\left(\frac{v_\theta}{r}\right) + \frac{1}{r}\frac{\partial v_r}{\partial \theta}\right] \tag{D}$$

$$\tau_{\theta z} = \tau_{z\theta} = -\mu\left(\frac{\partial v_\theta}{\partial z} + \frac{1}{r}\frac{\partial v_z}{\partial \theta}\right) \tag{E}$$

$$\tau_{zr} = \tau_{rz} = -\mu\left(\frac{\partial v_z}{\partial r} + \frac{\partial v_r}{\partial z}\right) \tag{F}$$

$$(\nabla \bullet \mathbf{v}) = \frac{1}{r}\frac{\partial}{\partial r}(rv_r) + \frac{1}{r}\frac{\partial v_\theta}{\partial \theta} + \frac{\partial v_z}{\partial z} \tag{G}$$

Table C.3. Components of the stress tensor for Newtonian fluids in spherical coordinates

$$\tau_{rr} = -\mu \left[2 \frac{\partial v_r}{\partial r} - \frac{2}{3} (\nabla \bullet \mathbf{v}) \right] \tag{A}$$

$$\tau_{\theta\theta} = -\mu \left[2 \left(\frac{1}{r} \frac{\partial v_\theta}{\partial \theta} + \frac{v_r}{r} \right) - \frac{2}{3} (\nabla \bullet \mathbf{v}) \right] \tag{B}$$

$$\tau_{\phi\phi} = -\mu \left[2 \left(\frac{1}{r \sin\theta} \frac{\partial v_\phi}{\partial \phi} + \frac{v_r}{r} + \frac{v_\theta \cot\theta}{r} \right) - \frac{2}{3} (\nabla \bullet \mathbf{v}) \right] \tag{C}$$

$$\tau_{r\theta} = \tau_{\theta r} = -\mu \left[r \frac{\partial}{\partial r} \left(\frac{v_\theta}{r} \right) + \frac{1}{r} \frac{\partial v_r}{\partial \theta} \right] \tag{D}$$

$$\tau_{\theta\phi} = \tau_{\phi\theta} = -\mu \left[\frac{\sin\theta}{r} \frac{\partial}{\partial \theta} \left(\frac{v_\phi}{\sin\theta} \right) + \frac{1}{r \sin\theta} \frac{\partial v_\theta}{\partial \phi} \right] \tag{E}$$

$$\tau_{\phi r} = \tau_{r\phi} = -\mu \left[\frac{1}{r \sin\theta} \frac{\partial v_r}{\partial \phi} + r \frac{\partial}{\partial r} \left(\frac{v_\phi}{r} \right) \right] \tag{F}$$

$$(\nabla \bullet \mathbf{v}) = \frac{1}{r^2} \frac{\partial}{\partial r} \left(r^2 v_r \right) + \frac{1}{r \sin\theta} \frac{\partial}{\partial \theta} (v_\theta \sin\theta) + \frac{1}{r \sin\theta} \frac{\partial v_\phi}{\partial \phi} \tag{G}$$

Table C.4. Flux expressions for energy transport in rectangular coordinates

Total Flux	Molecular Flux	Convective Flux	Constraint
e_x	$q_x = -k \dfrac{\partial T}{\partial x}$ $q_x = -\alpha \dfrac{\partial (\rho \widehat{C}_P T)}{\partial x}$	$(\rho \widehat{C}_P T) v_x$	None $\rho \widehat{C}_P = \text{constant}$
e_y	$q_y = -k \dfrac{\partial T}{\partial y}$ $q_y = -\alpha \dfrac{\partial (\rho \widehat{C}_P T)}{\partial y}$	$(\rho \widehat{C}_P T) v_y$	None $\rho \widehat{C}_P = \text{constant}$
e_z	$q_z = -k \dfrac{\partial T}{\partial z}$ $q_z = -\alpha \dfrac{\partial (\rho \widehat{C}_P T)}{\partial z}$	$(\rho \widehat{C}_P T) v_z$	None $\rho \widehat{C}_P = \text{constant}$

Table C.5. Flux expressions for energy transport in cylindrical coordinates

Total Flux	Molecular Flux	Convective Flux	Constraint
e_r	$q_r = -k \dfrac{\partial T}{\partial r}$	$(\rho \widehat{C}_P T) v_r$	None
	$q_r = -\alpha \dfrac{\partial (\rho \widehat{C}_P T)}{\partial r}$		$\rho \widehat{C}_P = \text{constant}$
e_θ	$q_\theta = -\dfrac{k}{r} \dfrac{\partial T}{\partial \theta}$	$(\rho \widehat{C}_P T) v_\theta$	None
	$q_\theta = -\dfrac{\alpha}{r} \dfrac{\partial (\rho \widehat{C}_P T)}{\partial \theta}$		$\rho \widehat{C}_P = \text{constant}$
e_z	$q_z = -k \dfrac{\partial T}{\partial z}$	$(\rho \widehat{C}_P T) v_z$	None
	$q_z = -\alpha \dfrac{\partial (\rho \widehat{C}_P T)}{\partial z}$		$\rho \widehat{C}_P = \text{constant}$

Table C.6. Flux expressions for energy transport in spherical coordinates

Total Flux	Molecular Flux	Convective Flux	Constraint
e_r	$q_r = -k \dfrac{\partial T}{\partial r}$	$(\rho \widehat{C}_P T) v_r$	None
	$q_r = -\alpha \dfrac{\partial (\rho \widehat{C}_P T)}{\partial r}$		$\rho \widehat{C}_P = \text{constant}$
e_θ	$q_\theta = -\dfrac{k}{r} \dfrac{\partial T}{\partial \theta}$	$(\rho \widehat{C}_P T) v_\theta$	None
	$q_\theta = -\dfrac{\alpha}{r} \dfrac{\partial (\rho \widehat{C}_P T)}{\partial \theta}$		$\rho \widehat{C}_P = \text{constant}$
e_ϕ	$q_\phi = -\dfrac{k}{r \sin \theta} \dfrac{\partial T}{\partial \phi}$	$(\rho \hat{C}_P T) v_\phi$	None
	$q_z = -\dfrac{\alpha}{r \sin \theta} \dfrac{\partial (\rho \hat{C}_P T)}{\partial \phi}$		$\rho \hat{C}_P = \text{constant}$

Table C.7. Flux expressions for mass transport in rectangular coordinates

Total Flux	Molecular Flux	Convective Flux	Constraint
$\mathcal{W}_{A_x}$	$j_{A_x} = -\rho \mathcal{D}_{AB} \dfrac{\partial \omega_A}{\partial x}$	$\rho_A v_x$	None
	$j_{A_x} = -\mathcal{D}_{AB} \dfrac{\partial \rho_A}{\partial x}$		$\rho = \text{constant}$
$\mathcal{W}_{A_y}$	$j_{A_y} = -\rho \mathcal{D}_{AB} \dfrac{\partial \omega_A}{\partial y}$	$\rho_A v_y$	None
	$j_{A_y} = -\mathcal{D}_{AB} \dfrac{\partial \rho_A}{\partial y}$		$\rho = \text{constant}$
$\mathcal{W}_{A_z}$	$j_{A_z} = -\rho \mathcal{D}_{AB} \dfrac{\partial \omega_A}{\partial z}$	$\rho_A v_z$	None
	$j_{A_z} = -\mathcal{D}_{AB} \dfrac{\partial \rho_A}{\partial z}$		$\rho = \text{constant}$
N_{A_x}	$J^*_{A_x} = -c \mathcal{D}_{AB} \dfrac{\partial x_A}{\partial x}$	$c_A v^*_x$	None
	$J^*_{A_x} = -\mathcal{D}_{AB} \dfrac{\partial c_A}{\partial x}$		$c = \text{constant}$
N_{A_y}	$J^*_{A_y} = -c \mathcal{D}_{AB} \dfrac{\partial x_A}{\partial y}$	$c_A v^*_y$	None
	$J^*_{A_y} = -\mathcal{D}_{AB} \dfrac{\partial c_A}{\partial y}$		$c = \text{constant}$
N_{A_z}	$J^*_{A_z} = -c \mathcal{D}_{AB} \dfrac{\partial x_A}{\partial z}$	$c_A v^*_z$	None
	$J^*_{A_z} = -\mathcal{D}_{AB} \dfrac{\partial c_A}{\partial z}$		$c = \text{constant}$

Table C.8. Flux expressions for mass transport in cylindrical coordinates

Total Flux	Molecular Flux	Convective Flux	Constraint
$\mathcal{W}_{A_r}$	$j_{A_r} = -\rho\,\mathcal{D}_{AB}\,\dfrac{\partial \omega_A}{\partial r}$	$\rho_A v_r$	None
	$j_{A_r} = -\mathcal{D}_{AB}\,\dfrac{\partial \rho_A}{\partial r}$		$\rho = \text{constant}$
$\mathcal{W}_{A_\theta}$	$j_{A_\theta} = -\dfrac{\rho\,\mathcal{D}_{AB}}{r}\,\dfrac{\partial \omega_A}{\partial \theta}$	$\rho_A v_\theta$	None
	$j_{A_\theta} = -\dfrac{\mathcal{D}_{AB}}{r}\,\dfrac{\partial \rho_A}{\partial \theta}$		$\rho = \text{constant}$
$\mathcal{W}_{A_z}$	$j_{A_z} = -\rho\,\mathcal{D}_{AB}\,\dfrac{\partial \omega_A}{\partial z}$	$\rho_A v_z$	None
	$j_{A_z} = -\mathcal{D}_{AB}\,\dfrac{\partial \rho_A}{\partial z}$		$\rho = \text{constant}$
N_{A_r}	$J^*_{A_r} = -c\,\mathcal{D}_{AB}\,\dfrac{\partial x_A}{\partial r}$	$c_A v^*_r$	None
	$J^*_{A_r} = -\mathcal{D}_{AB}\,\dfrac{\partial c_A}{\partial r}$		$c = \text{constant}$
N_{A_θ}	$J^*_{A_\theta} = -\dfrac{c\,\mathcal{D}_{AB}}{r}\,\dfrac{\partial x_A}{\partial \theta}$	$c_A v^*_\theta$	None
	$J^*_{A_\theta} = -\dfrac{\mathcal{D}_{AB}}{r}\,\dfrac{\partial c_A}{\partial \theta}$		$c = \text{constant}$
N_{A_z}	$J^*_{A_z} = -c\,\mathcal{D}_{AB}\,\dfrac{\partial x_A}{\partial z}$	$c_A v^*_z$	None
	$J^*_{A_z} = -\mathcal{D}_{AB}\,\dfrac{\partial c_A}{\partial z}$		$c = \text{constant}$

Table C.9. Flux expressions for mass transport in spherical coordinates

Total Flux	Molecular Flux	Convective Flux	Constraint
$\mathcal{W}_{A_r}$	$j_{A_r} = -\rho\, \mathcal{D}_{AB} \dfrac{\partial \omega_A}{\partial r}$	$\rho_A v_r$	None
	$j_{A_r} = -\mathcal{D}_{AB} \dfrac{\partial \rho_A}{\partial r}$		$\rho = $ constant
$\mathcal{W}_{A_\theta}$	$j_{A_\theta} = -\dfrac{\rho\, \mathcal{D}_{AB}}{r} \dfrac{\partial \omega_A}{\partial \theta}$	$\rho_A v_\theta$	None
	$j_{A_\theta} = -\dfrac{\mathcal{D}_{AB}}{r} \dfrac{\partial \rho_A}{\partial \theta}$		$\rho = $ constant
$\mathcal{W}_{A_\phi}$	$j_{A_\phi} = -\dfrac{\rho\, \mathcal{D}_{AB}}{r \sin\theta} \dfrac{\partial \omega_A}{\partial \phi}$	$\rho_A v_\phi$	None
	$j_{A_\phi} = -\dfrac{\mathcal{D}_{AB}}{r \sin\theta} \dfrac{\partial \rho_A}{\partial \phi}$		$\rho = $ constant
N_{A_r}	$J^*_{A_r} = -c\, \mathcal{D}_{AB} \dfrac{\partial x_A}{\partial r}$	$c_A v_r^*$	None
	$J^*_{A_r} = -\mathcal{D}_{AB} \dfrac{\partial c_A}{\partial r}$		$c = $ constant
N_{A_θ}	$J^*_{A_\theta} = -\dfrac{c\, \mathcal{D}_{AB}}{r} \dfrac{\partial x_A}{\partial \theta}$	$c_A v_\theta^*$	None
	$J^*_{A_\theta} = -\dfrac{\mathcal{D}_{AB}}{r} \dfrac{\partial c_A}{\partial \theta}$		$c = $ constant
N_{A_ϕ}	$J^*_{A_z} = -\dfrac{c\, \mathcal{D}_{AB}}{r \sin\theta} \dfrac{\partial x_A}{\partial \phi}$	$c_A v_\phi^*$	None
	$J^*_{A_z} = -\dfrac{\mathcal{D}_{AB}}{r \sin\theta} \dfrac{\partial c_A}{\partial \phi}$		$c = $ constant

Appendix D

PHYSICAL PROPERTIES

This appendix contains physical properties of some frequently encountered materials in the transport of momentum, energy, and mass. The reader should refer to either Perry's Chemical Engineers' Handbook (1997) or CRC Handbook of Chemistry and Physics (2001) for a more extensive list of physical properties.

Table D.1 contains viscosities of gases and liquids, as taken from Reid *et al.* (1977). Table D.2 contains thermal conductivities of gases, liquids, and solids. While gas and liquid thermal conductivities are compiled from Reid *et al.* (1977), solid thermal conductivity values are taken from Perry's Chemical Engineers' Handbook (1997). The values of the diffusion coefficients given in Table D.3 are compiled from Reid *et al.* (1977), Perry's Chemical Engineers' Handbook (1997), and Geankoplis (1972).

Table D.4 contains the physical properties of dry air at standard atmospheric pressure. The values are taken from Kays and Crawford (1980) who obtained the data from the three volumes of Touloukian *et al.* (1970). The physical properties of saturated liquid water, given in Table D.5, are taken from Incropera and DeWitt (1996) who adapted the data from Liley (1984).

Table D.1. Viscosities of various substances

Substance	T K	$\mu \times 10^4$ kg/m·s
Gases		
Ammonia	273	0.9
	373	1.31
Carbon dioxide	303	1.51
	373.5	1.81
Ethanol	383	1.11
	423	1.23
Sulfur dioxide	313	1.35
	373	1.63
Liquids		
Benzene	313	4.92
	353	3.18
Carbon tetrachloride	303	8.56
	343	5.34
Ethanol	313	8.26
	348	4.65

Table D.2. Thermal conductivities of various substances

Substance	T K	k W/m·K
Gases		
Ammonia	273	0.0221
	373	0.0320
Carbon dioxide	300	0.0167
	473	0.0283
Ethanol	293	0.0150
	375	0.0222
Sulfur dioxide	273	0.0083
Liquids		
Benzene	293	0.148
	323	0.137
Carbon tetrachloride	293	0.103
Ethanol	293	0.165
	313	0.152
Solids		
Aluminum	300	273
Brick	300	0.72
Copper	300	398
Glass Fiber	300	0.036
Steel	300	45

Table D.3. Experimental values of binary diffusion coefficients at 101.325 kPa

Substance	T K	$\mathcal{D}_{AB}$ m²/s
Gases		
Air–CO_2	317.2	1.77×10^{-5}
Air–Ethanol	313	1.45×10^{-5}
Air–Naphthalene	300	0.62×10^{-5}
Air–H_2O	313	2.88×10^{-5}
H_2–Acetone	296	4.24×10^{-5}
N_2–SO_2	263	1.04×10^{-5}
Liquids		
NH_3–H_2O	288	1.77×10^{-9}
Benzoic acid–H_2O	298	1.21×10^{-9}
CO_2–H_2O	298	1.92×10^{-9}
Ethanol–H_2O	283	0.84×10^{-9}
Solids		
Bi–Pb	293	1.1×10^{-20}
H_2–Nickel	358	1.16×10^{-12}
O_2–Vulc. Rubber	298	0.21×10^{-9}

Table D.4. Properties of air at $P = 101.325$ kPa

T K	ρ kg/m³	$\mu \times 10^6$ kg/m·s	$\nu \times 10^6$ m²/s	$\widehat{C}_P$ kJ/kg·K	$k \times 10^3$ W/m·K	Pr
100	3.5985	7.060	1.962	1.028	9.220	0.787
150	2.3673	10.38	4.385	1.011	13.75	0.763
200	1.7690	13.36	7.552	1.006	18.10	0.743
250	1.4119	16.06	11.37	1.003	22.26	0.724
263	1.3421	16.70	12.44	1.003	23.28	0.720
273	1.2930	17.20	13.30	1.004	24.07	0.717
275	1.2836	17.30	13.48	1.004	24.26	0.716
280	1.2607	17.54	13.92	1.004	24.63	0.715
283	1.2473	17.69	14.18	1.004	24.86	0.714
285	1.2385	17.79	14.36	1.004	25.00	0.714
288	1.2256	17.93	14.63	1.004	25.22	0.714
290	1.2172	18.03	14.81	1.004	25.37	0.714
293	1.2047	18.17	15.08	1.004	25.63	0.712
295	1.1966	18.27	15.27	1.005	25.74	0.713
298	1.1845	18.41	15.54	1.005	25.96	0.712
300	1.1766	18.53	15.75	1.005	26.14	0.711
303	1.1650	18.64	16.00	1.005	26.37	0.710
305	1.1573	18.74	16.19	1.005	26.48	0.711
308	1.1460	18.88	16.47	1.005	26.70	0.711
310	1.1386	18.97	16.66	1.005	26.85	0.710
313	1.1277	19.11	16.95	1.005	27.09	0.709
315	1.1206	19.20	17.14	1.006	27.22	0.709
320	1.1031	19.43	17.62	1.006	27.58	0.709
323	1.0928	19.57	17.91	1.006	27.80	0.708
325	1.0861	19.66	18.10	1.006	27.95	0.708
330	1.0696	19.89	18.59	1.006	28.32	0.707
333	1.0600	20.02	18.89	1.007	28.51	0.707
343	1.0291	20.47	19.89	1.008	29.21	0.706
350	1.0085	20.81	20.63	1.008	29.70	0.706
353	1.0000	20.91	20.91	1.008	29.89	0.705
363	0.9724	21.34	21.95	1.009	30.58	0.704
373	0.9463	21.77	23.01	1.010	31.26	0.703
400	0.8825	22.94	26.00	1.013	33.05	0.703
450	0.7844	24.93	31.78	1.020	36.33	0.700
500	0.7060	26.82	37.99	1.029	39.51	0.699
550	0.6418	28.60	44.56	1.039	42.60	0.698
600	0.5883	30.30	51.50	1.051	45.60	0.699
650	0.5431	31.93	58.80	1.063	48.40	0.701
700	0.5043	33.49	66.41	1.075	51.30	0.702

A widely used vapor pressure correlation over limited temperature ranges is the *Antoine equation* expressed in the form

$$\ln P^{sat} = A - \frac{B}{T + C}$$

where P^{sat} is in mmHg and T is in degrees Kelvin. The Antoine constants A, B, and C, given in Table D.6 for various substances, are taken from Reid *et al.* (1977).

Table D.5. Properties of saturated liquid water

T	P^{sat}	$\widehat{V} \times 10^3$	$\widehat{\lambda}$	$\widehat{C}_P$	$\mu \times 10^6$	$k \times 10^3$	Pr
273	0.00611	1.000	2502	4.217	1750	569	12.99
275	0.00697	1.000	2497	4.211	1652	574	12.22
280	0.00990	1.000	2485	4.198	1422	582	10.26
285	0.01387	1.000	2473	4.189	1225	590	8.70
288	0.01703	1.001	2466	4.186	1131	595	7.95
290	0.01917	1.001	2461	4.184	1080	598	7.56
293	0.02336	1.001	2454	4.182	1001	603	6.94
295	0.02617	1.002	2449	4.181	959	606	6.62
298	0.03165	1.003	2442	4.180	892	610	6.11
300	0.03531	1.003	2438	4.179	855	613	5.83
303	0.04240	1.004	2430	4.178	800	618	5.41
305	0.04712	1.005	2426	4.178	769	620	5.20
308	0.05620	1.006	2418	4.178	721	625	4.82
310	0.06221	1.007	2414	4.178	695	628	4.62
313	0.07373	1.008	2407	4.179	654	632	4.32
315	0.08132	1.009	2402	4.179	631	634	4.16
320	0.1053	1.011	2390	4.180	577	640	3.77
325	0.1351	1.013	2378	4.182	528	645	3.42
330	0.1719	1.016	2366	4.184	489	650	3.15
335	0.2167	1.018	2354	4.186	453	656	2.88
340	0.2713	1.021	2342	4.188	420	660	2.66
345	0.3372	1.024	2329	4.191	389	664	2.45
350	0.4163	1.027	2317	4.195	365	668	2.29
355	0.5100	1.030	2304	4.199	343	671	2.14
360	0.6209	1.034	2291	4.203	324	674	2.02
365	0.7514	1.038	2278	4.209	306	677	1.91
370	0.9040	1.041	2265	4.214	289	679	1.80
373	1.0133	1.044	2257	4.217	279	680	1.76
375	1.0815	1.045	2252	4.220	274	681	1.70
380	1.2869	1.049	2239	4.226	260	683	1.61
385	1.5233	1.053	2225	4.232	248	685	1.53
390	1.794	1.058	2212	4.239	237	686	1.47
400	2.455	1.067	2183	4.256	217	688	1.34

$T = \text{K}$; $P^{sat} = \text{bar}$; $\widehat{V} = \text{m}^3/\text{kg}$; $\widehat{\lambda} = \text{kJ/kg}$; $\widehat{C}_P = \text{kJ/kg·K}$; $\mu = \text{kg/m·s}$; $k = \text{W/m·K}$

Table D.6. Antoine equation constants

Substance	Range (K)	A	B	C
Acetone	241–350	16.6513	2940.46	−35.93
Benzene	280–377	15.9008	2788.51	−52.36
Benzoic acid	405–560	17.1634	4190.70	−125.2
Chloroform	260–370	15.9732	2696.79	−46.16
Ethanol	270–369	18.9119	3803.98	−41.68
Methanol	257–364	18.5875	3626.55	−34.29
Naphthalene	360–525	16.1426	3992.01	−71.29

REFERENCES

Geankoplis, C.J., 1972, Mass Transport Phenomena, Holt, Rinehart and Winston, New York.

Incropera, F.P. and D.P. DeWitt, 2002, Fundamentals of Heat and Mass Transfer, 5th Ed., Wiley, New York.

Kays, W.M. and M.E. Crawford, 1980, Convective Heat and Mass Transfer, 2nd Ed., McGraw-Hill, New York.

Lide, D.R., Ed., 2001, CRC Handbook of Chemistry and Physics, 82nd Ed., CRC Press, Boca Raton, Florida.

Liley, P.E., 1984, Steam Tables in SI Units, School of Mechanical Engineering, Purdue University, West Lafayette, Indiana.

Perry, R.H., D.W. Green and J.O. Maloney, Eds., 1997, Perry's Chemical Engineers' Handbook, 7th Ed., McGraw-Hill, New York.

Reid, R.C., J.M. Prausnitz and T.K. Sherwood, 1977, The Properties of Gases and Liquids, 3rd Ed., McGraw-Hill, New York.

Touloukian, Y.S., P.E. Liley and S.C. Saxena, 1970, Thermophysical Properties of Matter, Vol. 3: Thermal Conductivity. Nonmetallic Liquids and Gases, IFI/Plenum, New York.

Touloukian, Y.S., P.E. Liley and S.C. Saxena, 1970, Thermophysical Properties of Matter, Vol. 6: Specific Heat. Nonmetallic Liquids and Gases, IFI/Plenum, New York.

Touloukian, Y.S., P.E. Liley and S.C. Saxena, 1970, Thermophysical Properties of Matter, Vol. 11: Viscosity. Nonmetallic Liquids and Gases, IFI/Plenum, New York.

Appendix E

CONSTANTS AND CONVERSION FACTORS

PHYSICAL CONSTANTS

Gas constant ($\mathcal{R}$)
$$= 82.05 \text{ cm}^3 \cdot \text{atm/mol} \cdot \text{K}$$
$$= 0.08205 \text{ m}^3 \cdot \text{atm/kmol} \cdot \text{K}$$
$$= 1.987 \text{ cal/mol} \cdot \text{K}$$
$$= 8.314 \text{ J/mol} \cdot \text{K}$$
$$= 8.314 \times 10^{-3} \text{ kPa} \cdot \text{m}^3/\text{mol} \cdot \text{K}$$
$$= 8.314 \times 10^{-5} \text{ bar} \cdot \text{m}^3/\text{mol} \cdot \text{K}$$
$$= 8.314 \times 10^{-2} \text{ bar} \cdot \text{m}^3/\text{kmol} \cdot \text{K}$$
$$= 8.314 \times 10^{-6} \text{ MPa} \cdot \text{m}^3/\text{mol} \cdot \text{K}$$

Acceleration of gravity (g)
$$= 9.8067 \text{ m/s}^2$$
$$= 32.1740 \text{ ft/s}^2$$

Stefan-Boltzmann constant (σ)
$$= 5.67051 \times 10^{-8} \text{ W/m}^2 \cdot \text{K}^4$$
$$= 0.1713 \times 10^{-8} \text{ Btu/h} \cdot \text{ft}^2 \cdot {}^\circ\text{R}^4$$

CONVERSION FACTORS

Density
$$1 \text{ kg/m}^3 = 10^{-3} \text{ g/cm}^3 = 10^{-3} \text{ kg/L}$$
$$1 \text{ kg/m}^3 = 0.06243 \text{ lb/ft}^3$$

Diffusivity
(Kinematic, Mass, Thermal)
$$1 \text{ m}^2/\text{s} = 10^4 \text{ cm}^2/\text{s}$$
$$1 \text{ m}^2/\text{s} = 10.7639 \text{ ft}^2/\text{s} = 3.875 \times 10^4 \text{ ft}^2/\text{h}$$

Energy, Heat, Work
$$1 \text{ J} = 1 \text{ W} \cdot \text{s} = 1 \text{ N} \cdot \text{m} = 10^{-3} \text{ kJ}$$
$$1 \text{ cal} = 4.184 \text{ J}$$
$$1 \text{ kJ} = 2.7778 \times 10^{-4} \text{ kW} \cdot \text{h} = 0.94783 \text{ Btu}$$

Heat capacity
$$1 \text{ kJ/kg} \cdot \text{K} = 0.239 \text{ cal/g} \cdot \text{K}$$
$$1 \text{ kJ/kg} \cdot \text{K} = 0.239 \text{ Btu/lb} \cdot {}^\circ\text{R}$$

Force
$$1 \text{ N} = 1 \text{ kg} \cdot \text{m/s}^2 = 10^5 \text{ g} \cdot \text{cm/s}^2 \text{ (dyne)}$$
$$1 \text{ N} = 0.2248 \text{ lbf} = 7.23275 \text{ lb} \cdot \text{ft/s}^2 \text{ (poundals)}$$

| Heat flux | $1 \text{ W/m}^2 = 1 \text{ J/s·m}^2$ |
| | $1 \text{ W/m}^2 = 0.31709 \text{ Btu/h·ft}^2$ |

Heat transfer coefficient	$1 \text{ W/m}^2\text{·K} = 1 \text{ J/s·m}^2\text{·K}$
	$1 \text{ W/m}^2\text{·K} = 2.39 \times 10^{-5} \text{ cal/s·cm}^2\text{·K}$
	$1 \text{ W/m}^2\text{·K} = 0.1761 \text{ Btu/h·ft}^2\text{·°R}$

| Length | $1 \text{ m} = 100 \text{ cm} = 10^6 \text{ } \mu\text{m}$ |
| | $1 \text{ m} = 39.370 \text{ in} = 3.2808 \text{ ft}$ |

| Mass | $1 \text{ kg} = 1000 \text{ g}$ |
| | $1 \text{ kg} = 2.2046 \text{ lb}$ |

| Mass flow rate | $1 \text{ kg/s} = 2.2046 \text{ lb/s} = 7936.6 \text{ lb/h}$ |

| Mass flux | $1 \text{ kg/s·m}^2 = 0.2048 \text{ lb/s·ft}^2 = 737.3 \text{ lb/h·ft}^2$ |

| Mass transfer coefficient | $1 \text{ m/s} = 3.2808 \text{ ft/s}$ |

| Power | $1 \text{ W} = 1 \text{ J/s} = 10^{-3} \text{ kW}$ |
| | $1 \text{ kW} = 3412.2 \text{ Btu/h} = 1.341 \text{ hp}$ |

Pressure	$1 \text{ Pa} = 1 \text{ N/m}^2$
	$1 \text{ kPa} = 10^3 \text{ Pa} = 10^{-3} \text{ MPa}$
	$1 \text{ atm} = 101.325 \text{ kPa} = 1.01325 \text{ bar} = 760 \text{ mmHg}$
	$1 \text{ atm} = 14.696 \text{ lbf/in}^2$

| Temperature | $1 \text{ K} = 1.8 \text{ °R}$ |
| | $T(\text{°F}) = 1.8 T(\text{°C}) + 32$ |

| Thermal Conductivity | $1 \text{ W/m·K} = 1 \text{ J/s·m·K} = 2.39 \times 10^{-3} \text{ cal/s·cm·K}$ |
| | $1 \text{ W/m·K} = 0.5778 \text{ Btu/h·ft·°F}$ |

| Velocity | $1 \text{ m/s} = 3.60 \text{ km/h}$ |
| | $1 \text{ m/s} = 3.2808 \text{ ft/s} = 2.237 \text{ mi/h}$ |

Viscosity	$1 \text{ kg/m·s} = 1 \text{ Pa·s}$
	$1 \text{ P (poise)} = 1 \text{ g/cm·s}$
	$1 \text{ kg/m·s} = 10 \text{ P} = 10^3 \text{ cP}$
	$1 \text{ P (poise)} = 241.9 \text{ lb/ft·h}$

| Volume | $1 \text{ m}^3 = 1000 \text{ L}$ |
| | $1 \text{ m}^3 = 6.1022 \times 10^4 \text{ in}^3 = 35.313 \text{ ft}^3 = 264.17 \text{ gal}$ |

| Volumetric flow rate | $1 \text{ m}^3\text{/s} = 1000 \text{ L/s}$ |
| | $1 \text{ m}^3\text{/s} = 35.313 \text{ ft}^3\text{/s} = 1.27127 \times 10^5 \text{ ft}^3\text{/h}$ |

INDEX

Printed and bound by CPI Group (UK) Ltd, Croydon, CR0 4YY

03/10/2024

01040330-0017